An Introduction to Stochastic Mode

An Introduction to Stochastic Modeling

Fourth Edition

Mark A. Pinsky

Department of Mathematics
Northwestern University
Evanston, Illinois

Samuel Karlin

Department of Mathematics
Stanford University
Stanford, California

AMSTERDAM • BOSTON • HEIDELBERG • LONDON
NEW YORK • OXFORD • PARIS • SAN DIEGO
SAN FRANCISCO • SINGAPORE • SYDNEY • TOKYO
Academic Press is an imprint of Elsevier

Academic Press is an imprint of Elsevier
30 Corporate Drive, Suite 400, Burlington, MA 01803, USA
The Boulevard, Langford Lane, Kidlington, Oxford, OX5 1GB, UK

Notices
Knowledge and best practice in this field are constantly changing. As new research and experience broaden our understanding, changes in research methods, professional practices, or medical treatment may become necessary.

Practitioners and researchers must always rely on their own experience and knowledge in evaluating and using any information, methods, compounds, or experiments described herein. In using such information or methods they should be mindful of their own safety and the safety of others, including parties for whom they have a professional responsibility.

To the fullest extent of the law, neither the Publisher nor the authors, contributors, or editors, assume any liability for any injury and/or damage to persons or property as a matter of products' liability, negligence or otherwise, or from any use or operation of any methods, products, instructions, or ideas contained in the material herein.

Library of Congress Cataloging-in-Publication Data
Application submitted.

British Library Cataloguing-in-Publication Data
A catalogue record for this book is available from the British Library.

ISBN: 978-0-12-381416-6

For information on all Academic Press publications,
visit our website: *www.elsevierdirect.com*

Typeset by: diacriTech, India

Printed in the United States of America
10 11 12 13   8 7 6 5 4 3 2 1

Contents

Preface to the Fourth Edition

Since the publication of the third edition in 1998, some new developments have occurred. Samuel Karlin died in 2007, leaving a gap at the authorship level and the new designation of authors.

In the fourth edition, we have added two new chapters: Chapter 10 on random evolution and Chapter 11 on characteristic functions. *Random Evolution* denotes a set of stochastic models, which describe continuous motion with piecewise linear sample functions. Explicit formulas are available in the simplest cases. In the general case, one has a central limit theorem, which is pursued more generally in Chapter 11, "Characteristic Functions and Their Applications." Here the necessary tools from Fourier Analysis are developed and applied when necessary. Many theorems are proved in full detail, while other proofs are sketched—in the spirit of the earlier Chapters 1–9. Complete proofs may be found by consulting the intermediate textbooks listed in the section on further reading. Instructors who have taught from the third edition may be reassured that Chapters 1–9 of the new edition are identical to the corresponding chapters of the new book.

We express our thanks to Michael Perlman of the University of Washington and Russell Lyons of Indiana University for sharing their lists of errata from the third edition. We would also like to thank Craig Evans for useful advice on partial differential equations.

Biographical Note

Samuel Karlin earned his undergraduate degree from the Illinois Institute of Technology and his doctorate from Princeton University in 1947 at age 22. He served on the faculty of Caltech from 1948–1956 before joining the faculty of Stanford University, where he spent the remainder of his career. Karlin made fundamental contributions to mathematical economics, bioinformatics, game theory, evolutionary theory, biomolecular sequence analysis, mathematical population genetics, and total positivity.

Karlin authored 10 books and more than 450 articles. He was a member of the American Academy of Arts and Sciences and the National Academy of Sciences. In 1989, he received the National Medal of Science for his broad and remarkable researches in mathematical analysis, probability theory, and mathematical statistics and in the application of these ideas to mathematical economics, mechanics, and genetics. He died on December 18, 2007.

Mark A. Pinsky
Department of Mathematics
Northwestern University
Evanston, IL 60208-2730

Preface to the Third Edition

The purposes, level, and style of this new edition conform to the tenets set forth in the original preface. We continue with our objective of introducing some theory and applications of stochastic processes to students having a solid foundation in calculus and in calculus-level probability, but who are not conversant with the "epsilon–delta" definitions of mathematical analysis. We hope to entice students toward the deeper study of mathematics that is prerequisite to further work in stochastic processes by showing the myriad and interesting ways in which stochastic models can help us understand the real world.

We have removed some topics and added others. We added a small section on martingales that includes an example suggesting the martingale concept as appropriate for modeling the prices of assets traded in a perfect market. A new chapter introduces the Brownian motion process and includes several applications of it and its variants in financial modeling. In this chapter the Black–Scholes formula for option pricing is evaluated and compared with some reported prices of options. A Poisson process whose intensity is itself a stochastic process is described in another new section.

Some treatments have been updated. The law of rare events is presented via an inequality that measures the accuracy of a Poisson approximation for the distribution of the sum of independent, not necessarily identically distributed, Bernoulli random variables. We have added the shot noise model and related it to a random sum.

The text contains more than 250 exercises and 350 problems. Exercises are elementary drills intended to promote active learning and to develop familiarity with concepts through use. They often simply involve the substitution of numbers into given formulas or reasoning one or two steps away from a definition. They are the kinds of simple questions that we, as instructors, hope that students would pose and answer for themselves as they read a text. Answers to the exercises are given at the end of the book so that students may gauge their understanding as they go along.

Problems are more difficult. Some involve extensive algebraic or calculus manipulation. Many are "word problems" wherein the student is asked, in effect, to model some described scenario. As in formulating a model, the first step in the solution of a word problem is often a sentence of the form "Let $x = \ldots$." A manual containing the solutions to the problems is available from the publisher.

A reasonable strategy on the part of the teacher might be to hold students responsible for all of the exercises, but to require submitted solutions only to selected problems. Every student should attempt a representative selection of the problems in order to develop his or her ability to carry out stochastic modeling in his or her area of interest.

A small number of problems are labeled "Computer Challenges." These call for more than pencil and paper for their analyses, and either simulation, numerical exploration, or symbol manipulation may prove helpful. Computer Challenges are meant to be open-ended, intended to explore what constitutes an answer in today's world of computing power. They might be appropriate as part of an honors requirement.

Because our focus is on stochastic modeling, in some instances, we have omitted a proof and contented ourselves with a precise statement of a result and examples of its application. All such omitted proofs may be found in *A First Course in Stochastic Processes*, by the present authors. In this more advanced text, the ambitious student will also find additional material on martingales, Brownian motion, and renewal processes, and presentations of several other classes of stochastic processes.

Preface to the First Edition

Stochastic processes are ways of quantifying the dynamic relationships of sequences of random events. Stochastic models play an important role in elucidating many areas of the natural and engineering sciences. They can be used to analyze the variability inherent in biological and medical processes, to deal with uncertainties affecting managerial decisions and with the complexities of psychological and social interactions, and to provide new perspectives, methodology, models, and intuition to aid in other mathematical and statistical studies.

This book is intended as a beginning text in stochastic processes for students familiar with elementary probability calculus. Its aim is to bridge the gap between basic probability know-how and an intermediate-level course in stochastic processes—for example, *A First Course in Stochastic Processes*, by the present authors.

The objectives of this book are as follows: (1) to introduce students to the standard concepts and methods of stochastic modeling; (2) to illustrate the rich diversity of applications of stochastic processes in the sciences; and (3) to provide exercises in the application of simple stochastic analysis to appropriate problems.

The chapters are organized around several prototype classes of stochastic processes featuring Markov chains in discrete and continuous time, Poisson processes and renewal theory, the evolution of branching events, and queueing models. After the concluding Chapter 9, we provide a list of books that incorporate more advanced discussions of several of the models set forth in this text.

To the Instructor

If possible, we recommend having students skim the first two chapters, referring as necessary to the probability review material, and starting the course with Chapter 3, on Markov chains. A one-quarter course adapted to the junior–senior level could consist of a cursory (1-week) review of Chapters 1 and 2, followed in order by Chapters 3 through 6. For interested students, Chapters 7, 8, and 9 discuss other currently active areas of stochastic modeling. Starred sections contain material of a more advanced or specialized nature.

Acknowledgments

Many people helped to bring this text into being. We gratefully acknowledge the help of Anna Karlin, Shelley Stevens, Karen Larsen, and Laurieann Shoemaker. Chapter 9 was enriched by a series of lectures on queueing networks given by Ralph Disney at The Johns Hopkins University in 1982. Alan Karr, Ivan Johnstone, Luke Tierney, Bob Vanderbei, and others besides ourselves have taught from the text, and we have profited from their criticisms. Finally, we are grateful for improvements suggested by the several generations of students who have used the book over the past few years and have given us their reactions and suggestions.

1 Introduction

1.1 Stochastic Modeling

A quantitative description of a natural phenomenon is called a mathematical model of that phenomenon. Examples abound, from the simple equation $S = \frac{1}{2}gt^2$ describing the distance S traveled in time t by a falling object starting at rest to a complex computer program that simulates a biological population or a large industrial system.

In the final analysis, a model is judged using a single, quite pragmatic, factor, the model's *usefulness*. Some models are useful as detailed quantitative prescriptions of behavior, e.g., an inventory model that is used to determine the optimal number of units to stock. Another model in a different context may provide only general qualitative information about the relationships among and relative importance of several factors influencing an event. Such a model is useful in an equally important but quite different way. Examples of diverse types of stochastic models are spread throughout this book.

Such often mentioned attributes, such as realism, elegance, validity, and reproducibility, are important in evaluating a model only insofar as they bear on that model's ultimate usefulness. For instance, it is both unrealistic and quite inelegant to view the sprawling city of Los Angeles as a geometrical point, a mathematical object of no size or dimension. Yet, it is quite useful to do exactly that when using spherical geometry to derive a minimum-distance great circle air route from New York City, another "point."

There is no such thing as the best model for a given phenomenon. The pragmatic criterion of usefulness often allows the existence of two or more models for the same event, but serving distinct purposes. Consider light. The wave form model, in which light is viewed as a continuous flow, is entirely adequate for designing eyeglass and telescope lenses. In contrast, for understanding the impact of light on the retina of the eye, the photon model, which views light as tiny discrete bundles of energy, is preferred. Neither model supersedes the other; both are relevant and useful.

The word "stochastic" derives from a Greek word ($\sigma\tau o\chi\acute{\alpha}\zeta\epsilon\sigma\theta\alpha\iota$: to aim, to guess) and means "random" or "chance." The antonym is "sure," "deterministic," or "certain." A deterministic model predicts a single outcome from a given set of circumstances. A stochastic model predicts a set of possible outcomes weighted by their likelihoods or probabilities. A coin flipped into the air will surely return to earth somewhere. Whether it lands heads or tails is random. For a "fair" coin, we consider these alternatives equally likely and assign to each the probability $\frac{1}{2}$.

However, phenomena are not in and of themselves inherently stochastic or deterministic. Rather, to model a phenomenon as stochastic or deterministic is the choice of the observer. The choice depends on the observer's purpose; the criterion for judging the choice is usefulness. Most often the proper choice is quite clear, but controversial

An Introduction to Stochastic Modeling

situations do arise. If the coin once fallen is quickly covered by a book so that the outcome "heads" or "tails" remains unknown, two participants may still usefully employ probability concepts to evaluate what is a fair bet between them; i.e., they may usefully view the coin as random, even though most people would consider the outcome now to be fixed or deterministic. As a less mundane example of the converse situation, changes in the level of a large population are often usefully modeled deterministically, in spite of the general agreement among observers that many chance events contribute to their fluctuations.

Scientific modeling has three components: (1) a natural phenomenon under study, (2) a logical system for deducing implications about the phenomenon, and (3) a connection linking the elements of the natural system under study to the logical system used to model it. If we think of these three components in terms of the great-circle air route problem, the natural system is the earth with airports at Los Angeles and New York; the logical system is the mathematical subject of spherical geometry; and the two are connected by viewing the airports in the physical system as points in the logical system.

The modern approach to stochastic modeling is in a similar spirit. Nature does not dictate a unique definition of "probability," in the same way that there is no nature-imposed definition of "point" in geometry. "Probability" and "point" are terms in pure mathematics, defined only through the properties invested in them by their respective sets of axioms. (See Section 1.2.8 for a review of axiomatic probability theory.) There are, however, three general principles that are often useful in relating or connecting the abstract elements of mathematical probability theory to a real or natural phenomenon that is to be modeled. These are (1) the principle of equally likely outcomes, (2) the principle of long run relative frequency, and (3) the principle of odds making or subjective probabilities. Historically, these three concepts arose out of largely unsuccessful attempts to define probability in terms of physical experiences. Today, they are relevant as guidelines for the assignment of probability values in a model, and for the interpretation of the conclusions of a model in terms of the phenomenon under study.

We illustrate the distinctions between these principles with a long experiment. We will pretend that we are part of a group of people who decide to toss a coin and observe the event that the coin will fall heads up. This event is denoted by H, and the event of tails, by T.

Initially, everyone in the group agrees that $\Pr\{H\} = \frac{1}{2}$. When asked why, people give two reasons: Upon checking the coin construction, they believe that the two possible outcomes, heads and tails, are equally likely; and extrapolating from past experience, they also believe that if the coin is tossed many times, the fraction of times that heads is observed will be close to one-half.

The equally likely interpretation of probability surfaced in the works of Laplace in 1812, where the attempt was made to define the probability of an event A as the ratio of the total number of ways that A could occur to the total number of possible outcomes of the experiment. The equally likely approach is often used today to assign probabilities that reflect some notion of a total lack of knowledge about the outcome of a chance phenomenon. The principle requires judicious application if it is to be useful, however.

In our coin tossing experiment, for instance, merely introducing the *possibility* that the coin could land on its edge (E) instantly results in $\Pr\{H\} = \Pr\{T\} = \Pr\{E\} = \frac{1}{3}$.

The next principle, the long run relative frequency interpretation of probability, is a basic building block in modern stochastic modeling, made precise and justified within the axiomatic structure by the *law of large numbers*. This law asserts that the relative fraction of times, in which an event occurs in a sequence of independent similar experiments, approaches, in the limit, the probability of the occurrence of the event on any single trial.

The principle is not relevant in all situations, however. When the surgeon tells a patient that he has an 80–20 chance of survival, the surgeon means, most likely, that 80% of similar patients facing similar surgery will survive it. The patient at hand is not concerned with the long run, but in vivid contrast, he is vitally concerned only in the outcome of his, the next, trial.

Returning to the group experiment, we will suppose next that the coin is flipped into the air and, upon landing, is quickly covered so that no one can see the outcome. What is $\Pr\{H\}$ now? Several in the group argue that the outcome of the coin is no longer random, that $\Pr\{H\}$ is either 0 or 1, and that although we do not know which it is, probability theory does not apply.

Others articulate a different view, that the distinction between "random" and "lack of knowledge" is fuzzy, at best, and that a person with a sufficiently large computer and sufficient information about such factors as the energy, velocity, and direction used in tossing the coin could have predicted the outcome, heads or tails, with certainty before the toss. Therefore, even before the coin was flipped, the problem was a lack of knowledge and not some inherent randomness in the experiment.

In a related approach, several people in the group are willing to bet with each other, at even odds, on the outcome of the toss. That is, they are willing to *use* the calculus of probability to determine what is a fair bet, without considering whether the event under study is random or not. The usefulness criterion for judging a model has appeared.

While the rest of the mob were debating "random" versus "lack of knowledge," one member, Karen, looked at the coin. Her probability for heads is now different from that of everyone else. Keeping the coin covered, she announces the outcome "Tails," whereupon everyone mentally assigns the value $\Pr\{H\} = 0$. But then her companion, Mary, speaks up and says that Karen has a history of prevarication.

The last scenario explains why there are horse races; different people assign different probabilities to the same event. For this reason, probabilities used in odds making are often called *subjective* probabilities. Then, odds making forms the third principle for assigning probability values in models and for interpreting them in the real world.

The modern approach to stochastic modeling is to divorce the definition of probability from any particular type of application. Probability theory is an axiomatic structure (see Section 1.2.8), a part of pure mathematics. Its use in modeling stochastic phenomena is part of the broader realm of science and parallels the use of other branches of mathematics in modeling deterministic phenomena.

To be useful, a stochastic model must reflect all those aspects of the phenomenon under study that are relevant to the question at hand. In addition, the model must be amenable to calculation and must allow the deduction of important predictions or implications about the phenomenon.

1.1.1 Stochastic Processes

A *stochastic process* is a family of random variables X_t, where t is a parameter running over a suitable index set T. (Where convenient, we will write $X(t)$ instead of X_t.) In a common situation, the index t corresponds to discrete units of time, and the index set is $T = \{0, 1, 2, \ldots\}$. In this case, X_t might represent the outcomes at successive tosses of a coin, repeated responses of a subject in a learning experiment, or successive observations of some characteristics of a certain population. Stochastic processes for which $T = [0, \infty)$ are particularly important in applications. Here t often represents time, but different situations also frequently arise. For example, t may represent distance from an arbitrary origin, and X_t may indicate the number of defects in the interval $(0, t]$ along a thread, or the number of cars in the interval $(0, t]$ along a highway.

Stochastic processes are distinguished by their *state space*, or by the range of possible values for the random variables X_t, by their index set T, and by the dependence relations among the random variables X_t. The most widely used classes of stochastic processes are systematically and thoroughly presented for study in the following chapters, along with the mathematical techniques for calculation and analysis that are most useful with these processes. The use of these processes as models is taught by example. Sample applications from many and diverse areas of interest are an integral part of the exposition.

1.2 Probability Review*

This section summarizes the necessary background material and establishes the book's terminology and notation. It also illustrates the level of the exposition in the following chapters. Readers who find the major part of this section's material to be familiar and easily understood should have no difficulty with what follows. Others might wish to review their probability background before continuing.

In this section, statements frequently are made without proof. The reader desiring justification should consult any elementary probability text as the need arises.

1.2.1 Events and Probabilities

The reader is assumed to be familiar with the intuitive concept of an *event*. (Events are defined rigorously in Section 1.2.8, which reviews the axiomatic structure of probability theory.)

Let A and B be events. The event that at least one of A or B occurs is called the *union* of A and B and is written $A \cup B$; the event that both occur is called the

---

* Many readers will prefer to omit this review and move directly to Chapter 3, on Markov chains. They can then refer to the background material that is summarized in the remainder of this chapter and in Chapter 2 only as needed.

intersection of A and B and is written $A \cap B$, or simply AB. This notation extends to finite and countable sequences of events. Given events $A_1, A_2, \ldots$, the event that at least one occurs is written $A_1 \cup A_2 \cup \cdots = \bigcup_{i=1}^{\infty} A_i$, the event that all occur is written $A_1 \cap A_2 \cap \cdots = \bigcap_{i=1}^{\infty} A_i$.

The probability of an event A is written $\Pr\{A\}$. The *certain* event, denoted by Ω, always occurs, and $\Pr\{\Omega\} = 1$. The *impossible* event, denoted by $\emptyset$, never occurs, and $\Pr\{\emptyset\} = 0$. It is always the case that $0 \leq \Pr\{A\} \leq 1$ for any event A.

Events A and B are said to be *disjoint* if $A \cap B = \emptyset$, i.e., if A and B both cannot occur. For disjoint events A and B, we have the *addition law* $\Pr\{A \cup B\} = \Pr\{A\} + \Pr\{B\}$. A stronger form of the addition law is as follows: Let $A_1, A_2, \ldots$ be events with A_i and A_j disjoint whenever $i \neq j$. Then, $\Pr\left\{\bigcup_{i=1}^{\infty} A_i\right\} = \Sigma_{i=1}^{\infty} \Pr\{A_i\}$. The addition law leads directly to the *law of total probability*: Let $A_1, A_2, \ldots$ be disjoint events for which $\Omega = A_1 \cup A_2 \cup \cdots$. Equivalently, exactly one of the events $A_1, A_2, \ldots$ will occur. The law of total probability asserts that $\Pr\{B\} = \Sigma_{i=1}^{\infty} \Pr\{B \cap A_i\}$ for any event B. The law enables the calculation of the probability of an event B from the sometimes more easily determined probabilities $\Pr\{B \cap A_i\}$, where $i = 1, 2, \ldots$. Judicious choice of the events A_i is prerequisite to the profitable application of the law.

Events A and B are said to be *independent* if $\Pr\{A \cap B\} = \Pr\{A\} \times \Pr\{B\}$. Events $A_1, A_2, \ldots$ are *independent* if

$$\Pr\left\{A_{i_1} \cap A_{i_2} \cap \cdots \cap A_{i_n}\right\} = \Pr\left\{A_{i_1}\right\} \Pr\left\{A_{i_2}\right\} \cdots \Pr\left\{A_{i_n}\right\}$$

for every finite set of distinct indices $i_1, i_2, \ldots, i_n$.

1.2.2 Random Variables

An old-fashioned but very useful and highly intuitive definition describes a *random variable* as a variable that takes on its values by chance. In Section 1.2.8, we sketch the modern axiomatic structure for probability theory and random variables. The older definition just given serves quite adequately, however, in virtually all instances of stochastic modeling. Indeed, this older definition was the only approach available for well over a century of meaningful progress in probability theory and stochastic processes.

Most of the time we adhere to the convention of using capital letters such as X, Y, Z to denote random variables, and lowercase letters such as x, y, z for real numbers. The expression $\{X \leq x\}$ is the event that the random variable X assumes a value that is less than or equal to the real number x. This event may or may not occur, depending on the outcome of the experiment or phenomenon that determines the value for the random variable X. The probability that the event occurs is written $\Pr\{X \leq x\}$. Allowing x to vary, this probability defines a function

$$F(x) = \Pr\{X \leq x\}, \qquad -\infty < x < +\infty,$$

called the *distribution function* of the random variable X. Where several random variables appear in the same context, we may choose to distinguish their distribution functions with subscripts, writing, e.g., $F_X(\xi) = \Pr\{X \leq \xi\}$ and $F_Y(\xi) = \Pr\{Y \leq \xi\}$, defining the distribution functions of the random variables X and Y, respectively, as functions of the real variable ξ.

The distribution function contains all the information available about a random variable before its value is determined by experiment. We have, for instance, $\Pr\{X > a\} = 1 - F(a)$, $\Pr\{a < X \leq b\} = F(b) - F(a)$, and $\Pr\{X = x\} = F(x) - \lim_{\epsilon \downarrow 0} F(x - \epsilon) = F(x) - F(x-)$.

A random variable X is called *discrete* if there is a finite or denumerable set of distinct values $x_1, x_2, \ldots$ such that $a_i = \Pr\{X = x_i\} > 0$ for $i = 1, 2, \ldots$ and $\Sigma_i a_i = 1$. The function

$$p(x_i) = p_X(x_i) = a_i \quad \text{for } i = 1, 2, \ldots \tag{1.1}$$

is called the *probability mass function* for the random variable X and is related to the distribution function via

$$p(x_i) = F(x_i) - F(x_i-) \quad \text{and} \quad F(x) = \sum_{x_i \leq x} p(x_i).$$

The distribution function for a discrete random variable is a step function, which increases only in jumps, the size of the jump at x_i being $p(x_i)$.

If $\Pr\{X = x\} = 0$ for every value of x, then the random variable X is called *continuous* and its distribution function $F(x)$ is a continuous function of x. If there is a nonnegative function $f(x) = f_X(x)$ defined for $-\infty < x < \infty$ such that

$$\Pr\{a < X \leq b\} = \int_a^b f(x)\mathrm{d}x \quad \text{for} -\infty < a < b < \infty, \tag{1.2}$$

then $f(x)$ is called the *probability density function* for the random variable X. If X has a probability density function $f(x)$, then X is continuous and

$$F(x) = \int_{-\infty}^x f(\xi)\mathrm{d}\xi, \quad -\infty < x < \infty.$$

If $F(x)$ is differentiable in x, then X has a probability density function given by

$$f(x) = \frac{\mathrm{d}}{\mathrm{d}x}F(x) = F'(x), \quad -\infty < x < \infty. \tag{1.3}$$

In differential form, (1.3) leads to the informal statement

$$\Pr\{x < X \leq x + \mathrm{d}x\} = F(x + \mathrm{d}x) - F(x) = \mathrm{d}F(x) = f(x)\mathrm{d}x. \tag{1.4}$$

We consider (1.4) to be a shorthand version of the more precise statement

$$\Pr\{x < X \leq x + \Delta x\} = f(x)\Delta x + o(\Delta x), \quad \Delta x \downarrow 0, \tag{1.5}$$

where $o(\Delta x)$ is a generic remainder term of order less than Δx as $\Delta x \downarrow 0$. That is, $o(\Delta x)$ represents any term for which $\lim_{\Delta x \downarrow 0} o(\Delta x)/\Delta x = 0$. By the fundamental theorem of calculus, equation (1.5) is valid whenever the probability density function is continuous at x.

While examples are known of continuous random variables that do not possess probability density functions, they do not arise in stochastic models of common natural phenomena.

1.2.3 Moments and Expected Values

If X is a discrete random variable, then its mth *moment* is given by

$$E[X^m] = \sum_i x_i^m \Pr\{X = x_i\} \tag{1.6}$$

[where the x_i are specified in (1.1)], provided that the infinite sum converges absolutely. Where the infinite sum diverges, the moment is said not to exist. If X is a continuous random variable with probability density function $f(x)$, then its mth moment is given by

$$E[X^m] = \int_{-\infty}^{+\infty} x^m f(x)dx, \tag{1.7}$$

provided that this integral converges absolutely.

The *first moment*, corresponding to $m = 1$, is commonly called the *mean* or *expected value* of X and written m_X or μ_X. The mth *central moment* of X is defined as the mth moment of the random variable $X - \mu_X$, provided that μ_X exists. The first central moment is zero. The second central moment is called the *variance* of X and written σ_x^2 or Var[X]. We have the equivalent formulas $\text{Var}[X] = E\left[(X - \mu)^2\right] = E\left[X^2\right] - \mu^2$.

The *median* of a random variable X is any value v with the property that

$$\Pr\{X \geq v\} \geq \frac{1}{2} \quad \text{and} \quad \Pr\{X \leq v\} \geq \frac{1}{2}.$$

If X is a random variable and g is a function, then $Y = g(X)$ is also a random variable. If X is a discrete random variable with possible values $x_1, x_2, \ldots$, then the expectation of $g(X)$ is given by

$$E[g(X)] = \sum_{i=1}^{\infty} g(x_i) \Pr\{X = x_i\}, \tag{1.8}$$

provided that the sum converges absolutely. If X is continuous and has the probability density function f_X, then the expected value of $g(X)$ is evaluated from

$$E[g(X)] = \int g(x)f_X(x)dx. \tag{1.9}$$

The general formula, covering both the discrete and continuous cases, is

$$E[g(X)] = \int g(x)dF_X(x), \tag{1.10}$$

where F_X is the distribution function of the random variable X. Technically speaking, the integral in (1.10) is a Lebesgue–Stieltjes integral. We do not require knowledge of such integrals in this text, but interpret (1.10) to signify (1.8) when X is a discrete random variable, and to represent (1.9) when X possesses a probability density f_X.

Let $F_Y(y) = \Pr\{Y \leq y\}$ denote the distribution function for $Y = g(X)$. When X is a discrete random variable, then

$$E[Y] = \sum_j y_j \Pr\{Y = y_j\}$$
$$= \sum_i g(x_i) \Pr\{X = x_i\}$$

if $y_i = g(x_i)$ and provided that the second sum converges absolutely. In general,

$$E[Y] = \int y \, dF_Y(y)$$
$$= \int g(x) dF_X(x). \tag{1.11}$$

If X is a discrete random variable, then so is $Y = g(X)$. It may be, however, that X is a continuous random variable, while Y is discrete (the reader should provide an example). Even so, one may compute $E[Y]$ from either form in (1.11) with the same result.

1.2.4 Joint Distribution Functions

Given a pair (X, Y) of random variables, their *joint distribution function* is the function F_{XY} of two real variables given by

$$F_{XY}(x, y) = F(x, y) = \Pr\{X \leq x \quad \text{and} \quad Y \leq y\}.$$

Usually, the subscripts X, Y will be omitted, unless ambiguity is possible. A joint distribution function F_{XY} is said to possess a (joint) probability density if there exists a function f_{XY} of two real variables for which

$$F_{XY}(x, y) = \int_{-\infty}^{X} \int_{-\infty}^{y} f_{XY}(\xi, \eta) d\eta \, d\xi \quad \text{for all } x, y.$$

The function $F_X(x) = \lim_{y \to \infty} F(x, y)$ is a distribution function, called the *marginal distribution function* of X. Similarly, $F_Y(y) = \lim_{y \to \infty} F(x, y)$ is the marginal distribution function of Y. If the distribution function F possesses the joint density function f, then the marginal density functions for X and Y are given, respectively, by

$$f_X(x) = \int\limits_{-\infty}^{+\infty} f(x, y) \, dy \quad \text{and} \quad f_Y(y) = \int\limits_{-\infty}^{+\infty} f(x, y) \, dx.$$

If X and Y are jointly distributed, then $E[X + Y] = E[X] + E[Y]$, provided only that all these moments exist.

Independence

If it happens that $F(x, y) = F_X(x) \times F_Y(y)$ for every choice of x, y, then the random variables X and Y are said to be *independent*. If X and Y are independent and possess a joint density function $f(x, y)$, then necessarily $f(x, y) = f_X(x) f_Y(y)$ for all x, y.

Given jointly distributed random variables X and Y having means μ_X and μ_Y and finite variances, the *covariance* of X and Y, written σ_{XY} or Cov$[X, Y]$, is the product moment $\sigma_{XY} = E[(X - \mu_X)(Y - \mu_Y)] = E[XY] - \mu_X \mu_Y$, and X and Y are said to be *uncorrelated* if their covariance is zero, i.e., $\sigma_{XY} = 0$. Independent random variables having finite variances are uncorrelated, but the converse is not true; there are uncorrelated random variables that are not independent.

Dividing the covariance σ_{XY} by the standard deviations σ_X and σ_Y defines the *correlation coefficient* $\rho = \sigma_{XY}/\sigma_X \sigma_Y$ for which $-1 \leq \rho \leq +1$.

The joint distribution function of any finite collection $X_1, \ldots, X_n$ of random variables is defined as the function

$$F(x_1, \ldots, x_n) = F_{X_1, \ldots, X_n}(x_1, \ldots, x_n)$$
$$= \Pr\{X_1 \leq x_1, \ldots, X_n \leq x_n\}.$$

If $F(x_1, \ldots, x_n) = F_{X_1}(x_1) \cdots F_{X_n}(x_n)$ for all values of $x_1, \ldots, x_n$, then the random variables $X_1, \ldots, X_n$ are said to be independent.

A joint distribution function $F(x_1, \ldots, x_n)$ is said to have a probability density function $f(\xi_1, \ldots, \xi_n)$ if

$$F(x_1, \ldots, x_n) = \int\limits_{-\infty}^{x_1} \cdots \int\limits_{-\infty}^{x_n} f(\xi_1, \ldots, \xi_n) \, d\xi_n \ldots d\xi_1,$$

for all values of $x_1, \ldots, x_n$.

Expectation

For jointly distributed random variables $X_1, \ldots, X_n$ and arbitrary functions $h_1, \ldots, h_m$ of n variables each,

$$E\left[\sum_{j=1}^{m} h_j(X_1, \ldots, X_n)\right] = \sum_{j=1}^{m} E[h_j(X_1, \ldots, X_n)],$$

provided only that all these moments exist.

1.2.5   Sums and Convolutions

If X and Y are independent random variables having distribution functions F_X and F_Y, respectively, then the distribution function of their sum $Z = X + Y$ is the *convolution* of F_X and F_Y:

$$F_Z(z) = \int_{-\infty}^{+\infty} F_X(z - \xi) dF_Y(\xi) = \int_{-\infty}^{+\infty} F_Y(z - \eta) dF_X(\eta). \tag{1.12}$$

If we specialize to the situation where X and Y have the probability densities f_X and f_Y, respectively, then the density function f_Z of the sum $Z = X + Y$ is the convolution of the densities f_X and f_Y:

$$f_Z(z) = \int_{-\infty}^{\infty} f_X(z - \eta) f_Y(\eta) d\eta = \int_{-\infty}^{+\infty} f_Y(z - \xi) f_X(\xi) d\xi. \tag{1.13}$$

Where X and Y are nonnegative random variables, the range of integration is correspondingly reduced to

$$f_Z(z) = \int_0^z f_X(z - \eta) f_Y(\eta) d\eta = \int_0^z f_Y(z - \xi) f_X(\xi) d\xi \quad \text{for } z \geq 0. \tag{1.14}$$

If X and Y are independent and have respective variances σ_X^2 and σ_Y^2, then the variance of the sum $Z = X + Y$ is the sum of the variances: $\sigma_Z^2 = \sigma_X^2 + \sigma_Y^2$. More generally, if $X_1, \ldots, X_n$ are independent random variables having variances $\sigma_1^2, \ldots, \sigma_n^2$, respectively, then the variance of the sum $Z = X_1 + \cdots + X_n$ is $\sigma_Z^2 = \sigma_1^2 + \cdots + \sigma_n^2$.

1.2.6   Change of Variable

Suppose that X is a random variable with probability density function f_X and that g is a strictly increasing differentiable function. Then, $Y = g(X)$ defines a random variable, and the event $\{Y \leq y\}$ is the same as the event $\{X \leq g^{-1}(y)\}$, where g^{-1} is the inverse function to g; i.e., $y = g(x)$ if and only if $x = g^{-1}(y)$. Thus, we obtain the correspondence $F_Y(y) = \Pr\{Y \leq y\} = \Pr\{X \leq g^{-1}(y)\} = F_X(g^{-1}(y))$ between the distribution function of Y and that of X. Recall the differential calculus formula

$$\frac{dg^{-1}}{dy} = \frac{1}{g'(x)} = \frac{1}{dg/dx}, \quad \text{where } y = g(x),$$

and use this in the chain rule of differentiation to obtain

$$f_Y(y) = \frac{dF_Y(y)}{dy} = \frac{dF_X(g^{-1}(y))}{dy} = f_X(x) \frac{1}{g'(x)}, \quad \text{where } y = g(x).$$

The formula

$$f_Y(y) = \frac{1}{g'(x)} f_X(x), \quad \text{where } y = g(x),$$ (1.15)

expresses the density function for Y in terms of the density for X when g is strictly increasing and differentiable.

1.2.7 Conditional Probability

For any events A and B, the *conditional probability* of A given B is written $\Pr\{A|B\}$ and defined by

$$\Pr\{A|B\} = \frac{\Pr\{A \cap B\}}{\Pr\{B\}} \quad \text{if } \Pr\{B\} > 0,$$ (1.16)

and is left undefined if $\Pr\{B\} = 0$. [When $\Pr\{B\} = 0$, the right side of (1.16) is the indeterminate quantity $\frac{0}{0}$.]

In stochastic modeling, conditional probabilities are rarely procured via (1.16) but instead are dictated as primary data by the circumstances of the application, and then (1.16) is applied in its equivalent multiplicative form

$$\Pr\{A \cap B\} = \Pr\{A|B\}\Pr\{B\}$$ (1.17)

to compute other probabilities. (An example follows shortly.) Central in this role is the *law of total probability*, which results from substituting $\Pr\{A \cap B_i\} = \Pr\{A|B_i\}\Pr\{B_i\}$ into $\Pr\{A\} = \sum_{i=1}^{\infty} \Pr\{A \cap B_i\}$, where $\Omega = B_1 \cup B_2 \cup \cdots$ and $B_i \cap B_j = \emptyset$ if $i \neq j$ (see Section 1.2.1), to yield

$$\Pr\{A\} = \sum_{i=1}^{\infty} \Pr\{A|B_i\}\Pr\{B_i\}.$$ (1.18)

Example Gold and silver coins are allocated among three urns labeled I, II, III according to the following table:

Urn	Number of Gold Coins	Number of Silver Coins
I	4	8
II	3	9
III	6	6

An urn is selected at random, all urns being equally likely, and then a coin is selected at random from that urn. Using the notation I, II, III for the events of selecting urns

I, II, and III, respectively, and G for the event of selecting a gold coin, then the problem description provides the following probabilities and conditional probabilities as data:

$$\Pr\{I\} = \frac{1}{3}, \qquad \Pr\{G|I\} = \frac{4}{12},$$

$$\Pr\{II\} = \frac{1}{3}, \qquad \Pr\{G|II\} = \frac{3}{12},$$

$$\Pr\{III\} = \frac{1}{3}, \qquad \Pr\{G|III\} = \frac{6}{12},$$

and we *calculate* the probability of selecting a gold coin according to (1.18), via

$$\Pr\{G\} = \Pr\{G|I\}\Pr\{I\} + \Pr\{G|II\}\Pr\{II\} + \Pr\{G|III\}\Pr\{III\}$$

$$= \frac{4}{12}\left(\frac{1}{3}\right) + \frac{3}{12}\left(\frac{1}{3}\right) + \frac{6}{12}\left(\frac{1}{3}\right) = \frac{13}{36}.$$

As seen here, more often than not conditional probabilities are given as data and are not the end result of calculation.

Discussion of conditional distributions and conditional expectation merits an entire chapter (Chapter 2).

1.2.8 Review of Axiomatic Probability Theory*

For the most part, this book studies random variables only through their distributions. In this spirit, we defined a random variable as a variable that takes on its values by chance. For some purposes, however, a little more precision and structure are needed.

Recall that the basic elements of probability theory are

1. the *sample space*, a set Ω whose elements ω correspond to the possible outcomes of an experiment;
2. the family of events, a collection $\mathcal{F}$ of subsets A of Ω: we say that the event A *occurs* if the outcome ω of the experiment is an element of A; and
3. the *probability measure*, a function P defined on $\mathcal{F}$ and satisfying
   (a)

$$0 = P[\emptyset] \le P[A] \le P[\Omega] = 1 \quad \text{for } A \in \mathcal{F}$$

$$(\emptyset = \text{the empty set})$$

   and
   (b)

$$P\left[\bigcup_{n=1}^{\infty} A_n\right] = \sum_{n=1}^{\infty} P[A_n], \tag{1.19}$$

---

* The material included in this review of axiomatic probability theory is not used in the remainder of the book. It is included in this review chapter only for the sake of completeness.

if the events $A_1, A_2, \ldots$ are disjoint, i.e., if $A_i \cap A_j = \emptyset$ when $i \neq j$. The triple $(\Omega, \mathcal{F}, P)$ is called a *probability space*.

Example When there are only a denumerable number of possible outcomes, say $\Omega = \{\omega_1, \omega_2, \ldots\}$, we may take $\mathcal{F}$ to be the collection of all subsets of Ω. If $p_1, p_2, \ldots$ are nonnegative numbers with $\Sigma_n p_n = 1$, the assignment

$$P[A] = \sum_{\omega_i \in A} p_i$$

determines a probability measure defined on $\mathcal{F}$.

It is not always desirable, consistent, or feasible to take the family of events as the collection of *all* subsets of Ω. Indeed, when Ω is nondenumerably infinite, it may not be possible to define a probability measure on the collection of all subsets maintaining the properties of (1.19). In whatever way we prescribe $\mathcal{F}$ such that (1.19) holds, the family of events $\mathcal{F}$ should satisfy

(a) $\emptyset$ is in $\mathcal{F}$ and Ω is in $\mathcal{F}$;

(b) A^c is in $\mathcal{F}$ whenever A is in $\mathcal{F}$, where $A^c = \{\omega \in \Omega; \omega \notin A\}$ (1.20)
        is the complement of A; and

(c) $\bigcup_{n=1}^{\infty} A_n$ is in $\mathcal{F}$ whenever A_n is in $\mathcal{F}$ for $n = 1, 2, \ldots$.

A collection $\mathcal{F}$ of subsets of a set Ω satisfying (1.20) is called a *σ-algebra*. If $\mathcal{F}$ is a σ-algebra, then

$$\bigcap_{n=1}^{\infty} A_n = \left(\bigcup_{n=1}^{\infty} A_n^c \right)^c$$

is in $\mathcal{F}$ whenever A_n is in $\mathcal{F}$ for $n = 1, 2, \ldots$. Manifestly, as a consequence, we find that finite unions and finite intersections of members of $\mathcal{F}$ are maintained in $\mathcal{F}$.

In this framework, a real random variable X is a real-valued function defined on Ω fulfilling certain "measurability" conditions given here. The distribution function of the random variable X is formally given by

$$\Pr\{a < X \leq b\} = P[\{\omega; a < X(\omega) \leq b\}].$$ (1.21)

In words, the probability that the random variable X takes a value in $(a, b]$ is calculated as the probability of the set of outcomes ω for which $a < X(\omega) \leq b$. If relation (1.21) is to have meaning, X cannot be an arbitrary function on Ω, but must satisfy the condition that

$$\{\omega; a < X(\omega) \leq b\} \text{ is in } \mathcal{F} \text{ for all real } a < b,$$

since $\mathcal{F}$ embodies the only sets A for which $P[A]$ is defined. In fact, by exploiting the properties (1.20) of the σ-algebra $\mathcal{F}$, we find that it is enough to require

$$\{\omega; X(\omega) \leq x\} \text{ is in } \mathcal{F} \text{ for all real } x.$$

Let $\mathcal{A}$ be any σ-algebra of subsets of Ω. We say that X is *measurable with respect to* $\mathcal{A}$, or more briefly $\mathcal{A}$-*measurable*, if

$$\{\omega; X(\omega) \leq x\} \text{ is in } \mathcal{A} \text{ for all real } x.$$

Thus, every real random variable is by definition $\mathcal{F}$-measurable. There may, in general, be smaller σ-algebras with respect to which X is also measurable.

The σ-algebra *generated* by a random variable X is defined to be the smallest σ-algebra with respect to which X is measurable. It is denoted by $\mathcal{F}(X)$ and consists exactly of those sets $\mathcal{A}$ that are in every σ-algebra $\mathcal{A}$ for which X is $\mathcal{A}$-measurable. For example, if X has only denumerably many possible values $x_1, x_2, \ldots$, the sets

$$A_i = \{\omega; X(\omega) = x_i\}, \quad i = 1, 2, \ldots,$$

form a countable *partition* of Ω, i.e.,

$$\Omega = \bigcup_{i=1}^{\infty} A_i,$$

and

$$A_i \cap A_j = \emptyset \quad \text{if } i \neq j,$$

and then $\mathcal{F}(X)$ includes precisely $\emptyset$, Ω, and every set that is the union of some of the A_i's.

Example For the reader completely unfamiliar with this framework, the following simple example will help illustrate the concepts. The experiment consists in tossing a nickel and a dime and observing "heads" or "tails." We take Ω to be

$$\Omega = \{(H,H), (H,T), (T,H), (T,T)\},$$

where, e.g., (H, T) stands for the outcome "nickel = heads, and dime = tails." We will take the collection of all subsets of Ω as the family of events. Assuming each outcome in Ω to be equally likely, we arrive at the probability measure:

$A \in \mathscr{F}$	$P[A]$	$A \in \mathscr{F}$	$P[A]$
$\emptyset$	0	Ω	1
$\{(H,H)\}$	$\frac{1}{4}$	$\{(H,T),(T,H),(T,T)\}$	$\frac{3}{4}$
$\{(H,T)\}$	$\frac{1}{4}$	$\{(H,H),(T,H),(T,T)\}$	$\frac{3}{4}$
$\{(T,H)\}$	$\frac{1}{4}$	$\{(H,H),(H,T),(T,T)\}$	$\frac{3}{4}$
$\{(T,T)\}$	$\frac{1}{4}$	$\{(H,H),(H,T),(T,H)\}$	$\frac{3}{4}$
$\{(H,H),(H,T)\}$	$\frac{1}{2}$	$\{(T,H),(T,T)\}$	$\frac{1}{2}$
$\{(H,H),(T,H)\}$	$\frac{1}{2}$	$\{(H,T),(T,T)\}$	$\frac{1}{2}$
$\{(H,H),(T,T)\}$	$\frac{1}{2}$	$\{(H,T),(T,H)\}$	$\frac{1}{2}$

The event "nickel is heads" is $\{(H,H),(H,T)\}$ and has, according to the table, probability $\frac{1}{2}$, as it should.

Let X_n be 1 if the nickel is heads, and 0 otherwise; let X_d be the corresponding random variable for the dime; and let $Z = X_n + X_d$ be the total number of heads. As functions on Ω, we have

$\omega \in \Omega$	$X_n(\omega)$	$X_d(\omega)$	$Z(\omega)$
(H,H)	1	1	2
(H,T)	1	0	1
(T,H)	0	1	1
(T,T)	0	0	0

Finally, the σ-algebras generated by X_n and Z are

$$\mathscr{F}(X_n) = \emptyset, \Omega, \{(H,H),(H,T)\}, \{(T,H),(T,T)\},$$

and

$$\mathscr{F}(Z) = \emptyset, \Omega, \{(H,H)\}, \{(H,T),(T,H)\}, \{(T,T)\},$$
$$\{(H,T),(T,H),(T,T)\}, \{(H,H),(T,T)\},$$
$$\{(H,H),(H,T),(T,H)\}.$$

$\mathscr{F}(X_n)$ contains four sets and $\mathscr{F}(Z)$ contains eight. Is X_n measurable with respect to $\mathscr{F}(Z)$, or vice versa?

Every pair X, Y of random variables determines a σ-algebra called the σ-algebra generated by X, Y. It is the smallest σ-algebra with respect to which both X and Y are measurable. This σ-algebra comprises exactly those sets A that are in every σ-algebra $\mathscr{A}$ for which X and Y are both $\mathscr{A}$-measurable. If both X and Y assume only denumerably many possible values, say $x_1, x_2, \ldots$ and $y_1, y_2, \ldots$, respectively, then the sets

$$A_{ij} = \{\omega; X(\omega) = x_j, Y(\omega) = y_j\}, \quad i, j = 1, 2, \ldots,$$

present a countable partition of Ω, and $\mathscr{F}(X, Y)$ consists precisely of $\emptyset$, Ω, and every set that is the union of some of the A_{ij}'s. Observe that X is measurable with respect to $\mathscr{F}(X, Y)$, and thus $\mathscr{F}(X) \subset \mathscr{F}(X, Y)$.

More generally, let $\{X(t); t \in T\}$ be any family of random variables. Then, the σ-algebra generated by $\{X(t); t \in T\}$ is the smallest σ-algebra with respect to which every random variable $X(t), t \in T$, is measurable. It is denoted by $\mathscr{F}\{X(t); t \in T\}$.

A special role is played by a distinguished σ-algebra of sets of real numbers. The σ-algebra of *Borel sets* is the σ-algebra generated by the identity function $f(x) = x$, for $x \in (-\infty, \infty)$. Alternatively, the σ-algebra of Borel sets is the smallest σ-algebra containing every interval of the form $(a, b], -\infty \leq a \leq b < +\infty$. A real-valued function of a real variable is said to be *Borel measurable* if it is measurable with respect to the σ-algebra of Borel sets.

Exercises

1.2.1 Let A and B be arbitrary, not necessarily disjoint, events. Use the law of total probability to verify the formula

$$\Pr\{A\} = \Pr\{AB\} + \Pr\{AB^c\},$$

where B^c is the complementary event to B (i.e., B^c occurs if and only if B does not occur).

1.2.2 Let A and B be arbitrary, not necessarily disjoint, events. Establish the general addition law

$$\Pr\{A \cup B\} = \Pr\{A\} + \Pr\{B\} - \Pr\{AB\}.$$

Hint: Apply the result of Exercise 1.2.1 to evaluate $\Pr\{AB^c\} = \Pr\{A\} - \Pr\{AB\}$. Then, apply the addition law to the disjoint events AB and AB^c, noting that $A = (AB) \cup (AB^c)$.

1.2.3 (a) Plot the distribution function

$$F(x) = \begin{cases} 0 & \text{for } x \leq 0, \\ x^3 & \text{for } 0 < x < 1, \\ 1 & \text{for } x \geq 1. \end{cases}$$

(b) Determine the corresponding density function $f(x)$ in the three regions (1) $x \leq 0$, (2) $0 < x < 1$, and (3) $1 \leq x$.

(c) What is the mean of the distribution?

(d) If X is a random variable following the distribution specified in (a), evaluate $\Pr\left\{\frac{1}{4} \le X \le \frac{3}{4}\right\}$.

1.2.4 Let Z be a discrete random variable having possible values $0, 1, 2$, and 3 and probability mass function

$$p(0) = \frac{1}{4}, \quad p(2) = \frac{1}{8},$$
$$p(1) = \frac{1}{2}, \quad p(3) = \frac{1}{8}.$$

(a) Plot the corresponding distribution function.
(b) Determine the mean $E[Z]$.
(c) Evaluate the variance $\text{Var}[Z]$.

1.2.5 Let A, B, and C be arbitrary events. Establish the addition law

$$\Pr\{A \cup B \cup C\} = \Pr\{A\} + \Pr\{B\} + \Pr\{C\} - \Pr\{AB\} - \Pr\{AC\} - \Pr\{BC\} + \Pr\{ABC\}.$$

1.2.6 Let X and Y be independent random variables having distribution functions F_X and F_Y, respectively.
(a) Define $Z = \max\{X, Y\}$ to be the larger of the two. Show that $F_Z(z) = F_X(z)F_Y(z)$ for all z.
(b) Define $W = \min\{X, Y\}$ to be the smaller of the two. Show that $F_W(w) = 1 - [1 - F_X(w)][1 - F_Y(w)]$ for all w.

1.2.7 Suppose X is a random variable having the probability density function

$$f(x) = \begin{cases} Rx^{R-1} & \text{for } 0 \le x \le 1, \\ 0 & \text{elsewhere,} \end{cases}$$

where $R > 0$ is a fixed parameter.
(a) Determine the distribution function $F_X(x)$.
(b) Determine the mean $E[X]$.
(c) Determine the variance $\text{Var}[X]$.

1.2.8 A random variable V has the distribution function

$$F(v) = \begin{cases} 0 & \text{for } v < 0, \\ 1 - (1 - v)^A & \text{for } 0 \le v \le 1, \\ 1 & \text{for } v > 1, \end{cases}$$

where $A > 0$ is a parameter. Determine the density function, mean, and variance.

1.2.9 Determine the distribution function, mean, and variance corresponding to the triangular density.

$$f(x) = \begin{cases} x & \text{for } 0 \le x \le 1, \\ 2 - x & \text{for } 1 \le x \le 2, \\ 0 & \text{elsewhere.} \end{cases}$$

1.2.10 Let 1_A be the indicator random variable associated with an event A, defined to be one if A occurs, and zero otherwise. Define A^c, the complement of event A, to be the event that occurs when A does not occur. Show

(a) $1_{A^c} = 1 - 1_A$.

(b) $1_{A \cap B} = 1_A 1_B = \min\{1_A, 1_B\}$.

(c) $1_{A \cup B} = \max\{1_A, 1_B\}$.

Problems

1.2.1 Thirteen cards numbered $1, \ldots, 13$ are shuffled and dealt one at a time. Say a *match* occurs on deal k if the kth card revealed is card number k. Let N be the total number of matches that occur in the thirteen cards. Determine $E[N]$.

Hint: Write $N = 1\{A_1\} + \cdots + 1\{A_{13}\}$ where A_k is the event that a match occurs on deal k.

1.2.2 Let N cards carry the distinct numbers $x_1, \ldots, x_n$. If two cards are drawn at random without replacement, show that the correlation coefficient ρ between the numbers appearing on the two cards is $-1/(N-1)$.

1.2.3 A population having N distinct elements is sampled with replacement. Because of repetitions, a random sample of size r may contain fewer than r distinct elements. Let S_r be the sample size necessary to get r distinct elements. Show that

$$E[S_r] = N\left(\frac{1}{N} + \frac{1}{N-1} + \cdots + \frac{1}{N-r+1}\right).$$

1.2.4 A fair coin is tossed until the first time that the same side appears twice in succession. Let N be the number of tosses required.

(a) Determine the probability mass function for N.

(b) Let A be the event that N is even and B be the event that $N \leq 6$. Evaluate $\Pr\{A\}$, $\Pr\{B\}$, and $\Pr\{AB\}$.

1.2.5 Two players, A and B, take turns on a gambling machine until one of them scores a success, the first to do so being the winner. Their probabilities for success on a single play are p for A and q for B, and successive plays are independent.

(a) Determine the probability that A wins the contest given that A plays first.

(b) Determine the mean number of plays required, given that A wins.

1.2.6 A pair of dice is tossed. If the two outcomes are equal, the dice are tossed again, and the process repeated. If the dice are unequal, their sum is recorded. Determine the probability mass function for the sum.

1.2.7 Let U and W be jointly distributed random variables. Show that U and W are independent if

$$\Pr\{U > u \text{ and } W > w\} = \Pr\{U > u\}\Pr\{W > w\} \quad \text{for all } u, w.$$

1.2.8 Suppose X is a random variable with finite mean μ and variance σ^2, and $Y = a + bX$ for certain constants $a, b \neq 0$. Determine the mean and variance for Y.

1.2.9 Determine the mean and variance for the probability mass function

$$p(k) = \frac{2(n-k)}{n(n-1)} \quad \text{for } k = 1, 2, \ldots, n.$$

1.2.10 Random variables X and Y are independent and have the probability mass functions

$$p_X(0) = \frac{1}{2}, \qquad p_Y(1) = \frac{1}{6},$$

$$p_X(3) = \frac{1}{2}, \qquad p_Y(2) = \frac{1}{3},$$

$$p_Y(3) = \frac{1}{2}.$$

Determine the probability mass function of the sum $Z = X + Y$.

1.2.11 Random variables U and V are independent and have the probability mass functions

$$p_U(0) = \frac{1}{3}, \qquad p_V(1) = \frac{1}{2},$$

$$p_U(1) = \frac{1}{3}, \qquad p_V(2) = \frac{1}{2}.$$

$$p_U(2) = \frac{1}{3},$$

Determine the probability mass function of the sum $W = U + V$.

1.2.12 Let $U, V,$ and W be independent random variables with equal variances σ^2. Define $X = U + W$ and $Y = V - W$. Find the covariance between X and Y.

1.2.13 Let X and Y be independent random variables each with the uniform probability density function

$$f(x) = \begin{cases} 1 & \text{for } 0 < x < 1, \\ 0 & \text{elsewhere.} \end{cases}$$

Find the joint probability density function of U and V, where $U = \max\{X, Y\}$ and $V = \min\{X, Y\}$.

1.3 The Major Discrete Distributions

The most important discrete probability distributions and their relevant properties are summarized in this section. The exposition is brief, since most readers will be familiar with this material from an earlier course in probability.

1.3.1 Bernoulli Distribution

A random variable X following the Bernoulli distribution with parameter p has only two possible values, 0 and 1, and the probability mass function is $p(1) = p$ and $p(0) = 1 - p$, where $0 < p < 1$, and the mean and variance are $E[X] = p$ and $\text{Var}[X] = p(1 - p)$, respectively.

Bernoulli random variables occur frequently as indicators of events. The *indicator* of an event A is the random variable

$$1(A) = 1_A = \begin{cases} 1 & \text{if } A \text{ occurs,} \\ 0 & \text{if } A \text{ does not occur.} \end{cases} \tag{1.22}$$

Then, 1_A is a Bernoulli random variable with parameter $p = E[1_A] = \Pr\{A\}$.

The simple expedient of using indicators often reduces formidable calculations into trivial ones. For example, let $\alpha_1, \alpha_2, \ldots, \alpha_n$ be arbitrary real numbers and $A_1, A_2, \ldots, A_n$ be events, and consider the problem of showing that

$$\sum_{i=1}^{n} \sum_{i=1}^{n} \alpha_j \alpha_j \Pr\{A_i \cap A_j\} \geq 0. \tag{1.23}$$

Attacked directly, the problem is difficult. But bringing in the indicators $1(A_i)$ and observing that

$$0 \leq \left\{ \sum_{i=1}^{n} \alpha_i 1(A_i) \right\}^2 = \left\{ \sum_{i=1}^{n} \alpha_i 1(A_i) \right\} \left\{ \sum_{j=1}^{n} \alpha_j 1(A_j) \right\}$$

$$= \sum_{i=1}^{n} \sum_{j=1}^{n} \alpha_i \alpha_j 1(A_i) 1(A_j) = \sum_{i=1}^{n} \sum_{j=1}^{n} \alpha_i \alpha_j 1(A_i \cap A_j)$$

gives, after taking expectations,

$$0 \leq E\left[\left\{ \sum_{i=1}^{n} \alpha_i 1(A_i) \right\}^2 \right] = \sum_{i=1}^{n} \sum_{j=1}^{n} \alpha_i \alpha_j E[1(A_i \cap A_j)]$$

$$= \sum_{i=1}^{n} \sum_{j=1}^{n} \alpha_i \alpha_j \Pr\{A_i \cap A_j\},$$

and the demonstration of (1.23) is complete.

1.3.2 Binomial Distribution

Consider independent events $A_1, A_2, \ldots, A_n$, all having the same probability $p = \Pr\{A_i\}$ of occurrence. Let Y count the total number of events among $A_1, \ldots, A_n$ that occur.

Then, Y has a binomial distribution with parameters n and p. The probability mass function is

$$p_Y(k) = \Pr\{Y = k\}$$
$$= \frac{n!}{k!\,(n-k)!}p^k(1-p)^{n-k} \quad \text{for } k = 0, 1, \ldots, n. \tag{1.24}$$

Writing Y as a sum of indicators in the form $Y = \mathbf{1}(A_1) + \cdots + \mathbf{1}(A_n)$ makes it easy to determine the moments

$$E[Y] = E[\mathbf{1}(A_1)] + \cdots + E[\mathbf{1}(A_n)] = np,$$

and using independence, we can also determine that

$$\text{Var}[Y] = \text{Var}[\mathbf{1}(A_1)] + \cdots + \text{Var}[\mathbf{1}(A_n)] = np(1-p).$$

Briefly, we think of a binomial random variable as counting the number of "successes" in n independent trials where there is a constant probability p of success on any single trial.

1.3.3 Geometric and Negative Binominal Distributions

Let $A_1, A_2, \ldots$ be independent events having a common probability $p = \Pr\{A_i\}$ of occurrence. Say that trial k is a success (S) or failure (F), depending on whether A_k occurs or not, and let Z count the number of *failures* prior to the first success. To be precise, $Z = k$ if and only if $\mathbf{1}(A_1) = 0, \ldots, \mathbf{1}(A_k) = 0$, and $\mathbf{1}(A_{k+1}) = 1$. Then, Z has a geometric distribution with parameter p. The probability mass function is

$$p_Z(k) = p(1-p)^k \quad \text{for } k = 0, 1, \ldots, \tag{1.25}$$

and the first two moments are

$$E[Z] = \frac{1-p}{p}; \quad \text{Var}[Z] = \frac{1-p}{p^2}.$$

Sometimes the term "geometric distribution" is used in referring to the probability mass function

$$p_{Z'}(k) = p(1-p)^{k-1} \quad \text{for } k = 1, 2, \ldots. \tag{1.26}$$

This is merely the distribution of the random variable $Z' = 1 + Z$, the number of *trials* until the first success. Hence $E[Z'] = 1 + E[Z] = 1/p$, and $\text{Var}[Z'] = \text{Var}[Z] = (1-p)/p^2$.

Now fix an integer $r \geq 1$ and let W_r count the number of failures observed before the rth success in $A_1, A_2, \ldots$. Then, W_r has a *negative binominal* distribution with parameters r and p. The event $W_r = k$ calls for (A) exactly $r-1$ successes in the first

$k + r - 1$ trials, followed by (B) a success on trial $k + r$. The probability for (A) is obtained from a binomial distribution, and the probability for (B) is simply p, which leads to the following probability mass function for W_r:

$$p(k) = \Pr\{W_r = k\} = \frac{(k + r - 1)!}{(r - 1)!k!} p^r (1 - p)^k, \quad k = 0, 1, \ldots . \tag{1.27}$$

Another way of writing W_r is as the sum $W_r = Z_1 + \cdots + Z_r$, where $Z_1, \ldots, Z_r$ are independent random variables each having the geometric distribution of (1.25). This formulation readily yields the moments

$$E[W_r] = \frac{r(1 - p)}{p}; \quad \text{Var}[W_r] = \frac{r(1 - p)}{p^2}. \tag{1.28}$$

1.3.4 The Poisson Distribution

If distributions were graded on a scale of one to ten, the Poisson clearly merits a 10. It plays a role in the class of discrete distributions that parallels in some sense that of the normal distribution in the continuous class. The Poisson distribution occurs often in natural phenomena, for powerful and convincing reasons (the law of rare events, see later in this section). At the same time, the Poisson distribution has many elegant and surprising mathematical properties that make analysis a pleasure.

The Poisson distribution with parameter $\lambda > 0$ has the probability mass function

$$p(k) = \frac{\lambda^k e^{-\lambda}}{k!} \quad \text{for } k = 0, 1, \ldots . \tag{1.29}$$

Using this series expansion

$$e^\lambda = 1 + \lambda + \frac{\lambda^2}{2!} + \frac{\lambda^3}{3!} + \cdots \tag{1.30}$$

we see that $\Sigma_{k \geq 0} p(k) = 1$. The same series helps calculate the mean via

$$\sum_{k=0}^{\infty} k p(k) = \sum_{k=1}^{\infty} k \frac{\lambda^k e^{-\lambda}}{k!} = \lambda e^{-\lambda} \sum_{k=1}^{\infty} \frac{\lambda^{k-1}}{(k-1)!} = \lambda.$$

The same trick works on the variance, beginning with

$$\sum_{k=0}^{\infty} k(k-1) p(k) = \sum_{k=2}^{\infty} k(k-1) \frac{\lambda^k e^{-\lambda}}{k!} = \lambda^2 e^{-\lambda} \sum_{k=2}^{\infty} \frac{\lambda^{k-2}}{(k-2)!} = \lambda^2.$$

Written in terms of a random variable X having the Poisson distribution with parameter λ, we have just calculated $E[X] = \lambda$ and $E[X(X - 1)] = \lambda^2$, whence $E[X^2] = E[X(X - 1)] + E[X] = \lambda^2 + \lambda$ and $\text{Var}[X] = E[X^2] - \{E[X]\}^2 = \lambda$. That is, the mean and variance are both the same and equal to the parameter λ of the Poisson distribution.

The simplest form of the law of rare events asserts that the binomial distribution with parameters n and p converges to the Poisson with parameter λ if $n \to \infty$ and $p \to 0$ in such a way that $\lambda = np$ remains constant. In words, given an indefinitely large number of independent trials, where success on each trial occurs with the same arbitrarily small probability, then the total number of successes will follow, approximately, a Poisson distribution.

The proof is a relatively simple manipulation of limits. We begin by writing the binomial distribution in the form

$$\Pr\{X = k\} = \frac{n!}{k!\,(n-k)!}p^k(1-p)^{n-k}$$

$$= n(n-1)\cdots(n-k+1)\frac{p^k(1-p)^n}{k!\,(1-p)^k}$$

and then substitute $p = \lambda/n$ to get

$$\Pr\{X = k\} = n(n-1)\cdots(n-k+1)\frac{\left(\frac{\lambda}{n}\right)^k\left(1-\frac{\lambda}{n}\right)^n}{k!\,\left(1-\frac{\lambda}{n}\right)^k}$$

$$= 1\left(1-\frac{1}{n}\right)\cdots\left(1-\frac{k-1}{n}\right)\frac{\lambda^k\left(1-\frac{\lambda}{n}\right)^n}{k!\,\left(1-\frac{\lambda}{n}\right)^k}.$$

Now let $n \to \infty$ and observe that

$$1\left(1-\frac{1}{n}\right)\cdots\left(1-\frac{k-1}{n}\right) \to 1 \quad \text{as } n \to \infty;$$

$$\left(1-\frac{\lambda}{n}\right)^n \to e^{-\lambda} \quad \text{as } n \to \infty;$$

and

$$\left(1-\frac{\lambda}{n}\right)^k \to 1 \quad \text{as } n \to \infty;$$

to obtain the Poisson distribution

$$\Pr\{X = k\} = \frac{\lambda^k e^{-\lambda}}{k!} \quad \text{for } k = 0, 1, \ldots$$

in the limit. Extended forms of the law of rare events are presented in Chapter 5.

Example *You Be the Judge*   In a purse-snatching incident, a woman described her assailant as being seven feet tall and wearing an orange hat, red shirt, green trousers, and yellow shoes. A short while later and a few blocks away a person fitting that description was seen and charged with the crime.

In court, the prosecution argued that the characteristics of the assailant were so rare as to make the evidence overwhelming that the defendant was the criminal.

The defense argued that the description of the assailant was rare, and that, therefore, the number of people fitting the description should follow a Poisson distribution. Since one person fitting the description was found, the best estimate for the parameter is $\lambda = 1$. Finally, they argued that the relevant computation is the conditional probability that there is at least one other person at large fitting the description given that one was observed. The defense calculated

$$
\begin{aligned}
\Pr\{X \geq 2 | X \geq 1\} &= \frac{1 - \Pr\{X = 0\} - \Pr\{X = 1\}}{1 - \Pr\{X = 0\}} \\
&= \frac{1 - e^{-1} - e^{-1}}{1 - e^{-1}} = 0.4180,
\end{aligned}
$$

and since this figure is rather large, they argued that the circumstantial evidence arising out of the unusual description was too weak to satisfy the "beyond a reasonable doubt" criterion for guilt in criminal cases.

1.3.5   The Multinomial Distribution

This is a joint distribution of r variables in which only nonnegative integer values $0, \ldots, n$ are possible. The joint probability mass function is

$$
\begin{aligned}
&\Pr\{X_1 = k_1, \ldots, X_r = k_r\} \\
&= \begin{cases} \dfrac{n!}{k_1! \cdots k_r!} p_1^{k_1} \cdots p_r^{k_r} & \text{if } k_1 + \cdots + k_r = n, \\ 0 & \text{otherwise,} \end{cases}
\end{aligned}
\tag{1.31}
$$

where $p_i > 0$ for $i = 1, \ldots, r$ and $p_1 + \cdots + p_r = 1$.

Some moments are $E[X_i] = np_i$, $\text{Var}[X_i] = np_i(1 - p_i)$, and $\text{Cov}[X_i X_j] = -np_i p_j$.

The multinomial distribution generalizes the binomial. Consider an experiment having a total of r possible outcomes, and let the corresponding probabilities be $p_1, \ldots, p_r$, respectively. Now perform n independent replications of the experiment and let X_i record the total number of times that the ith type outcome is observed in the n trials. Then, $X_1, \ldots, X_r$ has the multinomial distribution given in (1.31).

Exercises

1.3.1 Consider tossing a fair coin five times and counting the total number of heads that appear. What is the probability that this total is three?

1.3.2 A fraction $p = 0.05$ of the items coming off a production process are defective. If a random sample of 10 items is taken from the output of the process, what is the probability that the sample contains exactly one defective item? What is the probability that the sample contains one or fewer defective items?

1.3.3 A fraction $p = 0.05$ of the items coming off of a production process are defective. The output of the process is sampled, one by one, in a random manner. What is the probability that the first defective item found is the tenth item sampled?

1.3.4 A Poisson distributed random variable X has a mean of $\lambda = 2$. What is the probability that X equals 2? What is the probability that X is less than or equal to 2?

1.3.5 The number of bacteria in a prescribed area of a slide containing a sample of well water has a Poisson distribution with parameter 5. What is the probability that the slide shows 8 or more bacteria?

1.3.6 The discrete uniform distribution on $\{1, \ldots, n\}$ corresponds to the probability mass function

$$p(k) = \begin{cases} \dfrac{1}{n} & \text{for } k = 1, \ldots, n, \\ 0 & \text{elsewhere.} \end{cases}$$

(a) Determine the mean and variance.

(b) Suppose X and Y are independent random variables, each having the discrete uniform distribution on $\{0, \ldots, n\}$. Determine the probability mass function for the sum $Z = X + Y$.

(c) Under the assumptions of (b), determine the probability mass function for the minimum $U = \min\{X, Y\}$.

Problems

1.3.1 Suppose that X has a discrete uniform distribution on the integers $0, 1, \ldots, 9$, and Y is independent and has the probability distribution $\Pr\{Y = k\} = a_k$ for $k = 0, 1, \ldots$. What is the distribution of $Z = X + Y \pmod{10}$, their sum modulo 10?

1.3.2 The *mode* of a probability mass function $p(k)$ is any value k^* for which $p(k^*) \geq p(k)$ for all k. Determine the mode(s) for
(a) The Poisson distribution with parameter $\lambda > 0$.
(b) The binomial distribution with parameters n and p.

1.3.3 Let X be a Poisson random variable with parameter λ. Determine the probability that X is odd.

1.3.4 Let U be a Poisson random variable with mean μ. Determine the expected value of the random variable $V = 1/(1 + U)$.

1.3.5 Let $Y = N - X$ where X has a binomial distribution with parameters N and p. Evaluate the product moment $E[XY]$ and the covariance $\mathrm{Cov}[X, Y]$.

1.3.6 Suppose (X_1, X_2, X_3) has a multinomial distribution with parameters M and $\pi_i > 0$ for $i = 1, 2, 3$, with $\pi_1 + \pi_2 + \pi_3 = 1$.
(a) Determine the marginal distribution for X_1.
(b) Find the distribution for $N = X_1 + X_2$.
(c) What is the conditional probability $\Pr\{X_1 = k | N = n\}$ for $0 \leq k \leq n$?

1.3.7 Let X and Y be independent Poisson distributed random variables having means μ and ν, respectively. Evaluate the convolution of their mass functions to determine the probability distribution of their sum $Z = X + Y$.

1.3.8 Let X and Y be independent binomial random variables having parameters (N, p) and (M, p), respectively. Let $Z = X + Y$.

(a) Argue that Z has a binomial distribution with parameters $(N+M, p)$ by writing X and Y as appropriate sums of Bernoulli random variables.

(b) Validate the result in (a) by evaluating the necessary convolution.

1.3.9 Suppose that X and Y are independent random variables with the geometric distribution

$$p(k) = (1 - \pi)\pi^k \quad \text{for } k = 0, 1, \ldots.$$

Perform the appropriate convolution to identify the distribution of $Z = X + Y$ as a negative binomial.

1.3.10 Determine numerical values to three decimal places for $\Pr\{X = k\}, k = 0, 1, 2,$ when

(a) X has a binomial distribution with parameters $n = 10$ and $p = 0.1$.

(b) X has a binomial distribution with parameters $n = 100$ and $p = 0.01$.

(c) X has a Poisson distribution with parameter $\lambda = 1$.

1.3.11 Let X and Y be independent random variables sharing the geometric distribution whose mass function is

$$p(k) = (1 - \pi)\pi^k \quad \text{for } k = 0, 1, \ldots,$$

where $0 < \pi < 1$. Let $U = \min\{X, Y\}$, $V = \max\{X, Y\}$, and $W = V - U$. Determine the joint probability mass function for U and W and show that U and W are independent.

1.3.12 Suppose that the telephone calls coming into a certain switchboard during a one-minute time interval follow a Poisson distribution with mean $\lambda = 4$. If the switchboard can handle at most 6 calls per minute, what is the probability that the switchboard will receive more calls than it can handle during a specified one-minute interval?

1.3.13 Suppose that a sample of 10 is taken from a day's output of a machine that produces parts of which 5% are normally defective. If 100% of a day's production is inspected whenever the sample of 10 gives 2 or more defective parts, then what is the probability that 100% of a day's production will be inspected? What assumptions did you make?

1.3.14 Suppose that a random variable Z has the geometric distribution

$$p_Z(k) = p(1 - p)^k \quad \text{for } k = 0, 1, \ldots,$$

where $p = 0.10$.

(a) Evaluate the mean and variance of Z.

(b) What is the probability that Z strictly exceeds 10?

1.3.15 Suppose that X is a Poisson distributed random variable with mean $\lambda = 2$. Determine $\Pr\{X \le \lambda\}$.

1.3.16 Consider the generalized geometric distribution defined by

$$p_k = b(1 - p)^k \quad \text{for } k = 1, 2, \ldots,$$

and

$$p_0 = 1 - \sum_{k=1}^{\infty} p_k,$$

where $0 < p < 1$ and $p \leq b \leq p/(1-p)$.

(a) Evaluate p_0 in terms of b and p.

(b) What does the generalized geometric distribution reduce to when $b = p$? When $b = p/(1-p)$?

(c) Show that $N = X + Z$ has the generalized geometric distribution when X is a Bernoulli random variable for which $\Pr\{X = 1\} = \alpha$, $0 < \alpha < 1$, and Z independently has the usual geometric distribution given in (1.25).

1.4   Important Continuous Distributions

For future reference, this section catalogs several continuous distributions and some of their properties.

1.4.1   The Normal Distribution

The *normal distribution* with parameters μ and $\sigma^2 > 0$ is given by the familiar bell-shaped probability density function

$$\phi\left(x; \mu, \sigma^2\right) = \frac{1}{\sqrt{2\pi}\sigma} e^{-(x-\mu)^2/2\sigma^2}, \quad -\infty < x < \infty. \tag{1.32}$$

The density function is symmetric about the point μ, and the parameter σ^2 is the variance of the distribution. The case $\mu = 0$ and $\sigma^2 = 1$ is referred to as the *standard normal distribution*. If X is normally distributed with mean μ and variance σ^2, then $Z = (X - \mu)/\sigma$ has a standard normal distribution. By this means, probability statements about arbitrary normal random variables can be reduced to equivalent statements about standard normal random variables. The standard normal density and distribution functions are given respectively by

$$\phi(\xi) = \frac{1}{\sqrt{2\pi}} e^{-\xi^2/2}, \quad -\infty < \xi < \infty, \tag{1.33}$$

and

$$\Phi(x) = \int_{-\infty}^{x} \phi(\xi) d\xi, \quad -\infty < x < \infty. \tag{1.34}$$

The *central limit theorem* explains in part the wide prevalence of the normal distribution in nature. A simple form of this aptly named result concerns the partial sums

$S_n = \xi_1 + \cdots + \xi_n$ of independent and identically distributed summands $\xi_1, \xi_2, \ldots$ having finite means $\mu = E[\xi_k]$ and finite variances $\sigma^2 = \text{Var}[\xi_k]$. In this case, the central limit theorem asserts that

$$\lim_{n \to \infty} \Pr\left\{\frac{S_n - n\mu}{\sigma\sqrt{n}} \leq x\right\} = \Phi(x) \quad \text{for all } x. \tag{1.35}$$

The precise statement of the theorem's conclusion is given by equation (1.35). Intuition is sometimes enhanced by the looser statement that, for large n, the sum S_n is approximately normally distributed with mean $n\mu$ and variance $n\sigma^2$.

In practical terms we expect the normal distribution to arise whenever the numerical outcome of an experiment results from numerous small additive effects, all operating independently, and where no single or small group of effects is dominant.

The Lognormal Distribution

If the natural logarithm of a nonnegative random variable V is normally distributed, then V is said to have a lognormal distribution. Conversely, if X is normally distributed with mean μ and variance σ^2, then $V = e^X$ defines a lognormally distributed random variable. The change-of-variable formula (1.15) applies to give the density function for V to be

$$f_V(v) = \frac{1}{\sqrt{2\pi}\sigma v}\exp\left\{-\frac{1}{2}\left(\frac{\ln v - \mu}{\sigma}\right)^2\right\}, \quad v \geq 0. \tag{1.36}$$

The mean and variance are, respectively,

$$E[V] = \exp\left\{\mu + \frac{1}{2}\sigma^2\right\},$$
$$\text{Var}[V] = \exp\left\{2\left(\mu + \frac{1}{2}\sigma^2\right)\right\}\left[\exp\left\{\sigma^2\right\} - 1\right]. \tag{1.37}$$

1.4.2 The Exponential Distribution

A nonnegative random variable T is said to have an exponential distribution with parameter $\lambda > 0$ if the probability density function is

$$f_T(t) = \begin{cases} \lambda e^{-\lambda t} & \text{for } t \geq 0, \\ 0 & \text{for } t < 0. \end{cases} \tag{1.38}$$

The corresponding distribution function is

$$F_T(t) = \begin{cases} 1 - e^{-\lambda t} & \text{for } t \geq 0, \\ 0 & \text{for } t < 0, \end{cases} \tag{1.39}$$

and the mean and variance are given, respectively, by

$$E[T] = \frac{1}{\lambda} \quad \text{and} \quad \text{Var}[T] = \frac{1}{\lambda^2}.$$

Note that the parameter is the reciprocal of the mean and *not* the mean itself.

The exponential distribution is fundamental in the theory of continuous-time Markov chains (see Chapter 5), due in major part to its *memoryless property*, as now explained. Think of T as a lifetime and, given that the unit has survived up to time t, ask for the conditional distribution of the remaining life $T - t$. Equivalently, for $x > 0$ determine the conditional probability $\Pr\{T - t > x | T > t\}$. Directly applying the definition of conditional probability (see Section 1.2.7), we obtain

$$\Pr\{T - t > x | T > t\} = \frac{\Pr\{T > t + x, T > t\}}{\Pr\{T > t\}}$$

$$= \frac{\Pr\{T > t + x\}}{\Pr\{T > t\}} \quad \text{(because } x > 0\text{)} \tag{1.40}$$

$$= \frac{e^{-\lambda(t+x)}}{e^{-\lambda t}} \quad \text{[from (1.39)]}$$

$$= e^{-\lambda x}.$$

There is no memory in the sense that $\Pr\{T - t > x | T > t\} = e^{-\lambda x} = \Pr\{T > x\}$, and an item that has survived for t units of time has a remaining lifetime that is statistically the same as that for a new item.

To view the memoryless property somewhat differently, we introduce the *hazard rate* or *failure rate* $r(s)$ associated with a nonnegative random variable S having continuous density $g(s)$ and distribution function $G(s) < 1$. The failure rate is defined by

$$r(s) = \frac{g(s)}{1 - G(s)} \quad \text{for } s > 0. \tag{1.41}$$

We obtain the interpretation by calculating (see Section 1.2.2)

$$\Pr\{s < S \leq s + \Delta s | s < S\} = \frac{\Pr\{s < S \leq s + \Delta s\}}{\Pr\{s < S\}}$$

$$= \frac{g(s)\Delta s}{1 - G(s)} + o(\Delta s) \quad \text{[from (1.5)]}$$

$$= r(s)\Delta s + o(\Delta s).$$

An item that has survived to time s will then fail in the interval $(s, s + \Delta s]$ with conditional probability $r(s)\Delta s + o(\Delta s)$, thus motivating the name "failure rate."

We can invert (4.10) by integrating

$$-r(s) = \frac{-g(s)}{1 - G(s)} = \frac{d[1 - G(s)]/ds}{1 - G(s)} = \frac{d\{\ln[1 - G(s)]\}}{ds}$$

to obtain

$$-\int_0^t r(s)ds = \ln[1 - G(t)],$$

or

$$G(t) = 1 - \exp\left\{ -\int_0^t r(s)ds \right\}, \quad t \geq 0,$$

which gives the distribution function explicitly in terms of the hazard rate.

The exponential distribution is uniquely the continuous distribution with the *constant* failure rate $r(t) \equiv \lambda$. (See Exercise 1.4.8 for the discrete analog.) The failure rate does not vary in time, another reflection of the memoryless property.

Section 1.5 contains several exercises concerning the exponential distribution. In addition to providing practice in relevant algebraic and calculus manipulations, these exercises are designed to enhance the reader's intuition concerning the exponential law.

1.4.3 The Uniform Distribution

A random variable U is uniformly distributed over the interval $[a, b]$, where $a < b$, if it has the probability density function

$$f_U(u) = \begin{cases} \dfrac{1}{b-a} & \text{for } a \leq u \leq b, \\ 0 & \text{elsewhere.} \end{cases} \tag{1.42}$$

The uniform distribution extends the notion of "equally likely" to the continuous case. The distribution function is

$$F_U(x) = \begin{cases} 0 & \text{for } u \leq a, \\ \dfrac{x-a}{b-a} & \text{for } a < x \leq b, \\ 1 & \text{for } x > b, \end{cases} \tag{1.43}$$

and the mean and variance are, respectively,

$$E[U] = \frac{1}{2}(a+b) \quad \text{and} \quad \text{Var}[U] = \frac{(b-a)^2}{12}.$$

The uniform distribution on the unit interval $[0, 1]$, for which $a = 0$ and $b = 1$, is most prevalent.

1.4.4 The Gamma Distribution

The gamma distribution with parameters $\alpha > 0$ and $\lambda > 0$ has probability density function

$$f(x) = \frac{\lambda}{\Gamma(\alpha)}(\lambda x)^{\alpha-1} e^{-\lambda x} \quad \text{for } x > 0. \tag{1.44}$$

Given an integer number α of independent exponentially distributed random variables $Y_1, \ldots, Y_\alpha$ having common parameter λ, then their sum $X_\alpha = Y_1 + \cdots + Y_\alpha$ has the gamma density of (1.44), from which we obtain the moments

$$E[X_\alpha] = \frac{\alpha}{\lambda} \quad \text{and} \quad \text{Var}[X_\alpha] = \frac{\alpha}{\lambda^2},$$

with these moment formulas holding for noninteger α as well.

1.4.5 The Beta Distribution

The beta density with parameters $\alpha > 0$ and $\beta > 0$ is given by

$$f(x) = \begin{cases} \dfrac{\Gamma(\alpha + \beta)}{\Gamma(\alpha)\Gamma(\beta)} x^{\alpha-1}(1-x)^{\beta-1} & \text{for } 0 < x < 1, \\ 0 & \text{elsewhere.} \end{cases} \tag{1.45}$$

The mean and variance are, respectively,

$$E[X] = \frac{\alpha}{\alpha + \beta} \quad \text{and} \quad \text{Var}[X] = \frac{\alpha\beta}{(\alpha + \beta)^2(\alpha + \beta + 1)}.$$

(The gamma and beta functions are defined and briefly discussed in Section 1.6.)

1.4.6 The Joint Normal Distribution

Let $\sigma_X, \sigma_Y, \mu_X, \mu_Y$, and ρ be real constants subject to $\sigma_X > 0, \sigma_Y > 0$, and $-1 < \rho < 1$. For real variables x and y, define

$$Q(x, y) = \frac{1}{1 - \rho^2} \left\{ \left(\frac{x - \mu_X}{\sigma_X} \right)^2 - 2\rho \left(\frac{x - \mu_X}{\sigma_X} \right) \left(\frac{y - \mu_Y}{\sigma_Y} \right) + \left(\frac{y - \mu_Y}{\sigma_Y} \right)^2 \right\}. \tag{1.46}$$

The joint normal (or bivariate normal) distribution for random variables X, Y is defined by the density function

$$\phi_{X,Y}(x, y) = \frac{1}{2\pi \sigma_X \sigma_Y \sqrt{1 - \rho^2}}$$

$$\times \exp\left\{ -\frac{1}{2} Q(x, y) \right\}, \quad -\infty < x, y < \infty. \tag{1.47}$$

The moments are

$$E[X] = \mu_X, \quad E[Y] = \mu_Y,$$
$$\text{Var}[X] = \sigma_X^2, \quad \text{Var}[Y] = \sigma_Y^2,$$

and

$$\text{Cov}[X, Y] = E[(X - \mu_X)(Y - \mu_Y)] = \rho \sigma_X \sigma_Y.$$

The dimensionless parameter ρ is called the *correlation coefficient*. When ρ is positive, then positive values of X are (stochastically) associated with positive values of Y. When ρ is negative, then positive values of X are associated with negative values of Y. If $\rho = 0$, then X and Y are independent random variables.

Linear Combinations of Normally Distributed Random Variables

Suppose X and Y have the bivariate normal density (1.47), and let $Z = aX + bY$ for arbitrary constants a, b. Then Z is normally distributed with mean

$$E[Z] = a\mu_X + b\mu_Y$$

and variance

$$\text{Var}[X] = a^2\sigma_X^2 + 2ab\rho\sigma_X\sigma_Y + b^2\sigma_Y^2.$$

A random vector $X_1, \ldots, X_n$, is said to have a *multivariate normal distribution*, or a *joint normal distribution*, if every linear combination $\alpha_1 X_1 + \cdots + \alpha_n X_n$, α_i real has a univariate normal distribution. Obviously, if $X_1, \ldots, X_n$ has a joint normal distribution, then so does the random vector $Y_1, \ldots, Y_m$, defined by the linear transformation in which

$$Y_j = \alpha_{j1}X_1 + \cdots + \alpha_{jn}X_n, \quad \text{for } j = 1, \ldots, m,$$

for arbitrary constants α_{ji}.

Exercises

1.4.1 The lifetime, in years, of a certain class of light bulbs has an exponential distribution with parameter $\lambda = 2$. What is the probability that a bulb selected at random from this class will last more than 1.5 years? What is the probability that a bulb selected at random will last exactly 1.5 years?

1.4.2 The median of a random variable X is any value a for which $\Pr\{X \le a\} \ge \frac{1}{2}$ and $\Pr\{X \ge a\} \ge \frac{1}{2}$. Determine the median of an exponentially distributed random variable with parameter λ. Compare the median to the mean.

1.4.3 The lengths, in inches, of cotton fibers used in a certain mill are exponentially distributed random variables with parameter λ. It is decided to convert all measurements in this mill to the metric system. Describe the probability distribution of the length, in centimeters, of cotton fibers in this mill.

1.4.4 Twelve independent random variables, each uniformly distributed over the interval $(0, 1]$, are added, and 6 is subtracted from the total. Determine the mean and variance of the resulting random variable.

1.4.5 Let X and Y have the joint normal distribution described in equation (1.47). What value of α minimizes the variance of $Z = \alpha X + (1 - \alpha)Y$? Simplify your result when X and Y are independent.

1.4.6 Suppose that U has a uniform distribution on the interval $[0, 1]$. Derive the density function for the random variables

(a) $Y = -\ln(1 - U)$.

(b) $W_n = U^n$ for $n \geq 1$.

Hint: Refer to Section 1.2.6.

1.4.7 Given independent exponentially distributed random variables S and T with common parameter λ, determine the probability density function of the sum $R = S + T$ and identify its type by name.

1.4.8 Let Z be a random variable with the geometric probability mass function

$$p(k) = (1 - \pi)\pi^k, \quad k = 0, 1, \ldots,$$

where $0 < \pi < 1$.

(a) Show that Z has a constant failure rate in the sense that $\Pr\{Z = k | Z \geq k\} = 1 - \pi$ for $k = 0, 1, \ldots$.

(b) Suppose Z' is a discrete random variable whose possible values are $0, 1, \ldots$, and for which $\Pr\{Z' = k | Z' \geq k\} = 1 - \pi$ for $k = 0, 1, \ldots$. Show that the probability mass function for Z' is $p(k)$.

Problems

1.4.1 Evaluate the moment $E\left[e^{\lambda Z}\right]$, where λ is an arbitrary real number and Z is a random variable following a standard normal distribution, by integrating

$$E[e^{\lambda Z}] \int\limits_{-\infty}^{+\infty} e^{\lambda z} \frac{1}{\sqrt{2\pi}} e^{-z^2/2} dz.$$

Hint: Complete the square $-\frac{1}{2}z^2 + \lambda z = -\frac{1}{2}\left[(z - \lambda)^2 - \lambda^2\right]$ and use the fact that

$$\int\limits_{-\infty}^{+\infty} \frac{1}{\sqrt{2\pi}} e^{-(z-\lambda)^2/2} dz = 1.$$

1.4.2 Let W be an exponentially distributed random variable with parameter θ and mean $\mu = 1/\theta$.

(a) Determine $\Pr\{W > \mu\}$.

(b) What is the mode of the distribution?

1.4.3 Let X and Y be independent random variables uniformly distributed over the interval $\left[\theta - \frac{1}{2}, \theta + \frac{1}{2}\right]$ for some fixed θ. Show that $W = X - Y$ has a distribution that is independent of θ with density function

$$f_W(w) = \begin{cases} 1 + w & \text{for } -1 \leq w < 0, \\ 1 - w & \text{for } 0 \leq w \leq 1, \\ 0 & \text{for } |w| > 1. \end{cases}$$

1.4.4 Suppose that the diameters of bearings are independent normally distributed random variables with mean $\mu_B = 1.005$ inch and variance $\sigma_B^2 = (0.003)^2$ inch^2. The diameters of shafts are independent normally distributed random variables having mean $\mu_S = 0.995$ inch and variance $\sigma_S^2 = (0.004)^2$ inch^2.

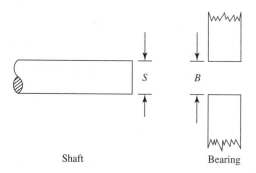

Shaft   Bearing

Let S be the diameter of a shaft taken at random and let B be the diameter of a bearing.

(a) What is the probability $\Pr\{S > B\}$ of interference?

(b) What is the probability of one or fewer interferences in 20 random shaft-bearing pairs?

Hint: The clearance, defined by $C = B - S$, is normally distributed (why?), and interference occurs only if $C < 0$.

1.4.5 If X follows an exponential distribution with parameter $\alpha = 2$, and independently, Y follows an exponential distribution with parameter $\beta = 3$, what is the probability that $X < Y$?

1.5 Some Elementary Exercises

We have collected in this section a number of exercises that go beyond what is usually covered in a first course in probability.

1.5.1 Tail Probabilities

In mathematics, what is a "trick" upon first encounter becomes a basic tool when familiarity through use is established. In dealing with nonnegative random variables, we can often simplify the analysis by the trick of approaching the problem through the upper tail probabilities of the form $\Pr\{X > x\}$. Consider the following example.

A jar has n chips numbered $1, 2, \ldots, n$. A person draws a chip, returns it, draws another, returns it, and so on, until a chip is drawn that has been drawn before. Let X be the number of drawings. Find the probability distribution for X.

It is easier to compute $\Pr\{X > k\}$ first. Then, $\Pr\{X > 1\} = 1$, since at least two draws are always required. The event $\{X > 2\}$ occurs when distinct numbers appear on the first two draws, whence $\Pr\{X > 2\} = (n/n)[(n-1)/n]$. Continuing in this manner, we obtain

$$\Pr\{X > k\} = 1\left(1 - \frac{1}{n}\right)\left(1 - \frac{2}{n}\right)\cdots\left(1 - \frac{k-1}{n}\right),$$

$$\text{for } k = 1, \ldots, n-1. \tag{1.48}$$

Finally,

$$\Pr\{X = k\} = \Pr\{X > k-1\} - \Pr\{X > k\}$$

$$= \left[\left(1 - \frac{1}{n}\right)\cdots\left(1 - \frac{k-2}{n}\right)\right]$$

$$- \left[\left(1 - \frac{1}{n}\right)\cdots\left(1 - \frac{k-2}{n}\right)\left(1 - \frac{k-1}{n}\right)\right]$$

$$= \left(1 - \frac{1}{n}\right)\cdots\left(1 - \frac{k-2}{n}\right)\left[1 - \left(1 - \frac{k-1}{n}\right)\right]$$

$$= \frac{k-1}{n}\left(1 - \frac{1}{n}\right)\cdots\left(1 - \frac{k-2}{n}\right),$$

$$\text{for } k = 2, \ldots, n+1.$$

Now try deriving $\Pr\{X = k\}$ directly, for comparison with the "trick" approach. The usefulness of the upper tail probabilities is enhanced by the formula

$$E[X] = \sum_{k=0}^{\infty} \Pr\{X > k\} = \sum_{k=1}^{\infty} \Pr\{X \geq k\}, \tag{1.49}$$

valid for nonnegative integer-valued random variables X. To establish (1.49), abbreviate the notation by using $p(k) = \Pr\{X = k\}$, and rearrange the terms in $E[X] = \Sigma_{k-\geq 0} k p(k)$ as follows:

$$E[X] = 0p(0) + 1p(1) + 2p(2) + 3p(3) + \cdots$$

$$= p(1) + p(2) + p(3) + p(4) + \cdots$$

$$+ p(2) + p(3) + p(4) + \cdots$$

$$+ p(3) + p(4) + \cdots$$

$$+ p(4) + \cdots$$

$$\vdots$$

$$= \Pr\{X \geq 1\} + \Pr\{X \geq 2\} + \Pr\{X \geq 3\} + \cdots$$

$$= \sum_{k=1}^{\infty} \Pr\{X \geq k\},$$

thus establishing (1.49).

For the chip drawing problem, the mean number of draws required is

$$E[X] = \Pr\{X > 0\} + \Pr\{X > 1\} + \cdots + \Pr\{X > n\},$$

since $\Pr\{X > k\} = 0$ for $k > n$. Substituting (1.48) into (1.49) leads directly to

$$E[X] = 2 + \left(1 - \frac{1}{n}\right) + \left(1 - \frac{1}{n}\right)\left(1 - \frac{2}{n}\right) + \cdots$$
$$+ \left(1 - \frac{1}{n}\right)\left(1 - \frac{2}{n}\right) \cdots \left(1 - \frac{n-1}{n}\right).$$

Now let X be a *nonnegative* continuous random variable with density $f(x)$ and distribution function $F(x)$. The analog to (1.49) is

$$E[X] = \int_0^\infty [1 - F(z)]dz, \tag{1.50}$$

obtained by interchanging an order of integration as follows:

$$E[X] = \int_0^\infty x f(x)dx = \int_0^\infty \left(\int_0^x dz\right) f(x)dx$$
$$= \int_0^\infty \left[\int_z^\infty f(x)dx\right] dz = \int_0^\infty [1 - F(z)]dz.$$

Interchanging the order of integration where the limits are variables often proves difficult for many students. The trick of using indicator functions to make the limits of integration constant may simplify matters. In the preceding interchange, let

$$1_z(x) := 1 \text{ if } 0 \le z < x \text{ and } 1_z(x) := 0 \text{ otherwise.}$$

and then

$$\int_0^\infty \left[\int_0^x dz\right] f(x)dx = \int_0^\infty \left[\int_0^\infty 1_z(x)f(x)dz\right] dx$$
$$= \int_0^\infty \left[\int_0^\infty 1_z(x)f(x)dx\right] dz = \int_0^\infty \left[\int_z^\infty f(x)dx\right] dz.$$

As an application of (1.50), let $X_c = \min\{c, X\}$ for some positive constant c. For example, suppose X is the failure time of a certain piece of equipment. A planned replacement policy is put in use that calls for replacement of the equipment upon its failure or upon its reaching age c, whichever occurs first. Then,

$$X_c = \min\{c, X\} = \begin{cases} X & \text{if } X \leq c, \\ c & \text{if } X > c \end{cases}$$

is the time for replacement.

Now

$$\Pr\{X_c > z\} = \begin{cases} 1 - F(z) & \text{if } 0 \leq z < c, \\ 0 & \text{if } c \leq z, \end{cases}$$

whence we obtain

$$E[X_c] = \int_0^c [1 - F(z)] \, dz,$$

which is decidedly shorter than

$$E[X_c] = \int_0^c x f(x) \, dx + c[1 - F(c)].$$

Observe that X_c is a random variable whose distribution is partly continuous and partly discrete, thus establishing by example that such distributions do occur in practical applications.

1.5.2 The Exponential Distribution

This exercise is designed to foster intuition about the exponential distribution, as well as to provide practice in algebraic and calculus manipulations relevant to stochastic modeling.

Let X_0 and X_1 be independent exponentially distributed random variables with respective parameters λ_0 and λ_1, so that

$$\Pr\{X_i > t\} = e^{-\lambda_i t} \quad \text{for } t \geq 0, i = 0, 1.$$

Let

$$N = \begin{cases} 0 & \text{if } X_0 \leq X_1, \\ 1 & \text{if } X_1 \leq X_0; \end{cases}$$

$$U = \min\{X_0, X_1\} = X_N;$$

$$M = 1 - N;$$

$$V = \max\{X_0, X_1\} = X_M;$$

and

$$W = V - U = |X_0 - X_1|.$$

In this context, we derive the following:

(a) $\Pr\{N = 0 \text{ and } U > t\} = e^{-(\lambda_0 + \lambda_1)t}\left(\dfrac{\lambda_0}{\lambda_0 + \lambda_1}\right).$

The event $\{N = 0 \text{ and } U > t\}$ is exactly the event $\{t < X_0 \leq X_1\}$, whence

$$\Pr\{N = 0, U > t\} = \Pr\{t < X_0 < X_1\}$$

$$= \iint\limits_{t < x_0 < x_l} \lambda_0 e^{-\lambda_0 x_0} \lambda_1 e^{-\lambda_1 x_1}\, dx_1\, dx_0$$

$$= \int_t^\infty \left(\int_{x_0}^\infty \lambda_1 e^{-\lambda_1 X_1}\, dx_1\right) \lambda_0 e^{-\lambda_0 x_0}\, dx_0$$

$$= \int_t^\infty e^{-\lambda_1 X_0} \lambda_0 e^{-\lambda_0 X_0}\, dx_0$$

$$= \frac{\lambda_0}{\lambda_0 + \lambda_1} \int_t^\infty (\lambda_0 + \lambda_1) e^{-(\lambda_0 + \lambda_1)x_0}\, dx_0$$

$$= \frac{\lambda_0}{\lambda_0 + \lambda_1} e^{-(\lambda_0 + \lambda_1)t}.$$

(b) $\Pr\{N = 0\} = \dfrac{\lambda_0}{\lambda_0 + \lambda_1}$ and $\Pr\{N = 1\} = \dfrac{\lambda_1}{\lambda_0 + \lambda_1}.$

We use the result in (a) as follows:

$$\Pr\{N = 0\} = \Pr\{N = 0, U > 0\} = \frac{\lambda_0}{\lambda_0 + \lambda_1} \quad \text{from (a).}$$

Obviously, $\Pr\{N = 1\} = 1 - \Pr\{N = 0\} = \lambda_1/(\lambda_0 + \lambda_1).$

(c) $\Pr\{U > t\} = e^{-(\lambda_0 + \lambda_1)t}, \quad t \geq 0.$

Upon adding the result in (a),

$$\Pr\{N = 0 \text{ and } U > t\} = e^{-(\lambda_0 + \lambda_1)t} \frac{\lambda_0}{\lambda_0 + \lambda_1},$$

to the corresponding quantity associated with $N = 1$,

$$\Pr\{N = 1 \text{ and } U > t\} = e^{-(\lambda_0 + \lambda_1)t} \frac{\lambda_1}{\lambda_0 + \lambda_1},$$

we obtain the desired result via

$$\Pr\{U > t\} = \Pr\{N = 0, U > t\} + \Pr\{N = 1, U > t\}$$

$$= e^{-(\lambda_0 + \lambda_1)t}\left(\frac{\lambda_0}{\lambda_0 + \lambda_1} + \frac{\lambda_1}{\lambda_0 + \lambda_1}\right)$$

$$= e^{-(\lambda_0 + \lambda_1)t}.$$

At this point observe that U and N are independent random variables. This follows because (a), (b), and (c) together give

$$\Pr\{N = 0 \text{ and } U > t\} = \Pr\{N = 0\} \times \Pr\{U > t\}.$$

Think about this remarkable result for a moment. Suppose X_0 and X_1 represent lifetimes, and $\lambda_0 = 0.001$, while $\lambda_1 = 1$. The mean lifetimes are $E[X_0] = 1000$ and $E[X_1] = 1$. Suppose we observe that the time of the first death is rather small, say, $U = \min\{X_0, X_1\} = \frac{1}{2}$. In spite of vast disparity between the mean lifetimes, the observation that $U = \frac{1}{2}$ provides no information about which of the two units, 0 or 1, was first to die! This apparent paradox is yet another, more subtle, manifestation of the memoryless property unique to the exponential density.

We continue with the exercise.

(d) $\Pr\{W > t | N = 0\} = e^{-\lambda_1 t}, \quad t \geq 0.$

The event $\{W > t \text{ and } N = 0\}$ for $t \geq 0$ corresponds exactly to the event $\{t < X_1 - X_0\}$. Thus,

$$\Pr\{W > t \text{ and } N = 0\} = \Pr\{X_1 - X_0 > t\}$$

$$= \iint\limits_{x_1 - x_0 > t} \lambda_0 e^{-\lambda_0 x_0} \lambda_1 e^{-\lambda_1 x_1} \, dx_0 dx_1$$

$$= \int_0^\infty \left(\int_{x_0 + t}^\infty \lambda_1 e^{-\lambda_1 x_1} \, dx_1 \right) \lambda_0 e^{-\lambda_0 x_0} \, dx_0$$

$$= \int_0^\infty e^{-\lambda_1 (x_0 + t)} \lambda_0 e^{-\lambda_0 x_0} \, dx_0$$

$$= \frac{\lambda_0}{\lambda_0 + \lambda_1} e^{-\lambda_1 t} \int_0^\infty (\lambda_0 + \lambda_1) e^{-(\lambda_0 + \lambda_1) x_0} \, dx_0$$

$$= \frac{\lambda_0}{\lambda_0 + \lambda_1} e^{-\lambda_1 t}$$

$$= \Pr\{N = 0\} e^{-\lambda_1 t} \quad \text{[from (b)]}.$$

Then, using the basic definition of conditional probability (Section 1.2.7), we obtain

$$\Pr\{W > t | N = 0\} = \frac{\Pr\{W > t, N = 0\}}{\Pr\{N = 0\}} = e^{-\lambda_1 t}, \quad t \geq 0,$$

as desired.

Of course a parallel formula holds conditional on $N = 1$:

$$\Pr\{W > t | N = 1\} = e^{-\lambda_0 t}, \quad t \geq 0,$$

and using the law of total probability we obtain the distribution of W in the form

$$\Pr\{W > t\} = \Pr\{W > t, N = 0\} + \Pr\{W > t, N = 1\}$$

$$= \frac{\lambda_0}{\lambda_0 + \lambda_1} e^{-\lambda_1 t} + \frac{\lambda_1}{\lambda_0 + \lambda_1} e^{-\lambda_0 t}, \quad t \geq 0,$$

(e) U and $W = V - U$ are independent random variables.

To establish this final consequence of the memoryless property, it suffices to show that

$$\Pr\{U > u \text{ and } W > w\} = \Pr\{U > u\}\Pr\{W > w\} \quad \text{for all } u \ge 0, w \ge 0.$$

Determining first

$$\Pr\{N = 0, U > u, W > w\} = \Pr\{u < X_0 < X_1 - w\}$$

$$= \iint_{u < x_0 < x_1 - w} \lambda_0 e^{-\lambda_0 x_0} \lambda_1 e^{-\lambda_1 x_1} \, dx_0 dx_1$$

$$= \int_u^\infty \left(\int_{x_0 + w}^\infty \lambda_1 e^{-\lambda_1 x_1} \, dx_1 \right) \lambda_0 e^{-\lambda_0 x_0} \, dx_0$$

$$= \int_u^\infty e^{-\lambda_1 (x_0 + w)} \lambda_0 e^{-\lambda_0 x_0} \, dx_0$$

$$= \left(\frac{\lambda_0}{\lambda_0 + \lambda_1} \right) e^{-\lambda_1 w} \int_u^\infty (\lambda_0 + \lambda_1) e^{-(\lambda_0 + \lambda_1) x_0} \, dx_0$$

$$= \left(\frac{\lambda_0}{\lambda_0 + \lambda_1} \right) e^{-\lambda_1 w} e^{-(\lambda_0 + \lambda_1) u},$$

and then, by symmetry,

$$\Pr\{N = 1, U > u, W > w\} = \left(\frac{\lambda_1}{\lambda_0 + \lambda_1} \right) e^{-\lambda_0 w} e^{-(\lambda_0 + \lambda_1) u},$$

and finally adding the two expressions, we obtain

$$\Pr\{U > u, W > w\} = \left[\left(\frac{\lambda_0}{\lambda_0 + \lambda_1} \right) e^{-\lambda_1 w} + \left(\frac{\lambda_1}{\lambda_0 + \lambda_1} \right) e^{-\lambda_0 w} \right] e^{-(\lambda_0 + \lambda_1) u}$$

$$= \Pr\{W > w\}\Pr\{U > u\}, \quad u, w \ge 0.$$

The calculation is complete.

Exercises

1.5.1 Let X have a binomial distribution with parameters $n = 4$ and $p = \frac{1}{4}$. Compute the probabilities $\Pr\{X \ge k\}$ for $k = 1, 2, 3, 4$, and sum these to verify that the mean of the distribution is 1.

1.5.2 A jar has four chips colored red, green, blue, and yellow. A person draws a chip, observes its color, and returns it. Chips are now drawn repeatedly, without replacement, until the first chip drawn is selected again. What is the mean number of draws required?

1.5.3 Let X be an exponentially distributed random variable with parameter λ. Determine the mean of X

(a) by integrating by parts in the definition in equation (1.7) with $m = 1$;

(b) by integrating the upper tail probabilities in accordance with equation (1.50).

Which method do you find easier?

1.5.4 A system has two components: A and B. The operating times until failure of the two components are independent and exponentially distributed random variables with parameter 2 for component A, and 3 for B. The system fails at the first component failure.

(a) What is the mean time to failure for component A? For component B?

(b) What is the mean time to system failure?

(c) What is the probability that it is component A that causes system failure?

(d) Suppose that it is component A that fails first. What is the mean remaining operating life of component B?

1.5.5 Consider a post office with two clerks. John, Paul, and Naomi enter simultaneously. John and Paul go directly to the clerks, while Naomi must wait until either John or Paul is finished before she begins service.

(a) If all of the service times are independent exponentially distributed random variables with the same mean $1/\lambda$, what is the probability that Naomi is still in the post office after the other two have left?

(b) How does your answer change if the two clerks have different service rates, say $\lambda_1 = 3$ and $\lambda_2 = 47$?

(c) The mean time that Naomi spends in the post office is less than that for John or Paul provided that $\max\{\lambda_1, \lambda_2\} > c \min\{\lambda_1, \lambda_2\}$ for a certain constant c. What is the value of this constant?

Problems

1.5.1 Let $X_1, X_2, \ldots$ be independent and identically distributed random variables having the cumulative distribution function $F(x) = \Pr\{X \leq x\}$. For a fixed number ξ, let N be the first index k for which $X_k > \xi$. That is, $N = 1$ if $X_1 > \xi$; $N = 2$ if $X_1 \leq \xi$ and $X_2 > \xi$; etc. Determine the probability mass function for N.

1.5.2 Let $X_1, X_2, \ldots, X_n$ be independent random variables, all exponentially distributed with the same parameter λ. Determine the distribution function for the minimum $Z = \min\{X_1, \ldots, X_n\}$.

1.5.3 Suppose that X is a discrete random variable having the geometric distribution whose probability mass function is

$$p(k) = p(1 - p)^k \quad \text{for } k = 0, 1, \ldots.$$

(a) Determine the upper tail probabilities $\Pr\{X > k\}$ for $k = 0, 1, \ldots$.

(b) Evaluate the mean via $E[X] = \Sigma_{k \geq 0} \Pr\{X > k\}$.

1.5.4 Let V be a continuous random variable taking both positive and negative values and whose mean exists. Derive the formula

$$E[V] = \int_0^\infty [1 - F_V(v)] dv - \int_{-\infty}^0 F_V(v) dv.$$

1.5.5 Show that

$$E[W^2] = \int_0^\infty 2y[1 - F_W(y)]dy$$

for a nonnegative random variable W.

1.5.6 Determine the upper tail probabilities $\Pr\{V > t\}$ and mean $E[V]$ for a random variable V having the exponential density

$$f_V(v) = \begin{cases} 0 & \text{for } v < 0, \\ \lambda e^{-\lambda v} & \text{for } v \geq 0, \end{cases}$$

where λ is a fixed positive parameter.

1.5.7 Let $X_1, X_2, \ldots, X_n$ be independent random variables that are exponentially distributed with respective parameters $\lambda_1, \lambda_2, \ldots, \lambda_n$. Identify the distribution of the minimum $V = \min\{X_1, X_2, \ldots, X_n\}$.

Hint: For any real number v, the event $\{V > v\}$ is equivalent to $\{X_1 > v, X_2 > v, \ldots, X_n > v\}$.

1.5.8 Let $U_1, U_2, \ldots, U_n$ be independent uniformly distributed random variables on the unit interval $[0, 1]$. Define the minimum $V_n = \min\{U_1, U_2, \ldots, U_n\}$.

(a) Show that $\Pr\{V_n > v\} = (1 - v)^n$ for $0 \leq v \leq 1$.

(b) Let $W_n = nV_n$. Show that $\Pr\{W_n > w\} = [1 - (w/n)]^n$ for $0 \leq w \leq n$, and thus

$$\lim_{n \to \infty} \Pr\{W_n > w\} = e^{-w} \quad \text{for } w \geq 0.$$

1.5.9 A flashlight requires two good batteries in order to shine. Suppose, for the sake of this academic exercise, that the lifetimes of batteries in use are independent random variables that are exponentially distributed with parameter $\lambda = 1$. Reserve batteries do not deteriorate. You begin with five fresh batteries. On average, how long can you shine your light?

1.6 Useful Functions, Integrals, and Sums

Collected here for later reference are some calculations and formulas that are especially pertinent in probability modeling.

We begin with several exponential integrals, the first and simplest being

$$\int e^{-x}dx = -e^{-x}. \tag{1.51}$$

When we use integration by parts, the second integral that we introduce reduces to the first in the manner

$$\int xe^{-x}dx = -xe^{-x} + \int e^{-x}dx = -e^{-x}(1 + x). \tag{1.52}$$

Then, (1.51) and (1.52) are the special cases of $\alpha = 1$ and $\alpha = 2$, respectively, in the general formula, valid for any real number α for which the integrals are defined, given by

$$\int x^{\alpha-1} e^{-x} dx = -x^{\alpha-1} e^{-x} + (\alpha - 1) \int x^{\alpha-2} e^{-x} dx. \tag{1.53}$$

Fixing the limits of integration leads to the gamma function, defined by

$$\Gamma(\alpha) = \int_0^\infty x^{\alpha-1} e^{-x} dx, \quad \text{for } \alpha > 0. \tag{1.54}$$

From (1.53), it follows that

$$\Gamma(\alpha) = (\alpha - 1)\Gamma(\alpha - 1), \tag{1.55}$$

and therefore, for any integers k,

$$\Gamma(k) = (k-1)(k-2)\cdots 2 \cdot \Gamma(1). \tag{1.56}$$

An easy consequence of (1.51) is the evaluation $\Gamma(1) = 1$, which with (1.55) shows that the gamma function at integral arguments is a generalization of the factorial function, and

$$\Gamma(k) = (k-1)! \quad \text{for } k = 1, 2, \ldots . \tag{1.57}$$

A more difficult integration shows that

$$\Gamma\left(\frac{1}{2}\right) = \sqrt{\pi}, \tag{1.58}$$

which with (1.56) provides

$$\Gamma\left(n + \frac{1}{2}\right) = \frac{1 \times 3 \times 5 \times \cdots \times (2n-1)}{2^n} \sqrt{\pi}, \quad \text{for } n = 0, 1, \ldots . \tag{1.59}$$

Stirling's formula is the following important asymptotic evaluation of the factorial function:

$$n! = n^n e^{-n} (2\pi n)^{1/2} e^{r(n)/12n}, \tag{1.60}$$

in which

$$1 - \frac{1}{12n+1} < r(n) < 1. \tag{1.61}$$

We sometimes write this in the looser form

$$n! \sim n^n e^{-n} (2\pi n)^{1/2} \quad \text{as } n \to \infty, \tag{1.62}$$

the symbol "$\sim$" signifying that the ratio of the two sides in (1.62) approaches 1 as $n \to \infty$. For the binomial coefficient $\binom{n}{k} = n!/[k!(n-k)!]$, we then obtain

$$\binom{n}{k} \sim \frac{(n-k)^k}{k!} \quad \text{as } n \to \infty, \tag{1.63}$$

as a consequence of (1.62) and the exponential limit

$$e^{-k} = \lim_{n \to \infty} \left(1 - \frac{k}{n}\right)^n.$$

The integral

$$B(m,n) = \int_0^1 x^{m-1}(1-x)^{n-1} dx, \tag{1.64}$$

which converges when m and n are positive, defines the *beta* function, related to the gamma function by

$$B(m,n) = \frac{\Gamma(m)\Gamma(n)}{\Gamma(m+n)} \quad \text{for } m > 0, n > 0. \tag{1.65}$$

For nonnegative integral values m and n,

$$B(m+1,n+1) = \int_0^1 x^m (1-x)^n dx = \frac{m!n!}{(m+n+1)!}. \tag{1.66}$$

For $n = 1, 2, \ldots$, the binomial theorem provides the evaluation

$$(1-x)^n = \sum_{k=0}^n (-1)^k \binom{n}{k} x^k, \quad \text{for } -\infty < x < \infty. \tag{1.67}$$

The formula may be generalized for nonintegral n by appropriately generalizing the binomial coefficient, defining for any real number α,

$$\binom{\alpha}{k} = \begin{cases} \dfrac{\alpha(\alpha-1)\cdots(\alpha-k+1)}{k!} & \text{for } k = 1, 2, \ldots, \\ 1 & \text{for } k = 0. \end{cases} \tag{1.68}$$

As a special case, for any positive integer n,

$$\binom{-n}{k} = (-1)^k \frac{n(n+1)\cdots(n+k-1)}{k!}$$

$$= (-1)^k \binom{n+k-1}{k}. \tag{1.69}$$

The general binomial theorem, valid for all real α, is

$$(1-x)^\alpha = \sum_{k=0}^{\infty} (-1)^k \binom{\alpha}{k} x^k \quad \text{for } -1 < x < 1. \tag{1.70}$$

When $\alpha = -n$ for a positive integer n, we obtain a group of formulas useful in dealing with geometric series. For a positive integer n, in view of (1.69) and (1.70), we have

$$(1-x)^{-n} = \sum_{k=0}^{\infty} \binom{n+k-1}{k} x^k \quad \text{for } |x| < 1. \tag{1.71}$$

The familiar formula

$$\sum_{k=0}^{\infty} x^k = 1 + x + x^2 + \cdots = \frac{1}{1-x} \quad \text{for } |x| < 1 \tag{1.72}$$

for the sum of a geometric series results from (1.71) with $n = 1$. The cases $n = 2$ and $n = 3$ yield the formulas

$$\sum_{k=0}^{\infty} (k+1)x^k = 1 + 2x + 3x^2 + \cdots$$

$$= \frac{1}{(1-x)^2} \quad \text{for } |x| < 1, \tag{1.73}$$

$$\sum_{k=0}^{\infty} (k+2)(k+1)x^k = \frac{2}{(1-x)^3} \quad \text{for } |x| < 1. \tag{1.74}$$

Sums of Numbers

The following sums of powers of integers have simple expressions:

$$1 + 2 + \cdots + n = \frac{n(n+1)}{2},$$

$$1 + 2^2 + \cdots + n^2 = \frac{n(n+1)(2n+1)}{6},$$

$$1 + 2^3 + \cdots + n^3 = \frac{n^2(n+1)^2}{4}.$$

2 Conditional Probability and Conditional Expectation

2.1 The Discrete Case

The conditional probability $\Pr\{A|B\}$ of the event A given the event B is defined by

$$\Pr\{A|B\} = \frac{\Pr\{A \text{ and } B\}}{\Pr\{B\}} \quad \text{if } \Pr\{B\} > 0, \tag{2.1}$$

and is not defined, or is assigned an arbitrary value, when $\Pr\{B\} = 0$. Let X and Y be random variables that can attain only countably many different values, say $0, 1, 2, \ldots$. The *conditional probability mass function* $p_{X|Y}(x|y)$ of X given $Y = y$ is defined by

$$p_{X|Y}(x|y) = \frac{\Pr\{X = x \text{ and } Y = y\}}{\Pr\{Y = y\}} \quad \text{if } \Pr\{Y = y\} > 0,$$

and is not defined, or is assigned an arbitrary value, whenever $\Pr\{Y = y\} = 0$. In terms of the joint and marginal probability mass functions $p_{XY}(x, y)$ and $p_Y(y) = \sum_x p_{XY}(x, y)$, respectively, the definition is

$$p_{X|Y}(x|y) = \frac{p_{XY}(x, y)}{p_Y(y)} \quad \text{if } p_Y(y) > 0; \quad x, y = 0, 1, \ldots. \tag{2.2}$$

Observe that $p_{X|Y}(x|y)$ is a probability mass function in x for each fixed y, i.e., $p_{X|Y}(x|y) \geq 0$ and $\sum_\xi p_{X|Y}(\xi|y) = 1$, for all x, y.

The law of total probability takes the form

$$\Pr\{X = x\} = \sum_{y=0}^{\infty} p_{X|Y}(x|y) p_Y(y). \tag{2.3}$$

Notice in (2.3) that the points y where $p_{X|Y}(x|y)$ is not defined are exactly those values for which $p_Y(y) = 0$, and hence, do not affect the computation. The lack of a complete prescription for the conditional probability mass function, a nuisance in some instances, is always consistent with subsequent calculations.

Example Let X have a binomial distribution with parameters p and N, where N has a binomial distribution with parameters q and M. What is the marginal distribution of X?

An Introduction to Stochastic Modeling

We are given the conditional probability mass function

$$p_{X|N}(k|n) = \binom{n}{k} p^k (1-p)^{n-k}, \quad k = 0, 1, \dots, n,$$

and the marginal distribution

$$p_N(n) = \binom{M}{n} q^n (1-q)^{M-n}, \quad n = 0, 1, \dots, M.$$

We apply the law of total probability in the form of (2.3) to obtain

$$\Pr\{X = k\} = \sum_{n=0}^{M} p_{X|N}(k|n) p_N(n)$$

$$= \sum_{n=k}^{M} \frac{n!}{k!\,(n-k)!} p^k (1-p)^{n-k} \frac{M!}{n!\,(M-n)!} q^n (1-q)^{M-n}$$

$$= \frac{M!}{k!} p^k (1-q)^M \left(\frac{q}{1-q}\right)^k \sum_{n=k}^{M} \frac{1}{(n-k)!\,(M-n)!} (1-p)^{n-k}$$

$$\times \left(\frac{q}{1-q}\right)^{n-k}$$

$$= \frac{M!}{k!\,(M-k)!} (pq)^k (1-q)^{M-k} \left[1 + \frac{q(1-p)}{1-q}\right]^{M-k}$$

$$= \frac{M!}{k!\,(M-k)!} (pq)^k (1-pq)^{M-k}, \quad k = 0, 1, \dots, M.$$

In words, X has a binomial distribution with parameters M and pq.

Example Suppose X has a binomial distribution with parameters p and N, where N has a Poisson distribution with mean λ. What is the marginal distribution for X?

Proceeding as in the previous example but now using

$$p_N(n) = \frac{\lambda^n e^{-\lambda}}{n!}, \quad n = 0, 1, \dots,$$

we obtain

$$\Pr\{X = k\} = \sum_{n=0}^{\infty} p_{X|N}(k|n) p_N(n)$$

$$= \sum_{n=k}^{\infty} \frac{n!}{k!\,(n-k)!} p^k (1-p)^{n-k} \frac{\lambda^n e^{-\lambda}}{n!}$$

$$= \frac{\lambda^k e^{-\lambda} p^k}{k!} \sum_{n=k}^{\infty} \frac{[\lambda(1-p)]^{n-k}}{(n-k)!}$$

$$= \frac{(\lambda p)^k e^{-\lambda}}{k!} e^{\lambda(1-p)}$$

$$= \frac{(\lambda p)^k e^{-\lambda p}}{k!} \quad \text{for } k = 0, 1, \dots.$$

In words, X has a Poisson distribution with mean λp.

Example Suppose X has a negative binomial distribution with parameters p and N, where N has the geometric distribution

$$p_N(n) = (1-\beta)\beta^{n-1} \quad \text{for } n = 1, 2, \dots.$$

What is the marginal distribution for X?

We are given the conditional probability mass function

$$p_{X|N}(k|n) = \binom{n+k-1}{k} p^n (1-p)^k, \quad k = 0, 1, \dots.$$

Using the law of total probability, we obtain

$$\Pr\{X = k\} = \sum_{n=0}^{\infty} p_{X|N}(k|n) p_N(n)$$

$$= \sum_{n=1}^{\infty} \frac{(n+k-1)!}{k!\,(n-1)!} p^n (1-p)^k (1-\beta)\beta^{n-1}$$

$$= (1-\beta)(1-p)^k p \sum_{n=1}^{\infty} \binom{n+k-1}{k} (\beta p)^{n-1}$$

$$= (1-\beta)(1-p)^k p (1-\beta p)^{-k-1}$$

$$= \left(\frac{p-\beta p}{1-\beta p}\right) \left(\frac{1-p}{1-\beta p}\right)^k \quad \text{for } k = 0, 1, \dots.$$

We recognize the marginal distribution of X as being of geometric form.

Let g be a function for which the expectation of $g(X)$ is finite. We define the *conditional* expected value of $g(X)$ given $Y = y$ by the formula

$$E[g(X)|Y = y] = \sum_{x} g(x) p_{X|Y}(x|y) \quad \text{if } p_Y(y) > 0, \tag{2.4}$$

and the conditional mean is not defined at values y for which $p_Y(y) = 0$. The law of total probability for conditional expectation reads

$$E[g(X)] = \sum_y E[g(X)|Y = y]p_Y(y). \tag{2.5}$$

The conditional expected value $E[g(X)|Y = y]$ is a function of the real variable y. If we evaluate this function at the random variable Y, we obtain a random variable that we denote by $E[g(X)|Y]$. The law of total probability in (2.5) now may be written in the form

$$E[g(X)] = E\{E[g(X)|Y]\}. \tag{2.6}$$

Since the conditional expectation of $g(X)$ given $Y = y$ is the expectation with respect to the conditional probability mass function $p_{X|Y}(x|y)$, conditional expectations behave in many ways like ordinary expectations. The following list summarizes some properties of conditional expectations. In this list, with or without affixes, X and Y are jointly distributed random variables; c is a real number; g is a function for which $E[|g(X)|] < \infty$; h is a bounded function; and v is a function of two variables for which $E[|v(X, Y)|] < \infty$. The properties are

1. $E[c_1 g_1(X_1) + c_2 g_2(X_2)|Y = y]$

$$= c_1 E[g_1(X_1)|Y = y] + c_2 E[g_2(X_2)|Y = y]. \tag{2.7}$$

2. if $g \geq 0$, then $E[g(X)|Y = y] \geq 0$. $\tag{2.8}$

3. $E[v(X, Y)|Y = y] = E[v(X, y)|Y = y].$ $\tag{2.9}$

4. $E[g(X)|Y = y] = E[g(X)]$ if X and Y are independent. $\tag{2.10}$

5. $E[g(X)h(Y)|Y = y] = h(y)E[g(X)|Y = y].$ $\tag{2.11}$

6. $E[g(X)h(Y)] = \sum_y h(y)E[g(X)|Y = y]p_Y(y)$

$$= E\{h(Y)E[g(X)|Y]\}. \tag{2.12}$$

As a consequence of (2.7), (2.11), and (2.12), with either $g \equiv 1$ or $h \equiv 1$, we obtain

$$E[c|Y = y] = c, \tag{2.13}$$

$$E[h(Y)|Y = y] = h(y), \tag{2.14}$$

$$E[g(X)] = \sum_y E[g(X)|Y = y]p_Y(y) = E\{E[g(X)|Y]\}. \tag{2.15}$$

Exercises

2.1.1 I roll a six-sided die and observe the number N on the uppermost face. I then toss a fair coin N times and observe X, the total number of heads to appear. What is the probability that $N = 3$ and $X = 2$? What is the probability that $X = 5$? What is $E[X]$, the expected number of heads to appear?

2.1.2 Four nickels and six dimes are tossed, and the total number N of heads is observed. If $N = 4$, what is the conditional probability that exactly two of the nickels were heads?

2.1.3 A poker hand of five cards is dealt from a normal deck of 52 cards. Let X be the number of aces in the hand. Determine $\Pr\{X > 1 | X \geq 1\}$. This is the probability that the hand contains more than one ace, given that it has at least one ace. Compare this with the probability that the hand contains more than one ace, given that it contains the ace of spades.

2.1.4 A six-sided die is rolled, and the number N on the uppermost face is recorded. From a jar containing 10 tags numbered $1, 2, \ldots, 10$, we then select N tags at random without replacement. Let X be the smallest number on the drawn tags. Determine $\Pr\{X = 2\}$.

2.1.5 Let X be a Poisson random variable with parameter λ. Find the conditional mean of X given that X is odd.

2.1.6 Suppose U and V are independent and follow the geometric distribution

$$p(k) = \rho(1 - \rho)^k \quad \text{for } k = 0, 1, \ldots.$$

Define the random variable $Z = U + V$.
(a) Determine the joint probability mass function $p_{U,Z}(u, z) = \Pr\{U = u, Z = z\}$.
(b) Determine the conditional probability mass function for U given that $Z = n$.

Problems

2.1.1 Let M have a binomial distribution with parameters N and p. Conditioned on M, the random variable X has a binomial distribution with parameters M and π.
(a) Determine the marginal distribution for X.
(b) Determine the covariance between X and $Y = M - X$.

2.1.2 A card is picked at random from N cards labeled $1, 2, \ldots, N$, and the number that appears is X. A second card is picked at random from cards numbered $1, 2, \ldots, X$ and its number is Y. Determine the conditional distribution of X given $Y = y$, for $y = 1, 2, \ldots$.

2.1.3 Let X and Y denote the respective outcomes when two fair dice are thrown. Let $U = \min\{X, Y\}$, $V = \max\{X, Y\}$, and $S = U + V$, $T = V - U$.
(a) Determine the conditional probability mass function for U given $V = v$.
(b) Determine the joint mass function for S and T.

2.1.4 Suppose that X has a binomial distribution with parameters $p = \frac{1}{2}$ and N, where N is also random and follows a binomial distribution with parameters $q = \frac{1}{4}$ and $M = 20$. What is the mean of X?

2.1.5 A nickel is tossed 20 times in succession. Every time that the nickel comes up heads, a dime is tossed. Let X count the number of heads appearing on tosses of the dime. Determine $\Pr\{X = 0\}$.

2.1.6 A dime is tossed repeatedly until a head appears. Let N be the trial number on which this first head occurs. Then, a nickel is tossed N times. Let X count the number of times that the nickel comes up tails. Determine $\Pr\{X=0\}, \Pr\{X=1\}$, and $E[X]$.

2.1.7 The probability that an airplane accident that is due to structural failure is correctly diagnosed is 0.85, and the probability that an airplane accident that is not due to structural failure is incorrectly diagnosed as being due to structural failure is 0.35. If 30% of all airplane accidents are due to structural failure, then find the probability that an airplane accident is due to structural failure given that it has been diagnosed as due to structural failure.

2.1.8 Initially an urn contains one red and one green ball. A ball is drawn at random from the urn, observed, and then replaced. If this ball is red, then an additional red ball is placed in the urn. If the ball is green, then a green ball is added. A second ball is drawn. Find the conditional probability that the first ball was red given that the second ball drawn was red.

2.1.9 Let N have a Poisson distribution with parameter $\lambda = 1$. Conditioned on $N = n$, let X have a uniform distribution over the integers $0, 1, \ldots, n + 1$. What is the marginal distribution for X?

2.1.10 *Do men have more sisters than women have?* In a certain society, all married couples use the following strategy to determine the number of children that they will have: If the first child is a girl, they have no more children. If the first child is a boy, they have a second child. If the second child is a girl, they have no more children. If the second child is a boy, they have exactly one additional child. (We ignore twins, assume sexes are equally likely, and the sex of distinct children are independent random variables, etc.) (a) What is the probability distribution for the number of children in a family? (b) What is the probability distribution for the number of girl children in a family? (c) A male child is chosen at random from all of the male children in the population. What is the probability distribution for the number of sisters of this child? What is the probability distribution for the number of his brothers?

2.2 The Dice Game Craps

An analysis of the dice game known as craps provides an educational example of the use of conditional probability in stochastic modeling. In craps, two dice are rolled and the sum of their uppermost faces is observed. If the sum has value 2, 3, or 12, the player loses immediately. If the sum is 7 or 11, the player wins. If the sum is 4, 5, 6, 8, 9, or 10, then further rolls are required to resolve the game. In the case where the sum is 4, e.g., the dice are rolled repeatedly until either a sum of 4 reappears or a sum of 7 is observed. If the sum of 4 appears first, the roller wins; if the sum of 7 appears first, he or she loses.

Consider repeated rolls of the pair of dice and let Z_n for $n = 0, 1, \ldots$ be the sum observed on the nth roll. Then, $Z_0, Z_1, \ldots$ are independent identically distributed random variables. If the dice are fair, the probability mass function is

$$
\begin{aligned}
& p_Z(2) = \frac{1}{36}, && p_Z(8) = \frac{5}{36}, \\
& p_Z(3) = \frac{2}{36}, && p_Z(9) = \frac{4}{36}, \\
& p_Z(4) = \frac{3}{36}, && p_Z(10) = \frac{3}{36}, \\
& p_Z(5) = \frac{4}{36}, && p_Z(11) = \frac{2}{36}, \\
& p_Z(6) = \frac{5}{36}, && p_Z(12) = \frac{1}{36}. \\
& p_Z(7) = \frac{6}{36},
\end{aligned}
\tag{2.16}
$$

Let A denote the event that the player wins the game. By the law of total probability,

$$
\Pr\{A\} = \sum_{k=2}^{12} \Pr\{A|Z_0 = k\} p_Z(k).
\tag{2.17}
$$

Because $Z_0 = 2, 3$, or 12 calls for an immediate loss, then $\Pr\{A|Z_0 = k\} = 0$ for $k = 2, 3$, or 12. Similarly, $Z_0 = 7$ or 11 results in an immediate win, and thus $\Pr\{A|Z_0 = 7\} = \Pr\{A|Z_0 = 11\} = 1$. It remains to consider the values $Z_0 = 4, 5, 6, 8, 9$, and 10, which call for additional rolls. Since the logic remains the same in each of these cases, we will argue only the case in which $Z_0 = 4$. Abbreviate with $\alpha = \Pr\{A|Z_0 = 4\}$. Then, α is the probability that in successive rolls $Z_1, Z_2, \ldots$ of a pair of dice, a sum of 4 appears before a sum of 7. Denote this event by B, and again bring in the law of total probability. Then,

$$
\alpha = \Pr\{B\} = \sum_{k=2}^{12} \Pr\{B|Z_1 = k\} p_Z(k).
\tag{2.18}
$$

Now $\Pr\{B|Z_1 = 4\} = 1$, while $\Pr\{B|Z_1 = 7\} = 0$. If the first roll results in anything other than a 4 or a 7, the problem is repeated in a statistically identical setting. That is, $\Pr\{B|Z_1 = k\} = \alpha$ for $k \neq 4$ or 7. Substitution into (2.18) results in

$$
\begin{aligned}
\alpha &= p_Z(4) \times 1 + p_Z(7) \times 0 + \sum_{k \neq 4, 7} p_Z(k) \times \alpha \\
&= p_Z(4) + [1 - p_Z(4) - p_Z(7)]\alpha,
\end{aligned}
$$

or

$$\alpha = \frac{p_Z(4)}{p_Z(4) + p_Z(7)}. \tag{2.19}$$

The same result may be secured by means of a longer, more computational, method. One may partition the event B into disjoint elemental events by writing

$$B = \{Z_1 = 4\} \cup \{Z_1 \neq 4 \text{ or } 7, Z_2 = 4\}$$
$$\cup \{Z_1 \neq 4 \text{ or } 7, Z_2 \neq 4 \text{ or } 7, Z_3 = 4\} \cup \cdots,$$

and then

$$\Pr\{B\} = \Pr\{Z_1 = 4\} + \Pr\{Z_1 \neq 4 \text{ or } 7, Z_2 = 4\}$$
$$+ \Pr\{Z_1 \neq 4 \text{ or } 7, Z_2 \neq 4 \text{ or } 7, Z_3 = 4\} + \cdots.$$

Now use the independence of $Z_1, Z_2, \ldots$ and sum a geometric series to secure

$$\Pr\{B\} = p_Z(4) + [1 - p_Z(4) - p_Z(7)]p_Z(4)$$
$$+ [1 - p_Z(4) - p_Z(7)]^2 p_Z(4) + \cdots$$
$$= \frac{p_Z(4)}{p_Z(4) + p_Z(7)}$$

in agreement with (2.19).

Extending the result just obtained to the other cases having more than one roll, we have

$$\Pr\{A|Z_0 = k\} = \frac{p_Z(k)}{p_Z(k) + p_Z(7)} \quad \text{for } k = 4, 5, 6, 8, 9, 10.$$

Finally, substitution into (2.17) yields the total win probability

$$\Pr\{A\} = p_Z(7) + p_Z(11) + \sum_{k=4,5,6,8,9,10} \frac{p_Z(k)^2}{p_Z(k) + p_Z(7)}. \tag{2.20}$$

The numerical values for $p_Z(k)$ given in (2.16), together with (2.20), determine the win probability

$$\Pr\{A\} = 0.49292929\cdots.$$

Having explained the computations, let us go on to a more interesting question. Suppose that the dice are not perfect cubes but are shaved so as to be slightly thinner in one dimension than in the other two. The numbers that appear on opposite faces on a single die always sum to 7. That is, 1 is opposite 6, 2 is opposite 5, and 3 is opposite 4. Suppose it is the 3-4 dimension that is smaller than the other two. See Figure 2.1. This

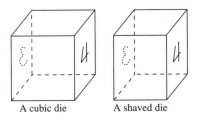

A cubic die   A shaved die

Figure 2.1 A cubic die versus a die that has been shaved down in one dimension.

will cause 3 and 4 to appear more frequently than the other faces, 1, 2, 5, and 6. To see this, think of the extreme case in which the 3-4 dimension is very thin, leading to a 3 or 4 on almost all tosses. Letting Y denote the result of tossing a single shaved die, we postulate that the probability mass function is given by

$$p_Y(3) = p_Y(4) = \frac{1}{6} + 2\varepsilon \equiv p_+,$$

$$p_Y(1) = p_Y(2) = p_Y(5) = p_Y(6) = \frac{1}{6} - \varepsilon \equiv p_-,$$

where $\varepsilon > 0$ is a small quantity depending on the amount by which the die has been biased.

If both dice are shaved in the same manner, the mass function for their sum can be determined in a straightforward manner from the following joint table:

				Die #1			
Die #2	1 $p-$	2 $p-$	3 $p+$	4 $p+$	5 $p-$	6 $p-$
1 $p-$	p_-^2	p_-^2	p_+p_-	p_+p_-	p_-^2	p_-^2
2 $p-$	p_-^2	p_-^2	p_+p_-	p_+p_-	p_-^2	p_-^2
3 $p+$	p_+p_-	p_+p_-	p_+^2	p_+^2	p_+p_-	p_+p_-
4 $p+$	p_+p_-	p_+p_-	p_+^2	p_+^2	p_+p_-	p_+p_-
5 $p-$	p_-^2	p_-^2	p_+p_-	p_+p_-	p_-^2	p_-^2
6 $p-$	p_-^2	p_-^2	p_+p_-	p_+p_-	p_-^2	p_-^2

It is easily seen that the probability mass function for the sum of the dice is

$$p(2) = p_-^2 = p(12),$$
$$p(3) = 2p_-^2 = p(11),$$
$$p(4) = p_-(p_- + 2p_+) = p(10),$$
$$p(5) = 4p_+p_- = p(9),$$
$$p(6) = p_-^2 + (p_+ + p_-)^2 = p(8),$$
$$p(7) = 4p_-^2 + 2p_+^2.$$

To obtain a numerical value to compare to the win probability $0.492929\cdots$ associated with fair dice, let us arbitrarily set $\varepsilon = 0.02$ so that $p_- = 0.146666\cdots$ and $p_+ = 0.206666\cdots$. Then, routine substitutions according to the table lead to

$$
\begin{aligned}
&p(2) = p(12) = 0.02151111, \quad p(5) = p(9) = 0.12124445, \\
&p(3) = p(11) = 0.04302222, \quad p(6) = p(8) = 0.14635556, \qquad (2.21) \\
&p(4) = p(10) = 0.08213333, \quad p(7) = 0.17146667,
\end{aligned}
$$

and the win probability becomes $\Pr\{A\} = 0.5029237$.

The win probability of 0.4929293 with fair dice is unfavorable, i.e., is less than $\frac{1}{2}$. With shaved dice, the win probability is favorable, now being 0.5029237. What appears to be a slight change becomes, in fact, quite significant when a large number of games are played. See Chapter 3, Section 3.5.

Exercises

2.2.1 A red die is rolled a single time. A green die is rolled repeatedly. The game stops the first time that the sum of the two dice is either 4 or 7. What is the probability that the game stops with a sum of 4?

2.2.2 Verify the win probability of 0.5029237 by substituting from (2.21) into (2.20).

2.2.3 Determine the win probability when the dice are shaved on the 1–6 faces and $p_+ = 0.206666\cdots$ and $p_- = 0.146666\cdots$.

Problems

2.2.1 Let $X_1, X_2, \ldots$ be independent identically distributed positive random variables whose common distribution function is F. We interpret $X_1, X_2, \ldots$ as successive bids on an asset offered for sale. Suppose that the policy is followed of accepting the first bid that exceeds some prescribed number A. Formally, the accepted bid is X_N, where

$$
N = \min\{k \geq 1; X_k > A\}.
$$

Set $\alpha = \Pr\{X_1 > A\}$ and $M = E[X_N]$.

(a) Argue the equation

$$
M = \int_A^\infty x \, \mathrm{d}F(x) + (1 - \alpha)M
$$

by considering the possibilities, either the first bid is accepted or it is not.

(b) Solve for M, thereby obtaining

$$M = \alpha^{-1} \int_A^\infty x \, dF(x).$$

(c) When X_1 has an exponential distribution with parameter λ, use the memoryless property to deduce $M = A + \lambda^{-1}$.

(d) Verify this result by calculation in (b).

2.2.2 Consider a pair of dice that are unbalanced by the addition of weights in the following manner: Die #1 has a small piece of lead placed near the four side, causing the appearance of the outcome 3 more often than usual, while die #2 is weighted near the three side, causing the outcome 4 to appear more often than usual. We assign the probabilities

Die #1

$$p(1) = p(2) = p(5) = p(6) = 0.166667,$$
$$p(3) = 0.186666,$$
$$p(4) = 0.146666;$$

Die #2

$$p(1) = p(2) = p(5) = p(6) = 0.166667,$$
$$p(3) = 0.146666,$$
$$p(4) = 0.186666.$$

Determine the win probability if the game of craps is played with these loaded dice.

2.3 Random Sums

Sums of the form $X = \xi_1 + \cdots + \xi_N$, where N is random, arise frequently and in varied contexts. Our study of random sums begins with a crisp definition and a precise statement of the assumptions effective in this section, followed by some quick examples.

We postulate a sequence $\xi_1, \xi_2, \ldots$ of independent and identically distributed random variables. Let N be a discrete random variable, independent of $\xi_1, \xi_2, \ldots$ and having the probability mass function $p_N(n) = \Pr\{N = n\}$ for $n = 0, 1, \ldots$. Define the random sum X by

$$X = \begin{cases} 0 & \text{if } N = 0, \\ \xi_1 + \cdots + \xi_N & \text{if } N > 0. \end{cases} \tag{2.22}$$

We save space by abbreviating (2.22) to simply $X = \xi_1 + \cdots + \xi_N$, understanding that $X = 0$ whenever $N = 0$.

Examples

(a) *Queueing*   Let N be the number of customers arriving at a service facility in a specified period of time, and let ξ_i be the service time required by the ith customer. Then, $X = \xi_1 + \cdots + \xi_N$ is the total demand for service time.

(b) *Risk Theory*   Suppose that a total of N claims arrives at an insurance company in a given week. Let ξ_i be the amount of the ith claim. Then, the total liability of the insurance company is $X = \xi_1 + \cdots + \xi_N$.

(c) *Population Models*   Let N be the number of plants of a given species in a specified area, and let ξ_i be the number of seeds produced by the ith plant. Then, $X = \xi_1 + \cdots + \xi_N$ gives the total number of seeds produced in the area.

(d) *Biometrics*   A wildlife sampling scheme traps a random number N of a given species. Let ξ_i be the weight of the ith specimen. Then, $X = \xi_1 + \cdots + \xi_N$ is the total weight captured.

The necessary background in conditional probability was covered in Section 2.1 for when $\xi_1, \xi_2, \ldots$ are discrete random variables. In order to study the random sum $X = \xi_1 + \cdots + \xi_N$ when $\xi_1, \xi_2, \ldots$ are continuous random variables, we need to extend our knowledge of conditional distributions.

2.3.1   Conditional Distributions: The Mixed Case

Let X and N be jointly distributed random variables and suppose that the possible values for N are the discrete set $n = 0, 1, 2, \ldots$. Then, the elementary definition of conditional probability (2.1) applies to define the *conditional distribution function* $F_{X|N}(x|n)$ of the random variable X, given that $N = n$, to be

$$F_{X|N}(x|n) = \frac{\Pr\{X \le x \text{ and } N = n\}}{\Pr\{N = n\}} \quad \text{if } \Pr\{N = n\} > 0, \tag{2.23}$$

and the conditional distribution function is not defined at values of n for which $\Pr\{N = n\} = 0$. It is elementary to verify that $F_{X|N}(x|n)$ is a probability distribution function in x at each value of n for which it is defined.

   The case in which X is a discrete random variable was covered in Section 2.1. Now let us suppose that X is continuous and that $F_{X|N}(x|n)$ is differentiable in x at each value of n for which $\Pr\{N = n\} > 0$. We define the *conditional probability density function* $f_{X|N}(x|n)$ for the random variable X given that $N = n$ by setting

$$f_{X|N}(x|n) = \frac{\mathrm{d}}{\mathrm{d}x} F_{X|N}(x|n) \quad \text{if } \Pr\{N = n\} > 0. \tag{2.24}$$

Again, $f_{X|N}(x|n)$ is a probability density function in x at each value of n for which it is defined. Moreover, the conditional density as defined in (2.24) has the appropriate properties, e.g.,

$$\Pr\{a \le X < b, N = n\} = \int_a^b f_{X|N}(x|n)p_N(n)\mathrm{d}x \tag{2.25}$$

for $a < b$ and where $p_N(n) = \Pr\{N = n\}$. The law of total probability leads to the marginal probability density function for X via

$$f_X(x) = \sum_{n=0}^{\infty} f_{X|N}(x|n)p_N(n). \tag{2.26}$$

Suppose that g is a function for which $E[|g(X)|] < \infty$. The conditional expectation of $g(X)$ given that $N = n$ is defined by

$$E[g(X)|N = n] = \int g(x)f_{X|N}(x|n)\,\mathrm{d}x. \tag{2.27}$$

Stipulated thus, $E[g(X)|N = n]$ satisfies the properties listed in (2.7) to (2.15) for the joint discrete case. For example, the law of total probability is

$$E[g(X)] = \sum_{n=0}^{\infty} E[g(X)|N = n]p_N(n) = E\{E[g(X)|N]\}. \tag{2.28}$$

2.3.2 The Moments of a Random Sum

Let us assume that ξ_k and N have the finite moments

$$\begin{aligned} E[\xi_k] &= \mu, \quad \mathrm{Var}[\xi_k] = \sigma^2, \\ E[N] &= v, \quad \mathrm{Var}[N] = \tau^2, \end{aligned} \tag{2.29}$$

and determine the mean and variance for $X = \xi_1 + \cdots + \xi_N$ as defined in (2.22). The derivation provides practice in manipulating conditional expectations, and the results,

$$E[X] = \mu v, \quad \mathrm{Var}[X] = v\sigma^2 + \mu^2\tau^2, \tag{2.30}$$

are useful and important. The properties of conditional expectation listed in (2.7) to (2.15) justify the steps in the determination.

If we begin with the mean $E[X]$, then

$$
\begin{aligned}
E[X] &= \sum_{n=0}^{\infty} E[X|N=n] p_N(n) && \text{[by (2.15)]} \\
&= \sum_{n=1}^{\infty} E[\xi_1 + \cdots + \xi_N | N=n] p_N(n) && \text{(definition of X)} \\
&= \sum_{n=1}^{\infty} E[\xi_1 + \cdots + \xi_n | N=n] p_N(n) && \text{[by (2.9)]} \\
&= \sum_{n=1}^{\infty} E[\xi_1 + \cdots + \xi_n] p_N(n) && \text{[by (2.10)]} \\
&= \mu \sum_{n=1}^{\infty} n p_N(n) = \mu v.
\end{aligned}
$$

To determine the variance, we begin with the elementary step

$$
\begin{aligned}
\text{Var}[X] &= E\left[(X - \mu v)^2\right] = E\left[(X - N\mu + N\mu - v\mu)^2\right] \\
&= E\left[(X - N\mu)^2\right] + E\left[\mu^2(N - v)^2\right] \\
&\quad + 2E[\mu(X - N\mu)(N - v)].
\end{aligned}
\tag{2.31}
$$

Then,

$$
\begin{aligned}
E\left[(X - N\mu)^2\right] &= \sum_{n=0}^{\infty} E\left[(X - N\mu)^2 | N=n\right] p_N(n) \\
&= \sum_{n=1}^{\infty} E\left[(\xi_1 + \cdots + \xi_n - n\mu)^2 | N=n\right] p_N(n) \\
&= \sigma^2 + \sum_{n=1}^{\infty} n p_N(n) = v\sigma^2,
\end{aligned}
$$

and

$$
E\left[\mu^2(N - v)^2\right] = \mu^2 E\left[(N - v)^2\right] = \mu^2 \tau^2,
$$

while

$$
\begin{aligned}
E[\mu(X - N\mu)(N - v)] &= \mu \sum_{n=0}^{\infty} E[(X - n\mu)(n - v)|N=n] p_N(n) \\
&= \mu \sum_{n=0}^{\infty} (n - v) E[(X - n\mu)|N=n] p_N(n) \\
&= 0
\end{aligned}
$$

(because $E[(X - n\mu)|N = n] = E[\xi_1 + \cdots + \xi_n - n\mu] = 0$). Then, (2.31) with the subsequent three calculations validates the variance of X as stated in (2.30).

Example The number of offspring of a given species is a random variable having probability mass function $p(k)$ for $k = 0, 1, \ldots$. A population begins with a single parent who produces a random number N of progeny, each of which independently produces offspring according to $p(k)$ to form a second generation. Then, the total number of descendants in the second generation may be written $X = \xi_1 + \cdots + \xi_N$, where ξ_k is the number of progeny of the kth offspring of the original parent. Let $E[N] = E[\xi_k] = \mu$ and $\text{Var}[N] = \text{Var}[\xi_k] = \sigma^2$. Then,

$$E[X] = \mu^2 \quad \text{and} \quad \text{Var}[X] = \mu\sigma^2(1 + \mu).$$

2.3.3 The Distribution of a Random Sum

Suppose that the summands $\xi_1, \xi_2, \ldots$ are continuous random variables having a probability density function $f(z)$. For $n \geq 1$, the probability density function for the fixed sum $\xi_1 + \cdots + \xi_n$ is the n-fold convolution of the density $f(z)$, denoted by $f^{(n)}(z)$ and recursively defined by

$$f^{(1)}(z) = f(z)$$

and

$$f^{(n)}(z) = \int f^{(n-1)}(z - u)f(u)du \quad \text{for } n > 1. \tag{2.32}$$

(See Chapter 1, Section 1.2.5 for a discussion of convolutions.) Because N and $\xi_1, \xi_2, \ldots$ are independent, then $f^{(n)}(z)$ is also the conditional density function for $X = \xi_1 + \cdots + \xi_N$ given that $N = n \geq 1$. Let us suppose that $\Pr\{N = 0\} = 0$. Then, by the law of total probability as expressed in (2.26), X is continuous and has the marginal density function

$$f_X(x) = \sum_{n=1}^{\infty} f^{(n)}(x)p_N(n). \tag{2.33}$$

Remark When $N = 0$ can occur with positive probability, then $X = \xi_1 + \cdots + \xi_N$ is a random variable having both continuous and discrete components to its distribution. Assuming that $\xi_1, \xi_2, \ldots$ are continuous with probability density function $f(z)$, then

$$\Pr\{X = 0\} = \Pr\{N = 0\} = p_N(0),$$

while for $0 < a < b$ or $a < b < 0$, then

$$\Pr\{a < X < b\} = \int_a^b \left\{ \sum_{n=1}^{\infty} f^{(n)}(z)p_N(n) \right\} dz. \tag{2.34}$$

Example *A Geometric Sum of Exponential Random Variables* In the following computational example, suppose

$$f(z) = \begin{cases} \lambda e^{-\lambda z} & \text{for } z \geq 0, \\ 0 & \text{for } z < 0, \end{cases}$$

and

$$p_N(n) = \beta(1-\beta)^{n-1} \quad \text{for } n = 1, 2, \dots.$$

For $n \geq 1$, the n-fold convolution of $f(z)$ is the gamma density

$$f^{(n)}(z) = \begin{cases} \dfrac{\lambda^n}{(n-1)!} z^{n-1} e^{-\lambda z} & \text{for } z \geq 0, \\ 0 & \text{for } z < 0. \end{cases}$$

(See Chapter 1, Section 1.4.4 for discussion.)

The density for $X = \xi_1 + \cdots + \xi_N$ is given, according to (2.26), by

$$\begin{aligned}
f_X(z) &= \sum_{n=1}^{\infty} f^{(n)}(z) p_N(n) \\
&= \sum_{n=1}^{\infty} \frac{\lambda^n}{(n-1)!} z^{n-1} e^{-\lambda z} \beta(1-\beta)^{n-1} \\
&= \lambda \beta e^{-\lambda z} \sum_{n=1}^{\infty} \frac{[\lambda(1-\beta)z]^{n-1}}{(n-1)!} \\
&= \lambda \beta e^{-\lambda z} e^{\lambda(1-\beta)z} \\
&= \lambda \beta e^{-\lambda \beta z}, \quad z \geq 0.
\end{aligned}$$

Surprise! X has an exponential distribution with parameter $\lambda\beta$.

Example *Stock Price Changes* Stochastic models for price fluctuations of publicly traded assets were developed as early as 1900.

Let Z denote the difference in price of a single share of a certain stock between the close of one trading day and the close of the next day. For an actively traded stock, a large number of transactions take place in a single day, and the total daily price change is the sum of the changes over these individual transactions. If we assume that price changes over successive transactions are independent random variables having a common finite variance,* then the central limit theorem applies. The price change over a large number of transactions should follow a normal, or Gaussian, distribution.

---

* Rather strong economic arguments in support of these assumptions can be given. The independence follows from concepts of a "perfect market" and the common variance from notions of time stationarity.

A variety of empirical studies have supported this conclusion. For the most part, these studies involved price changes over a fixed number of transactions. Other studies found discrepancies in that both very small and very large price changes occurred more frequently in the data than suggested by normal theory. At the same time, intermediate-size price changes were under represented in the data. For the most part, these studies examined price changes over fixed durations containing a random number of transactions.

A natural question arises: Does the random number of transactions in a given day provide a possible explanation for the departures from normality that are observed in data of daily price changes? Let us model the daily price change in the form

$$Z = \xi_0 + \xi_1 + \cdots + \xi_N = \xi_0 + X, \tag{2.35}$$

where $\xi_0, \xi_1, \ldots$ are independent normally distributed random variables with common mean zero and variance σ^2, and N has a Poisson distribution with mean v.

We interpret N as the number of transactions during the day, ξ_i for $i \geq 1$ as the price change during the ith transaction, and ξ_0 as an initial price change arising between the close of the market on one day and the opening of the market on the next day. (An obvious generalization would allow the distribution of ξ_0 to differ from that of $\xi_1, \xi_2, \ldots$.)

Conditioned on $N = n$, the random variable $Z = \xi_0 + \xi_1 + \cdots + \xi_n$ is normally distributed with mean zero and variance $(n+1)\sigma^2$. The conditional density function is

$$\phi_n(z) = \frac{1}{\sqrt{2\pi(n+1)}\sigma} \exp\left\{ -\frac{1}{2} \frac{1z^2}{(n+1)\sigma^2} \right\}.$$

Since the probability mass function for N is

$$p_N(n) = \frac{\lambda^n e^{-\lambda}}{n!}, \quad n = 0, 1, \ldots,$$

using (2.33) we determine the probability density function for the daily price change to be

$$f_Z(z) = \sum_{n=0}^{\infty} \phi_n(z) \frac{\lambda^n e^{-\lambda}}{n!}.$$

The formula for the density $f_Z(z)$ does not simplify. Nevertheless, numerical calculations are possible. When $\lambda = 1$ and $\sigma^2 = \frac{1}{2}$, then (2.30) shows that the variance of the daily price change Z in the model (2.35) is $\text{Var}[Z] = (1+\lambda)\sigma^2 = 1$. Thus, comparing the density $f_Z(z)$ when $\lambda = 1$ and $\sigma^2 = \frac{1}{2}$ to a normal density with mean zero and variance, one sheds some light on the question at hand.

The calculations were carried out and are shown in Figure 2.2.

The departure from normality that is exhibited by the random sum in Figure 2.2 is consistent with the departure from normality shown by stock price changes over fixed time intervals. Of course, our calculations do not *prove* that the observed departure

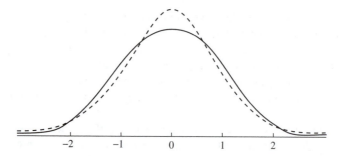

Figure 2.2 A standard normal density (solid line) as compared with a density for a random sum (dashed line). Both densities have zero mean and unit variance.

from normality is *caused* by the random number of transactions in a fixed time interval. Rather, the calculations show only that such an explanation is consistent with the data and is, therefore, a possible cause.

Exercises

2.3.1 A six-sided die is rolled, and the number N on the uppermost face is recorded. Then a fair coin is tossed N times, and the total number Z of heads to appear is observed. Determine the mean and variance of Z by viewing Z as a random sum of N Bernoulli random variables. Determine the probability mass function of Z, and use it to find the mean and variance of Z.

2.3.2 Six nickels are tossed, and the total number N of heads is observed. Then N dimes are tossed, and the total number Z of tails among the dimes is observed. Determine the mean and variance of Z. What is the probability that $Z = 2$?

2.3.3 Suppose that upon striking a plate a single electron is transformed into a number N of electrons, where N is a random variable with mean μ and standard deviation σ. Suppose that each of these electrons strikes a second plate and releases further electrons, independently of each other and each with the same probability distribution as N. Let Z be the total number of electrons emitted from the second plate. Determine the mean and variance of Z.

2.3.4 A six-sided die is rolled, and the number N on the uppermost face is recorded. From a jar containing 10 tags numbered $1, 2, \ldots, 10$ we then select N tags at random without replacement. Let X be the smallest number on the drawn tags. Determine $\Pr\{X = 2\}$ and $E[X]$.

2.3.5 The number of accidents occurring in a factory in a week is a Poisson random variable with mean 2. The number of individuals injured in different accidents is independently distributed, each with mean 3 and variance 4. Determine the mean and variance of the number of individuals injured in a week.

Problems

2.3.1 The following experiment is performed: An observation is made of a Poisson random variable N with parameter λ. Then N independent Bernoulli trials are performed, each with probability p of success. Let Z be the total number of successes observed in the N trials.

(a) Formulate Z as a random sum and thereby determine its mean and variance.

(b) What is the distribution of Z?

2.3.2 For each given p, let Z have a binomial distribution with parameters p and N. Suppose that N is itself binomially distributed with parameters q and M. Formulate Z as a random sum and show that Z has a binomial distribution with parameters pq and M.

2.3.3 Suppose that $\xi_1, \xi_2, \ldots$ are independent and identically distributed with $\Pr\{\xi_k = \pm 1\} = \frac{1}{2}$. Let N be independent of $\xi_1, \xi_2, \ldots$ and follow the geometric probability mass function

$$p_N(k) = \alpha(1-\alpha)^k \quad \text{for } k = 0, 1, \ldots,$$

where $0 < \alpha < 1$. Form the random sum $Z = \xi_1 + \cdots + \xi_N$.

(a) Determine the mean and variance of Z.

(b) Evaluate the higher moments $m_3 = E[Z^3]$ and $m_4 = E[Z^4]$.

Hint: Express Z^4 in terms of the ξ_i's where $\xi_i^2 = 1$ and $E[\xi_i \xi_j] = 0$.

2.3.4 Suppose $\xi_1, \xi_2, \ldots$ are independent and identically distributed random variables having mean μ and variance σ^2. Form the random sum $S_N = \xi_1 + \cdots + \xi_N$.

(a) Derive the mean and variance of S_N when N has a Poisson distribution with parameter λ.

(b) Determine the mean and variance of S_N when N has a geometric distribution with mean $\lambda = (1-p)/p$.

(c) Compare the behaviors in (a) and (b) as $\lambda \to \infty$.

2.3.5 To form a slightly different random sum, let $\xi_0, \xi_1, \ldots$ be independent identically distributed random variables and let N be a nonnegative integer-valued random variable, independent of $\xi_0, \xi_1, \ldots$. The first two moments are

$$E[\xi_k] = \mu, \quad \text{Var}[\xi_k] = \sigma^2,$$
$$E[N] = \nu, \quad \text{Var}[N] = \tau^2.$$

Determine the mean and variance of the random sum $Z = \xi_0 + \cdots + \xi_N$.

2.4 Conditioning on a Continuous Random Variable*

Let X and Y be jointly distributed continuous random variables with joint probability density function $f_{X,Y}(x,y)$. We define the conditional probability density function $f_{X|Y}(x|y)$ for the random variable X given that $Y = y$ by the formula

$$f_{X|Y}(x|y) = \frac{f_{X,Y}(x,y)}{f_Y(y)} \quad \text{if } f_Y(y) > 0, \tag{2.36}$$

---

* The reader may wish to defer reading this section until encountering Chapter 7, on renewal processes, where conditioning on a continuous random variable first appears.

and the conditional density is not defined at values y for which $f_Y(y) = 0$. The conditional distribution function for X given $Y = y$ is defined by

$$F_{X|Y}(x|y) = \int_{-\infty}^{x} f_{X|Y}(\xi|y)d\xi \quad \text{if } f_Y(y) > 0. \tag{2.37}$$

Finally, given a function g for which $E[|g(X)|] < \infty$, the conditional expectation of $g(X)$ given that $Y = y$ is defined to be

$$E[g(X)|Y = y] = \int g(x)f_{X|Y}(x|y)dx \quad \text{if } f_Y(y) > 0. \tag{2.38}$$

The definitions given in (2.36) to (2.38) are a significant extension of our elementary notions of conditional probability because they allow us to condition on certain events having zero probability. To understand the distinction, try to apply the elementary formula

$$\Pr\{A|B\} = \frac{\Pr\{A \text{ and } B\}}{\Pr\{B\}} \quad \text{if } \Pr\{B\} > 0 \tag{2.39}$$

to evaluate the conditional probability $\Pr\{a < X \leq b | Y = y\}$. We set $A = \{a < X \leq b\}$ and $B = \{Y = y\}$. But Y is a continuous random variable, and thus, $\Pr\{B\} = \Pr\{Y = y\} = 0$, and (2.39) cannot be applied. Equation (2.37) saves the day, yielding

$$\Pr\{a < X \leq b | Y = y\} = F_{X|Y}(b|y) - F_{X|Y}(a|y) = \int_{a}^{b} f_{X|Y}(\xi|y)d\xi, \tag{2.40}$$

provided only that the density $f_Y(y)$ is strictly positive at the point y.

To emphasize the important advance being made, we consider the following simple problem. A woman arrives at a bus stop at a time Y that is uniformly distributed between 0 (noon) and 1. Independently, the bus arrives at a time Z that is also uniformly distributed between 0 and 1. Given that the woman arrives at time $Y = 0.20$, what is the probability that she misses the bus?

On the one hand, the answer $\Pr\{Z < Y | Y = 0.20\} = 0.20$ is obvious. On the other hand, this elementary question cannot be answered by the elementary conditional probability formula (2.39) because the event $\{Y = 0.20\}$ has zero probability. To apply (2.36), start with the joint density function

$$f_{Z,Y}(z,y) = \begin{cases} 1 & \text{for } 0 \leq z, y \leq 1, \\ 0 & \text{elsewhere,} \end{cases}$$

and change variables according to $X = Y - Z$. Then,

$$f_{X,Y}(x,y) = 1 \quad \text{for } 0 \leq y \leq 1, y - 1 \leq x \leq y,$$

and, applying (2.36), we find that

$$f_{X|Y}(x|0.20) = \frac{f_{X,Y}(x, 0.20)}{f_Y(0.20)} = 1 \quad \text{for } -0.80 \le x \le 0.20.$$

Finally,

$$\Pr\{Z < Y | Y = 0.20\} = \Pr\{X > 0 | Y = 0.20\} = \int_0^\infty f_{X|Y}(x|0.20)dx = 0.20.$$

We see that the definition in (2.36) leads to the intuitively correct answer.

The conditional density function that is prescribed by (2.36) possesses all of the properties that are called for by our intuition and the basic concept of conditional probability. In particular, one can calculate the probability of joint events by the formula

$$\Pr\{a < X < b, c < Y < d\} = \int_c^d \left\{ \int_a^b f_{X|Y}(x|y)dx \right\} f_Y(y)dy, \tag{2.41}$$

which becomes the law of total probability by setting $c = -\infty$ and $d = +\infty$;

$$\Pr\{a < X < b\} = \int_{-\infty}^{+\infty} \left\{ \int_a^b f_{X|Y}(x|y)dx \right\} f_Y(y)dy. \tag{2.42}$$

For the same reasons, the conditional expectation as defined in (2.38) satisfies the requirements listed in (2.7) to (2.11). The property (2.12), adapted to a continuous random variable Y, is written

$$E[g(X)h(Y)] = E\{h(Y)E[g(X)|Y]\}$$
$$= \int h(y)E[g(X)|Y = y]f_Y(y)dy, \tag{2.43}$$

valid for any bounded function h, and assuming $E[|g(X)|] < \infty$. When $h(y) \equiv 1$, we recover the law of total probability in the form

$$E[g(X)] = E\{E[g(X)|Y]\} = \int E[g(X)|Y = y]f_Y(y)dy. \tag{2.44}$$

Both the discrete and continuous cases of (2.43) and (2.44) are contained in the expressions

$$E[g(X)h(Y)] = E\{h(Y)E[g(X)|Y]\} = \int h(y)E[g(X)|Y = y]dF_Y(y), \tag{2.45}$$

and

$$E[g(X)] = E\{E[g(X)|Y]\} = \int E[g(X)|Y=y] \, dF_Y(y). \qquad (2.46)$$

[See the discussion following (1.9) in Chapter 1 for an explanation of the symbolism in (2.45) and (2.46).]

The following exercises provide practice in deriving conditional probability density functions and in manipulating the law of total probability.

Example Suppose X and Y are jointly distributed random variables having the density function

$$f_{XY}(x,y) = \frac{1}{y} e^{-(x/y)-y} \quad \text{for } x, y > 0.$$

We first determine the marginal density for y, obtaining

$$f_Y(y) = \int_0^\infty f_{XY}(x,y) \, dx$$

$$= e^{-y} \int_0^\infty y^{-1} e^{-(x/y)} \, dx = e^{-y} \quad \text{for } y > 0.$$

Then,

$$f_{X|Y}(x|y) = \frac{f_{XY}(x,y)}{f_Y(y)} = y^{-1} e^{-(x/y)} \quad \text{for } x, y > 0.$$

That is, conditional on $Y = y$, the random variable X has an exponential distribution with parameter $1/y$. It is easily seen that $E[X|Y=y] = y$.

Example For each given p, let X have a binomial distribution with parameters p and N. Suppose that p is uniformly distributed on the interval $[0, 1]$. What is the resulting distribution of X?

We are given the marginal distribution for p and the conditional distribution for X. Applying the law of total probability and the beta integral in Chapter 1 (1.66), we obtain

$$\Pr\{X = k\} = \int_0^1 \Pr\{X = k | p = \xi\} f_p(\xi) \, d\xi$$

$$= \int_0^1 \frac{N!}{k!\,(N-k)!} \xi^k (1-\xi)^{N-k} \, d\xi$$

$$= \frac{N!}{k!\,(N-k)!} \frac{k!\,(N-k)!}{(N+1)!}$$

$$= \frac{1}{N+1} \quad \text{for } k = 0, 1, \ldots, N.$$

That is, X is uniformly distributed on the integers $0, 1, \ldots, N$.

When p has the beta distribution with parameters r and s, then similar calculations give

$$\Pr\{X = k\} = \frac{N!}{k!\,(N-k)!}\,\frac{\Gamma(r+s)}{\Gamma(r)\Gamma(s)}\int_0^1 \xi^{r-1}(1-\xi)^{s-1}\xi^k(1-\xi)^{N-k}\,d\xi$$

$$= \binom{N}{k}\frac{\Gamma(r+s)\Gamma(r+k)\Gamma(s+N-k)}{\Gamma(r)\Gamma(s)\Gamma(N+r+s)}\quad \text{for } k = 0, 1, \ldots, N.$$

Example A random variable Y follows the exponential distribution with parameter θ. Given that $Y = y$, the random variable X has a Poisson distribution with mean y. Applying the law of total probability then yields

$$\Pr\{X = k\} = \int_0^\infty \frac{y^k e^{-y}}{k!}\theta e^{-\theta y}\,dy$$

$$= \frac{\theta}{k!}\int_0^\infty y^k e^{-(1+\theta)y}\,dy$$

$$= \frac{\theta}{k!\,(1+\theta)^{k+1}}\int_0^\infty u^k e^{-u}\,du$$

$$= \frac{\theta}{(1+\theta)^{k+1}}\quad \text{for } k = 0, 1, \ldots.$$

Suppose that Y has the gamma density

$$f_Y(y) = \frac{\theta}{\Gamma(\alpha)}(\theta y)^{\alpha-1}e^{-\theta y}, \quad y \geq 0.$$

Then, similar calculations yield

$$\Pr\{X = k\} = \int_0^\infty \frac{y^k e^{-y}}{k!}\frac{\theta}{\Gamma(\alpha)}(\theta y)^{\alpha-1}e^{-\theta y}\,dy$$

$$= \frac{\theta^\alpha}{k!\,\Gamma(\alpha)(1+\theta)^{k+\alpha}}\int_0^\infty u^{k+\alpha-1}e^{-u}\,du$$

$$= \frac{\Gamma(k+\alpha)}{k!\,\Gamma(\alpha)}\left(\frac{\theta}{1+\theta}\right)^\alpha\left(\frac{1}{1+\theta}\right)^k, \quad k = 0, 1, \ldots.$$

This is the negative binomial distribution.

Exercises

2.4.1 Suppose that three contestants on a quiz show are each given the same question and that each answers it correctly, independently of the others, with probability p. But the difficulty of the question is itself a random variable, so let us suppose, for the sake of illustration, that p is uniformly distributed over the interval $(0, 1]$. What is the probability that exactly two of the contestants answer the question correctly?

2.4.2 Suppose that three components in a certain system each function with probability p and fail with probability $1 - p$, each component operating or failing independently of the others. But the system is in a random environment so that p is itself a random variable. Suppose that p is uniformly distributed over the interval $(0, 1]$. The system operates if at least two of the components operate. What is the probability that the system operates?

2.4.3 A random variable T is selected that is uniformly distributed over the interval $(0, 1]$. Then, a second random variable U is chosen, uniformly distributed on the interval $(0, T]$. What is the probability that U exceeds $\frac{1}{2}$?

2.4.4 Suppose X and Y are independent random variables, each exponentially distributed with parameter λ. Determine the probability density function for $Z = X/Y$.

2.4.5 Let U be uniformly distributed over the interval $[0, L]$ where L follows the gamma density $f_L(x) = xe^{-x}$ for $x \geq 0$. What is the joint density function of U and $V = L - U$?

Problems

2.4.1 Suppose that the outcome X of a certain chance mechanism depends on a parameter p according to $\Pr\{X = 1\} = p$ and $\Pr\{X = 0\} = 1 - p$, where $0 \leq p \leq 1$. Suppose that p is chosen at random, uniformly distributed over the unit interval $[0, 1]$, and then, that two independent outcomes X_1 and X_2 are observed. What is the unconditional correlation coefficient between X_1 and X_2?

Note: Conditionally independent random variables may become dependent if they share a common parameter.

2.4.2 Let N have a Poisson distribution with parameter $\lambda > 0$. Suppose that, conditioned on $N = n$, the random variable X is binomially distributed with parameters $N = n$ and p. Set $Y = N - X$. Show that X and Y have Poisson distributions with respective parameters λp and $\lambda(1 - p)$ and that X and Y are independent.

Note: Conditionally dependent random variables may become independent through randomization.

2.4.3 Let X have a Poisson distribution with parameter $\lambda > 0$. Suppose λ itself is random, following an exponential density with parameter θ.
   (a) What is the marginal distribution of X?
   (b) Determine the conditional density for λ given $X = k$.

2.4.4 Suppose X and Y are independent random variables having the same Poisson distribution with parameter λ, but where λ is also random, being exponentially distributed with parameter θ. What is the conditional distribution for X given that $X + Y = n$?

2.4.5 Let X and Y be jointly distributed random variables whose joint probability mass function is given in the following table:

		x		
		-1	0	1
y	-1	$\dfrac{1}{9}$	$\dfrac{2}{9}$	0
	0	0	$\dfrac{1}{9}$	$\dfrac{2}{9}$
	1	$\dfrac{2}{9}$	0	$\dfrac{1}{9}$

$$p(x, y) = \Pr\{X = x, Y = y\}$$

Show that the covariance between X and Y is zero even though X and Y are not independent.

2.4.6 Let $X_0, X_1, X_2, \ldots$ be independent identically distributed nonnegative random variables having a continuous distribution. Let N be the first index k for which $X_k > X_0$. That is, $N = 1$ if $X_1 > X_0, N = 2$ if $X_1 \le X_0$ and $X_2 > X_0$, etc. Determine the probability mass function for N and the mean $E[N]$. (Interpretation: $X_0, X_1, \ldots$ are successive offers or bids on a car that you are trying to sell. Then, N is the index of the first bid that is better than the initial bid.)

2.4.7 Suppose that X and Y are independent random variables, each having the same exponential distribution with parameter α. What is the conditional probability density function for X, given that $Z = X + Y = z$?

2.4.8 Let X and Y have the normal density given in Chapter 1, in (1.47). Show that the conditional density function for X, given that $Y = y$, is normal with moments

$$\mu_{X|Y} = \mu_X + \frac{\rho \sigma_X}{\sigma_Y}(y - \mu_Y)$$

and

$$\sigma_{X|Y} = \sigma_X \sqrt{1 - \rho^2}.$$

2.5 Martingales*

Stochastic processes are characterized by the dependence relationships that exist among their variables. The martingale property is one such relationship that captures a notion of a game being fair. The martingale property is a restriction solely on the

* Some problems scattered throughout the text call for the student to identify certain stochastic processes as martingales. Otherwise, the material of this section is not used in the sequel.

conditional means of some of the variables, given values of others, and does not otherwise depend on the actual distribution of the random variables in the stochastic process. Despite the apparent weakness of the martingale assumption, the consequences are striking, as we hope to suggest.

2.5.1 The Definition

We begin the presentation with the simplest definition.

Definition A stochastic process $\{X_n; n = 0, 1, \ldots\}$ is a martingale if for $n = 0, 1, \ldots$,

(a) $E[|X_n|] < \infty$,
and
(b) $E[X_{n+1}|X_0, \ldots, X_n] = X_n$.

Taking expectations on both sides of (b),

$$E\{E[X_{n+1}|X_0, \ldots, X_n]\} = E\{X_n\},$$

and using the law of total probability in the form

$$E\{E[X_{n+1}|X_1, \ldots, X_n]\} = E[X_{n+1}]$$

shows that

$$E[X_{n+1}] = E[X_n],$$

and consequently, a martingale has constant mean:

$$E[X_0] = E[X_k] = E[X_n], \quad 0 \le k \le n. \tag{2.47}$$

A similar conditioning (see Problem 2.5.1) verifies that the martingale equality (b) extends to future times in the form

$$E[X_m|X_0, \ldots, X_n] = X_n \quad \text{for } m \ge n. \tag{2.48}$$

To relate the martingale property to concepts of fairness in gambling, consider X_n to be a certain player's fortune after the nth play of a game. The game is "fair" if on average, the player's fortune neither increases nor decreases at each play. The martingale property (b) requires the player's fortune after the next play to equal, on average, his current fortune and not be otherwise affected by previous history. Some early work in martingale theory was motivated in part by problems in gambling. For example, *martingale systems theorems* consider whether an astute choice of betting strategy can turn a fair game into a favorable one, and the name "martingale" derives from a French term for the particular strategy of doubling one's bets until a win is secured. While it remains popular to illustrate martingale concepts with gambling examples, today, martingale theory has such broad scope and diverse applications that to think of it purely in terms of gambling would be unduly restrictive and misleading.

Example *Stock Prices in a Perfect Market* Let X_n be the closing price at the end of day n of a certain publicly traded security such as a share of stock. While daily prices may fluctuate, many scholars believe that, in a perfect market, these price sequences should be martingales. In a perfect market freely open to all, they argue, it should not be possible to predict with any degree of accuracy whether a future price X_{n+1} will be higher or lower than the current price X_n. For example, if a future price could be expected to be higher, then a number of buyers would enter the market, and their demand would raise the current price X_n. Similarly, if a future price could be predicted as lower, a number of sellers would appear and tend to depress the current price. Equilibrium obtains where the future price cannot be predicted, on average, as higher or lower, i.e., where price sequences are martingales.

2.5.2 The Markov Inequality

What does the mean of a random variable tell us about its distribution? For a nonnegative random variable X, Markov's inequality is $\lambda \Pr\{X \geq \lambda\} \leq E[X]$, for any positive constant λ. For example, if $E[X] = 1$, then $\Pr\{X \geq 4\} \leq \frac{1}{4}$, no matter what the actual distribution of X is. The proof uses two properties: (i) $X \geq 0$ (X is a nonnegative random variable), and (ii) $E[X\mathbf{1}\{X \geq \lambda\}] \geq \lambda \Pr\{X \geq \lambda\}$. (Recall that $\mathbf{1}(A)$ is the *indicator* of an event A and is one if A occurs and zero otherwise. See Chapter 1, Section 1.3.1.) Then, by the law of total probability,

$$
\begin{aligned}
E[X] &= E\left[X\mathbf{1}_{[\lambda,\infty)}(X)\right] + E\left[X\mathbf{1}_{(-\infty,\lambda)}(X)\right] \\
&\geq E\left[X\mathbf{1}_{[\lambda,\infty)}(X)\right] \\
&\geq \lambda \Pr[X \geq \lambda]
\end{aligned}
$$

and Markov's inequality results.

2.5.3 The Maximal Inequality for Nonnegative Martingales

Because a martingale has constant mean, Markov's inequality applied to a nonnegative martingale immediately yields

$$
\Pr\{X_n \geq \lambda\} \leq \frac{E[X_0]}{\lambda}, \quad \lambda > 0.
$$

We will extend the reasoning behind Markov's inequality to achieve an inequality of far greater power:

$$
\Pr\left\{\max_{0 \leq n \leq m} X_n \geq \lambda\right\} \leq \frac{E[X_0]}{\lambda}. \tag{2.49}
$$

Instead of limiting the probability of a large value for a single observation X_n, the *maximal inequality* (2.49) limits the probability of observing a large value anywhere in the time interval $0, \ldots, m$, and since the right side of (2.49) does not depend on the length of the interval, the maximal inequality limits the probability of observing a large value at any time in the infinite future of the martingale!

In order to prove the maximal inequality for nonnegative martingales, we need but a single additional fact: If X and Y are jointly distributed random variables and B is an arbitrary set, then

$$E[X \mathbf{1}_B(Y)] = E[E(X|Y) \mathbf{1}_B(Y)] \tag{2.50}$$

But (2.50) follows from the conditional expectation property (2.12), $E[g(X)h(Y)] = E\{h(Y)E[g(X)|Y]\}$, with $g(x) = x$ and $h(y) = \mathbf{1}(y \text{ in } B)$. We will have need of (2.50) with $X = X_m$ and $Y = (X_0, \ldots, X_n)$, whereupon (2.50) followed by (2.48) then justifies

$$E[X_m \mathbf{1}\{X_0 < \lambda, \ldots, X_{n-1} < \lambda, X_n \geq \lambda\}]$$
$$= E[E\{X_m|X_0, \ldots, X_n\} \mathbf{1}\{X_0 < \lambda, \ldots, X_{n-1} < \lambda, X_n \geq \lambda\}] \tag{2.51}$$
$$= E[X_n \mathbf{1}\{X_0 < \lambda, \ldots, X_{n-1} < \lambda, X_n \geq \lambda\}].$$

Theorem 2.1. *Let* $X_0, X_1, \ldots$ *be a martingale with nonnegative values; i.e.,* $\Pr\{X_n \geq 0\} = 1$ *for* $n = 0, 1, \ldots$ *For any* $\lambda > 0$,

$$\Pr\left\{\max_{0 \leq n \leq m} X_n \geq \lambda\right\} \leq \frac{E[X_0]}{\lambda}, \quad \text{for } 0 \leq n \leq m \tag{2.52}$$

and

$$\Pr\left\{\max_{n \geq 0} X_n > \lambda\right\} \leq \frac{E[X_0]}{\lambda}, \quad \text{for all } n. \tag{2.53}$$

Proof. Inequality (2.53) follows from (2.52) because the right side of (2.52) does not depend on m. We begin with the law of total probability, as in Chapter 1, Section 1.2.1. Either the $\{X_0, \ldots, X_m\}$ sequence rises above λ for the first time at some index n or else it remains always below λ. As these possibilities are mutually exclusive and exhaustive, we apply the law of total probability to obtain

$$E[X_m] = \sum_{n=0}^{m} E[X_m \mathbf{1}\{X_0 < \lambda, \ldots, X_{n-1} < \lambda, X_n \geq \lambda\}$$
$$+ E[X_m \mathbf{1}\{X_0 < \lambda, \ldots, X_m < \lambda\}]$$
$$\geq \sum_{n=0}^{m} E[X_m \mathbf{1}\{X_0 < \lambda, \ldots, X_{n-1} < \lambda, X_n \geq \lambda\}] \quad (X_m \geq 0)$$

$$= \sum_{n=0}^{m} E[X_n \mathbf{1}\{X_0 < \lambda, \ldots, X_{n-1} < \lambda, X_n \geq \lambda\}] \quad \text{[using (2.51)]}$$

$$\geq \lambda \sum_{n=0}^{m} \Pr\{X_0 < \lambda, \ldots, X_{n-1} < \lambda, X_n \geq \lambda\}$$

$$= \lambda \Pr\left\{\max_{0 \leq n \leq m} X_n \geq \lambda\right\}.$$

∎

Example A gambler begins with a unit amount of money and faces a series of independent fair games. Beginning with $X_0 = 1$, the gambler bets the amount p, $0 < p < 1$. If the first game is a win, which occurs with probability $\frac{1}{2}$, the gambler's fortune is $X_1 = 1 + pX_0 = 1 + p$. If the first game is a loss, then $X_1 = 1 - pX_0 = 1 - p$. After the nth play and with a current fortune of X_n, the gambler wagers pX_n, and

$$X_{n+1} = \begin{cases} (1+p)X_n & \text{with probability } \dfrac{1}{2}, \\[2mm] (1-p)X_n & \text{with probability } \dfrac{1}{2}. \end{cases}$$

Then, $\{X_n\}$ is a nonnegative martingale, and the maximal inequality (2.52) with $\lambda = 2$, e.g., asserts that *the probability that the gambler ever doubles his money is less than or equal to $\frac{1}{2}$, and this holds no matter what the game is, as long as it is fair, and no matter what fraction p of his fortune is wagered at each play.* Indeed, the fraction wagered may vary from play to play, as long as it is chosen without knowledge of the next outcome.

As amply demonstrated by this example, the maximal inequality is a very strong statement. Indeed, more elaborate arguments based on the maximal and other related martingale inequalities are used to show that a nonnegative martingale converges: If $\{X_n\}$ is a nonnegative martingale, then there exists a random variable, let us call it X_∞, for which $\lim_{n \to \infty} X_n = X_\infty$. We cannot guarantee the equality of the expectations in the limit, but the inequality $E[X_0] \geq E[X_\infty] \geq 0$ can be established.

Example In Chapter 3, Section 3.8, we will introduce the branching process model for population growth. In this model, X_n is the number of individuals in the population in the nth generation, and $\mu > 0$ is the mean family size or expected number of offspring of any single individual. The mean population size in the nth generation is $X_0 \mu^n$. In this branching process model, X_n/μ^n is a nonnegative martingale (see Chapter 3, Problem 3.8.4), and the maximal inequality implies that the probability of the actual population ever exceeding 10 times the mean size is less than or equal to 1/10. The nonnegative martingale convergence theorem asserts that the evolution of such a population after many generations may be described by a single random variable X_∞ in the form

$$X_n \approx X_\infty \mu^n, \quad \text{for large } n.$$

Example *How NOT to generate a uniformly distributed random variable*   An urn initially contains one red and one green ball. A ball is drawn at random and it is returned to the urn, together with another ball of the same color. This process is repeated indefinitely. After the nth play, there will be a total of $n + 2$ balls in the urn. Let R_n be the number of these balls that are red, and $X_n = R_n/(n + 2)$ the fraction of red balls. We claim that $\{X_n\}$ is a martingale. First, observe that

$$R_{n+1} = \begin{cases} R_n + 1 & \text{with probability } X_n \\ R_n & \text{with probability } 1 - X_n \end{cases}$$

so that

$$E[R_{n+1}|X_n] = R_n + X_n = X_n(2 + n + 1),$$

and finally,

$$E[X_{n+1}|X_n] = \frac{1}{n+3} E[R_{n+1}|X_n] = \frac{2+n+1}{n+3} X_n = X_n.$$

This verifies the martingale property, and because such a fraction is always nonnegative, indeed, between 0 and 1, there must be a random variable X_∞ to which the martingale converges. We will derive the probability distribution of the random limit. It is immediate that R_1 is equally likely to be 1 or 2, since the first ball chosen is equally likely to be red or green. Continuing,

$$Pr\{R_2 = 3\} = Pr\{R_2 = 3|R_1 = 2\} Pr\{R_1 = 2\}$$

$$= \left(\frac{2}{3}\right)\left(\frac{1}{2}\right) = \frac{1}{3};$$

$$Pr\{R_2 = 2\} = Pr\{R_2 = 2|R_1 = 1\} Pr\{R_1 = 1\}$$
$$+ Pr\{R_2 = 2|R_1 = 2\} Pr\{R_1 = 2\}$$

$$= \left(\frac{1}{3}\right)\left(\frac{1}{2}\right) + \left(\frac{1}{3}\right)\left(\frac{1}{2}\right) = \frac{1}{3};$$

and since the probabilities must sum to 1,

$$Pr\{R_2 = 1\} = \frac{1}{3}.$$

By repeating these simple calculations, it is easy to see that

$$Pr\{R_n = k\} = \frac{1}{n+1} \quad \text{for } k = 1, 2, \ldots, n+1,$$

and that, therefore, X_n is uniformly distributed over the values $1/(n+2)$, $2/(n+2), \ldots, (n+1)/(n+2)$. This uniform distribution must prevail in the limit, which leads to

$$\Pr\{X_\infty \le x\} = x \quad \text{for } 0 < x < 1.$$

Think about this remarkable result for a minute! If you sit down in front of such an urn and play this game, eventually the fraction of red balls in your urn will stabilize in the near vicinity of some value, call it U. If I play the game, the fraction of red balls in my urn will stabilize also, but at another value, U'. Anyone who plays the game will find the fraction of red balls in the urn tending toward some limit, but everyone will experience a different limit. In fact, each play of the game generates a fresh, uniformly distributed random variable, in the limit. Of course, there may be faster and simpler ways to generate uniformly distributed random variables.

Martingale implications include many more inequalities and convergence theorems. As briefly mentioned at the start, there are so-called *systems theorems* that delimit the conditions under which a gambling system, such as doubling the bets until a win is secured, can turn a fair game into a winning game. A deeper discussion of martingale theory would take us well beyond the scope of this introductory text, and our aim must be limited to building an enthusiasm for further study. Nevertheless, a large variety of important martingales will be introduced in the Problems at the end of each section in the remainder of the book.

Exercises

2.5.1 Let X be an exponentially distributed random variable with mean $E[X] = 1$. For $x = 0.5, 1$, and 2, compare $\Pr\{X > x\}$ with the Markov inequality bound $E[X]/x$.

2.5.2 Let X be a Bernoulli random variable with parameter p. Compare $\Pr\{X \ge 1\}$ with the Markov inequality bound.

2.5.3 Let ξ be a random variable with mean μ and standard deviation σ. Let $X = (\xi - \mu)^2$. Apply Markov's inequality to X to deduce Chebyshev's inequality:

$$\Pr\{|\xi - \mu| \ge \varepsilon\} \le \frac{\sigma^2}{\varepsilon^2} \quad \text{for any } \varepsilon > 0.$$

Problems

2.5.1 Use the law of total probability for conditional expectations $E[E\{X|Y,Z\}|Z] = E[X|Z]$ to show

$$E[X_{n+2}|X_0, \ldots, X_n] = E[E\{X_{n+2}|X_0, \ldots, X_{n+1}\}|X_0, \ldots, X_n].$$

Conclude that when X_n is a martingale,

$$E[X_{n+2}|X_0, \ldots, X_n] = X_n.$$

2.5.2 Let $U_1, U_2, \ldots$ be independent random variables each uniformly distributed over the interval $(0, 1]$. Show that $X_0 = 1$ and $X_n = 2^n U_1 \cdots U_n$ for $n = 1, 2, \ldots$ defines a martingale.

2.5.3 Let $S_0 = 0$, and for $n \geq 1$, let $S_n = \varepsilon_1 + \cdots + \varepsilon_n$ be the sum of n independent random variables, each exponentially distributed with mean $E[\varepsilon] = 1$. Show that

$$X_n = 2^n \exp(-S_n), \quad n \geq 0$$

defines a martingale.

2.5.4 Let $\xi_1, \xi_2, \ldots$ be independent Bernoulli random variables with parameter $p, 0 < p < 1$. Show that $X_0 = 1$ and $X_n = p^{-n} \xi_1 \cdots \xi_n, n = 1, 2, \ldots$, defines a nonnegative martingale. What is the limit of X_n as $n \to \infty$?

2.5.5 Consider a stochastic process that evolves according to the following laws: If $X_n = 0$, then $X_{n+1} = 0$, whereas if $X_n > 0$, then

$$X_{n+1} = \begin{cases} X_n + 1 & \text{with probability } \dfrac{1}{2} \\[2ex] X_n - 1 & \text{with probability } \dfrac{1}{2}. \end{cases}$$

(a) Show that X_n is a nonnegative martingale.

(b) Suppose that $X_0 = i > 0$. Use the maximal inequality to bound

$$\Pr\{X_n \geq N \quad \text{for some } n \geq 0 | X_0 = i\}.$$

Note: X_n represents the fortune of a player of a fair game who wagers \$1 at each bet and who is forced to quit if all money is lost ($X_n = 0$). This *gambler's ruin* problem is discussed fully in Chapter 3, Section 3.5.3.

3 Markov Chains: Introduction

3.1 Definitions

A *Markov process* $\{X_t\}$ is a stochastic process with the property that, given the value of X_t, the values of X_s for $s > t$ are not influenced by the values of X_u for $u < t$. In words, the probability of any particular future behavior of the process, when its current state is known exactly, is not altered by additional knowledge concerning its past behavior. A *discrete-time Markov chain* is a Markov process whose state space is a finite or countable set, and whose (time) index set is $T = (0, 1, 2, \ldots)$. In formal terms, the Markov property is that

$$\Pr\{X_{n+1} = j | X_0 = i_0, \ldots, X_{n-1} = i_{n-1}, X_n = i\}$$
$$= \Pr\{X_{n+1} = j | X_n = i\} \tag{3.1}$$

for all time points n and all states $i_0, \ldots, i_{n-1}, i, j$.

It is frequently convenient to label the state space of the Markov chain by the non-negative integers $\{0, 1, 2, \ldots\}$, which we will do unless the contrary is explicitly stated, and it is customary to speak of X_n as being in state i if $X_n = i$.

The probability of X_{n+1} being in state j given that X_n is in state i is called the *one-step transition probability* and is denoted by $P_{ij}^{n,n+1}$. That is,

$$P_{ij}^{n,n+1} = \Pr\{X_{n+1} = j | X_n = i\}. \tag{3.2}$$

The notation emphasizes that in general the transition probabilities are functions not only of the initial and final states but also of the time of transition as well. When the one-step transition probabilities are independent of the time variable n, we say that the Markov chain has *stationary transition probabilities*. Since the vast majority of Markov chains that we shall encounter have stationary transition probabilities, we limit our discussion to this case. Then, $P_{ij}^{n,n+1} = P_{ij}$ is independent of n, and P_{ij} is the conditional probability that the state value undergoes a transition from i to j in one trial. It is customary to arrange these numbers P_{ij} in a *matrix*, in the infinite square array

$$\mathbf{P} = \begin{Vmatrix} P_{00} & P_{01} & P_{02} & P_{03} & \cdots \\ P_{10} & P_{11} & P_{12} & P_{13} & \cdots \\ P_{20} & P_{21} & P_{22} & P_{23} & \cdots \\ \vdots & \vdots & \vdots & \vdots & \\ P_{i0} & P_{i1} & P_{i2} & P_{i3} & \cdots \\ \vdots & \vdots & \vdots & \vdots & \end{Vmatrix}$$

An Introduction to Stochastic Modeling

and refer to $\mathbf{P} = \|P_{ij}\|$ as the Markov matrix or *transition probability matrix* of the process.

The ith row of $\mathbf{P}$, for $i = 0, 1, \ldots$, is the probability distribution of the values of X_{n+1} under the condition that $X_n = i$. If the number of states is finite, then $\mathbf{P}$ is a finite square matrix whose order (the number of rows) is equal to the number of states. Clearly, the quantities P_{ij} satisfy the conditions

$$P_{ij} \geq 0 \quad \text{for } i, j = 0, 1, 2, \ldots, \tag{3.3}$$

$$\sum_{j=0}^{\infty} P_{ij} = 1 \quad \text{for } i = 0, 1, 2, \ldots. \tag{3.4}$$

The condition (3.4) merely expresses the fact that some transition occurs at each trial. (For convenience, one says that a transition has occurred even if the state remains unchanged.)

A Markov process is completely defined once its transition probability matrix and initial state X_0 (or, more generally, the probability distribution of X_0) are specified. We shall now prove this fact.

Let $\Pr\{X_0 = i\} = p_i$. It is enough to show how to compute the quantities

$$\Pr\{X_0 = i_0, X_1 = i_1, X_2 = i_2, \ldots, X_n = i_n\}, \tag{3.5}$$

since any probability involving $X_{j_1}, \ldots, X_{j_k}$, for $j_1 < \cdots < j_k$, can be obtained, according to the axiom of total probability, by summing terms of the form (3.5).

By the definition of conditional probabilities, we obtain

$$\begin{aligned}
\Pr\{X_0 &= i_0, X_1 = i_1, X_2 = i_2, \ldots, X_n = i_n\} \\
&= \Pr\{X_0 = i_0, X_1 = i_1, \ldots, X_{n-1} = i_{n-1}\} \\
&\quad \times \Pr\{X_n = i_n | X_0 = i_0, X_1 = i_1, \ldots, X_{n-1} = i_{n-1}\}.
\end{aligned} \tag{3.6}$$

Now, by the definition of a Markov process,

$$\begin{aligned}
\Pr\{X_n &= i_n | X_0 = i_0, X_1 = i_1, \ldots, X_{n-1} = i_{n-1}\} \\
&= \Pr\{X_n = i_n | X_{n-1} = i_{n-1}\} = P_{i_{n-1}, i_n}.
\end{aligned} \tag{3.7}$$

Substituting (3.7) into (3.6) gives

$$\begin{aligned}
\Pr\{X_0 &= i_0, X_1 = i_1, \ldots, X_n = i_n\} \\
&= \Pr\{X_0 = i_0, X_1 = i_1, \ldots, X_{n-1} = i_{n-1}\} P_{i_{n-1}, i_n}.
\end{aligned}$$

Then, upon repeating the argument $n - 1$ additional times, (3.5) becomes

$$\begin{aligned}
\Pr\{X_0 &= i_0, X_1 = i_1, \ldots, X_n = i_n\} \\
&= p_{i_0} P_{i_0, i_1} \cdots P_{i_{n-2}, i_{n-1}} P_{i_{n-1}, i_n}.
\end{aligned} \tag{3.8}$$

This shows that all finite-dimensional probabilities are specified once the transition probabilities and initial distribution are given, and in this sense, the process is defined by these quantities.

Related computations show that (3.1) is equivalent to the Markov property in the form

$$\Pr\{X_{n+1} = j_1, \ldots, X_{n+m} = j_m | X_0 = i_0, \ldots, X_n = i_n\}$$
$$= \Pr\{X_{n+1} = j_1, \ldots, X_{n+m} = j_m | X_n = i_n\} \tag{3.9}$$

for all time points n, m and all states $i_0, \ldots, i_n, j_1, \ldots, j_m$. In other words, once (3.9) is established for the value $m = 1$, it holds for all $m \geq 1$ as well.

Exercises

3.1.1 A Markov chain $X_0, X_1, \ldots$ on states 0, 1, 2 has the transition probability matrix

$$\mathbf{P} = \begin{array}{c} 0 \\ 1 \\ 2 \end{array} \begin{array}{|ccc|} 0 & 1 & 2 \\ \hline 0.1 & 0.2 & 0.7 \\ 0.9 & 0.1 & 0 \\ 0.1 & 0.8 & 0.1 \end{array}$$

and initial distribution $p_0 = \Pr\{X_0 = 0\} = 0.3, p_1 = \Pr\{X_0 = 1\} = 0.4$, and $p_2 = \Pr\{X_0 = 2\} = 0.3$. Determine $\Pr\{X_0 = 0, X_1 = 1, X_2 = 2\}$.

3.1.2 A Markov chain $X_0, X_1, X_2, \ldots$ has the transition probability matrix

$$\mathbf{P} = \begin{array}{c} 0 \\ 1 \\ 2 \end{array} \begin{array}{|ccc|} 0 & 1 & 2 \\ \hline 0.7 & 0.2 & 0.1 \\ 0 & 0.6 & 0.4 \\ 0.5 & 0 & 0.5 \end{array}$$

Determine the conditional probabilities

$$\Pr\{X_2 = 1, X_3 = 1 | X_1 = 0\} \quad \text{and} \quad \Pr\{X_1 = 1, X_2 = 1 | X_0 = 0\}.$$

3.1.3 A Markov chain $X_0, X_1, X_2, \ldots$ has the transition probability matrix

$$\mathbf{P} = \begin{array}{c} 0 \\ 1 \\ 2 \end{array} \begin{array}{|ccc|} 0 & 1 & 2 \\ \hline 0.6 & 0.3 & 0.1 \\ 0.3 & 0.3 & 0.4 \\ 0.4 & 0.1 & 0.5 \end{array}$$

If it is known that the process starts in state $X_0 = 1$, determine the probability $\Pr\{X_0 = 1, X_1 = 0, X_2 = 2\}$.

3.1.4 A Markov chain $X_0, X_1, X_2, \ldots$ has the transition probability matrix

$$
\mathbf{P} = \begin{array}{c} \\ 0 \\ 1 \\ 2 \end{array}
\begin{array}{ccc}
0 & 1 & 2 \\
\left\|\begin{array}{ccc}
0.1 & 0.1 & 0.8 \\
0.2 & 0.2 & 0.6 \\
0.3 & 0.3 & 0.4
\end{array}\right\|
\end{array}.
$$

Determine the conditional probabilities

$$\Pr\{X_1 = 1, X_2 = 1 | X_0 = 0\} \quad \text{and} \quad \Pr\{X_2 = 1, X_3 = 1 | X_1 = 0\}.$$

3.1.5 A Markov chain $X_0, X_1, X_2, \ldots$ has the transition probability matrix

$$
\mathbf{P} = \begin{array}{c} \\ 0 \\ 1 \\ 2 \end{array}
\begin{array}{ccc}
0 & 1 & 2 \\
\left\|\begin{array}{ccc}
0.3 & 0.2 & 0.5 \\
0.5 & 0.1 & 0.4 \\
0.5 & 0.2 & 0.3
\end{array}\right\|
\end{array}
$$

and initial distribution $p_0 = 0.5$ and $p_1 = 0.5$. Determine the probabilities

$$\Pr\{X_0 = 1, X_1 = 1, X_2 = 0\} \quad \text{and} \quad \Pr\{X_1 = 1, X_2 = 1, X_3 = 0\}.$$

Problems

3.1.1 A simplified model for the spread of a disease goes this way: The total population size is $N = 5$, of which some are diseased and the remainder are healthy. During any single period of time, two people are selected at random from the population and assumed to interact. The selection is such that an encounter between any pair of individuals in the population is just as likely as between any other pair. If one of these persons is diseased and the other not, with probability $\alpha = 0.1$ the disease is transmitted to the healthy person. Otherwise, no disease transmission takes place. Let X_n denote the number of diseased persons in the population at the end of the nth period. Specify the transition probability matrix.

3.1.2 Consider the problem of sending a binary message, 0 or 1, through a signal channel consisting of several stages, where transmission through each stage is subject to a fixed probability of error α. Suppose that $X_0 = 0$ is the signal that is sent and let X_n be the signal that is received at the nth stage. Assume that $\{X_n\}$ is a Markov chain with transition probabilities $P_{00} = P_{11} = 1 - \alpha$ and $P_{01} = P_{10} = \alpha$, where $0 < \alpha < 1$.

(a) Determine $\Pr\{X_0 = 0, X_1 = 0, X_2 = 0\}$, the probability that no error occurs up to stage $n = 2$.

(b) Determine the probability that a correct signal is received at stage 2.

Hint: This is $\Pr\{X_0 = 0, X_1 = 0, X_2 = 0\} + \Pr\{X_0 = 0, X_1 = 1, X_2 = 0\}$.

3.1.3 Consider a sequence of items from a production process, with each item being graded as good or defective. Suppose that a good item is followed by another good item with probability α and is followed by a defective item with probability $1 - \alpha$. Similarly, a defective item is followed by another defective item with probability β and is followed by a good item with probability $1 - \beta$. If the first item is good, what is the probability that the first defective item to appear is the fifth item?

3.1.4 The random variables $\xi_1, \xi_2, \ldots$ are independent and with the common probability mass function

$k =$	0	1	2	3
$\Pr\{\xi = k\} =$	0.1	0.3	0.2	0.4

Set $X_0 = 0$, and let $X_n = \max\{\xi_1, \ldots, \xi_n\}$ be the largest ξ observed to date. Determine the transition probability matrix for the Markov chain $\{X_n\}$.

3.2 Transition Probability Matrices of a Markov Chain

A Markov chain is completely defined by its one-step transition probability matrix and the specification of a probability distribution on the state of the process at time 0. The analysis of a Markov chain concerns mainly the calculation of the probabilities of the possible realizations of the process.

Central in these calculations are the n-step transition probability matrices $\mathbf{P}^{(n)} = \|P_{ij}^{(n)}\|$. Here, $P_{ij}^{(n)}$ denotes the probability that the process goes from state i to state j in n transitions. Formally,

$$P_{ij}^{(n)} = \Pr\{X_{m+n} = j | X_m = i\}. \tag{3.10}$$

Observe that we are dealing only with temporally homogeneous processes having stationary transition probabilities, since otherwise the left side of (3.10) would also depend on m.

The Markov property allows us to express (3.10) in terms of $\|P_{ij}\|$ as stated in the following theorem.

Theorem 3.1. *The n-step transition probabilities of a Markov chain satisfy*

$$P_{ij}^{(n)} = \sum_{k=0}^{\infty} P_{ik} P_{kj}^{(n-1)}, \tag{3.11}$$

where we define

$$P_{ij}^{(0)} = \begin{cases} 1 & \text{if } i = j, \\ 0 & \text{if } i \neq j. \end{cases}$$

From the theory of matrices, we recognize the relation (3.11) as the formula for matrix multiplication so that $\mathbf{P}^{(n)} = \mathbf{P} \times \mathbf{P}^{(n-1)}$. By iterating this formula, we obtain

$$\mathbf{P}^{(n)} = \underbrace{\mathbf{P} \times \mathbf{P} \times \cdots \times \mathbf{P}}_{n \text{ factors}} = \mathbf{P}^n; \tag{3.12}$$

in other words, the n-step transition probabilities $P_{ij}^{(n)}$ are the entries in the matrix $\mathbf{P}^n$, the nth power of $\mathbf{P}$.

Proof. The proof proceeds via a *first step analysis*, a breaking down, or analysis, of the possible transitions on the first step, followed by an application of the Markov property. The event of going from state i to state j in n transitions can be realized in the mutually exclusive ways of going to some intermediate state $k(k = 0, 1, \ldots)$ in the first transition, and then going from state k to state j in the remaining $(n - 1)$ transitions. Because of the Markov property, the probability of the second transition is $P_{kj}^{(n-1)}$ and that of the first is clearly P_{ik}. If we use the law of total probability, then (3.11) follows. The steps are

$$P_{ij}^{(n)} = \Pr\{X_n = j | X_0 = i\} = \sum_{k=0}^{\infty} \Pr\{X_n = j, X_1 = k | X_0 = i\}$$

$$= \sum_{k=0}^{\infty} \Pr\{X_1 = k | X_0 = i\} \Pr\{X_n = j | X_0 = i, X_1 = k\}$$

$$= \sum_{k=0}^{\infty} P_{ik} P_{kj}^{(n-1)}.$$

If the probability of the process initially being in state j is p_j, i.e., the distribution law of X_0 is $\Pr\{X_0 = j\} = p_j$, then the probability of the process being in state k at time n is

$$p_k^{(n)} = \sum_{j=0}^{\infty} p_j P_{jk}^{(n)} = \Pr\{X_n = k\}. \tag{3.13}$$

■

Exercises

3.2.1 A Markov chain $\{X_n\}$ on the states $0, 1, 2$ has the transition probability matrix

$$\mathbf{P} = \begin{array}{c} \\ 0 \\ 1 \\ 2 \end{array} \begin{array}{ccc} 0 & 1 & 2 \\ \left\| \begin{array}{ccc} 0.1 & 0.2 & 0.7 \\ 0.2 & 0.2 & 0.6 \\ 0.6 & 0.1 & 0.3 \end{array} \right\| \end{array}.$$

(a) Compute the two-step transition matrix P^2.

(b) What is $\Pr\{X_3 = 1 | X_1 = 0\}$?

(c) What is $\Pr\{X_3 = 1 | X_0 = 0\}$?

3.2.2 A particle moves among the states $0, 1, 2$ according to a Markov process whose transition probability matrix is

$$
\mathbf{P} =
\begin{array}{c}
\\
0 \\
1 \\
2
\end{array}
\begin{array}{ccc}
0 & 1 & 2 \\
\left\| \begin{array}{ccc}
0 & \dfrac{1}{2} & \dfrac{1}{2} \\
\dfrac{1}{2} & 0 & \dfrac{1}{2} \\
\dfrac{1}{2} & \dfrac{1}{2} & 0
\end{array} \right\|
\end{array}.
$$

Let X_n denote the position of the particle at the nth move. Calculate $\Pr\{X_n = 0 | X_0 = 0\}$ for $n = 0, 1, 2, 3, 4$.

3.2.3 A Markov chain $X_0, X_1, X_2, \ldots$ has the transition probability matrix

$$
\mathbf{P} =
\begin{array}{c}
\\
0 \\
1 \\
2
\end{array}
\begin{array}{ccc}
0 & 1 & 2 \\
\left\| \begin{array}{ccc}
0.7 & 0.2 & 0.1 \\
0 & 0.6 & 0.4 \\
0.5 & 0 & 0.5
\end{array} \right\|
\end{array}.
$$

Determine the conditional probabilities

$$\Pr\{X_3 = 1 | X_0 = 0\} \quad \text{and} \quad \Pr\{X_4 = 1 | X_0 = 0\}.$$

3.2.4 A Markov chain $X_0, X_1, X_2, \ldots$ has the transition probability matrix

$$
\mathbf{P} =
\begin{array}{c}
\\
0 \\
1 \\
2
\end{array}
\begin{array}{ccc}
0 & 1 & 2 \\
\left\| \begin{array}{ccc}
0.6 & 0.3 & 0.1 \\
0.3 & 0.3 & 0.4 \\
0.4 & 0.1 & 0.5
\end{array} \right\|
\end{array}.
$$

If it is known that the process starts in state $X_0 = 1$, determine the probability $\Pr\{X_2 = 2\}$.

3.2.5 A Markov chain $X_0, X_1, X_2, \ldots$ has the transition probability matrix

$$
\mathbf{P} =
\begin{array}{c}
\\
0 \\
1 \\
2
\end{array}
\begin{array}{ccc}
0 & 1 & 2 \\
\left\| \begin{array}{ccc}
0.1 & 0.1 & 0.8 \\
0.2 & 0.2 & 0.6 \\
0.3 & 0.3 & 0.4
\end{array} \right\|
\end{array}.
$$

Determine the conditional probabilities

$$\Pr\{X_3 = 1 | X_1 = 0\} \quad \text{and} \quad \Pr\{X_2 = 1 | X_0 = 0\}.$$

3.2.6 A Markov chain $X_0, X_1, X_2, \ldots$ has the transition probability matrix

$$
\mathbf{P} =
\begin{array}{c c}
 & \begin{array}{ccc} 0 & 1 & 2 \end{array} \\
\begin{array}{c} 0 \\ 1 \\ 2 \end{array} &
\left\|
\begin{array}{ccc}
0.3 & 0.2 & 0.5 \\
0.5 & 0.1 & 0.4 \\
0.5 & 0.2 & 0.3
\end{array}
\right\|
\end{array}
$$

and initial distribution $p_0 = 0.5$ and $p_1 = 0.5$. Determine the probabilities $\Pr\{X_2 = 0\}$ and $\Pr\{X_3 = 0\}$.

Problems

3.2.1 Consider the Markov chain whose transition probability matrix is given by

$$
\mathbf{P} =
\begin{array}{c c}
 & \begin{array}{cccc} 0 & 1 & 2 & 3 \end{array} \\
\begin{array}{c} 0 \\ 1 \\ 2 \\ 3 \end{array} &
\left\|
\begin{array}{cccc}
0.4 & 0.3 & 0.2 & 0.1 \\
0.1 & 0.4 & 0.3 & 0.2 \\
0.3 & 0.2 & 0.1 & 0.4 \\
0.2 & 0.1 & 0.4 & 0.3
\end{array}
\right\|
\end{array}.
$$

Suppose that the initial distribution is $p_i = \frac{1}{4}$ for $i = 0, 1, 2, 3$. Show that $\Pr\{X_n = k\} = \frac{1}{4}, k = 0, 1, 2, 3$, for all n. Can you deduce a general result from this example?

3.2.2 Consider the problem of sending a binary message, 0 or 1, through a signal channel consisting of several stages, where transmission through each stage is subject to a fixed probability of error α. Let X_0 be the signal that is sent, and let X_n be the signal that is received at the nth stage. Suppose X_n is a Markov chain with transition probabilities $P_{00} = P_{11} = 1 - \alpha$ and $P_{01} = P_{10} = \alpha$, $(0 < \alpha < 1)$. Determine $\Pr\{X_5 = 0 | X_0 = 0\}$, the probability of correct transmission through five stages.

3.2.3 Let X_n denote the quality of the nth item produced by a production system with $X_n = 0$ meaning "good" and $X_n = 1$ meaning "defective." Suppose that X_n evolves as a Markov chain whose transition probability matrix is

$$
\mathbf{P} =
\begin{array}{c c}
 & \begin{array}{cc} 0 & 1 \end{array} \\
\begin{array}{c} 0 \\ 1 \end{array} &
\left\|
\begin{array}{cc}
0.99 & 0.01 \\
0.12 & 0.88
\end{array}
\right\|
\end{array}.
$$

What is the probability that the fourth item is defective given that the first item is defective?

3.2.4 Suppose X_n is a two-state Markov chain whose transition probability matrix is

$$\mathbf{P} = \begin{matrix} & 0 & 1 \\ 0 & \alpha & 1-\alpha \\ 1 & 1-\beta & \beta \end{matrix}.$$

Then, $Z_n = (X_{n-1}, X_n)$ is a Markov chain having the four states $(0,0)$, $(0,1)$, $(1,0)$, and $(1,1)$. Determine the transition probability matrix.

3.2.5 A Markov chain has the transition probability matrix

$$\mathbf{P} = \begin{matrix} & 0 & 1 & 2 \\ 0 & 0.7 & 0.2 & 0.1 \\ 1 & 0.3 & 0.5 & 0.2 \\ 2 & 0 & 0 & 1 \end{matrix}.$$

The Markov chain starts at time zero in state $X_0 = 0$. Let

$$T = \min\{n \geq 0; X_n = 2\}$$

be the first time that the process reaches state 2. Eventually, the process will reach and be absorbed into state 2. If in some experiment we observed such a process and noted that absorption had not yet taken place, we might be interested in the conditional probability that the process is in state 0 (or 1), given that absorption had not yet taken place. Determine $\Pr\{X_3 = 0 | X_0, T > 3\}$.

Hint: The event $\{T > 3\}$ is exactly the same as the event $\{X_3 \neq 2\} = \{X_3 = 0\} \cup \{X_3 = 1\}$.

3.3 Some Markov Chain Models

Markov chains can be used to model and quantify a large number of natural physical, biological, and economic phenomena that can be described by them. This is enhanced by the amenability of Markov chains to quantitative manipulation. In this section, we give several examples of Markov chain models that arise in various parts of science. General methods for computing certain functionals on Markov chains are derived in the following section.

3.3.1 An Inventory Model

Consider a situation in which a commodity is stocked in order to satisfy a continuing demand. We assume that the replenishment of stock takes place at the end of periods labeled $n = 0, 1, 2, \ldots$, and we assume that the total aggregate demand for the commodity during period n is a random variable ξ_n whose distribution function is independent of the time period,

$$\Pr\{\xi_n = k\} = a_k \quad \text{for } k = 0, 1, 2, \ldots, \tag{3.14}$$

where $a_k \geq 0$ and $\sum_{k=0}^{\infty} a_k = 1$. The stock level is examined at the end of each period. A replenishment policy is prescribed by specifying two nonnegative critical numbers s and $S > s$ whose interpretation is, if the end-of-period stock quantity is not greater than s, then an amount sufficient to increase the quantity of stock on hand up to the level S is immediately procured. If, however, the available stock is in excess of s, then no replenishment of stock is undertaken. Let X_n denote the quantity on hand at the end of period n just prior to restocking. The states of the process $\{X_n\}$ consist of the possible values of stock size

$$S, S-1, \ldots, +1, 0, -1, -2, \ldots,$$

where a negative value is interpreted as an unfilled demand that will be satisfied immediately upon restocking.

The process $\{X_n\}$ is depicted in Figure 3.1.

According to the rules of the inventory policy, the stock levels at two consecutive periods are connected by the relation

$$X_{n+1} = \begin{cases} X_n - \xi_{n+1} & \text{if } s < X_n \leq S, \\ S - \xi_{n+1} & \text{if } X_n \leq s, \end{cases} \tag{3.15}$$

where ξ_n is the quantity demanded in the nth period, stipulated to follow the probability law (3.14). If we assume that the successive demands $\xi_1, \xi_2, \ldots$ are independent random variables, then the stock values $X_0, X_1, X_2, \ldots$ constitute a Markov chain whose transition probability matrix can be calculated in accordance with relation (3.15). Explicitly,

$$P_{ij} = \Pr\{X_{n+1} = j | X_n = i\}$$
$$= \begin{cases} \Pr\{\xi_{n+1} = i - j\} & \text{if } s < i \leq S, \\ \Pr\{\xi_{n+1} = S - j\} & \text{if } i \leq s. \end{cases}$$

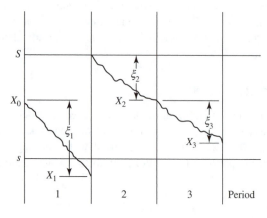

Figure 3.1 The inventory process.

Consider a spare parts inventory model as a numerical example in which either 0, 1, or 2 repair parts are demanded in any period, with

$$\Pr\{\xi_n = 0\} = 0.5, \quad \Pr\{\xi_n = 1\} = 0.4, \quad \Pr\{\xi_n = 2\} = 0.1,$$

and suppose $s = 0$, while $S = 2$. The possible values for X_n are $S = 2, 1, 0$, and -1. To illustrate the transition probability calculations, we will consider first the determination of $P_{10} = \Pr\{X_{n+1} = 0 | X_n = 1\}$. When $X_n = 1$, then no replenishment takes place and the next state $X_{n+1} = 0$ results when the demand $\xi_{n+1} = 1$, and this occurs with probability $P_{10} = 0.4$. To illustrate another case, if $X_n = 0$, then instantaneous replenishment to $S = 2$ ensues, and a next period level of $X_{n+1} = 0$ results from the demand quantity $\xi_{n+1} = 2$. The corresponding probability of this outcome yields $P_{00} = 0.1$. Continuing in this manner, we obtain the transition probability matrix

$$\mathbf{P} = \begin{array}{c} \\ -0 \\ 0 \\ +1 \\ +2 \end{array} \begin{array}{cccc} -1 & 0 & +1 & +2 \\ \left\| \begin{array}{cccc} 0 & 0.1 & 0.4 & 0.5 \\ 0 & 0.1 & 0.4 & 0.5 \\ 0.1 & 0.4 & 0.5 & 0 \\ 0 & 0.1 & 0.4 & 0.5 \end{array} \right\| \end{array}.$$

Important quantities of interest in inventory models of this type are the long-term fraction of periods in which demand is not met ($X_n < 0$) and long-term average inventory level. Using the notation $p_j^{(n)} = \Pr\{X_n = j\}$, we give these quantities, respectively, as $\lim_{n \to \infty} \Sigma_{j<0} p_j^{(n)}$ and $\lim_{n \to \infty} \Sigma_{j>0} j p_j^{(n)}$. This illustrates the importance of determining conditions under which the probabilities $p_j^{(n)}$ stabilize and approach limiting probabilities π_j as $n \to \infty$ and of determining methods for calculating the limiting probabilities π_j when they exist. These topics are the subject of Chapter 4.

3.3.2 The Ehrenfest Urn Model

A classical mathematical description of diffusion through a membrane is the famous Ehrenfest urn model. Imagine two containers containing a total of $2a$ balls (molecules). Suppose the first container, labeled A, holds k balls and the second container, B, holds the remaining $2a - k$ balls. A ball is selected at random (all selections are equally likely) from the totality of the $2a$ balls and moved to the other container. (A molecule diffuses at random through the membrane.) Each selection generates a transition of the process. Clearly, the balls fluctuate between the two containers with an average drift from the urn with the excess numbers to the one with the smaller concentration.

Let Y_n be the number of balls in urn A at the nth stage, and define $X_n = Y_n - a$. Then, $\{X_n\}$ is a Markov chain on the states $i = -a, -a+1, \ldots, -1, 0, +1, \ldots, a$ with transition probabilities

$$P_{ij} = \begin{cases} \dfrac{a - i}{2a} & \text{if } j = i + 1, \\ \dfrac{a + i}{2a} & \text{if } j = i - 1, \\ 0 & \text{otherwise.} \end{cases}$$

An important quantity in the Ehrenfest urn model is the long-term, or equilibrium, distribution of the number of balls in each urn.

3.3.3 Markov Chains in Genetics

The following idealized genetics model was introduced by S. Wright to investigate the fluctuation of gene frequency under the influence of mutation and selection. We begin by describing the so-called simple haploid model of random reproduction, disregarding mutation pressures and selective forces. We assume that we are dealing with a fixed population size of $2N$ genes composed of type-a and type-A individuals. The makeup of the next generation is determined by $2N$ independent Bernoulli trials as follows: If the parent population consists of j **a**-genes and $2N - j$ **A**-genes, then each trial results in **a** or **A** with probabilities

$$p_j = \frac{j}{2N}, \quad q_j = 1 - \frac{j}{2N},$$

respectively. Repeated selections are done with replacement. By this procedure, we generate a Markov chain $\{X_n\}$, where X_n is the number of **a**-genes in the nth generation among a constant population size of $2N$ individuals. The state space contains the $2N + 1$ values $\{0, 1, 2, \ldots, 2N\}$. The transition probability matrix is computed according to the binomial distribution as

$$\Pr\{X_{n+1} = k | X_n = j\} = P_{jk} = \binom{2N}{k} p_j^k q_j^{2N-k} \tag{3.16}$$

$$(j, k = 0, 1, \ldots, 2N).$$

For some discussion of the biological justification of these postulates, we refer the reader to Fisher.[*]

Notice that states 0 and $2N$ are completely absorbing in the sense that once $X_n = 0$ (or $2N$), then $X_{n+k} = 0$ (or $2N$, respectively) for all $k \geq 0$. One of the questions of interest is to determine the probability, under the condition $X_0 = i$, that the population will attain fixation, i.e., that it will become a pure population composed only of **a**-genes or **A**-genes. It is also pertinent to determine the rate of approach to fixation. We will examine such questions in our general analysis of absorption probabilities.

A more complete model takes account of mutation pressures. We assume that prior to the formation of the new generation, each gene has the possibility to mutate, i.e., to change into a gene of the other kind. Specifically, we assume that for each gene the mutation **a** $\to$ **A** occurs with probability α, and **A** $\to$ **a** occurs with probability β. Again we assume that the composition of the next generation is determined by $2N$ independent binomial trials. The relevant values of p_j and q_j when the parent

---

[*] R. A. Fisher, *The Genetical Theory of Natural Selection*, Oxford (Clarendon) Press, London and New York, 1962.

population consists of j **a**-genes are now taken to be

$$p_j = \frac{j}{2N}(1-\alpha) + \left(1 - \frac{j}{2N}\right)\beta$$

and (3.17)

$$q_j = \frac{j}{2N}\alpha + \left(1 - \frac{j}{2N}\right)(1-\beta).$$

The rationale is as follows: We assume that the mutation pressures operate first, after which a new gene is chosen by random selection from the population. Now, the probability of selecting an **a**-gene after the mutation forces have acted is just $1/(2N)$ times the number of **a**-genes present; hence, the average probability (averaged with respect to the possible mutations) is simply $1/(2N)$ times the average number of **a**-genes after mutation. But this average number is clearly $j(1-\alpha) + (2N-j)\beta$, which leads at once to (3.17).

The transition probabilities of the associated Markov chain are calculated by (3.16) using the values of p_j and q_j given in (3.17).

If $\alpha\beta > 0$, then fixation will not occur in any state. Instead, as $n \to \infty$, the distribution function of X_n will approach a steady-state distribution of a random variable ξ, where $\Pr\{\xi = k\} = \pi_k (k = 0, 1, 2, \ldots, 2N) \left(\sum_{k=0}^{n} \pi_k = 1, \pi_k > 0\right)$. The distribution function of ξ is called the steady-state gene frequency distribution.

We return to the simple random mating model and discuss the concept of a selection force operating in favor of, say, **a**-genes. Suppose we wish to impose a selective advantage for **a**-genes over **A**-genes so that the relative number of offspring have expectations proportional to $1+s$ and 1, respectively, where s is small and positive. We replace $p_j = j/(2N)$ and $q_j = 1 - j/(2N)$ by

$$p_j = \frac{(1+s)j}{2N+sj}, \quad q_j = 1 - p_j,$$

and build the next generation by binomial sampling as before. If the parent population consisted of j **a**-genes, then in the next generation the expected population sizes of **a**-genes and **A**-genes, respectively, are

$$2N\frac{(1+s)j}{2N+sj}, \quad 2N\frac{(2N-j)}{2N+sj}.$$

The ratio of expected population size of **a**-genes to **A**-genes at the $(n+1)$st generation is

$$\frac{1+s}{1} \times \frac{j}{2N-j} = \left(\frac{1+s}{1}\right)\left(\frac{\text{number of \textbf{a}-genes in the } n\text{th generation}}{\text{number of \textbf{A}-genes in the } n\text{th generation}}\right),$$

which explains the meaning of selection.

3.3.4 A Discrete Queueing Markov Chain

Customers arrive for service and take their place in a waiting line. During each period of time, a single customer is served, provided that at least one customer is present. If no customer awaits service, then during this period no service is performed. (We can imagine, e.g., a taxi stand at which a cab arrives at fixed time intervals to give service. If no one is present, the cab immediately departs.) During a service period, new customers may arrive. We suppose that the actual number of customers that arrive during the nth period is a random variable ξ_n whose distribution is independent of the period and is given by

$$\Pr\{k \text{ customers arrive in a service period}\} = \Pr\{\xi_n = k\} = a_k,$$

for $k = 0, 1, \ldots$, where $a_k \geq 0$ and $\sum_{k=0}^{\infty} a_k = 1$.

We also assume that $\xi_1, \xi_2, \ldots$ are independent random variables. The state of the system at the start of each period is defined to be the number of customers waiting in line for service. If the present state is i, then after the lapse of one period the state is

$$j = \begin{cases} i - 1 + \xi & \text{if } i \geq 1, \\ \xi & \text{if } i = 0, \end{cases} \tag{3.18}$$

where ξ is the number of new customers having arrived in this period while a single customer was served. In terms of the random variables of the process, we can express (3.18) formally as

$$X_{n+1} = (X_n - 1)^+ + \xi_n,$$

where $Y^+ = \max\{Y, 0\}$. In view of (3.18), the transition probability matrix may be calculated easily, and we obtain

$$\mathbf{P} = \begin{Vmatrix} a_0 & a_1 & a_2 & a_3 & a_4 & \cdots \\ a_0 & a_1 & a_2 & a_3 & a_4 & \cdots \\ 0 & a_0 & a_1 & a_2 & a_3 & \cdots \\ 0 & 0 & a_0 & a_1 & a_2 & \cdots \\ 0 & 0 & 0 & a_0 & a_1 & \cdots \\ \vdots & \vdots & \vdots & \vdots & \vdots & \end{Vmatrix}.$$

It is intuitively clear that if the expected number of new customers, $\sum_{k=0}^{\infty} k a_k$, who arrive during a service period exceeds one, then with the passage of time the length of the waiting line increases without limit. On the other hand, if $\sum_{k=0}^{\infty} k a_k < 1$, then the length of the waiting line approaches a statistical equilibrium that is described by a limiting distribution

$$\lim_{n \to \infty} \Pr\{X_n = k \mid X_0 = j\} = \pi_k > 0, \quad \text{for } k = 0, 1, \ldots,$$

where $\sum_{k=0}^{x} \pi_k = 1$. Important quantities to be determined by this model include the long run fraction of time that the service facility is idle, given by π_0, and the long run mean time that a customer spends in the system, given by $\sum_{k=0}^{\infty} (1 + k) \pi_k$.

Exercises

3.3.1 Consider a spare parts inventory model in which either 0, 1, or 2 repair parts are demanded in any period, with

$$\Pr\{\xi_n = 0\} = 0.4, \quad \Pr\{\xi_n = 1\} = 0.3, \quad \Pr\{\xi_n = 2\} = 0.3,$$

and suppose $s = 0$ and $S = 3$. Determine the transition probability matrix for the Markov chain $\{X_n\}$, where X_n is defined to be the quantity on hand at the end-of-period n.

3.3.2 Consider two urns A and B containing a total of N balls. An experiment is performed in which a ball is selected at random (all selections equally likely) at time $t(t = 1, 2, \ldots)$ from among the totality of N balls. Then, an urn is selected at random (A is chosen with probability p and B is chosen with probability q) and the ball previously drawn is placed in this urn. The state of the system at each trial is represented by the number of balls in A. Determine the transition matrix for this Markov chain.

3.3.3 Consider the inventory model of Section 3.3.1. Suppose that $S = 3$. Set up the corresponding transition probability matrix for the end-of-period inventory level X_n.

3.3.4 Consider the inventory model of Section 3.3.1. Suppose that $S = 3$ and that the probability distribution for demand is $\Pr\{\xi = 0\} = 0.1$, $\Pr\{\xi = 1\} = 0.4$, $\Pr\{\xi = 2\} = 0.3$, and $\Pr\{\xi = 3\} = 0.2$. Set up the corresponding transition probability matrix for the end-of-period inventory level X_n.

3.3.5 An urn initially contains a single red ball and a single green ball. A ball is drawn at random, removed, and replaced by a ball of the opposite color, and this process repeats so that there are always exactly two balls in the urn. Let X_n be the number of red balls in the urn after n draws, with $X_0 = 1$. Specify the transition probabilities for the Markov chain $\{X_n\}$.

Problems

3.3.1 An urn contains six tags, of which three are red and three are green. Two tags are selected from the urn. If one tag is red and the other is green, then the selected tags are discarded and two blue tags are returned to the urn. Otherwise, the selected tags are resumed to the urn. This process repeats until the urn contains only blue tags. Let X_n denote the number of red tags in the urn after the nth draw, with $X_0 = 3$. (This is an elementary model of a chemical reaction in which red and green atoms combine to form a blue molecule.) Give the transition probability matrix.

3.3.2 Three fair coins are tossed, and we let X_1 denote the number of heads that appear. Those coins that were heads on the first trial (there were X_1 of them) we pick up and toss again, and now we let X_2 be the total number of tails, including those left from the first toss. We toss again all coins showing tails, and let X_3 be the resulting total number of heads, including those left from the previous toss. We continue the process. The pattern is, count heads, toss heads, count tails, toss tails, count heads, toss heads, etc., and $X_0 = 3$. Then, $\{X_n\}$ is a Markov chain. What is the transition probability matrix?

3.3.3 Consider the inventory model of Section 3.3.1. Suppose that unfulfilled demand is not back ordered but is lost.

    (a) Set up the corresponding transition probability matrix for the end-of-period inventory level X_n.

    (b) Express the long run fraction of lost demand in terms of the demand distribution and limiting probabilities for the end-of-period inventory.

3.3.4 Consider the queueing model of Section 3.4. Now, suppose that at most a single customer arrives during a single period, but that the service time of a customer is a random variable Z with the geometric probability distribution

$$\Pr\{Z = k\} = \alpha(1 - \alpha)^{k-1} \quad \text{for } k = 1, 2, \dots.$$

Specify the transition probabilities for the Markov chain whose state is the number of customers waiting for service or being served at the start of each period. Assume that the probability that a customer arrives in a period is β and that no customer arrives with probability $1 - \beta$.

3.3.5 You are going to successively flip a quarter until the pattern *HHT* appears, that is, until you observe two successive heads followed by a tails. In order to calculate some properties of this game, you set up a Markov chain with the following states: $0, H, HH$, and HHT, where 0 represents the starting point, H represents a single observed head on the last flip, HH represents two successive heads on the last two flips, and HHT is the sequence that you are looking for. Observe that if you have just tossed a tails, followed by a heads, a next toss of a tails effectively starts you over again in your quest for the *HHT* sequence. Set up the transition probability matrix.

3.3.6 Two teams, A and B, are to play a best of seven series of games. Suppose that the outcomes of successive games are independent, and each is won by A with probability p and won by B with probability $1 - p$. Let the state of the system be represented by the pair (a, b), where a is the number of games won by A, and b is the number of games won by B. Specify the transition probability matrix. Note that $a + b \leq 7$ and that the entries end whenever $a = 4$ or $b = 4$.

3.3.7 A component in a system is placed into service, where it operates until its failure, whereupon it is replaced *at the end of the period* with a new component having statistically identical properties, and the process repeats. The probability that a component lasts for k periods is α_k, for $k = 1, 2, \dots$. Let X_n be the remaining life of the component in service *at the end-of-period n*. Then, $X_n = 0$ means that X_{n+1} will be the total operating life of the next component. Give the transition probabilities for the Markov chain $\{X_n\}$.

3.3.8 Two urns A and B contain a total of N balls. Assume that at time t, there were exactly k balls in A. At time $t + 1$, an urn is selected at random in proportion to its contents (i.e., A is chosen with probability k/N and B is chosen with probability $(N - k)/N$). Then, a ball is selected from A with probability p or from B with probability q and placed in the previously chosen urn. Determine the transition matrix for this Markov chain.

3.3.9 Suppose that two urns A and B contain a total of N balls. Assume that at time t, there are exactly k balls in A. At time $t + 1$, a ball and an urn are chosen with probability depending on the contents of the urn (i.e., a ball is chosen from A with probability k/N or from B with probability $(N - k)/N$). Then, the ball is placed into one of the urns, where urn A is chosen with probability k/N or urn B is chosen with probability $(N - k)/N$. Determine the transition matrix of the Markov chain with states represented by the contents of A.

3.3.10 Consider a discrete-time, periodic review inventory model and let ξ_n be the total demand in period n, and let X_n be the inventory quantity on hand at the end-of-period n. An (s, S) inventory policy is used: If the end-of-period stock is not greater than s, then a quantity is instantly procured to bring the level up to S. If the end-of-period stock exceeds s, then no replenishment takes place.

    (a) Suppose that $s = 1, S = 4$, and $X_0 = S = 4$. If the period demands turn out to be $\xi_1 = 2, \xi_2 = 3, \xi_3 = 4, \xi_4 = 0, \xi_5 = 2, \xi_6 = 1, \xi_7 = 2$, and $\xi_8 = 2$, what are the end-of-period stock levels X_n for periods $n = 1, 2, \ldots, 8$?

    (b) Suppose $\xi_1, \xi_2, \ldots$ are independent random variables where $\Pr\{\xi_n = 0\} = 0.1, \Pr\{\xi_n = 1\} = 0.3, \Pr\{\xi_n = 2\} = 0.3, \Pr\{\xi_n = 3\} = 0.2$, and $\Pr\{\xi_n = 4\} = 0.1$. Then, $X_0, X_1, \ldots$ is a Markov chain. Determine P_{41} and P_{04}.

3.4 First Step Analysis

A surprising number of functionals on a Markov chain can be evaluated by a technique that we call *first step analysis*. This method proceeds by analyzing, or breaking down, the possibilities that can arise at the end of the first transition, and then invoking the law of total probability coupled with the Markov property to establish a characterizing relationship among the unknown variables. We first applied this technique in Theorem 3.1. In this section, we develop a series of applications of the technique.

3.4.1 Simple First Step Analyses

Consider the Markov chain $\{X_n\}$ whose transition probability matrix is

$$
\mathbf{P} = \begin{array}{c} 0 \\ 1 \\ 2 \end{array} \begin{array}{ccc} 0 & 1 & 2 \\ \left\| \begin{array}{ccc} 1 & 0 & 0 \\ \alpha & \beta & \gamma \\ 0 & 0 & 1 \end{array} \right\|, \end{array}
$$

where $\alpha > 0, \beta > 0, \gamma > 0$, and $\alpha + \beta + \gamma = 1$. If the Markov chain begins in state 1, it remains there for a random duration and then proceeds either to state 0 or to state 2, where it is trapped or absorbed. That is, once in state 0, the process remains there for ever after, as it also does in state 2. Two questions arise: In which state, 0 or 2, is the process ultimately trapped, and how long, on the average, does it take to reach one of these states? Both questions are easily answered by instituting a first step analysis.

We begin by more precisely defining the questions. Let

$$T = \min\{n \geq 0; X_n = 0 \quad \text{or} \quad X_n = 2\}$$

be the time of absorption of the process. In terms of this random absorption time, the two questions ask us to find

$$u = \Pr\{X_T = 0 | X_0 = 1\}$$

and

$$v = E[T | X_0 = 1].$$

We proceed to institute a first step analysis, considering separately the three contingencies $X_1 = 0, X_1 = 1$, and $X_1 = 2$, with respective probabilities α, β, and γ. Consider $u = \Pr\{X_T = 0 | X_0 = 1\}$. If $X_1 = 0$, which occurs with probability α, then $T = 1$ and $X_T = 0$. If $X_1 = 2$, which occurs with probability γ, then again $T = 1$, but $X_T = 2$. Finally, if $X_1 = 1$, which occurs with probability β, then the process returns to state 1 and the problem repeats from the same state as before. In symbols, we claim that

$$\Pr\{X_T = 0 | X_1 = 0\} = 1,$$
$$\Pr\{X_T = 0 | X_1 = 2\} = 0,$$
$$\Pr\{X_T = 0 | X_1 = 1\} = u,$$

which inserted into the law of total probability gives

$$u = \Pr\{X_T = 0 | X_0 = 1\}$$
$$= \sum_{k=0}^{2} \Pr\{X_T = 0 | X_0 = 1, X_1 = k\} \Pr\{X_1 = k | X_0 = 1\}$$
$$= \sum_{k=0}^{2} \Pr\{X_T = 0 | X_1 = k\} \Pr\{X_1 = k | X_0 = 1\}$$

$$\text{(by the Markov property)}$$
$$= 1(\alpha) + u(\beta) + 0(\gamma).$$

Thus, we obtain the equation

$$u = \alpha + \beta u, \tag{3.19}$$

which gives

$$u = \frac{\alpha}{1 - \beta} = \frac{\alpha}{\alpha + \gamma}.$$

Observe that this quantity is the conditional probability of a transition to 0, given that a transition to 0 or 2 occurred. That is, the answer makes sense.

We turn to determining the mean time to absorption, again analyzing the possibilities arising on the first step. The absorption time T is always at least 1. If either $X_1 = 0$ or $X_1 = 2$, then no further steps are required. If, on the other hand, $X_1 = 1$, then the process is back at its starting point, and on the average, $v = E[T|X_0 = 1]$ *additional* steps are required for absorption. Weighting these contingencies by their respective probabilities, we obtain for $v = E[T|X_0 = 1]$,

$$v = 1 + \alpha(0) + \beta(v) + \gamma(0)$$
$$= 1 + \beta v, \tag{3.20}$$

which gives

$$v = \frac{1}{1 - \beta}.$$

In the example just studied, the reader is invited to verify that T has the geometric distribution in which

$$\Pr\{T > k | X_0 = 1\} = \beta^k \quad \text{for } k = 0, 1, \ldots,$$

and, therefore,

$$E[T|X_0 = 1] = \sum_{k=0}^{\infty} \Pr\{T > k | X_0 = 1\} = \frac{1}{1 - \beta}.$$

That is, a direct calculation verifies the result of the first step analysis. Unfortunately, in more general Markov chains, a direct calculation is rarely possible, and first step analysis provides the only solution technique.

A significant extension occurs when we move up to the four-state Markov chain whose transition probability matrix is

$$
\mathbf{P} = \begin{array}{c} \\ 0 \\ 1 \\ 2 \\ 3 \end{array}
\begin{array}{cccc} 0 & 1 & 2 & 3 \\ \left\| \begin{array}{cccc} 1 & 0 & 0 & 0 \\ P_{10} & P_{11} & P_{12} & P_{13} \\ P_{20} & P_{21} & P_{22} & P_{23} \\ 0 & 0 & 0 & 1 \end{array} \right\| \end{array}.
$$

Absorption now occurs in states 0 and 3, and states 1 and 2 are "transient." The probability of ultimate absorption in state 0, say, now depends on the transient state in which the process began. Accordingly, we must extend our notation to include the starting state. Let

$$T = \min\{n \geq 0; X_n = 0 \quad \text{or} \quad X_n = 3\},$$
$$u_i = \Pr\{X_T = 0 | X_0 = i\} \quad \text{for } i = 1, 2,$$

and

$$v_i = E[T|X_0 = i] \quad \text{for } i = 1, 2.$$

We may extend the definitions for u_i and v_i in a consistent and commonsense manner by prescribing $u_0 = 1, u_3 = 0$, and $v_0 = v_3 = 0$.

The first step analysis now requires us to consider the two possible starting states $X_0 = 1$ and $X_0 = 2$ separately. Considering $X_0 = 1$ and applying a first step analysis to $u_1 = \Pr\{X_T = 0|X_0 = 1\}$, we obtain

$$u_1 = P_{10} + P_{11}u_1 + P_{12}u_2. \tag{3.21}$$

The three terms on the right correspond to the contingencies $X_1 = 0, X_1 = 1$, and $X_1 = 2$, respectively, with the conditional probabilities

$$\Pr\{X_T = 0|X_1 = 0\} = 1,$$
$$\Pr\{X_T = 0|X_1 = 1\} = u_1,$$

and

$$\Pr\{X_T = 0|X_1 = 2\} = u_2.$$

The law of total probability then applies to give (3.21), just as it was used in obtaining (3.19). A similar equation is obtained for u_2:

$$u_2 = P_{20} + P_{21}u_1 + P_{22}u_2. \tag{3.22}$$

The two equations in u_1 and u_2 are now solved simultaneously. To give a numerical example, we will suppose

$$\mathbf{P} = \begin{array}{c} \\ 0 \\ 1 \\ 2 \\ 3 \end{array} \begin{array}{cccc} 0 & 1 & 2 & 3 \\ \left\| \begin{array}{cccc} 1 & 0 & 0 & 0 \\ 0.4 & 0.3 & 0.2 & 0.1 \\ 0.1 & 0.3 & 0.3 & 0.3 \\ 0 & 0 & 0 & 1 \end{array} \right\| \end{array}. \tag{3.23}$$

The first step analysis equations (3.21) and (3.22) for u_1 and u_2 are

$$u_1 = 0.4 + 0.3u_1 + 0.2u_2,$$
$$u_2 = 0.1 + 0.3u_1 + 0.3u_2,$$

or

$$0.7u_1 - 0.2u_2 = 0.4,$$
$$-0.3u_1 + 0.7u_2 = 0.1.$$

The solution is $u_1 = \frac{30}{43}$ and $u_2 = \frac{19}{43}$. Note that one cannot, in general, solve for u_1 without bringing in u_2, and vice versa. The result $u_2 = \frac{19}{43}$ tells us that once begun in state $X_0 = 2$, the Markov chain $\{X_n\}$ described by (3.23) will ultimately end up in state 0 with probability $u_2 = \frac{19}{43}$, and alternatively, will be absorbed in state 3 with probability $1 - u_2 = \frac{24}{43}$.

The mean time to absorption also depends on the starting state. The first step analysis equations for $v_i = E[T|X_0 = i]$ are

$$
\begin{aligned}
v_1 &= 1 + P_{11}v_1 + P_{12}v_2, \\
v_2 &= 1 + P_{21}v_1 + P_{22}v_2.
\end{aligned}
\tag{3.24}
$$

The right side of (3.24) asserts that at least one step is always taken. If the first move is to either $X_1 = 1$ or $X_1 = 2$, then additional steps are needed, and on the average, these are v_1 and v_2, respectively. Weighting the contingencies $X_1 = 1$ and $X_1 = 2$ by their respective probabilities and summing according to the law of total probability results in (3.24).

For the transition matrix given in (3.23), the equations are

$$
\begin{aligned}
v_1 &= 1 + 0.3v_1 + 0.2v_2, \\
v_2 &= 1 + 0.3v_1 + 0.3v_2,
\end{aligned}
$$

and their solutions are $v_1 = \frac{90}{43}$ and $v_2 = \frac{100}{43}$. Again, v_1 cannot be obtained without also considering v_2, and vice versa. For a process that begins in state $X_0 = 2$, on the average $v_2 = \frac{100}{43} = 2.33$ steps will transpire prior to absorption.

To study the method in a more general context, let $\{X_n\}$ be a finite-state Markov chain whose states are labeled $0, 1, \ldots, N$. Suppose that states $0, 1, \ldots, r - 1$ are *transient** in that $P_{ij}^{(n)} \to 0$ as $n \to \infty$ for $0 \leq i, j < r$, while states $r, \ldots, N$ are *absorbing* ($P_{ii} = 1$ for $r \leq i \leq N$). The transition matrix has the form

$$
\mathbf{P} = \begin{Vmatrix} \mathbf{Q} & \mathbf{R} \\ \mathbf{0} & \mathbf{I} \end{Vmatrix}
\tag{3.25}
$$

where $\mathbf{0}$ is an $(N - r + 1) \times r$ matrix all of whose entries are zero, $\mathbf{I}$ is an $(N - r + 1) \times (N - r + 1)$ identity matrix, and $Q_{ij} = P_{ij}$ for $0 \leq i, j < r$.

Started at one of the transient states $X_0 = i$, where $0 \leq i < r$, such a process will remain in the transient states for some random duration, but ultimately the process gets trapped in one of the absorbing states $i = r, \ldots, N$. Functionals of importance are the mean duration until absorption and the probability distribution over the states in which absorption takes place.

Let us consider the second question first and fix a state k among the absorbing states ($r \leq k \leq N$). The probability of ultimate absorption in state k, as opposed to some other absorbing state, depends on the initial state $X_0 = i$. Let $U_{ik} = u_i$ denote this probability, where we suppress the target state k in the notation for typographical convenience.

---

* The definition of a transient state is different for an infinite-state Markov chain. See Chapter 4, Section 4.3.

We begin a first step analysis by enumerating the possibilities in the first transition. Starting from state i, with probability P_{ik} the process immediately goes to state k, thereafter to remain, and this is the first possibility considered. Alternatively, the process could move on its first step to an absorbing state $j \neq k$, where $r \leq j \leq N$, in which case ultimate absorption in state k is precluded. Finally, the process could move to a transient state $j < r$. Because of the Markov property, once in state j, then the probability of ultimate absorption in state k is $u_j = U_{jk}$ by definition. Weighting the enumerated possibilities by their respective probabilities via the law of total probability, we obtain the relation

$$u_i = \Pr\{\text{Absorption in } k | X_0 = i\}$$

$$= \sum_{j=0}^{N} \Pr\{\text{Absorption in } k | X_0 = i, X_1 = j\} P_{ij}$$

$$= P_{ik} + \sum_{\substack{j=r \\ j \neq k}}^{N} P_{ij} \times 0 + \sum_{j=0}^{r-1} P_{ij} u_j.$$

To summarize, for a fixed absorbing state k, the quantities

$$u_i = U_{ik} = \Pr\{\text{Absorption in } k | X_0 = i\} \quad \text{for } 0 \leq i < r$$

satisfy the inhomogeneous system of linear equations

$$U_{ik} = P_{ik} + \sum_{j=0}^{r-1} P_{ij} U_{jk}, \quad i = 0, 1, \ldots, r-1. \tag{3.26}$$

Example *A Maze*   A white rat is put into the maze shown:

In the absence of learning, one might hypothesize that the rat would move through the maze at random; i.e., if there are k ways to leave a compartment, then the rat would choose each of these with probability $1/k$. Assume that the rat makes one change to some adjacent compartment at each unit of time and let X_n denote the compartment occupied at stage n. We suppose that compartment 7 contains food and compartment

8 contains an electrical shocking mechanism, and we ask the probablity that the rat, moving at random, encounters the food before being shocked. The appropriate transition probability matrix is

$$
\mathbf{P} = \begin{array}{c}
\\ 0 \\ 1 \\ 2 \\ 3 \\ 4 \\ 5 \\ 6 \\ 7 \\ 8
\end{array}
\begin{array}{|ccccccccc|}
0 & 1 & 2 & 3 & 4 & 5 & 6 & 7 & 8 \\
\frac{1}{2} & \frac{1}{2} & & & & & & & \\
\frac{1}{3} & & \frac{1}{3} & & & & & \frac{1}{3} & \\
\frac{1}{3} & & & \frac{1}{3} & & & & & \frac{1}{3} \\
& \frac{1}{4} & \frac{1}{4} & & \frac{1}{4} & \frac{1}{4} & & & \\
& & & \frac{1}{3} & & & \frac{1}{3} & \frac{1}{3} & \\
& & & \frac{1}{3} & & & \frac{1}{3} & & \frac{1}{3} \\
& & & & \frac{1}{2} & \frac{1}{2} & & & \\
& & & & & & & 1 & \\
& & & & & & & & 1
\end{array}.
$$

Let $u_i = u_i(7)$ denote the probability of absorption in the food compartment 7, given that the rat is dropped initially in compartment i. Then, equation (3.26) becomes, in this particular instance,

$$
u_0 = \frac{1}{2}u_1 + \frac{1}{2}u_2,
$$

$$
u_1 = \frac{1}{3} + \frac{1}{3}u_0 + \frac{1}{3}u_3,
$$

$$
u_2 = \frac{1}{3}u_0 + \frac{1}{3}u_3,
$$

$$
u_3 = \frac{1}{4}u_1 + \frac{1}{4}u_2 + \frac{1}{4}u_4 + \frac{1}{4}u_5,
$$

$$
u_4 = \frac{1}{3} + \frac{1}{3}u_3 + \frac{1}{3}u_6,
$$

$$
u_5 = \frac{1}{3}u_3 + \frac{1}{3}u_6,
$$

$$
u_6 = \frac{1}{2}u_4 + \frac{1}{2}u_5.
$$

Turning to the solution, we see that the symmetry of the maze implies that $u_0 = u_6$, $u_2 = u_5$, and $u_1 = u_4$. We also must have $u_3 = \frac{1}{2}$. With these simplifications, the equations for u_0, u_1, and u_2 become

$$u_0 = \frac{1}{2}u_1 + \frac{1}{2}u_2,$$

$$u_1 = \frac{1}{2} + \frac{1}{3}u_0,$$

$$u_2 = \frac{1}{6} + \frac{1}{3}u_0,$$

and the natural substitutions give $u_0 = \frac{1}{2}\left(\frac{1}{2} + \frac{1}{3}u_0\right) + \frac{1}{2}\left(\frac{1}{6} + \frac{1}{3}u_0\right)$, or $u_0 = \frac{1}{2}, u_1 = \frac{2}{3}$, and $u_2 = \frac{1}{3}$.

One might compare these theoretical values under random moves with actual observations as an indication of whether or not learning is taking place.

3.4.2 The General Absorbing Markov Chain

Let $\{X_n\}$ be a Markov chain whose transition probability matrix takes the form (3.25). We turn to a more general form of the first question by introducing the random absorption time T. Formally, we define

$$T = \min\{n \geq 0; X_n \geq r\}.$$

Let us suppose that associated with each transient state i is a rate $g(i)$ and that we wish to determine the mean total rate that is accumulated up to absorption. Let w_i be this mean total amount, where the subscript i denotes the starting position $X_0 = i$. To be precise, let

$$w_i = E\left[\sum_{n=0}^{T-1} g(X_n)|X_0 = i\right].$$

The choice $g(i) = 1$ for all i yields $\sum_{n=0}^{T-1} g(X_n) = \sum_{n=0}^{T-1} 1 = T$, and then w_i is identical to $v_i \equiv E[T|X_0 = i]$, the mean time until absorption. For a transient state k, the choice

$$g(i) = \begin{cases} 1 & \text{if } i = k, \\ 0 & \text{if } i \neq k, \end{cases}$$

gives $w_i = W_{ik}$, the mean number of visits to state $k(0 \leq k < r)$ prior to absorption.

We again proceed via a first step analysis. The sum $\sum_{n=0}^{T-1} g(X_n)$ always includes the first term $g(X_0) = g(i)$. In addition, if a transition is made from i to a transient state j, then the sum includes future terms as well. By invoking the Markov property, we deduce that this future sum proceeding from state j has an expected value equal to w_j.

Weighting this by the transition probability P_{ij} and then summing all contributions in accordance with the law of total probability, we obtain the joint relations

$$w_i = g(i) + \sum_{j=0}^{r-1} P_{ij} w_j \quad \text{for } i = 0, \ldots, r-1. \tag{3.27}$$

The special case in which $g(i) = 1$ for all i determines $v_i = E[T|X_0 = i]$ as solving

$$v_i = 1 + \sum_{j=0}^{r-1} P_{ij} v_j \quad \text{for } i = 0, 1, \ldots, r-1. \tag{3.28}$$

The case in which

$$g(i) = \delta_{ik} = \begin{cases} 1 & \text{if } i = k, \\ 0 & \text{if } i \neq k, \end{cases}$$

determines W_{ik}, the mean number of visits to state k prior to absorption starting from state i, as solving

$$W_{ik} = \delta_{ik} + \sum_{j=0}^{r-1} P_{ij} W_{jk} \quad \text{for } i = 0, 1, \ldots, r-1. \tag{3.29}$$

Example *A Model of Fecundity* Changes in sociological patterns such as increase in age at marriage, more remarriages after widowhood, and increased divorce rates have profound effects on overall population growth rates. Here, we attempt to model the lifespan of a female in a population in order to provide a framework for analyzing the effect of social changes on average fecundity.

The general model we propose has a large number of states delimiting the age and status of a typical female in the population. For example, we begin with the 12 age groups 0–4 years, 5–9 years, ..., 50–54 years, 55 years, and over. In addition, each of these age groups might be further subdivided according to marital status: single, married, separated, divorced, or widowed, and might also be subdivided according to the number of children. Each female would begin in the (0–4, single) category and end in a distinguished state Δ corresponding to death or emigration from the population. However, the duration spent in the various other states might differ among different females. Of interest is the mean duration spent in the categories of maximum fertility, or more generally, a mean sum of durations weighted by appropriate fecundity rates.

When there are a large number of states in the model, as just sketched, the relevant calculations require a computer. We turn to a simpler model which, while less realistic, will serve to illustrate the concepts and approach. We introduce the states

E_0: Prepuberty, E_3: Divorced,

E_1: Single, E_4: Widowed,

E_2: Married, E_5: Δ,

and we are interested in the mean duration spent in state E_2: Married, since this corresponds to the state of maximum fecundity. To illustrate the computations, we will suppose the transition probability matrix is

$$\mathbf{P} = \begin{array}{c} \\ E_0 \\ E_1 \\ E_2 \\ E_3 \\ E_4 \\ E_5 \end{array} \begin{array}{c} \begin{matrix} E_0 & E_1 & E_2 & E_3 & E_4 & E_5 \end{matrix} \\ \left\| \begin{matrix} 0 & 0.9 & 0 & 0 & 0 & 0.1 \\ 0 & 0.5 & 0.4 & 0 & 0 & 0.1 \\ 0 & 0 & 0.6 & 0.2 & 0.1 & 0.1 \\ 0 & 0 & 0.4 & 0.5 & 0 & 0.1 \\ 0 & 0 & 0.4 & 0 & 0.5 & 0.1 \\ 0 & 0 & 0 & 0 & 0 & 1.0 \end{matrix} \right\| \end{array}.$$

In practice, such a matrix would be estimated from demographic data.

Every person begins in state E_0 and ends in state E_5, but a variety of intervening states may be visited. We wish to determine the mean duration spent in state E_2: Married. The powerful approach of *first step analysis* begins by considering the slightly more general problem in which the initial state is varied. Let $w_i = W_{i2}$ be the mean duration in state E_2 given the initial state $X_0 = E_i$ for $i = 0, 1, \ldots, 5$. We are interested in w_0, the mean duration corresponding to the initial state E_0.

First step analysis breaks down, or analyzes, the possibilities arising in the first transition, and using the Markov property, an equation that relates $w_0, \ldots, w_5$ results.

We begin by considering w_0. From state E_0, a transition to one of the states E_1 or E_5 occurs, and the mean duration spent in E_2 starting from E_0 must be the appropriately weighted average of w_1 and w_5. That is,

$$w_0 = 0.9w_1 + 0.1w_5.$$

Proceeding in a similar manner, we obtain

$$w_1 = 0.5w_1 + 0.4w_2 + 0.1w_5.$$

The situation changes when the process begins in state E_2 because in counting the mean duration spent in E_2, we must count this initial visit plus any subsequent visits that may occur. Thus, for E_2, we have

$$w_2 = 1 + 0.6w_2 + 0.2w_3 + 0.1w_4 + 0.1w_5.$$

The other states give us

$$w_3 = 0.4w_2 + 0.5w_3 + 0.1w_5,$$
$$w_4 = 0.4w_2 + 0.5w_4 + 0.1w_5,$$
$$w_5 = w_5.$$

Since state E_5 corresponds to death, it is clear that we must have $w_5 = 0$. With this prescription, the reduced equations become, after elementary simplification,

$$
\begin{aligned}
-1.0w_0 + 0.9w_1 &= 0, \\
-0.5w_1 + 0.4w_2 &= 0, \\
-0.4w_2 + 0.2w_3 + 0.1w_4 &= -1, \\
0.4w_2 - 0.5w_3 &= 0, \\
0.4w_2 \qquad\quad -0.5w_4 &= 0.
\end{aligned}
$$

The unique solution is

$$w_0 = 4.5, \quad w_1 = 5.00, \quad w_2 = 6.25, \quad w_3 = w_4 = 5.00.$$

Each female, on the average, spends $w_0 = W_{02} = 4.5$ periods in the childbearing state E_2 during her lifetime.

Exercises

3.4.1 Find the mean time to reach state 3 starting from state 0 for the Markov chain whose transition probability matrix is

$$
\mathbf{P} = \begin{array}{c} \\ 0 \\ 1 \\ 2 \\ 3 \end{array}
\begin{array}{cccc}
0 & 1 & 2 & 3 \\
\left\|\begin{array}{cccc}
0.4 & 0.3 & 0.2 & 0.1 \\
0 & 0.7 & 0.2 & 0.1 \\
0 & 0 & 0.9 & 0.1 \\
0 & 0 & 0 & 1
\end{array}\right\|
\end{array}.
$$

3.4.2 Consider the Markov chain whose transition probablity matrix is given by

$$
\mathbf{P} = \begin{array}{c} \\ 0 \\ 1 \\ 2 \end{array}
\begin{array}{ccc}
0 & 1 & 2 \\
\left\|\begin{array}{ccc}
1 & 0 & 0 \\
0.1 & 0.6 & 0.3 \\
0 & 0 & 1
\end{array}\right\|
\end{array}.
$$

(a) Starting in state 1, determine the probability that the Markov chain ends in state 0.

(b) Determine the mean time to absorption.

3.4.3 Consider the Markov chain whose transition probability matrix is given by

$$\mathbf{P} = \begin{array}{c} \\ 0 \\ 1 \\ 2 \\ 3 \end{array} \begin{array}{|cccc|} 0 & 1 & 2 & 3 \\ \hline 1 & 0 & 0 & 0 \\ 0.1 & 0.6 & 0.1 & 0.2 \\ 0.2 & 0.3 & 0.4 & 0.1 \\ 0 & 0 & 0 & 1 \end{array}.$$

(a) Starting in state 1, determine the probability that the Markov chain ends in state 0.
(b) Determine the mean time to absorption.

3.4.4 A coin is tossed repeatedly until two successive heads appear. Find the mean number of tosses required.

Hint: Let X_n be the cumulative number of successive heads. The state space is $0, 1, 2$, and the transition probability matrix is

$$\mathbf{P} = \begin{array}{c} \\ 0 \\ 1 \\ 2 \end{array} \begin{array}{|ccc|} 0 & 1 & 2 \\ \hline \dfrac{1}{2} & \dfrac{1}{2} & 0 \\ \dfrac{1}{2} & 0 & \dfrac{1}{2} \\ 0 & 0 & 1 \end{array}.$$

Determine the mean time to reach state 2 starting from state 0 by invoking a first step analysis.

3.4.5 A coin is tossed repeatedly until either two successive heads appear or two successive tails appear. Suppose the first coin toss results in a head. Find the probability that the game ends with two successive tails.

3.4.6 Consider the Markov chain whose transition probability matrix is given by

$$\mathbf{P} = \begin{array}{c} \\ 0 \\ 1 \\ 2 \\ 3 \end{array} \begin{array}{|cccc|} 0 & 1 & 2 & 3 \\ \hline 1 & 0 & 0 & 0 \\ 0.1 & 0.4 & 0.1 & 0.4 \\ 0.2 & 0.1 & 0.6 & 0.1 \\ 0 & 0 & 0 & 1 \end{array}.$$

(a) Starting in state 1, determine the probability that the Markov chain ends in state 0.
(b) Determine the mean time to absorption.

3.4.7 Consider the Markov chain whose transition probability matrix is given by

$$P = \begin{array}{c} \\ 0 \\ 1 \\ 2 \\ 3 \end{array} \begin{array}{cccc} 0 & 1 & 2 & 3 \\ \left\| \begin{array}{cccc} 1 & 0 & 0 & 0 \\ 0.1 & 0.2 & 0.5 & 0.2 \\ 0.1 & 0.2 & 0.6 & 0.1 \\ 0 & 0 & 0 & 1 \end{array} \right\| \end{array}.$$

Starting in state 1, determine the mean time that the process spends in state 1 prior to absorption and the mean time that the process spends in state 2 prior to absorption. Verify that the sum of these is the mean time to absorption.

3.4.8 Consider the Markov chain whose transition probability matrix is given by

$$P = \begin{array}{c} \\ 0 \\ 1 \\ 2 \\ 3 \end{array} \begin{array}{cccc} 0 & 1 & 2 & 3 \\ \left\| \begin{array}{cccc} 1 & 0 & 0 & 0 \\ 0.5 & 0.2 & 0.1 & 0.2 \\ 0.2 & 0.1 & 0.6 & 0.1 \\ 0 & 0 & 0 & 1 \end{array} \right\| \end{array}.$$

Starting in state 1, determine the mean time that the process spends in state 1 prior to absorption and the mean time that the process spends in state 2 prior to absorption. Verify that the sum of these is the mean time to absorption.

3.4.9 Consider the Markov chain whose transition probability matrix is given by

$$P = \begin{array}{c} \\ 0 \\ 1 \\ 2 \\ 3 \end{array} \begin{array}{cccc} 0 & 1 & 2 & 3 \\ \left\| \begin{array}{cccc} 1 & 0 & 0 & 0 \\ 0.1 & 0.2 & 0.5 & 0.2 \\ 0.1 & 0.2 & 0.6 & 0.1 \\ 0 & 0 & 0 & 1 \end{array} \right\| \end{array}.$$

Starting in state 1, determine the probability that the process is absorbed into state 0. Compare this with the $(1,0)$th entry in the matrix powers $\mathbf{P}^2, \mathbf{P}^4, \mathbf{P}^8$, and $\mathbf{P}^{16}$.

Problems

3.4.1 Which will take fewer flips, on average: successively flipping a quarter until the pattern *HHT* appears, i.e., until you observe two successive heads followed by a tails; or successively flipping a quarter until the pattern *HTH* appears? Can you explain why these are different?

3.4.2 A zero-seeking device operates as follows: If it is in state m at time n, then at time $n + 1$, its position is uniformly distributed over the states $0, 1, \ldots, m - 1$. Find the expected time until the device first hits zero starting from state m.

Note: This is a highly simplified model for an algorithm that seeks a maximum over a finite set of points.

3.4.3 A zero-seeking device operates as follows: If it is in state j at time n, then at time $n + 1$, its position is 0 with probability $1/j$, and its position is k (where k is one of the states $1, 2, \ldots, j - 1$) with probability $2k/j^2$. Find the expected time until the device first hits zero starting from state m.

3.4.4 Consider the Markov chain whose transition probability matrix is given by

$$
\mathbf{P} = \begin{array}{c} \\ 0 \\ 1 \\ 2 \\ 3 \end{array}
\begin{array}{cccc}
0 & 1 & 2 & 3 \\
\left\| \begin{array}{cccc} 1 & 0 & 0 & 0 \\ 0.1 & 0.2 & 0.5 & 0.2 \\ 0.1 & 0.2 & 0.6 & 0.1 \\ 0.2 & 0.2 & 0.3 & 0.3 \end{array} \right\|
\end{array}.
$$

Starting in state $X_0 = 1$, determine the probability that the process never visits state 2. Justify your answer.

3.4.5 A white rat is put into compartment 4 of the maze shown here:

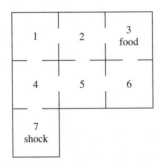

It moves through the compartments at random; i.e., if there are k ways to leave a compartment, it chooses each of these with probability $1/k$. What is the probability that it finds the food in compartment 3 before feeling the electric shock in compartment 7?

3.4.6 Consider the Markov chain whose transition matrix is

$$
\mathbf{P} = \begin{array}{c} \\ 0 \\ 1 \\ 2 \\ 3 \\ 4 \end{array}
\begin{array}{ccccc}
0 & 1 & 2 & 3 & 4 \\
\left\| \begin{array}{ccccc} q & p & 0 & 0 & 0 \\ q & 0 & p & 0 & 0 \\ q & 0 & 0 & p & 0 \\ q & 0 & 0 & 0 & p \\ 0 & 0 & 0 & 0 & 1 \end{array} \right\|
\end{array},
$$

where $p + q = 1$. Determine the mean time to reach state 4 starting from state 0. That is, find $E[T|X_0 = 0]$, where $T = \min\{n \geq 0; X_n = 4\}$.

Hint: Let $v_i = E[T|X_0 = i]$ for $i = 0, 1, \ldots, 4$. Establish equations for $v_0, v_1, \ldots, v_4$ by using a first step analysis and the boundary condition $v_4 = 0$. Then, solve for v_0.

3.4.7 Let X_n be a Markov chain with transition probabilities P_{ij}. We are given a "discount factor" β with $0 < \beta < 1$ and a cost function $c(i)$, and we wish to determine the total expected discounted cost starting from state i, defined by

$$h_i = E\left[\sum_{n=0}^{\infty} \beta^n c(X_n)|X_0 = i\right].$$

Using a first step analysis show that h_i satisfies the system of linear equations

$$h_i = c(i) + \beta \sum_j P_{ij} h_j \quad \text{for all states } i.$$

3.4.8 An urn contains five red and three green balls. The balls are chosen at random, one by one, from the urn. If a red ball is chosen, it is removed. Any green ball that is chosen is returned to the urn. The selection process continues until all of the red balls have been removed from the urn. What is the mean duration of the game?

3.4.9 An urn contains five red and three yellow balls. The balls are chosen at random, one by one, from the urn. Each ball removed is replaced in the urn by a yellow ball. The selection process continues until all of the red balls have been removed from the urn. What is the mean duration of the game?

3.4.10 You have five fair coins. You toss them all so that they randomly fall heads or tails. Those that fall tails in the first toss you pick up and toss again. You toss again those that show tails after the second toss, and so on, until all show heads. Let X be the number of coins involved in the *last* toss. Find $\Pr\{X = 1\}$.

3.4.11 An urn contains two red and two green balls. The balls are chosen at random, one by one, and removed from the urn. The selection process continues until all of the green balls have been removed from the urn. What is the probability that a single red ball is in the urn at the time that the last green ball is chosen?

3.4.12 A Markov chain $X_0, X_1, X_2, \ldots$ has the transition probability matrix

$$\mathbf{P} = \begin{array}{c} 0 \\ 1 \\ 2 \end{array} \begin{array}{ccc} 0 & 1 & 2 \\ \left\| \begin{array}{ccc} 0.3 & 0.2 & 0.5 \\ 0.5 & 0.1 & 0.4 \\ 0 & 0 & 1 \end{array} \right\| \end{array}$$

and is known to start in state $X_0 = 0$. Eventually, the process will end up in state 2. What is the probability that when the process moves into state 2, it does so from state 1?

Hint: Let $T = \min\{n \geq 0; X_n = 2\}$, and let

$$z_i = \Pr\{X_{T-1} = 1 | X_0 = i\} \quad \text{for } i = 0, 1.$$

Establish and solve the first step equations

$$z_0 = \qquad 0.3z_0 + 0.2z_1,$$
$$z_1 = 0.4 + 0.5z_0 + 0.1z_1.$$

3.4.13 A Markov chain $X_0, X_1, X_2, \ldots$ has the transition probability matrix

$$\mathbf{P} = \begin{array}{c} \\ 0 \\ 1 \\ 2 \end{array} \begin{array}{ccc} 0 & 1 & 2 \\ \left\| \begin{array}{ccc} 0.3 & 0.2 & 0.5 \\ 0.5 & 0.1 & 0.4 \\ 0 & 0 & 1 \end{array} \right\| \end{array}$$

and is known to start in state $X_0 = 0$. Eventually, the process will end up in state 2. What is the probability that the time $T = \min\{n \geq 0; X_n = 2\}$ is an odd number?

3.4.14 A single die is rolled repeatedly. The game stops the first time that the sum of two successive rolls is either 5 or 7. What is the probability that the game stops at a sum of 5?

3.4.15 A simplified model for the spread of a rumor goes this way: There are $N = 5$ people in a group of friends, of which some have heard the rumor and the others have not. During any single period of time, two people are selected at random from the group and assumed to interact. The selection is such that an encounter between any pair of friends is just as likely as between any other pair. If one of these persons has heard the rumor and the other has not, then with probability $\alpha = 0.1$ the rumor is transmitted. Let X_n denote the number of friends who have heard the rumor at the end of the nth period.

Assuming that the process begins at time 0 with a single person knowing the rumor, what is the mean time that it takes for everyone to hear it?

3.4.16 An urn contains five tags, of which three are red and two are green. A tag is randomly selected from the urn and replaced with a tag of the opposite color. This continues until only tags of a single color remain in the urn. Let X_n denote the number of red tags in the urn after the nth draw, with $X_0 = 3$. What is the probability that the game ends with the urn containing only red tags?

3.4.17 The *damage* X_n of a system subjected to wear is a Markov chain with the transition probability matrix

$$\mathbf{P} = \begin{array}{c} \\ 0 \\ 1 \\ 2 \end{array} \begin{array}{ccc} 0 & 1 & 2 \\ \left\| \begin{array}{ccc} 0.7 & 0.3 & 0 \\ 0 & 0.6 & 0.4 \\ 0 & 0 & 1 \end{array} \right\| \end{array}.$$

The system starts in state 0 and fails when it first reaches state 2. Let $T = \min\{n \geq 0; X_n = 2\}$ be the time of failure. Use a first step analysis to evaluate $\phi(s) = E[s^T]$ for a fixed number $0 < s < 1$. (This is called the *generating function* of T. See Section 3.9.)

3.4.18 *Time-dependent transition probabilities.* A well-disciplined man, who smokes exactly one half of a cigar each day, buys a box containing N cigars. He cuts a cigar in half, smokes half, and returns the other half to the box. In general, on a day in which his cigar box contains w whole cigars and h half cigars, he will pick one of the $w + h$ smokes at random, each whole and half cigar being equally likely, and if it is a half cigar, he smokes it. If it is a whole cigar, he cuts it in half, smokes one piece, and returns the other to the box. What is the expected value of T, the day on which the last whole cigar is selected from the box?

Hint: Let X_n be the number of whole cigars in the box after the nth smoke. Then, X_n is a Markov chain whose transition probabilities vary with n. Define $v_n(w) = E[T|X_n = w]$. Use a first step analysis to develop a recursion for $v_n(w)$ and show that the solution is

$$v_n(w) = \frac{2Nw + n + 2w}{w + 1} - \sum_{k=1}^{w} \frac{1}{k},$$

whence

$$E[T] = v_0(N) = 2N - \sum_{k=1}^{N} \frac{1}{k}.$$

3.4.19 *Computer Challenge.* Let N be a positive integer and let $Z_1, \ldots, Z_N$ be independent random variables, each having the geometric distribution

$$\Pr\{Z = k\} = \left(\frac{1}{2}\right)^k, \quad \text{for } k = 1, 2, \ldots.$$

Since these are discrete random variables, the maximum among them may be unique, or there may be ties for the maximum. Let p_N be the probability that the maximum is unique. How does p_N behave when N is large? (Alternative formulation: You toss N dimes. Those that are heads you set aside; those that are tails you toss again. You repeat this until all of the coins are heads. Then, p_N is the probability that the last toss was of a single coin.)

3.5 Some Special Markov Chains

We introduce several particular Markov chains that arise in a variety of applications.

3.5.1   The Two-State Markov Chain

Let

$$\mathbf{P} = \begin{matrix} & 0 & 1 \\ 0 \\ 1 \end{matrix} \left\| \begin{matrix} 1-a & a \\ b & 1-b \end{matrix} \right\|, \quad \text{where } 0 < a, b < 1, \tag{3.30}$$

be the transition matrix of a two-state Markov chain.

When $a = 1 - b$ so that the rows of $\mathbf{P}$ are the same, then the states $X_1, X_2, \ldots$ are independent identically distributed random variables with $\Pr\{X_n = 0\} = b$ and $\Pr\{X_n = 1\} = a$. When $a \neq 1 - b$, the probability distribution for X_n varies depending on the outcome X_{n-1} at the previous stage.

For the two-state Markov chain, it is readily verified by induction that the n-step transition matrix is given by

$$\mathbf{P}^n = \frac{1}{a+b} \left\| \begin{matrix} b & a \\ b & a \end{matrix} \right\| + \frac{(1-a-b)^n}{a+b} \left\| \begin{matrix} a & -a \\ -b & b \end{matrix} \right\|. \tag{3.31}$$

To verify this general formula, introduce the abbreviations

$$\mathbf{A} = \left\| \begin{matrix} b & a \\ b & a \end{matrix} \right\| \quad \text{and} \quad \mathbf{B} = \left\| \begin{matrix} a & -a \\ -b & b \end{matrix} \right\|$$

so that (3.31) can be written

$$\mathbf{P}^n = (a+b)^{-1} \left[\mathbf{A} + (1-a-b)^n \mathbf{B} \right].$$

Next, check the multiplications

$$\mathbf{AP} = \left\| \begin{matrix} b & a \\ b & a \end{matrix} \right\| \times \left\| \begin{matrix} 1-a & a \\ b & 1-b \end{matrix} \right\| = \left\| \begin{matrix} b & a \\ b & a \end{matrix} \right\| = \mathbf{A}$$

and

$$\mathbf{BP} = \left\| \begin{matrix} a & -a \\ -b & b \end{matrix} \right\| \times \left\| \begin{matrix} 1-a & a \\ b & 1-b \end{matrix} \right\|$$

$$= \left\| \begin{matrix} a-a^2-ab & a^2-a+ab \\ -b+ab+b^2 & -ab+b-b^2 \end{matrix} \right\| = (1-a-b)\mathbf{B}.$$

Now, (3.31) is easily seen to be true when $n = 1$, since then

$$\mathbf{P}^1 = \frac{1}{a+b} \begin{Vmatrix} b & a \\ b & a \end{Vmatrix} + \frac{(1-a-b)}{a+b} \begin{Vmatrix} a & -a \\ -b & b \end{Vmatrix}$$

$$= \frac{1}{a+b} \begin{Vmatrix} b+a-a^2-ab & a-a+a^2+ab \\ b-b+ab+b^2 & a+b-ab-b^2 \end{Vmatrix}$$

$$= \begin{Vmatrix} 1-a & a \\ b & 1-b \end{Vmatrix} = \mathbf{P}.$$

To complete an induction proof, assume that the formula is true for n. Then,

$$\mathbf{P}^n\mathbf{P} = (a+b)^{-1}\left[\mathbf{A} + (1-a-b)^n\mathbf{B}\right]\mathbf{P}$$

$$= (a+b)^{-1}\left[\mathbf{AP} + (1-a-b)^n\mathbf{BP}\right]$$

$$= (a+b)^{-1}\left[\mathbf{A} + (1-a-b)^{n+1}\mathbf{B}\right] = \mathbf{P}^{n+1}.$$

We have verified that the formula holds for $n+1$. It, therefore, is established for all n.
   Note that $|1-a-b| < 1$ when $0 < a, b < 1$, and thus $|1-a-b|^n \to 0$ as $n \to \infty$
and

$$\lim_{n\to\infty} \mathbf{P}^n = \begin{Vmatrix} \dfrac{b}{a+b} & \dfrac{a}{a+b} \\ \dfrac{b}{a+b} & \dfrac{a}{a+b} \end{Vmatrix}. \tag{3.32}$$

This tells us that such a system, in the long run, will be in state 0 with probability
$b/(a+b)$ and in state 1 with probability $a/(a+b)$, irrespective of the initial state in
which the system started.
   For a numerical example, suppose that the items produced by a certain worker are
graded as defective or not and that due to trends in raw material quality, whether or
not a particular item is defective depends in part on whether or not the previous item
was defective. Let X_n denote the quality of the nth item with $X_n = 0$ meaning "good"
and $X_n = 1$ meaning "defective." Suppose that $\{X_n\}$ evolves as a Markov chain whose
transition matrix is

$$\mathbf{P} = \begin{matrix} & 0 & 1 \\ 0 & \\ 1 & \end{matrix} \begin{Vmatrix} 0.99 & 0.01 \\ 0.12 & 0.88 \end{Vmatrix}.$$

Defective items would tend to appear in bunches in the output of such a system.
   In the long run, the probability that an item produced by this system is defective is
given by $a/(a+b) = 0.01/(0.01+0.12) = 0.077$.

3.5.2   Markov Chains Defined by Independent Random Variables

Let ξ denote a discrete-valued random variable whose possible values are the non-negative integers and where $\Pr\{\xi = i\} = a_i \geq 0$, for $i = 0, 1, \ldots$, and $\sum_{i=0}^{\infty} a_i = 1$. Let $\xi_1, \xi_2, \ldots, \xi_n, \ldots$ represent independent observations of ξ.

We shall now describe three different Markov chains connected with the sequence $\xi_1, \xi_2, \ldots$. In each case, the state space of the process is the set of nonnegative integers.

Example *Independent Random Variables*   Consider the process $X_n, n = 0, 1, 2, \ldots$, defined by $X_n = \xi_n$ ($X_0 = \xi_0$ prescribed). Its Markov matrix has the form

$$\mathbf{P} = \begin{Vmatrix} a_0 & a_1 & a_2 & \cdots \\ a_0 & a_1 & a_2 & \cdots \\ a_0 & a_1 & a_2 & \cdots \\ \vdots & \vdots & \vdots & \end{Vmatrix}. \tag{3.33}$$

That all rows are identical plainly expresses the fact that the random variable X_{n+1} is independent of X_n.

Example *Successive Maxima*   The partial maxima of $\xi_1, \xi_2, \ldots$ define a second important Markov chain. Let

$$\theta_n = \max\{\xi_1, \ldots, \xi_n\}, \quad \text{for } n = 1, 2, \ldots,$$

with $\theta_0 = 0$. The process defined by $X_n = \theta_n$ is readily seen to be a Markov chain, and the relation $X_{n+1} = \max\{X_n, \xi_{n+1}\}$ allows the transition probabilities to be computed to be

$$\mathbf{P} = \begin{Vmatrix} A_0 & a_1 & a_2 & a_3 & \cdots \\ 0 & A_1 & a_2 & a_3 & \cdots \\ 0 & 0 & A_2 & a_3 & \cdots \\ 0 & 0 & 0 & A_3 & \cdots \\ \vdots & \vdots & \vdots & \vdots & \end{Vmatrix}, \tag{3.34}$$

where $A_k = a_0 + \cdots + a_k$ for $k = 0, 1, \ldots$.

Suppose $\xi_1, \xi_2, \ldots$ represent successive bids on a certain asset that is offered for sale. Then, $X_n = \max\{\xi_1, \ldots, \xi_n\}$ is the maximum that is bid up to stage n. Suppose that the bid that is accepted is the first bid that equals or exceeds a prescribed level M. The time of sale is the random variable $T = \min\{n \geq 1; X_n \geq M\}$. A first step analysis shows that the mean $\mu = E[T]$ satisfies

$$\mu = 1 + \mu \Pr\{\xi_1 < M\}, \tag{3.35}$$

or $\mu = 1/\Pr\{\xi_1 \geq M\} = 1/(a_M + a_{M+1} + \cdots)$. The first step analysis invoked in establishing (3.35) considers the two possibilities $\{\xi_1 < M\}$ and $\{\xi_1 \geq M\}$. With this breakdown, the law of total probability justifies the sum

$$E[T] = E[T|\xi_1 \geq M]\Pr\{\xi_1 \geq M\} + E[T|\xi_1 < M]\Pr\{\xi_1 < M\}. \tag{3.36}$$

Clearly, $E[T|\xi_1 \geq M] = 1$, since no further bids are examined in this case. On the other hand, when $\xi_1 < M$, we have the first bid, which was not accepted, plus some future bids. The future bids $\xi_2, \xi_3, \ldots$ have the same probabilistic properties as in the original problem, and they are examined until the first acceptable bid appears. This reasoning leads to $E[T|\xi_1 < M] = 1 + \mu$. Substitution into (3.36) then yields (3.35) as follows:

$$\begin{aligned} E[T] &= 1 \times \Pr\{\xi_1 \geq M\} + (1 + \mu)\Pr\{\xi_1 < M\} \\ &= 1 + \mu\Pr\{\xi_1 < M\}. \end{aligned}$$

To restate the argument somewhat differently, one always examines the first bid ξ_1. If $\xi_1 < M$, then further bids are examined in a future that is probabilistically similar to the original problem. That is, when $\xi_1 < M$, then on the average μ bids in addition to ξ_1 must be examined before an acceptable bid appears. Equation (3.35) results.

Example *Partial Sums* Another important Markov chain arises from consideration of the successive partial sums η_n of the ξ_i, i.e.,

$$\eta_n = \xi_1 + \cdots + \xi_n, \quad n = 1, 2, \ldots,$$

and by definition, $\eta_0 = 0$. The process $X_n = \eta_n$ is readily seen to be a Markov chain via

$$\begin{aligned} \Pr\{X_{n+1} = j|X_1 = i_1, \ldots, X_{n-1} = i_{n-1}, X_n = i\} \\ = \Pr\{\xi_{n+1} = j - i|\xi_1 = i_1, \xi_2 = i_2 - i_1, \ldots, \xi_n = i - i_{n-1}\} \\ = \Pr\{\xi_{n+1} = j - i\} \quad \text{(independence of } \xi_1, \xi_2, \ldots) \\ = \Pr\{X_{n+1} = j|X_n = i\}. \end{aligned}$$

The transition probability matrix is determined by

$$\begin{aligned} \Pr\{X_{n+1} = j|X_n = i\} &= \Pr\{\xi_1 + \cdots + \xi_{n+1} = j|\xi_1 + \cdots + \xi_n = i\} \\ &= \Pr\{\xi_{n+1} = j - i\} \\ &= \begin{cases} a_{j-i} & \text{for } j \geq i, \\ 0 & \text{for } j < i, \end{cases} \end{aligned}$$

where we have used the independence of the ξ_i.

Schematically, we have

$$\mathbf{P} = \left\| \begin{matrix} a_0 & a_1 & a_2 & a_3 & \cdots \\ 0 & a_0 & a_1 & a_2 & \cdots \\ 0 & 0 & a_0 & a_1 & \cdots \\ \vdots & \vdots & \vdots & \vdots & \end{matrix} \right\|. \tag{3.37}$$

If the possible values of the random variable ξ are permitted to be the positive and negative integers, then the possible values of η_n for each n will be contained among the totality of all integers. Instead of labeling the states conventionally by means of the nonnegative integers, it is more convenient to identify the state space with the totality of integers, since the transition probability matrix will then appear in a more symmetric form. The state space consists then of the values $\ldots -2, -1, 0, 1, 2, \ldots$.

The transition probability matrix becomes

$$\mathbf{P} = \left\| \begin{matrix} & \vdots & \vdots & \vdots & \vdots & \vdots & \\ \cdots & a_{-1} & a_0 & a_1 & a_2 & a_3 & \cdots \\ \cdots & a_{-2} & a_{-1} & a_0 & a_1 & a_2 & \cdots \\ \cdots & a_{-3} & a_{-2} & a_{-1} & a_0 & a_1 & \cdots \\ & \vdots & \vdots & \vdots & \vdots & \vdots & \end{matrix} \right\|,$$

where $\Pr\{\xi = k\} = a_k$ for $k = 0, \pm 1, \pm 2, \ldots$, and $a_k \geq 0$, $\sum_{k=-\infty}^{+\infty} a_k = 1$.

3.5.3 One-Dimensional Random Walks

When we discuss random walks, it is an aid to intuition to speak about the state of the system as the position of a moving "particle."

A one-dimensional random walk is a Markov chain whose state space is a finite or infinite subset $a, a+1, \ldots, b$ of the integers, in which the particle, if it is in state i, can in a single transition either stay in i or move to one of the neighboring states $i-1, i+1$. If the state space is taken as the nonnegative integers, the transition matrix of a random walk has the form

$$\mathbf{P} = \begin{matrix}
 & \begin{matrix} 0 & \quad 1 & \quad\quad 2 & \quad\quad i-1 & \quad i & \; i+1 \end{matrix} \\
\begin{matrix} 0 \\ 1 \\ 2 \\ \vdots \\ i \end{matrix} & \left\| \begin{matrix} r_0 & p_0 & 0 & \cdots & 0 & \cdots & & \\ q_1 & r_1 & p_1 & \cdots & 0 & \cdots & & \\ 0 & q_2 & r_2 & \cdots & 0 & \cdots & & \\ & & \ddots & & & & & \\ & & 0 & & q_i & r_i & p_i & 0 \\ & & & & \ddots & & & \ddots \end{matrix} \right\|
\end{matrix}, \tag{3.38}$$

where $p_i > 0, q_i > 0, r_i \geq 0$, and $q_i + r_i + p_i = 1, i = 1, 2, \ldots (i \geq 1), p_0 \geq 0, r_0 \geq 0$, $r_0 + p_0 = 1$. Specifically, if $X_n = i$, then for $i \geq 1$,

$$\Pr\{X_{n+1} = i + 1 | X_n = i\} = p_i,$$
$$\Pr\{X_{n+1} = i - 1 | X_n = i\} = q_j,$$
$$\Pr\{X_{n+1} = i | X_n = i\} = r_i,$$

with the obvious modifications holding for $i = 0$.

The designation "random walk" seems apt, since a realization of the process describes the path of a person (suitably intoxicated) moving randomly one step forward or backward.

The fortune of a player engaged in a series of contests is often depicted by a random walk process. Specifically, suppose an individual (player A) with fortune k plays a game against an infinitely rich adversary and has probability p_k of winning one unit and probability $q_k = 1 - p_k (k \geq 1)$ of losing one unit in the next contest (the choice of the contest at each stage may depend on his fortune), and $r_0 = 1$. The process X_n, where X_n represents his fortune after n contests, is clearly a random walk. Note that once the state 0 is reached (i.e., player A is wiped out), the process remains in that state. The event of reaching state $k = 0$ is commonly known as the "gambler's ruin."

If the adversary, player B, also starts with a limited fortune l and player A has an initial fortune $k(k + l = N)$, then we may again consider the Markov chain process X_n representing player A's fortune. However, the states of the process are now restricted to the values $0, 1, 2, \ldots, N$. At any trial, $N - X_n$ is interpreted as player B's fortune. If we allow the possibility of neither player winning in a contest, the transition probability matrix takes the form

$$
\mathbf{P} =
\begin{array}{c c}
& \begin{array}{c c c c c c} 0 & 1 & 2 & 3 & & N \end{array} \\
\begin{array}{c} 0 \\ 1 \\ 2 \\ \\ \\ N \end{array} &
\left\| \begin{array}{c c c c c c}
1 & 0 & 0 & 0 & \cdots & \\
q_1 & r_1 & p_1 & 0 & \cdots & \\
0 & q_2 & r_2 & p_2 & \cdots & \\
& \ddots & & & & \\
& & q_{N-1} & r_{N-1} & p_{N-1} \\
0 & \cdots & & \cdots & 0 & 0 & 1
\end{array} \right\|
\end{array}.
\tag{3.39}
$$

Again $p_i(q_i), i = 1, 2, \ldots, N - 1$, denotes the probability of player A's fortune increasing (decreasing) by 1 at the subsequent trial when his present fortune is i, and r_i may be interpreted as the probability of a draw. Note that, in accordance with the Markov chain given in (3.39), when player A's fortune (the state of the process) reaches 0 or N, it remains in this same state forever. We say player A is ruined when the state of the process reaches 0, and player B is ruined when the state of the process reaches N.

The probability of gambler's ruin (for player A) is derived in the next section by solving a first step analysis. Some more complex functionals on random walk processes are also derived in the next section.

The random walk corresponding to $p_k = p, q_k = 1 - p = q$ for all $k \geq 1$ and $r_0 = 1$ describes the situation of identical contests. There is a definite advantage to player A in each individual trial if $p > q$, and conversely, an advantage to player B if $p < q$. A "fair" contest corresponds to $p = q = \frac{1}{2}$. Suppose the total of both players' fortunes is N. Then, the corresponding walk, where X_n is player A's fortune at stage n, has the transition probability matrix

$$
\mathbf{P} =
\begin{array}{c}
\\
0 \\
1 \\
2 \\
\vdots \\
N-1 \\
N
\end{array}
\begin{array}{c}
\begin{array}{cccccccc}
0 & 1 & 2 & 3 & & N-1 & N
\end{array} \\
\left\|
\begin{array}{ccccccc}
1 & 0 & 0 & 0 & \cdots & 0 & 0 \\
q & 0 & p & 0 & \cdots & 0 & 0 \\
0 & q & 0 & p & \cdots & 0 & 0 \\
\vdots & \vdots & \vdots & \vdots & & \vdots & \vdots \\
0 & 0 & 0 & 0 & \cdots & 0 & p \\
0 & 0 & 0 & 0 & \cdots & 0 & 1
\end{array}
\right\|
\end{array}
. \tag{3.40}
$$

Let $u_i = U_{i0}$ be the probability of gambler's ruin starting with the initial fortune i. Then, u_i is the probability that the random walk reaches state 0 before reaching state N, starting from $X_0 = i$. The first step analysis of Section 3.4, as used in deriving equation (3.26), shows that these ruin probabilities satisfy

$$
u_i = p u_{i+1} + q u_{i-1} \quad \text{for } i = 1, \ldots, N-1 \tag{3.41}
$$

together with the obvious boundary conditions

$$
u_0 = 1 \quad \text{and} \quad u_N = 0.
$$

These equations are solved in the next section following a straight-forward but arduous method. There it is shown that the gambler's ruin probabilities corresponding to the transition probability matrix given in (3.40) are

$$
u_i = \Pr\{X_n \text{ reaches state 0 before state } N | X_0 = i\}
$$

$$
= \begin{cases}
\dfrac{N-i}{N} & \text{when } p = q = \dfrac{1}{2}, \\[3mm]
\dfrac{(q/p)^i - (q/p)^N}{1 - (q/p)^N} & \text{when } p \neq q.
\end{cases} \tag{3.42}
$$

The ruin probabilities u_i given by (3.42) have the following interpretation. In a game in which player A begins with an initial fortune of i units and player B begins with $N - i$ units, the probability that player A loses all his money before player B goes broke is given by u_i, where p is the probability that player A wins in a single contest. If player B is infinitely rich ($N \to \infty$), then passing to the limit in (3.42) and using $(q/p)^N \to \infty$ as $N \to \infty$ if $p < q$, while $(q/p)^N \to 0$ if $p > q$, we see that the ruin probabilities become

$$
u_i = \begin{cases} 1 & \text{if } p \leq q, \\ \left(\dfrac{q}{p}\right)^i & \text{if } p > q. \end{cases} \tag{3.43}
$$

(In passing to the limit, the case $p = q = \frac{1}{2}$ must be treated separately.) We see that ruin is certain ($u_i = 1$) against an infinitely rich adversary when the game is unfavorable ($p < q$), and even when the game is fair ($p = q$). In a favorable game ($p > q$), starting with initial fortune i, then ruin occurs (player A goes broke) with probability $(q/p)^i$. This ruin probability decreases as the initial fortune i increases. In a favorable game against an infinitely rich opponent, with probability $1 - (q/p)^i$ player A's fortune increases, in the long run, without limit.

More complex gambler's-ruin-type problems find practical relevance in certain models describing the fluctuation of insurance company assets over time.

Random walks are not only useful in simulating situations of gambling but frequently serve as reasonable discrete approximations to physical processes describing the motion of diffusing particles. If a particle is subjected to collisions and random impulses, then its position fluctuates randomly, although the particle describes a continuous path. If the future position (i.e., its probability distribution) of the particle depends only on the present position, then the process X_t, where X_t is the position at time t, is Markov. A discrete approximation to such a continuous motion corresponds to a random walk. A classical discrete version of Brownian motion (VIII) is provided by the symmetric random walk. By a symmetric random walk on the integers (say all the integers) we mean a Markov chain with state space the totality of all integers and whose transition probability matrix has the elements

$$
P_{ij} = \begin{cases} p & \text{if } j = i+1, \\ p & \text{if } j = i-1, \\ r & \text{if } j = i, \\ 0 & \text{otherwise,} \end{cases} \quad i,j = 0, 1, 2, \ldots,
$$

where $p > 0$, $r \geq 0$, and $2p + r = 1$. Conventionally, "simple random walk" refers only to the case $r = 0$, $p = \frac{1}{2}$.

The classical simple random walk in n dimensions admits the following formulation. The state space is identified with the set of all integral lattice points in E^n (Euclidean n space); that is, a state is an n-tuple $k = (k_1, k_2, \ldots, k_n)$ of integers. The transition probability matrix is defined by

$$
\mathbf{P_{kl}} = \begin{cases} \dfrac{1}{2n} & \text{if } \sum_{i=0}^{n} |l_i - k_i| = 1, \\ 0 & \text{otherwise.} \end{cases}
$$

Analogous to the one-dimensional case, the simple random walk in E^n represents a discrete version of n-dimensional Brownian motion.

3.5.4 Success Runs

Consider a Markov chain on the nonnegative integers with transition probability matrix of the form

$$
\mathbf{P} = \begin{array}{c} \\ 0 \\ 1 \\ 2 \\ 3 \\ \\ \end{array}
\begin{array}{ccccc}
0 & 1 & 2 & 3 & 4 \\
\end{array}
\left\| \begin{array}{ccccc}
p_0 & q_0 & 0 & 0 & 0 & \cdots \\
p_1 & r_1 & q_1 & 0 & 0 & \cdots \\
p_2 & 0 & r_2 & q_2 & 0 & \cdots \\
p_3 & 0 & 0 & r_3 & q_3 & \cdots \\
\vdots & \vdots & \vdots & \vdots & \vdots &
\end{array} \right\|,
\tag{3.44}
$$

where $q_i > 0$, $p_i > 0$, and $p_i + q_i + r_i = 1$ for $i = 0, 1, 2, \ldots$. The zero state plays a distinguished role in that it can be reached in one transition from any other state, while state $i + 1$ can be reached only from state i.

This example arises surprisingly often in applications and at the same time is very easy to compute with. We will frequently illustrate concepts and results in terms of it.

A special case of this transition matrix arises when one is dealing with success runs resulting from repeated trials, each of which admits two possible outcomes, success S or failure F. More explicitly, consider a sequence of trials with two possible outcomes, S or F. Moreover, suppose that in each trial, the probability of S is α and the probability of F is $\beta = 1 - \alpha$. We say a success run of length r happened at trial n if the outcomes in the preceding $r + 1$ trials, including the present trial as the last, were respectively $F, S, S, \ldots, S$. Let us now label the present state of the process by the length of the success run currently under way. In particular, if the last trial resulted in a failure, then the state is zero. Similarly, when the preceding $r + 1$ trials in order have the outcomes $F, S, S, \ldots, S$, the state variable would carry the label r. The process is clearly Markov (since the individual trials were independent of each other), and its transition matrix has the form (3.44), where

$$
p_n = \beta, \quad r_n = 0, \quad \text{and} \quad q_n = \alpha \quad \text{for } n = 0, 1, 2, \ldots.
$$

A second example is furnished by the *current age* in a *renewal process*. Consider a light bulb whose lifetime, measured in discrete units, is a random variable ξ, where

$$
\Pr\{\xi = k\} = a_k > 0 \quad \text{for } k = 1, 2, \ldots, \sum_{k=1}^{\infty} a_k = 1.
$$

Let each bulb be replaced by a new one when it burns out. Suppose the first bulb lasts until time ξ_1, the second bulb until time $\xi_1 + \xi_2$, and the nth bulb until time $\xi_1 + \cdots + \xi_n$, where the individual lifetimes $\xi_1, \xi_2, \ldots$ are independent random variables each having the same distribution as ξ. Let X_n be the age of the bulb in service at time n. This current age process is depicted in Figure 3.2.

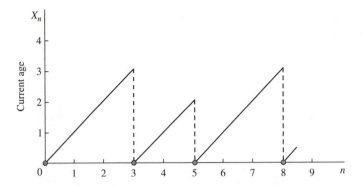

Figure 3.2 The current age X_n in a renewal process. Here, $\xi_1 = 3, \xi_2 = 2$, and $\xi_3 = 3$.

By convention, we set $X_n = 0$ at the time of a failure.

The current age is a success run Markov process for which

$$p_k = \frac{a_{k+1}}{a_{k+1} + a_{k+2} + \cdots}, \quad r_k = 0, q_k = 1 - p_k,$$

$$\text{for } k = 0, 1, \ldots. \tag{3.45}$$

We reason as follows: The age process reverts to zero upon failure of the item in service. Given that the age of the item in current service is k, then failure occurs in the next time period with *conditional probability* $p_k = a_{k+1}/(a_{k+1} + a_{k+2} + \cdots)$. Given that the item has survived k periods, it survives at least to the next period with the remaining probability $q_k = 1 - p_k$.

Renewal processes are extensively discussed in Chapter 7.

Exercises

3.5.1 The probability of the thrower winning in the dice game called "craps" is $p = 0.4929$. Suppose Player A is the thrower and begins the game with \$5, and Player B, his opponent, begins with \$10. What is the probability that Player A goes bankrupt before Player B? Assume that the bet is \$1 per round.

   Hint: Use equation (3.42).

3.5.2 Determine the gambler's ruin probability for Player A when both players begin with \$50, bet \$1 on each play, and where the win probability for Player A in each game is

   (a) $p = 0.49292929$

   (b) $p = 0.5029237$

   (See Chapter 2, Section 2.2.)

   What are the gambler's ruin probabilities when each player begins with \$500?

3.5.3 Determine $\mathbf{P}^n$ for $n = 2, 3, 4, 5$ for the Markov chain whose transition probability matrix is

$$\mathbf{P} = \begin{Vmatrix} 0.4 & 0.6 \\ 0.7 & 0.3 \end{Vmatrix}.$$

3.5.4 A coin is tossed repeatedly until three heads in a row appear. Let X_n record the current number of successive heads that have appeared. That is, $X_n = 0$ if the nth toss resulted in tails; $X_n = 1$ if the nth toss was heads and the $(n-1)$st toss was tails; and so on. Model X_n as a success runs Markov chain by specifying the probabilities p_i and q_i.

3.5.5 Suppose that the items produced by a certain process are each graded as defective or good and that whether or not a particular item is defective or good depends on the quality of the previous item. To be specific, suppose that a defective item is followed by another defective item with probability 0.80, whereas a good item is followed by another good item with probability 0.95. Suppose that the initial (zeroth) item is good. Using equation (3.31), determine the probability that the eighth item is good, and verify this by computing the eighth matrix power of the transition probability matrix.

3.5.6 A baseball trading card that you have for sale may be quite valuable. Suppose that the successive bids $\xi_1, \xi_2, \ldots$ that you receive are independent random variables with the geometric distribution

$$\Pr\{\xi = k\} = 0.01(0.99)^k \quad \text{for } k = 0, 1, \ldots.$$

If you decide to accept any bid over \$100, how many bids, on the average, will you receive before an acceptable bid appears?

Hint: Review the discussion surrounding equation (3.35).

3.5.7 Consider the random walk Markov chain whose transition probability matrix is given by

$$\mathbf{P} = \begin{array}{c c} & \begin{array}{c c c c} 0 & 1 & 2 & 3 \end{array} \\ \begin{array}{c} 0 \\ 1 \\ 2 \\ 3 \end{array} & \begin{Vmatrix} 1 & 0 & 0 & 0 \\ 0.3 & 0 & 0.7 & 0 \\ 0 & 0.3 & 0 & 0.7 \\ 0 & 0 & 0 & 1 \end{Vmatrix} \end{array}.$$

Starting in state 1, determine the probability that the process is absorbed into state 0. Do this first using the basic first step approach of equations (3.21) and (3.22) and second using the particular results for a random walk given in equation (3.42).

3.5.8 As a special case, consider a discrete-time queueing model in which at most a single customer arrives in any period and at most a single customer completes service. Suppose that in any single period, a single customer arrives with probability α, and no customers arrive with probability $1 - \alpha$. Provided that there are customers in the system, in a single period a single customer completes service with probability β, and no customers leave with probability $1 - \beta$. Then X_n, the number of customers in the system at the end-of-period n, is a random walk in the sense of Section 3.5.3. Referring to equation (3.38), specify the transition probabilities p_i, q_i, and r_i for $i = 0, 1, \ldots$.

3.5.9 In a simplified model of a certain television game show, suppose that the contestant, having won k dollars, will at the next play have $k + 1$ dollars with probability q and be put out of the game and leave with nothing with probability $p = 1 - q$. Suppose that the contestant begins with one dollar. Model her winnings after n plays as a success runs Markov chain by specifying the transition probabilities p_i, q_i, and r_i in equation (3.44).

Problems

3.5.1 As a special case of the successive maxima Markov chain whose transition probabilities are given in equation (3.34), consider the Markov chain whose transition probability matrix is given by

$$
\mathbf{P} = \begin{array}{c} \\ 0 \\ 1 \\ 2 \\ 3 \end{array}
\begin{array}{cccc}
0 & 1 & 2 & 3 \\
\left\| \begin{array}{cccc}
a_0 & a_1 & a_2 & a_3 \\
0 & a_0 + a_1 & a_2 & a_3 \\
0 & 0 & a_0 + a_1 + a_2 & a_3 \\
0 & 0 & 0 & 1
\end{array} \right\|
\end{array}.
$$

Starting in state 0, show that the mean time until absorption is $v_0 = 1/a_3$.

3.5.2 A component of a computer has an active life, measured in discrete units, that is a random variable T, where $\Pr\{T = k\} = a_k$ for $k = 1, 2, \ldots$. Suppose one starts with a fresh component, and each component is replaced by a new component upon failure. Let X_n be the age of the component in service at time n. Then, $\{X_n\}$ is a success runs Markov chain.

(a) Specify the probabilities p_i and q_i.

(b) A "planned replacement" policy calls for replacing the component upon its failure or upon its reaching age N, whichever occurs first. Specify the success runs probabilities p_i and q_i under the planned replacement policy.

3.5.3 *A Batch Processing Model.* Customers arrive at a facility and wait there until K customers have accumulated. Upon the arrival of the Kth customer, all are instantaneously served, and the process repeats. Let $\xi_0, \xi_1, \ldots$ denote the arrivals in successive periods, assumed to be independent random variables whose distribution is given by

$$
\Pr\{\xi_k = 0\} = \alpha, \quad \Pr\{\xi_k = 1\} = 1 - \alpha,
$$

where $0 < \alpha < 1$. Let X_n denote the number of customers in the system at time n. Then, $\{X_n\}$ is a Markov chain on the states $0, 1, \ldots, K - 1$. With $K = 3$, give the transition probability matrix for $\{X_n\}$. Be explicit about any assumptions you make.

3.5.4 Martha has a fair die with the usual six sides. She throws the die and records the number. She throws the die again and adds the second number to the first. She repeats this until the cumulative sum of all the tosses first exceeds 10. What is the probability that she stops at a cumulative sum of 13?

3.5.5 Let $\{X_n\}$ be a random walk for which zero is an absorbing state and such that from a positive state, the process is equally likely to go up or down one unit. The transition probability matrix is given by (3.38) with $r_0 = 1$ and $p_i = q_i = \frac{1}{2}$ for $i \geq 1$. (a) Show that $\{X_n\}$ is a nonnegative martingale. (b) Use the maximal inequality in Chapter 2, (2.53) to limit the probability that the process ever gets as high as $N > 0$.

3.6 Functionals of Random Walks and Success Runs

Consider first the random walk on $N + 1$ states whose transition probability matrix is given by

$$
\mathbf{P} = \begin{array}{c} \\ 0 \\ 1 \\ 2 \\ \vdots \\ N \end{array}
\begin{array}{cccccc}
0 & 1 & 2 & 3 & \cdots & N \\
\left\| \begin{array}{ccccc} 1 & 0 & 0 & 0 & \cdots & 0 \\ q & 0 & p & 0 & \cdots & 0 \\ 0 & q & 0 & p & \cdots & 0 \\ \vdots & \vdots & \vdots & \vdots & & \vdots \\ 0 & 0 & 0 & 0 & \cdots & 1 \end{array} \right\|
\end{array}.
$$

"Gambler's ruin" is the event that the process reaches state 0 before reaching state N. This event can be stated more formally if we introduce the concept of *hitting time*. Let T be the (random) time that the process first reaches, or hits, state 0 or N. In symbols,

$$T = \min\{n \geq 0; X_n = 0 \text{ or } X_n = N\}.$$

The random time T is shown in Figure 3.3 in a typical case.

In terms of T, the event written as $X_T = 0$ is the event of gambler's ruin, and the probability of this event starting from the initial state k is

$$u_k = \Pr\{X_T = 0 | X_0 = k\}.$$

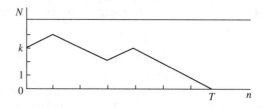

Figure 3.3 The hitting time to 0 or N. As depicted here, state 0 was reached first.

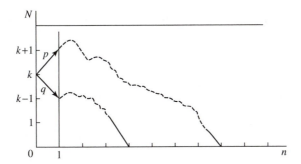

Figure 3.4 First step analysis for the gambler's ruin problem.

Figure 3.4 shows the first step analysis that leads to the equation

$$u_k = pu_{k+1} + qu_{k-1}, \quad \text{for } k = 1, \dots, N-1, \tag{3.46}$$

with the obvious boundary conditions

$$u_0 = 1, \quad u_N = 0.$$

Equation (3.46) yields to straightforward but tedious manipulations. Because the approach has considerable generality and arises frequently, it is well worth pursuing in this simplest case.

We begin the solution by introducing the differences $x_k = u_k - u_{k-1}$ for $k = 1, \dots, N$. Using $p + q = 1$ to write $u_k = (p+q)u_k = pu_k + qu_k$, equation (3.46) becomes

$$
\begin{aligned}
k &= 1; & 0 &= p(u_2 - u_1) - q(u_1 - u_0) = px_2 - qx_1; \\
k &= 2; & 0 &= p(u_3 - u_2) - q(u_2 - u_1) = px_3 - qx_2; \\
k &= 3; & 0 &= p(u_4 - u_3) - q(u_3 - u_2) = px_4 - qx_3; \\
&\;\;\vdots \\
k &= N-1; & 0 &= p(u_N - u_{N-1}) - q(u_{N-1} - u_{N-2}) = px_N - qx_{N-1};
\end{aligned}
$$

or

$$
\begin{aligned}
x_2 &= (q/p)x_1, \\
x_3 &= (q/p)x_2 = (q/p)^2 x_1, \\
x_4 &= (q/p)x_3 = (q/p)^3 x_1 \\
&\;\;\vdots \\
x_k &= (q/p)x_{k-1} = (q/p)^{k-1} x_1, \\
&\;\;\vdots \\
x_N &= (q/p)x_{N-1} = (q/p)^{N-1} x_1.
\end{aligned}
$$

We now recover $u_0, u_1, \ldots, u_N$ by invoking the conditions $u_0 = 1, u_N = 0$ and summing the x_k's:

$$x_1 = u_1 - u_0 = u_1 - 1,$$
$$x_2 = u_2 - u_1, \qquad\qquad\qquad x_1 + x_2 = u_2 - 1,$$
$$x_3 = u_3 - u_2, \qquad\qquad\qquad x_1 + x_2 + x_3 = u_3 - 1,$$
$$\vdots \qquad\qquad\qquad\qquad\qquad \vdots$$
$$x_k = u_k - u_{k-1}, \qquad\qquad\qquad x_1 + \cdots + x_k = u_k - 1,$$
$$\vdots \qquad\qquad\qquad\qquad\qquad \vdots$$
$$x_N = u_N - u_{N-1} = -u_{N-1}, \quad x_1 + \cdots + x_N = u_N - 1 = -1.$$

The equation for general k gives

$$
\begin{aligned}
u_k &= 1 + x_1 + x_2 + \cdots + x_k \\
&= 1 + x_1 + (q/p)x_1 + \cdots + (q/p)^{k-1}x_1 \\
&= 1 + [1 + (q/p) + \cdots + (q/p)^{k-1}]x_1,
\end{aligned}
\tag{3.47}
$$

which expresses u_k in terms of the as yet undetermined x_1. But $u_N = 0$ gives

$$0 = 1 + [1 + (q/p) + \cdots + (q/p)^{N-1}]x_1,$$

or

$$x_1 = -\frac{1}{1 + (q/p) + \cdots + (q/p)^{N-1}},$$

which substituted into (3.47) gives

$$u_k = 1 - \frac{1 + (q/p) + \cdots + (q/p)^{k-1}}{1 + (q/p) + \cdots + (q/p)^{N-1}}.$$

The geometric series sums to

$$
1 + (q/p) + \cdots + (q/p)^{k-1} =
\begin{cases}
k & \text{if } p = q = \dfrac{1}{2}, \\[2mm]
\dfrac{1 - (q/p)^k}{1 - (q/p)} & \text{if } p \neq q,
\end{cases}
$$

whence

$$
u_k =
\begin{cases}
1 - (k/N) = (N - k)/N & \text{when } p = q = \dfrac{1}{2}, \\[2mm]
1 - \dfrac{1 - (q/p)^k}{1 - (q/p)^N} = \dfrac{(q/p)^k - (q/p)^N}{1 - (q/p)^N} & \text{when } p \neq q.
\end{cases}
\tag{3.48}
$$

A similar approach works to evaluate the mean duration

$$v_i = E[T|X_0 = i]. \tag{3.49}$$

The time T is composed of a first step plus the remaining steps. With probability p, the first step is to state $i+1$, and then, the remainder, on the average, is v_{i+1} additional steps. With probability q, the first step is to $i-1$, and then, on the average, there are v_{i-1} further steps. Thus, for the mean duration, a first step analysis leads to the equation

$$v_i = 1 + pv_{i+1} + qv_{i-1} \quad \text{for } i = 1, \ldots, N-1. \tag{3.50}$$

Of course, the game ends in states 0 and N, and thus,

$$v_0 = 0, \quad v_N = 0.$$

We will solve equation (3.50) when $p = q = \frac{1}{2}$. The solution for other values of p proceeds in a similar manner, and the solution for a general random walk is given later in this section.

Again, we introduce the differences $x_k = v_k - v_{k-1}$ for $k = 1, \ldots, N$, writing (3.50) in the form

$$k = 1; \quad -1 = \frac{1}{2}(v_2 - v_1) - \frac{1}{2}(v_1 - v_0) = \frac{1}{2}x_2 - \frac{1}{2}x_1;$$

$$k = 2; \quad -1 = \frac{1}{2}(v_3 - v_2) - \frac{1}{2}(v_2 - v_1) = \frac{1}{2}x_3 - \frac{1}{2}x_2;$$

$$k = 3; \quad -1 = \frac{1}{2}(v_4 - v_3) - \frac{1}{2}(v_3 - v_2) = \frac{1}{2}x_4 - \frac{1}{2}x_3;$$

$$\vdots$$

$$k = N-1; \quad -1 = \frac{1}{2}(v_N - v_{N-1}) - \frac{1}{2}(v_{N-1} - v_{N-2}) = \frac{1}{2}x_N - \frac{1}{2}x_{N-1}.$$

The right side forms a collapsing sum. Upon adding, we obtain

$$k = 1; \quad -1 = \frac{1}{2}x_2 - \frac{1}{2}x_1;$$

$$k = 2; \quad -2 = \frac{1}{2}x_3 - \frac{1}{2}x_1;$$

$$k = 3; \quad -3 = \frac{1}{2}x_4 - \frac{1}{2}x_1;$$

$$\vdots$$

$$k = N-1; \quad -(N-1) = \frac{1}{2}x_N - \frac{1}{2}x_1.$$

The general line gives $x_k = x_1 - 2(k-1)$ for $k = 2, 3, \ldots, N$. We return to the v_k's by means of

$$
\begin{aligned}
x_1 &= v_1 - v_0 = v_1; \\
x_2 &= v_2 - v_1; & x_1 + x_2 &= v_2; \\
x_3 &= v_3 - v_2; & x_1 + x_2 + x_3 &= v_3; \\
&\;\;\vdots \\
x_k &= v_k - v_{k-1}; & x_1 + \cdots + x_k &= v_k;
\end{aligned}
$$

or

$$
v_k = kv_1 - 2[1 + 2 + \cdots + (k-1)] = kv_1 - k(k-1), \tag{3.51}
$$

which gives v_k in terms of the as yet unknown v_1. We impose the boundary condition $v_N = 0$ to obtain $0 = Nv_1 - N(N-1)$ or $v_1 = (N-1)$. Substituting this into (3.51), we obtain

$$
v_k = k(N-k), \quad k = 0, 1, \ldots, N, \tag{3.52}
$$

for the mean duration of the game. Note that the mean duration is greatest for initial fortunes k that are midway between the boundaries 0 and N, as we would expect.

3.6.1   The General Random Walk

We give the results of similar derivations on the random walk whose transition matrix is

$$
\mathbf{P} = \begin{array}{c@{}c}
 & \begin{array}{cccccc} 0 & 1 & 2 & 3 & \cdots & N \end{array} \\
\begin{array}{c} 0 \\ 1 \\ 2 \\ \vdots \\ N \end{array} &
\left\| \begin{array}{cccccc}
1 & 0 & 0 & 0 & \cdots & 0 \\
q_1 & r_1 & p_1 & 0 & \cdots & 0 \\
0 & q_2 & r_2 & p_2 & \cdots & 0 \\
\vdots & \vdots & \vdots & \vdots & & \vdots \\
0 & 0 & 0 & 0 & \cdots & 1
\end{array} \right\|
\end{array},
$$

where $q_k > 0$ and $p_k > 0$ for $k = 1, \ldots, N-1$. Let $T = \min\{n \geq 0; X_n = 0 \text{ or } X_n = N\}$ be the hitting time to states 0 and N.

Example As a sample calculation of these functionals, we consider the special case in which the transition probabilities are the same from row to row. That is, we study the random walk whose transition probability matrix is

$$
\mathbf{P} = \begin{array}{c} \\ 0 \\ 1 \\ 2 \\ \vdots \\ N \end{array}
\begin{array}{c} \begin{array}{ccccccc} 0 & 1 & 2 & 3 & \cdots & N \end{array} \\
\left\| \begin{array}{cccccc}
1 & 0 & 0 & 0 & \cdots & 0 \\
q & r & p & 0 & \cdots & 0 \\
0 & q & r & p & \cdots & 0 \\
\vdots & \vdots & \vdots & \vdots & & \vdots \\
0 & 0 & 0 & 0 & & 1
\end{array} \right\|
\end{array},
$$

with $p > 0$, $q > 0$, and $p + q + r = 1$. Let us abbreviate by setting $\theta = (q/p)$, and then ρ_k, as defined in (3.63), simplifies according to

$$
\rho_k = \frac{q_1 q_2 \cdots q_k}{p_1 p_2 \cdots p_k} = \left(\frac{q}{p} \right)^k = \theta^k \quad \text{for } k = 1, \ldots, N-1.
$$

The probability of gambler's ruin, as defined in (3.61) and evaluated in (3.62), becomes

$$
u_k = \Pr\{X_T = 0 | X_0 = k\}
$$

$$
= \frac{\theta^k + \cdots + \theta^{N-1}}{1 + \theta + \cdots + \theta^{N-1}}
$$

$$
= \begin{cases}
\dfrac{\theta^k - \theta^N}{1 - \theta^N} & \text{if } \theta \equiv (q/p) \neq 1, \\[2ex]
\dfrac{N - k}{N} & \text{if } \theta \equiv (q/p) = 1.
\end{cases}
$$

This, of course, agrees with the answer given in (3.48).

We turn to evaluating the mean time

$$
v_k = E[T | X_0 = k] \quad \text{for } k = 1, \ldots, N = 1
$$

by first substituting $\rho_i = \theta^i$ into (3.67) to obtain

$$
\Phi_i = \left(\frac{1}{q} + \frac{1}{q\theta} + \cdots + \frac{1}{q\theta^{i-1}} \right) \theta^i
$$

$$
= \frac{1}{q} (\theta^i + \theta^{i-1} + \cdots + \theta)
$$

$$
= \frac{1}{p} (1 + \theta + \cdots + \theta^{i-1})
$$

$$= \begin{cases} \dfrac{i}{p} & \text{when } p = q (\theta = 1), \\[2ex] \dfrac{1}{p}\left(\dfrac{1-\theta^i}{1-\theta}\right) & \text{when } p \neq q (\theta \neq 1). \end{cases}$$

Now observe that

$$1 + \rho_1 + \cdots + \rho_{i-1} = 1 + \theta + \cdots + \theta^{i-1}$$
$$= p\Phi_i$$

so that (3.66) reduces to

$$v_k = \frac{\Phi_k}{\Phi_N}(\Phi_1 + \cdots + \Phi_{N-1}) - (\Phi_1 + \cdots + \Phi_{k-1}). \tag{3.53}$$

In order to continue, we need to simplify the terms of the form $\Phi_1 + \cdots + \Phi_{j-1}$. We consider the two cases $\theta \equiv (q/p) = 1$ and $\theta \equiv (q/p) \neq 1$ separately.

When $p = q$, or equivalently, $\theta = 1$, then $\Phi_i = i/p$, whence

$$\Phi_1 + \cdots + \Phi_{j-1} = \frac{1 + \cdots + (j-1)}{p} = \frac{j(j-1)}{2p},$$

which inserted into (3.53) gives

$$v_i \equiv E[T \mid X_0 = i]$$
$$= \frac{i}{N}\left[\frac{N(N-1)}{2p}\right] - \frac{i(i-1)}{2p} \tag{3.54}$$
$$= \frac{i(N-i)}{2p} \quad \text{if } p = q.$$

When $p = \frac{1}{2}$, then $v_i = i(N-i)$ in agreement with (3.52).
When $p \neq q$, so that $\theta \equiv q/p \neq 1$, then

$$\Phi_i = \frac{1}{p}\left(\frac{1-\theta^i}{1-\theta}\right),$$

whence

$$\Phi_1 + \cdots + \Phi_{j-1} = \frac{1}{p(1-\theta)}\left[(j-1) - \left(\theta + \theta^2 + \cdots + \theta^{j-1}\right)\right]$$
$$= \frac{1}{p(1-\theta)}\left[(j-1) - \theta\left(\frac{1-\theta^{j-1}}{1-\theta}\right)\right],$$

and

$$v_i = E[T|X_0 = i]$$

$$= \left(\frac{1-\theta^i}{1-\theta^N}\right)\frac{1}{p(1-\theta)}\left[N - \left(\frac{1-\theta^N}{1-\theta}\right)\right] - \frac{1}{p(1-\theta)}\left[i - \left(\frac{1-\theta^i}{1-\theta}\right)\right]$$

$$= \frac{1}{p(1-\theta)}\left[N\left(\frac{1-\theta^i}{1-\theta^N}\right) - i\right],$$

when $\theta \equiv (q/p) \neq 1$.

Finally, we evaluate W_{ik}, expressed verbally as the mean number of visits to state k starting from $X_0 = i$ and defined formally in (3.68). Again, we consider the two cases $\theta \equiv (q/p) = 1$ and $\theta \equiv (q/p) \neq 1$.

When $\theta = 1$, then $\rho_j = \theta^j = 1$ and $1 + \cdots + \rho_{i-1} = i$, $\rho_k + \cdots + \rho_{N-1} = N - k$, and (3.69) simplifies to

$$W_{ik} = \begin{cases} \dfrac{i(N-k)}{qN} & \text{for } 0 < i \leq k < N, \\[3mm] \dfrac{1}{q}\left[\dfrac{i(N-k)}{N} - (i-k)\right] = \dfrac{k(N-i)}{qN} & \text{for } 0 < k < i < N, \end{cases}$$

$$= \frac{i(N-k)}{qN} - \frac{\max\{o, i-k\}}{q}. \tag{3.55}$$

When $\theta = (q/p) \neq 1$, then $\rho_j = \theta^j$ and

$$1 + \cdots + \rho_{i-1} = \frac{1-\theta^i}{1-\theta},$$

$$\rho_k + \cdots + \rho_{N-1} = \frac{\theta^k - \theta^N}{1-\theta},$$

and

$$q\rho_{k-1} = p\rho_k = p\theta^k.$$

In this case, (3.69) simplifies to

$$W_{ik} = \frac{\left(1-\theta^i\right)\left(\theta^k - \theta^N\right)}{(1-\theta)\left(1-\theta^N\right)}\left(\frac{1}{p\theta^k}\right) \quad \text{for } 0 < i \leq k < N,$$

and

$$W_{ik} = \frac{\left(1-\theta^k\right)\left(\theta^i - \theta^N\right)}{(1-\theta)\left(1-\theta^N\right)}\left(\frac{1}{p\theta^k}\right) \quad \text{for } 0 < k < i < N.$$

We may write the expression for W_{ik} in a single line by introducing the notation $(i - k)^+ = \max\{0, i - k\}$. Then,

$$W_{ik} = \frac{(1 - \theta^i)(1 - \theta^{N-k})}{p(1 - \theta)(1 - \theta^N)} - \frac{1 - \theta^{(i-k)^+}}{p(1 - \theta)}. \tag{3.56}$$

3.6.2 Cash Management

Short-term cash management is the review and control of a corporation's cash balances, short-term loan balances, and short-term marketable security holdings. The objective is to maintain the smallest cash balances that are adequate to meet future disbursements. The corporation cashier tries to eliminate idle cash balances (e.g., by reducing short-term loans or buying treasury bills) but to cover potential cash shortages (by selling treasury bills or increasing short-term loans). The analogous problem for an individual is to maintain an optimal balance between a checking and a savings account.

In the absence of intervention, the corporation's cash level fluctuates randomly as the result of many relatively small transactions. We model this by dividing time into successive, equal length periods, each of short duration, and by assuming that from period to period, the cash level moves up or down one unit, each with a probability of one-half. Let X_n be the cash on hand in period n. We are assuming that $\{X_n\}$ is the random walk in which

$$\Pr\{X_{n+1} = k \pm 1 | X_n = k\} = \frac{1}{2}.$$

The cashier's job is to intervene if the cash level ever gets too low or too high. We consider cash management strategies that are specified by two parameters, s and $\mathscr{S}$, where $0 < s < \mathscr{S}$. The policy is as follows: If the cash level ever drops to zero, then sell sufficient treasury bills to replenish the cash level up to s. If the cash level ever increases up to $\mathscr{S}$, then invest in treasury bills in order to reduce the cash level to s. A typical sequence of cash levels $\{X_n\}$ when $s = 2$ and $\mathscr{S} = 5$ is depicted in Figure 3.5.

We see that the cash level fluctuates in a series of statistically similar cycles, each cycle beginning with s units of cash on hand and ending at the next intervention,

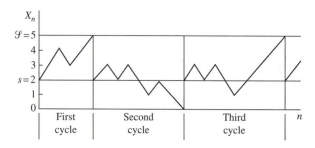

Figure 3.5 Several typical cycles in a cash inventory model.

whether a replenishment or reduction in cash. We begin our study by evaluating the mean length of a cycle and the mean total unit periods of cash on hand during a cycle. Later, we use these quantities to evaluate the long run performance of the model.

Let T denote the random time at which the cash on hand first reaches the level $\mathscr{S}$ or 0. That is, T is the time of the first transaction. Let $v_s = E[T|X_0 = s]$ be the mean time to the first transaction, or the mean cycle length. From (3.52), we have

$$v_s = s(\mathscr{S} - s).\tag{3.57}$$

Next, fix an arbitrary state $k(0 < k < \mathscr{S})$ and let W_{sk} be the mean number of visits to k up to time T for a process starting at $X_0 = s$. From (3.55), we have

$$W_{sk} = 2\left[\frac{s}{\mathscr{S}}(\mathscr{S} - k) - (s - k)^+\right].\tag{3.58}$$

Using this we obtain the mean total unit periods of cash on hand up to time T starting from $X_0 = s$ by weighting W_{sk} by k and summing according to

$$
\begin{aligned}
W_s &= \sum_{k=1}^{\mathscr{S}-1} k W_{sk} \\
&= 2\left\{\frac{s}{\mathscr{S}}\sum_{k=1}^{\mathscr{S}-1} k(\mathscr{S} - k) - \sum_{k=1}^{s-1} k(s - k)\right\} \\
&= 2\left\{\frac{s}{\mathscr{S}}\left[\frac{\mathscr{S}(\mathscr{S} - 1)(\mathscr{S} + 1)}{6}\right] - \frac{s(s - 1)(s + 1)}{6}\right\}^* \\
&= \frac{s}{3}\left[\mathscr{S}^2 - s^2\right].
\end{aligned}\tag{3.59}
$$

Having obtained these single cycle results, we will use them to evaluate the long run behavior of the model. Note that each cycle starts from the cash level s, and thus, the cycles are statistically independent. Let K be the fixed cost of each transaction. Let T_i be the duration of the ith cycle and let R_i be the total opportunity cost of holding cash on hand during that time. Over n cycles the average cost per unit time is

$$\text{Average cost} = \frac{nK + R_1 + \cdots + R_n}{T_1 + \cdots + T_n}.$$

Next, divide the numerator and denominator by n, let $n \to \infty$, and invoke the law of large numbers to obtain

$$\text{Long run average cost} = \frac{K + E[R_i]}{E[T_i]}.$$

* Use the sum $\sum_{k=1}^{a-1} k(a - k) = \frac{1}{6}a(a + 1)(a - 1)$.

Let r denote the opportunity cost per unit time of cash on hand. Then, $E[R_i] = rW_s$, while $E[T_i] = v_s$. Since these quantities were determined in (3.57) and (3.59), we have

$$\text{Long run average cost} = \frac{K + (1/3)rs\left(\mathcal{S}^2 - s^2\right)}{s(\mathcal{S} - s)}. \tag{3.60}$$

In order to use calculus to determine the cost-minimizing values for $\mathcal{S}$ and s, it simplifies matters if we introduce the new variable $x = s/\mathcal{S}$. Then, (3.60) becomes

$$\text{Long run average cost} = \frac{K + (1/3)r\mathcal{S}^3 x \left(1 - x^2\right)}{\mathcal{S}^2 x(1 - x)},$$

whence

$$\frac{d(\text{average cost})}{dx} = 0 = -\frac{K(1 - 2x)}{\mathcal{S}^2 x^2 (1 - x)^2} + \frac{1}{3}r\mathcal{S},$$

$$\frac{d(\text{average cost})}{d\mathcal{S}} = 0 = -\frac{2K}{\mathcal{S}^3 x(1 - x)} + \frac{r(1 + x)}{3},$$

which yield

$$x_{\text{opt}} = \frac{1}{3} \quad \text{and} \quad \mathcal{S}_{\text{opt}} = 3s_{\text{opt}} = 3\sqrt[3]{\frac{3K}{4r}}.$$

Implementing the cash management strategy with the values s_{opt} and $\mathcal{S}_{\text{opt}}$ results in the optimal balance between transaction costs and the opportunity cost of holding cash on hand.

3.6.3 The Success Runs Markov Chain

Consider the success runs Markov chain on $N + 1$ states whose transition matrix is

$$
\mathbf{P} =
\begin{array}{c@{}c}
 & \begin{array}{cccccc} 0 & 1 & 2 & 3 & \cdots & N \end{array} \\
\begin{array}{c} 0 \\ 1 \\ 2 \\ \vdots \\ N-1 \\ N \end{array} &
\left\|
\begin{array}{cccccc}
1 & 0 & 0 & 0 & \cdots & 0 \\
p_1 & r_1 & q_1 & 0 & \cdots & 0 \\
p_2 & 0 & r_2 & q_2 & \cdots & 0 \\
\vdots & \vdots & \vdots & \vdots & & \vdots \\
p_{N-1} & 0 & 0 & 0 & \cdots & q_{N-1} \\
0 & 0 & 0 & 0 & \cdots & 1
\end{array}
\right\|
\end{array}.
$$

Note that states 0 and N are absorbing; once the process reaches one of these two states it remains there.

Let T be the hitting time to states 0 or N,

$$T = \min\{n \geq 0; X_n = 0 \text{ or } X_n = N\}.$$

Exercises

3.6.1 A rat is put into the linear maze as shown:

0 shock	1	2	3	4	5 food

   (a) Assume that the rat is equally likely to move right or left at each step. What is the probability that the rat finds the food before getting shocked?

   (b) As a result of learning, at each step the rat moves to the right with probability $p > \frac{1}{2}$ and to the left with probability $q = 1 - p < \frac{1}{2}$. What is the probability that the rat finds the food before getting shocked?

3.6.2 Customer accounts receivable at Smith Company are classified each month according to

0: Current
1: 30–60 days past due
2: 60–90 days past due
3: Over 90 days past due

Consider a particular customer account and suppose that it evolves month to month as a Markov chain $\{X_n\}$ whose transition probability matrix is

$$\mathbf{P} = \begin{array}{c} \\ 0 \\ 1 \\ 2 \\ 3 \end{array} \begin{array}{cccc} 0 & 1 & 2 & 3 \\ \left\| \begin{array}{cccc} 0.9 & 0.1 & 0 & 0 \\ 0.5 & 0 & 0.5 & 0 \\ 0.3 & 0 & 0 & 0.7 \\ 0.2 & 0 & 0 & 0.8 \end{array} \right\| \end{array}.$$

Suppose that a certain customer's account is now in state 1: 30–60 days past due. What is the probability that this account will be paid (and thereby enter state 0: Current) before it becomes over 90 days past due? That is, let $T = \min\{n \geq 0; X_n = 0 \text{ or } X_n = 3\}$. Determine $\Pr\{X_T = 0 | X_0 = 1\}$.

3.6.3 Players A and B each have \$50 at the beginning of a game in which each player bets \$1 at each play, and the game continues until one player is broke. Suppose there is a constant probability $p = 0.492929\ldots$ that Player A wins on any given bet. What is the mean duration of the game?

3.6.4 Consider the random walk Markov chain whose transition probability matrix is given by

$$\mathbf{P} = \begin{array}{c} \\ 0 \\ 1 \\ 2 \\ 3 \end{array} \begin{array}{cccc} 0 & 1 & 2 & 3 \\ \left\| \begin{array}{cccc} 1 & 0 & 0 & 0 \\ 0.3 & 0 & 0.7 & 0 \\ 0 & 0.3 & 0 & 0.7 \\ 0 & 0 & 0 & 1 \end{array} \right\| \end{array}.$$

Starting in state 1, determine the mean time until absorption. Do this first using the basic first step approach of equation (3.24), and second using the particular formula for v_i that follows equation (3.54), which applies for a random walk in which $p \neq q$.

Problems

3.6.1 The probability of gambler's ruin

$$u_i = \Pr\{X_T = 0 | X_0 = i\} \tag{3.61}$$

satisfies the first step analysis equation

$$u_i = q_i u_{i-1} + r_i u_i + p_i u_{i+1} \quad \text{for } i = 1, \ldots, N-1,$$

and

$$u_0 = 1, \quad u_N = 0.$$

The solution is

$$u_i = \frac{\rho_i + \cdots + \rho_{N-1}}{1 + \rho_1 + \rho_2 + \cdots + \rho_{N-1}}, \quad i = 1, \ldots, N-1, \tag{3.62}$$

where

$$\rho_k = \frac{q_1 q_2 \cdots q_k}{p_1 p_2 \cdots p_k}, \quad k = 1, \ldots, N-1. \tag{3.63}$$

3.6.2 The mean hitting time

$$v_k = E[T | X_0 = k] \tag{3.64}$$

satisfies the equations

$$v_k = 1 + q_k v_{k-1} + r_k v_k + p_k v_{k+1} \quad \text{and} \quad v_0 = v_N = 0. \tag{3.65}$$

The solution is

$$v_k = \left(\frac{\Phi_1 + \cdots + \Phi_{N-1}}{1 + \rho_1 + \cdots + \rho_{N-1}} \right) (1 + \rho_1 + \cdots + \rho_{k-1})$$
$$- (\Phi_1 + \cdots + \Phi_{k-1}) \quad \text{for } k = 1, \ldots, N-1, \tag{3.66}$$

where ρ_i is given in (3.63) and

$$\Phi_i = \left(\frac{1}{q_1} + \frac{1}{q_2 \rho_1} + \cdots + \frac{1}{q_i \rho_{i-1}} \right) \rho_i$$
$$= \frac{q_2 \cdots q_i}{p_1 \cdots p_i} + \frac{q_3 \cdots q_i}{p_2 \cdots p_i} + \cdots + \frac{q_i}{p_{i-1} p_i} + \frac{1}{p_i} \quad \text{for } i = 1, \ldots, N-1. \tag{3.67}$$

3.6.3 Fix a state k, where $0 < k < N$, and let W_{ik} be the mean total visits to state k starting from i. Formally, the definition is

$$W_{ik} = E\left[\sum_{n=0}^{T-1} 1\{X_n = k\}|X_0 = i\right], \tag{3.68}$$

where

$$1\{X_n = k\} = \begin{cases} 1 & \text{if } X_n = k, \\ 0 & \text{if } X_n \neq k. \end{cases}$$

Then, W_{ik} satisfies the equations

$$W_{ik} = \delta_{ik} + q_i W_{i-1,k} + r_i W_{ik} + p_i W_{i+1,k} \quad \text{for } i = 1, \dots, N-1$$

and

$$W_{0k} = W_{Nk} = 0,$$

where

$$\delta_{ik} = \begin{cases} 1 & \text{if } i = k, \\ 0 & \text{if } i \neq k. \end{cases}$$

The solution is

$$W_{ik} = \begin{cases} \dfrac{(1 + \cdots + \rho_{i-1})(\rho_k + \cdots + \rho_{N-1})}{1 + \cdots + \rho_{N-1}} \left(\dfrac{1}{q_k \rho_{k-1}}\right) & \text{for } i \leq k, \\[4ex] \left[\dfrac{(1 + \cdots + \rho_{i-1})(\rho_k + \cdots + \rho_{N-1})}{1 + \cdots + \rho_{N-1}} \right. \\[3ex] \qquad \left. - (\rho_k + \cdots + \rho_{i-1})\right]\left(\dfrac{1}{q_k \rho_{k-1}}\right) & \text{for } i \geq k. \end{cases} \tag{3.69}$$

3.6.4 The probability of absorption at 0 starting from state k

$$u_k = \Pr\{X_T = 0|X_0 = k\} \tag{3.70}$$

satisfies the equation

$$u_k = p_k + r_k u_k + q_k u_{k+1},$$
$$\text{for } k = 1, \dots, N-1 \quad \text{and} \quad u_0 = 1, u_N = 0.$$

The solution is

$$u_k = 1 - \left(\frac{q_k}{p_k + q_k}\right) \cdots \left(\frac{q_{N-1}}{p_{N-1} + q_{N-1}}\right) \quad \text{for } k = 1, \ldots, N-1. \quad (3.71)$$

3.6.5 The mean hitting time

$$v_k = E[T|X_0 = k] \quad (3.72)$$

satisfies the equation

$$v_k = 1 + r_k v_k + q_k v_{k+1} \quad \text{for } k = 1, \ldots, N-1 \quad \text{and} \quad v_0 = v_N = 0.$$

The solution is

$$v_k = \frac{1}{p_k + q_k} + \frac{\pi_{k,k+1}}{p_{k+1} + q_{k+1}} + \cdots + \frac{\pi_{k,N-1}}{p_{N-1} + q_{N-1}}, \quad (3.73)$$

where

$$\pi_{kj} = \left(\frac{q_k}{p_k + q_k}\right)\left(\frac{q_{k+1}}{p_{k+1} + q_{k+1}}\right) \cdots \left(\frac{q_{j-1}}{p_{j-1} + q_{j-1}}\right) \quad (3.74)$$
$$\text{for } k < j.$$

3.6.6 Fix a state $j(0 < j < N)$ and let W_{ij} be the mean total visits to state j starting from state i [see equation (3.68)]. Then,

$$W_{iJ} = \begin{cases} \dfrac{1}{p_i + q_i} & \text{for } j = i, \\[2mm] \left(\dfrac{q_i}{p_i + q_i}\right) \cdots \left(\dfrac{q_{j-1}}{p_{j-1} + q_{j-1}}\right)\dfrac{1}{p_j + q_j} & \text{for } i < j, \\[2mm] 0 & \text{for } i > j. \end{cases} \quad (3.75)$$

3.6.7 Consider the random walk Markov chain whose transition probability matrix is given by

$$
\mathbf{P} = \begin{array}{c} \\ 0 \\ 1 \\ 2 \\ 3 \end{array}
\begin{array}{c} \begin{array}{cccc} 0 & 1 & 2 & 3 \end{array} \\ \left\|\begin{array}{cccc} 1 & 0 & 0 & 0 \\ 0.3 & 0 & 0.7 & 0 \\ 0 & 0.1 & 0 & 0.9 \\ 0 & 0 & 0 & 1 \end{array}\right\| \end{array}.
$$

Starting in state 1, determine the mean time until absorption. Do this first using the basic first step approach of equation (3.24) and second using the particular results for a random walk given in equation (3.66).

3.6.8 Consider the Markov chain $\{X_n\}$ whose transition matrix is

$$
\mathbf{P} = \begin{array}{c|cccc}
 & 0 & 1 & 2 & 3 \\
\hline
0 & \alpha & 0 & \beta & 0 \\
1 & \alpha & 0 & 0 & \beta \\
2 & \alpha & \beta & 0 & 0 \\
3 & 0 & 0 & 0 & 1
\end{array},
$$

where $\alpha > 0, \beta > 0$, and $\alpha + \beta = 1$. Determine the mean time to reach state 3 starting from state 0. That is, find $E[T|X_0 = 0]$, where $T = \min\{n \geq 0; X_n = 3\}$.

3.6.9 *Computer Challenge.* You have two urns: A and B, with a balls in A and b balls in B. You pick an urn at random, each urn being equally likely, and move a ball from it to the other urn. You do this repeatedly. The game ends when either of the urns becomes empty. The number of balls in A at the nth move is a simple random walk, and the expected duration of the game is $E[T] = ab$ [see equation (3.52)]. Now consider three urns, A, B, and C, with a, b, and c balls, respectively. You pick an urn at random, each being equally likely, and move a ball from it to one of the other two urns, each being equally likely. The game ends when one of the three urns becomes empty. What is the mean duration of the game? If you can guess the general form of this mean time by computing it in a variety of particular cases, it is not particularly difficult to verify it by a first step analysis. What about four urns?

3.7 Another Look at First Step Analysis*

In this section, we provide an alternative approach to evaluating the functionals treated in Section 3.4. The nth power of a transition probability matrix having both transient and absorbing states is directly evaluated. From these nth powers, it is possible to extract the mean number of visits to a transient state j prior to absorption, the mean time until absorption, and the probability of absorption in any particular absorbing state k. These functionals all depend on the initial state $X_0 = i$, and as a by-product of the derivation, we show that, as functions of this initial state i, these functionals satisfy their appropriate first step analysis equations.

Consider a Markov chain whose states are labeled $0, 1, \ldots, N$. States $0, 1, \ldots, r-1$ are transient in that $P_{ij}^{(n)} \to 0$ as $n \to \infty$ for $0 \leq i, j < r$, while states $r, \ldots, N$ are absorbing, or trap, and here $P_{ii} = 1$ for $r \leq i \leq N$. The transition matrix has the form

$$
\mathbf{P} = \begin{Vmatrix} \mathbf{Q} & \mathbf{R} \\ \mathbf{0} & \mathbf{I} \end{Vmatrix}, \tag{3.76}
$$

---

* This section contains material at a more difficult level. It is not prerequisite to what follows.

where $\mathbf{0}$ is an $(N - r + 1) \times r$ matrix all of whose components are zero, $\mathbf{I}$ is an $(N - r + 1) \times (N - r + 1)$ identity matrix, and $Q_{ij} = P_{ij}$ for $0 \leq i, j < r$.

To illustrate the calculations, begin with the four-state transition matrix

$$
\mathbf{P} = \begin{array}{c} \\ 0 \\ 1 \\ 2 \\ 3 \end{array} \begin{array}{cccc} 0 & 1 & 2 & 3 \\ \left\| \begin{array}{cccc} Q_{00} & Q_{01} & R_{02} & R_{03} \\ Q_{10} & Q_{11} & R_{12} & R_{13} \\ 0 & 0 & 1 & 0 \\ 0 & 0 & 0 & 1 \end{array} \right\| \end{array} .
\tag{3.77}
$$

Straightforward matrix multiplication shows the square of $\mathbf{P}$ to be

$$
\mathbf{P}^2 = \left\| \begin{array}{cc} \mathbf{Q}^2 & \mathbf{R} + \mathbf{QR} \\ \mathbf{0} & \mathbf{I} \end{array} \right\| .
\tag{3.78}
$$

Continuing on to the third power, we have

$$
\mathbf{P}^3 = \left\| \begin{array}{cc} \mathbf{Q} & \mathbf{R} \\ \mathbf{0} & \mathbf{I} \end{array} \right\| \times \left\| \begin{array}{cc} \mathbf{Q}^2 & \mathbf{R} + \mathbf{QR} \\ \mathbf{0} & \mathbf{I} \end{array} \right\| = \left\| \begin{array}{cc} \mathbf{Q}^3 & \mathbf{R} + \mathbf{QR} + \mathbf{Q}^2\mathbf{R} \\ \mathbf{0} & \mathbf{I} \end{array} \right\| ,
$$

and for higher values of n,

$$
\mathbf{P}^n = \left\| \begin{array}{cc} \mathbf{Q}^n & \left(\mathbf{I} + \mathbf{Q} + \cdots + \mathbf{Q}^{n-1} \right) \mathbf{R} \\ \mathbf{0} & \mathbf{I} \end{array} \right\| .
\tag{3.79}
$$

The consideration of four states was for typographical convenience only. It is straightforward to verify that the nth power of $\mathbf{P}$ is given by (3.79) for the general $(N + 1)$-state transition matrix of (3.76) in which states $0, 1, \ldots, r - 1$ are *transient* ($P_{ij}^{(n)} \to 0$ as $n \to \infty$ for $0 \leq i, j < r$) while states $r, \ldots, N$ are *absorbing* ($P_{ii} = 1$ for $r \leq i \leq N$).

We turn to the interpretation of (3.79). Let $W_{ij}^{(n)}$ be the mean number of visits to state j up to stage n for a Markov chain starting in state i. Formally,

$$
W_{ij}^{(n)} = E \left[\sum_{l=0}^{n} \mathbf{1}\{X_l = j\} | X_0 = i \right],
\tag{3.80}
$$

where

$$
\mathbf{1}\{X_l = j\} = \begin{cases} 1 & \text{if } X_l = j, \\ 0 & \text{if } X_l \neq j. \end{cases}
\tag{3.81}
$$

Now, $E[1\{X_l = j\}|X_0 = i] = \Pr\{X_l = j|X_0 = i\} = P_{ij}^{(l)}$, and since the expected value of a sum is the sum of the expected values, we obtain from (3.80) that

$$W_{ij}^{(n)} = \sum_{i=0}^{n} E[1\{X_l = j\}|X_0 = i]$$

$$= \sum_{l=0}^{n} P_{ij}^{(l)}.$$

(3.82)

Equation (3.82) holds for all states i, j, but it has the most meaning when i and j are transient. Because (3.79) asserts that $P_{ij}^{(l)} = Q_{ij}^{(l)}$ when $0 \leq i, j < r$, then

$$W_{ij}^{(n)} = Q_{ij}^{(0)} + Q_{ij}^{(1)} + \cdots + Q_{ij}^{(n)}, \quad 0 \leq i, j < r,$$

where

$$Q_{ij}^{(0)} = \begin{cases} 1 & \text{if } i = j, \\ 0 & \text{if } i \neq j. \end{cases}$$

In matrix notation, $\mathbf{Q}^{(0)} = \mathbf{I}$, and because $\mathbf{Q}^{(n)} = \mathbf{Q}^n$, the nth power of $\mathbf{Q}$, then

$$\begin{aligned} \mathbf{W}^{(n)} &= \mathbf{I} + \mathbf{Q} + \mathbf{Q}^2 + \cdots + \mathbf{Q}^n \\ &= \mathbf{I} + \mathbf{Q}\left(\mathbf{I} + \mathbf{Q} + \cdots + \mathbf{Q}^{n-1}\right) \\ &= \mathbf{I} + \mathbf{Q}\mathbf{W}^{(n-1)}. \end{aligned}$$

(3.83)

Upon writing out the matrix equation (3.83) in terms of the matrix entries, we recognize the results of a first step analysis. We have

$$\begin{aligned} W_{ij}^{(n)} &= \delta_{ij} + \sum_{k=0}^{r-1} Q_{ik} W_{kj}^{(n-1)} \\ &= \delta_{ij} + \sum_{k=0}^{r-1} P_{ik} W_{kj}^{(n-1)}. \end{aligned}$$

In words, the equation asserts that the mean number of visits to state j in the first n stages starting from the initial stage i includes the initial visit if $i = j (\delta_{ij})$ plus the future visits during the $n - 1$ remaining stages weighted by the appropriate transition probabilities.

We pass to the limit in (3.83) and obtain for

$$W_{ij} = \lim_{n \to \infty} W_{ij}^{(n)} = E[\text{Total visits to } j|X_0 = i], \quad 0 \leq i, j < r,$$

the matrix equations

$$W = I + Q + Q^2 + \cdots$$

and

$$W = I + QW. \tag{3.84}$$

In terms of its entries, (3.84) is

$$W_{ij} = \delta_{ij} + \sum_{l=0}^{r-1} P_{il} W_{lj} \quad \text{for } i, j = 0, \ldots, r-1. \tag{3.85}$$

Equation (3.85) is the same as equation (3.29), which was derived by a first step analysis.

Rewriting equation (3.84) in the form

$$W - QW = (I - Q)W = I, \tag{3.86}$$

we see that $W = (I - Q)^{-1}$, the inverse matrix to $I - Q$. The matrix W is often called the *fundamental* matrix associated with Q.

Let T be the time of absorption. Formally, since states $r, r+1, \ldots, N$ are the absorbing ones, the definition is

$$T = \min\{n \geq 0; r \leq X_n \leq N\}.$$

Then, the (i,j)th element W_{ij} of the fundamental matrix W evaluates

$$W_{ij} = E\left[\sum_{n=0}^{T-1} \mathbf{1}\{X = j\} | X_0 = i\right] \quad \text{for } 0 \leq i, j < r. \tag{3.87}$$

Let $v_i = E[T|X_0 = i]$ be the mean time to absorption starting from state i. The time to absorption is composed of sojourns in the transient states. Formally,

$$\sum_{j=0}^{r-1} \sum_{n=0}^{T-1} \mathbf{1}\{X_n = j\} = \sum_{n=0}^{T-1} \sum_{j=0}^{r-1} \mathbf{1}\{X_n = j\}$$

$$= \sum_{n=0}^{T-1} 1 = T.$$

It follows from (3.87), then, that

$$\sum_{j=0}^{r-1} W_{ij} = \sum_{j=0}^{r-1} E\left[\sum_{n=0}^{T-1} \mathbf{1}\{X_n = j\} | X_0 = i\right] \tag{3.88}$$

$$= E[T|X_0 = i] = v_i \quad \text{for } 0 \leq i < r.$$

Summing equation (3.85) over transient states j as follows,

$$\sum_{j=0}^{r-1} W_{ij} = \sum_{j=0}^{r-1} \delta_{ij} + \sum_{j=0}^{r-1}\sum_{k=0}^{r-1} P_{ij}W_{kj} \quad \text{for } i = 0, 1, \ldots, r-1,$$

and using the equivalence $v_i = \sum_{j=0}^{r-1} W_{ij}$ leads to

$$v_i = 1 + \sum_{k=0}^{r-1} P_{ij}v_k \quad \text{for } i = 0, 1, \ldots, r-1. \tag{3.89}$$

This equation is identical with that derived by first step analysis in (3.28). We turn to the hitting probabilities. Recall that states $k = r, \ldots, N$ are absorbing. Since such a state cannot be left once entered, the probability of absorption in a particular absorbing state k up to time n, starting from initial state i, is simply

$$\begin{aligned}
P_{ik}^{(n)} &= \Pr\{X_n = k|X_0 = i\} \\
&= \Pr\{T \le n \text{ and } X_T = k|X_0 = i\} \\
&\quad \text{for } i = 0, \ldots, r-1; k = r, \ldots, N,
\end{aligned} \tag{3.90}$$

where $T = \min\{n \ge 0 : r \le X_n \le N\}$ is the time of absorption. Let

$$\begin{aligned}
U_{ik}^{(n)} &= \Pr\{T \le n \text{ and } X_T = k|X_0 = i\} \\
&\quad \text{for } 0 \le i < r \text{ and } r \le k \le N.
\end{aligned} \tag{3.91}$$

Referring to (3.79) and (3.90), we give the matrix $\mathbf{U}^{(n)}$ by

$$\begin{aligned}
\mathbf{U}^{(n)} &= \left(\mathbf{I} + \mathbf{Q} + \cdots + \mathbf{Q}^{n-1}\right)\mathbf{R} \\
&= \mathbf{W}^{(n-1)}\mathbf{R} \quad \text{[by (3.83)]}.
\end{aligned} \tag{3.92}$$

If we pass to the limit in n, we obtain the hitting probabilities

$$U_{ik} = \lim_{n\to\infty} U_{ik}^{(n)} = \Pr\{X_T = k|X_0 = i\} \quad \text{for } 0 \le i < r \text{ and } r \le k \le N.$$

Equation (3.92) then leads to an expression of the hitting probability matrix $\mathbf{U}$ in terms of the fundamental matrix $\mathbf{W}$ as simply $\mathbf{U} = \mathbf{WR}$, or

$$U_{ik} = \sum_{j=0}^{r-1} W_{ij}R_{jk} \quad \text{for } 0 \le i < r \text{ and } r \le k \le N. \tag{3.93}$$

Equation (3.93) may be used in conjunction with (3.85) to verify the first step analysis equation for U_{ik}. We multiply (3.85) by R_{jk} and sum, obtaining thereby

$$\sum_{j=0}^{r-1} W_{ij}R_{jk} = \sum_{j=0}^{r-1} \delta_{ij}R_{jk} + \sum_{j=0}^{r-1}\sum_{l=0}^{r-1} P_{il}W_{lj}R_{jk},$$

which with (3.93) gives

$$U_{ik} = R_{ik} + \sum_{l=0}^{r-1} P_{il}U_{lk}$$

$$= P_{ik} + \sum_{l=0}^{r-1} P_{il}U_{lk} \quad \text{for } 0 \le i < r \text{ and } r \le k \le N.$$

This equation was derived earlier by first step analysis in (3.26).

Exercises

3.7.1 Consider the Markov chain whose transition probability matrix is given by

$$\mathbf{P} = \begin{array}{c} \\ 0 \\ 1 \\ 2 \\ 3 \end{array} \begin{array}{cccc} 0 & 1 & 2 & 3 \\ \left\| \begin{array}{cccc} 1 & 0 & 0 & 0 \\ 0.1 & 0.2 & 0.5 & 0.2 \\ 0.1 & 0.2 & 0.6 & 0.1 \\ 0 & 0 & 0 & 1 \end{array} \right\| \end{array}.$$

The transition probability matrix corresponding to the nonabsorbing states is

$$\mathbf{Q} = \begin{array}{c} \\ 0 \\ 1 \end{array} \begin{array}{cc} 1 & 2 \\ \left\| \begin{array}{cc} 0.2 & 0.5 \\ 0.2 & 0.6 \end{array} \right\| \end{array}.$$

Calculate the matrix inverse to $\mathbf{I} - \mathbf{Q}$, and from this determine
(a) the probability of absorption into state 0 starting from state 1;
(b) the mean time spent in each of states 1 and 2 prior to absorption.

3.7.2 Consider the random walk Markov chain whose transition probability matrix is given by

$$\mathbf{P} = \begin{array}{c} \\ 0 \\ 1 \\ 2 \\ 3 \end{array} \begin{array}{cccc} 0 & 1 & 2 & 3 \\ \left\| \begin{array}{cccc} 1 & 0 & 0 & 0 \\ 0.3 & 0 & 0.7 & 0 \\ 0 & 0.3 & 0 & 0.7 \\ 0 & 0 & 0 & 1 \end{array} \right\| \end{array}.$$

The transition probability matrix corresponding to the nonabsorbing states is

$$\mathbf{Q} = \begin{matrix} & \begin{matrix} 1 & \ \ 2 \end{matrix} \\ \begin{matrix} 0 \\ 1 \end{matrix} & \left\| \begin{matrix} 0 & 0.7 \\ 0.3 & 0 \end{matrix} \right\| \end{matrix}.$$

Calculate the matrix inverse to $\mathbf{I} - \mathbf{Q}$, and from this determine
(a) the probability of absorption into state 0 starting from state 1;
(b) the mean time spent in each of states 1 and 2 prior to absorption.

Problems

3.7.1 A zero-seeking device operates as follows: If it is in state m at time n, then at time $n+1$ its position is uniformly distributed over the states $0, 1, \ldots, m-1$. State 0 is absorbing. Find the inverse of the $\mathbf{I} - \mathbf{Q}$ matrix for the transient states $1, 2, \ldots, m$.

3.7.2 A zero-seeking device operates as follows: If it is in state j at time n, then at time $n+1$ its position is 0 with probability $1/j$, and its position is k (where k is one of the states $1, 2, \ldots, j-1$) with probability $2k/j^2$. State 0 is absorbing. Find the inverse of the $\mathbf{I} - \mathbf{Q}$ matrix.

3.7.3 Let X_n be an absorbing Markov chain whose transition probability matrix takes the form given in equation (3.76). Let $\mathbf{W}$ be the fundamental matrix, the matrix inverse of $\mathbf{I} - \mathbf{Q}$. Let

$$T = \min\{n \geq 0; r \leq n \leq N\}$$

be the random time of absorption (recall that states $r, r+1, \ldots, N$ are the absorbing states). Establish the joint distribution

$$\Pr\{X_{T-1} = j, X_T = k | X_0 = i\} = W_{ij} P_{jk} \quad \text{for } 0 \leq i, j < r; r \leq k \leq N,$$

whence

$$\Pr\{X_{T-1} = j | X_0 = i\} = \sum_{k=r}^{N} W_{ij} P_{jk} \quad \text{for } 0 \leq i, j < r.$$

3.7.4 The possible states for a Markov chain are the integers $0, 1, \ldots, N$, and if the chain is in state j, at the next step it is equally likely to be in any of the states $0, 1, \ldots, j-1$. Formally,

$$P_{ij} = \begin{cases} 1, & \text{if } i = j = 0, \\ 0 & \text{if } 0 < i \leq j \leq N, \\ 1/i, & \text{if } 0 \leq j < i \leq N. \end{cases}$$

(a) Determine the fundamental matrix for the transient states $1, 2, \ldots, N$.

(b) Determine the probability distribution for the last positive integer that the chain visits.

3.7.5 *Computer Challenge.* Consider the partial sums:

$$S_0 = k \quad \text{and} \quad S_m = k + \xi_1 + \cdots + \xi_m, \; k > 0,$$

where $\xi_1, \xi_2, \ldots$ are independent and identically distributed as

$$\Pr\{\xi = 0\} = 1 - \frac{2}{\pi}$$

and

$$\Pr\{\xi = \pm j\} = \frac{2}{\pi(4j^2 - 1)}, j = 1, 2, \ldots.$$

Can you find an explicit formula for the mean time v_k for the partial sums starting from $S_0 = k$ to exit the interval $[0, N] = \{0, 1, \ldots, N\}$? In another context, the answer was found by computing it in a variety of special cases.

Note: A simple random walk *on the* integer plane moves according to the rule: If $(X_n, Y_n) = (i, j)$, then the next position is equally likely to be any of the four points $(i+1, j)$, $(i-1, j)$, $(i, j+1)$, or $(i, j-1)$. Let us suppose that the process starts at the point $(X_0, Y_0) = (k, k)$ on the diagonal, and we observe the process only when it visits the diagonal. Formally, we define

$$\tau_1 = \min\{n > 0; X_n = Y_n\},$$

and

$$\tau_m = \min\{n > \tau_{m-1}; X_n = Y_n\}.$$

It is not hard to show that

$$S_0 = k, \; S_m = X_{\tau_m} = Y_{\tau_m}, \; m > 0,$$

is a version of the above partial sum process.

3.8 Branching Processes*

Suppose an organism at the end of its lifetime produces a random number ξ of offspring with probability distribution

$$\Pr\{\xi = k\} = p_k \quad \text{for } k = 0, 1, 2, \ldots, \tag{3.94}$$

* Branching processes are Markov chains of a special type. Sections 3.8 and 3.9 are not prerequisites to the later chapters.

where as usual, $p_k \geq 0$ and $\sum_{k=0}^{\infty} p_k = 1$. We assume that all offspring act independently of each other and at the ends of their lifetimes (for simplicity, the lifespans of all organisms are assumed to be the same) individually have progeny in accordance with the probability distribution (3.94), thus propagating their species. The process $\{X_n\}$, where X_n is the population size at the nth generation, is a Markov chain of special structure called a *branching process*.

The Markov property may be reasoned simply as follows. In the nth generation, the X_n individuals independently give rise to numbers of offspring $\xi_1^{(n)}, \xi_2^{(n)}, \ldots, \xi_{X_n}^{(n)}$, and hence the cumulative number produced for the $(n+1)$st generation is

$$X_{n+1} = \xi_1^{(n)} + \xi_2^{(n)} + \cdots + \xi_{X_n}^{(n)}. \tag{3.95}$$

3.8.1 Examples of Branching Processes

There are numerous examples of Markov branching processes that arise naturally in various scientific disciplines. We list some of the more prominent cases.

Electron Multipliers

An electron multiplier is a device that amplifies a weak current of electrons. A series of plates are set up in the path of electrons emitted by a source. Each electron, as it strikes the first plate, generates a random number of new electrons, which in turn strike the next plate and produce more electrons, and so forth. Let X_0 be the number of electrons initially emitted and X_1 be the number of electrons produced on the first plate by the impact due to the X_0 initial electrons; in general, let X_n be the number of electrons emitted from the nth plate due to electrons emanating from the $(n-1)$st plate. The sequence of random variables $X_0, X_1, X_2, \ldots, X_n, \ldots$ constitutes a branching process.

Neutron Chain Reaction

A nucleus is split by a chance collision with a neutron. The resulting fission yields a random number of new neutrons. Each of these secondary neutrons may hit some other nucleus, producing a random number of additional neutrons, and so forth. In this case, the initial number of neutrons is $X_0 = 1$. The first generation of neutrons comprises all those produced from the fission caused by the initial neutron. The size of the first generation is a random variable X_1. In general, the population X_n at the nth generation is produced by the chance hits of the X_{n-1} individual neutrons of the $(n-1)$st generation.

Survival of Family Names

The family name is inherited by sons only. Suppose that each individual has probability p_k of having k male offspring. Then, from one individual there result the 1st, 2nd, $\ldots$, nth, $\ldots$ generations of descendants. We may investigate the distribution of such random variables as the number of descendants in the nth generation, or the probability that the family name will eventually become extinct. Such questions will be dealt with beginning in Section 3.8.3.

Survival of Mutant Genes

Each individual gene has a chance to give birth to k offspring, $k = 1, 2, \ldots$, which are genes of the same kind. Any individual, however, has a chance to transform into a different type of mutant gene. This gene may become the first in a sequence of generations of a particular mutant gene. We may inquire about the chances of survival of the mutant gene within the population of the original genes. In this example, the number of offspring is often assumed to follow a Poisson distribution.

The rationale behind this choice of distribution is as follows. In many populations a large number of zygotes (fertilized eggs) are produced, only a small number of which grow to maturity. The events of fertilization and maturation of different zygotes obey the law of independent binomial trials. The number of trials (i.e., number zygotes) is large. The law of rare events then implies that the number of progeny that mature will approximately follow the Poisson distribution. The Poisson assumption seems quite appropriate in the model of population growth of a rare mutant gene. If the mutant gene carries a biological advantage (or disadvantage), then the probability distribution is taken to be the Poisson distribution with mean $\lambda > 1$ or (< 1).

All of the preceding examples possess the following structure. Let X_0 denote the size of the initial population. Each individual gives birth to k new individuals with probability p_k *independently of the others*. The totality of all the direct descendants of the initial population constitutes the first generation, whose size we denote by X_1. Each individual of the first generation independently bears a progeny set whose size is governed by the probability distribution (3.94). The descendants produced constitute the second generation, of size X_2. In general, the nth generation is composed of descendants of the $(n-1)$st generation, each of whose members independently produces k progeny with probability $p_k, k = 0, 1, 2, \ldots$. The population size of the nth generation is denoted by X_n. The X_n forms a sequence of integer-valued random variables that generate a Markov chain in the manner described by (3.95).

3.8.2 The Mean and Variance of a Branching Process

Equation (3.95) characterizes the evolution of the branching process as successive random sums of random variables. Random sums were studied in Chapter 2, Section 2.3, and we can use the moment formulas developed there to compute the mean and variance of the population size X_n. First some notation. Let $\mu = E[\xi]$ and $\sigma^2 = \text{Var}[\xi]$ be the mean and variance, respectively, of the offspring distribution (3.94). Let $M(n)$ and $V(n)$ be the mean and variance of X_n under the initial condition $X_0 = 1$. Then, direct application of Chapter 2, (2.30) with respect to the random sum (3.95) gives the recursions

$$M(n+1) = \mu M(n) \tag{3.96}$$

and

$$V(n+1) = \sigma^2 M(n) + \mu^2 V(n). \tag{3.97}$$

The initial condition $X_0 = 1$ starts the recursions (3.96) and (3.97) at $M(0) = 1$ and $V(0) = 0$. Then, from (3.96), we obtain $M(1) = \mu 1 = \mu, M(2) = \mu M(1) = \mu^2$, and, in general,

$$M(n) = \mu^n \quad \text{for } n = 0, 1, \ldots. \tag{3.98}$$

Thus, the mean population size increases geometrically when $\mu > 1$, decreases geometrically when $\mu < 1$, and remains constant when $\mu = 1$.

Next, substitution of $M(n) = \mu^n$ into (3.97) gives $V(n+1) = \sigma^2 \mu^n + \mu^2 V(n)$, which with $V(0) = 0$ yields

$$V(1) = \sigma^2,$$

$$V(2) = \sigma^2 \mu + \mu^2 V(1) = \sigma^2 \mu + \sigma^2 \mu^2,$$

$$V(3) = \sigma^2 \mu^2 + \mu^2 V(2)$$
$$= \sigma^2 \mu^2 + \sigma^2 \mu^3 + \sigma^2 \mu^4,$$

and, in general,

$$V(n) = \sigma^2 \left[\mu^{n-1} + \mu^n + \cdots + \mu^{2n-2} \right]$$
$$= \sigma^2 \mu^{n-1} \left[1 + \mu + \cdots + \mu^{n-1} \right] \tag{3.99}$$
$$= \sigma^2 \mu^{n-1} \times \begin{cases} n & \text{if } \mu = 1, \\ \dfrac{1 - \mu^n}{1 - \mu} & \text{if } \mu \neq 1. \end{cases}$$

Thus, the variance of the population size increases geometrically if $\mu > 1$, increases linearly if $\mu = 1$, and decreases geometrically if $\mu < 1$.

3.8.3 Extinction Probabilities

Population extinction occurs when and if the population size is reduced to zero. The random time of extinction N is thus the first time n for which $X_n = 0$, and then, obviously, $X_k = 0$ for all $k \geq N$. In Markov chain terminology, 0 is an absorbing state, and we may calculate the probability of extinction by invoking a first step analysis. Let

$$u_n = \Pr\{N \leq n\} = \Pr\{X_n = 0\} \tag{3.100}$$

be the probability of extinction at or prior to the nth generation, beginning with a single parent $X_0 = 1$. Suppose that the single parent represented by $X_0 = 1$ gives rise to $\xi_1^{(0)} = k$ offspring. In turn, each of these offspring will generate a population of its own descendants, and if the original population is to die out in n generations, then each of these k lines of descent must die out in $n - 1$ generations. The analysis is depicted in Figure 3.6.

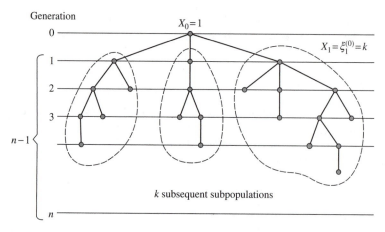

Figure 3.6 The diagram illustrates that if the original population is to die out by generation n, then the subpopulations generated by distinct initial offspring must all die out in $n-1$ generations.

Now, the k subpopulations generated by the distinct offspring of the original parent are independent, and they have the same statistical properties as the original population. Therefore, the probability that any particular one of them dies out in $n-1$ generations is u_{n-1} by definition, and the probability that all k subpopulations die out in $n-1$ generations is the kth power $(u_{n-1})^k$ because they are independent. Upon weighting this factor by the probability of k offspring and summing according to the law of total probability, we obtain

$$u_n = \sum_{k=0}^{\infty} p_k(u_{n-1})^k, \quad n = 1, 2, \ldots. \tag{3.101}$$

Of course $u_0 = 0$, and $u_1 = p_0$, the probability that the original parent had no offspring.

Example Suppose a parent has no offspring with probability $\frac{1}{4}$ and two offspring with probability $\frac{3}{4}$. Then, the recursion (3.101) specializes to

$$u_n = \frac{1}{4} + \frac{3}{4}(u_{n-1})^2 = \frac{1 + 3(u_{n-1})^2}{4}.$$

Beginning with $u_0 = 0$, we successively compute

$$
\begin{aligned}
u_1 &= 0.2500, & u_6 &= 0.3313, \\
u_2 &= 0.2969, & u_7 &= 0.3323, \\
u_3 &= 0.3161, & u_8 &= 0.3328, \\
u_4 &= 0.3249, & u_9 &= 0.3331, \\
u_5 &= 0.3292, & u_{10} &= 0.3332.
\end{aligned}
$$

We see that the chances are very nearly $\frac{1}{3}$ that such a population will die out by the tenth generation.

Exercises

3.8.1 A population begins with a single individual. In each generation, each individual in the population dies with probability $\frac{1}{2}$ or doubles with probability $\frac{1}{2}$. Let X_n denote the number of individuals in the population in the nth generation. Find the mean and variance of X_n.

3.8.2 The number of offspring of an individual in a population is 0, 1, or 2 with respective probabilities $a > 0$, $b > 0$, and $c > 0$, where $a + b + c = 1$. Express the mean and variance of the offspring distribution in terms of b and c.

3.8.3 Suppose a parent has no offspring with probability $\frac{1}{2}$ and has two offspring with probability $\frac{1}{2}$. If a population of such individuals begins with a single parent and evolves as a branching process, determine u_n, the probability that the population is extinct by the nth generation, for $n = 1, 2, 3, 4, 5$.

3.8.4 At each stage of an electron multiplier, each electron, upon striking the plate, generates a Poisson distributed number of electrons for the next stage. Suppose the mean of the Poisson distribution is λ. Determine the mean and variance for the number of electrons in the nth stage.

Problems

3.8.1 Each adult individual in a population produces a fixed number M of offspring and then dies. A fixed number L of these remain at the location of the parent. These local offspring will either all grow to adulthood, which occurs with a fixed probability β, or all will die, which has probability $1 - \beta$. Local mortality is catastrophic in that it affects the entire local population. The remaining $N = M - L$ offspring disperse. Their successful growth to adulthood will occur statistically independently of one another, but at a lower probability $\alpha = p\beta$, where p may be thought of as the probability of successfully surviving the dispersal process. Define the random variable ξ to be the number of offspring of a single parent that survive to reach adulthood in the next generation. According to our assumptions, we may write ξ as

$$\xi = v_1 + v_2 + \cdots + v_N + (M - N)\Theta,$$

where $\Theta, v_1, v_2, \ldots, v_N$ are independent with $\Pr\{v_k = 1\} = \alpha$, $\Pr\{v_k = 0\} = 1 - \alpha$, and with $\Pr\{\Theta = 1\} = \beta$ and $\Pr\{\Theta = 0\} = 1 - \beta$. Show that the mean number of offspring reaching adulthood is $E[\xi] = \alpha N + \beta(M - N)$, and since $\alpha < \beta$, the mean number of surviving offspring is maximized by dispersing none ($N = 0$). Show that the probability of having no offspring surviving to adulthood is

$$\Pr\{\xi = 0\} = (1 - \alpha)^N (1 - \beta)$$

and that this probability is made smallest by making N large.

3.8.2 Let $Z = \sum_{n=0}^{x} X_n$ be the total family size in a branching process whose offspring distribution has a mean $\mu = E[\xi] < 1$. Assuming that $X_0 = 1$, show that $E[Z] = 1/(1-\mu)$.

3.8.3 Families in a certain society choose the number of children that they will have according to the following rule: If the first child is a girl, they have exactly one more child. If the first child is a boy, they continue to have children until the first girl, and then cease childbearing.

   (a) For $k = 0, 1, 2, \ldots$, what is the probability that a particular family will have k children in total?

   (b) For $k = 0, 1, 2, \ldots$, what is the probability that a particular family will have exactly k male children among their offspring?

3.8.4 Let $\{X_n\}$ be a branching process with mean family size μ. Show that $Z_n = X_n/\mu^n$ is a nonnegative martingale. Interpret the maximal inequality as applied to $\{Z_n\}$.

3.9   Branching Processes and Generating Functions*

Consider a nonnegative integer-valued random variable ξ whose probability distribution is given by

$$\Pr\{\xi = k\} = p_k \quad \text{for } k = 0, 1, \ldots. \tag{3.102}$$

The *generating function* $\phi(s)$ associated with the random variable ξ (or equivalently, with the distribution $\{p_k\}$) is defined by

$$\phi(s) = E\left[s^{\xi}\right] = \sum_{k=0}^{\infty} p_k s^k \quad \text{for } 0 \leq s \leq 1. \tag{3.103}$$

Much of the importance of generating functions derives from the following three results.

First, the relation between probability mass functions (3.102) and generating functions (3.103) is one-to-one. Thus, knowing the generating function is equivalent, in some sense, to knowing the distribution. The relation that expresses the probability mass function $\{p_k\}$ in terms of the generating function $\phi(s)$ is

$$p_k = \frac{1}{k!} \frac{d^k \phi(s)}{ds^k} \bigg|_{s=0}. \tag{3.104}$$

For example,

$$\phi(s) = p_0 + p_1 s + p_2 s^2 + \cdots,$$

---

* This topic is not prerequisite to what follows.

whence

$$p_0 = \phi(0),$$

and

$$\frac{d\phi(s)}{ds} = p_1 + 2p_2 s + 3p_3 s^2 + \cdots,$$

whence

$$p_1 = \left.\frac{d\phi(s)}{ds}\right|_{s=0}.$$

Second, if $\xi_1, \ldots, \xi_n$ are independent random variables having generating functions $\phi_1(s), \ldots, \phi_n(s)$, respectively, then the generating function of their sum $X = \xi_1 + \cdots + \xi_n$ is simply the product

$$\phi_X(s) = \phi_1(s)\phi_2(s)\cdots\phi_n(s). \tag{3.105}$$

This simple result makes generating functions extremely helpful in dealing with problems involving sums of independent random variables. It is to be expected, then, that generating functions might provide a major tool in the analysis of branching processes.

Third, the moments of a nonnegative integer-valued random variable may be found by differentiating the generating function. For example, the first derivative is

$$\frac{d\phi(s)}{ds} = p_1 + 2p_2 s + 3p_3 s^2 + \cdots,$$

whence

$$\left.\frac{d\phi(s)}{ds}\right|_{s=1} = p_1 + 2p_2 + 3p_3 + \cdots = E[\xi], \tag{3.106}$$

and the second derivative is

$$\frac{d^2\phi(s)}{ds^2} = 2p_2 + 3(2)p_3 s + 4(3)p_4 s^2 + \cdots,$$

whence

$$\left.\frac{d^2\phi(s)}{ds^2}\right|_{s=1} = 2p_2 + 3(2)p_3 + 4(3)p_4 + \cdots$$

$$= \sum_{k=2}^{\infty} k(k-1)p_k = E[\xi(\xi-1)]$$

$$= E[\xi^2] - E[\xi].$$

Thus

$$E\left[\xi^2\right] = \frac{d^2\phi(s)}{ds^2}\bigg|_{s=1} + E[\xi]$$

$$= \frac{d^2\phi(s)}{ds^2}\bigg|_{s=1} + \frac{d\phi(s)}{ds}\bigg|_{s=1}$$

and

$$\text{Var}[\xi] = E\left[\xi^2\right] - \{E[\xi]\}^2$$

$$= \frac{d^2\phi(s)}{ds^2}\bigg|_{s=1} + \frac{d\phi(s)}{ds}\bigg|_{s=1} - \left\{\frac{d\phi(s)}{ds}\bigg|_{s=1}\right\}^2. \tag{3.107}$$

Example If ξ has a Poisson distribution with mean λ for which

$$p_k = \Pr\{\xi = k\} = \frac{\lambda^k e^{-\lambda}}{k!} \quad \text{for } k = 0, 1, \ldots,$$

then,

$$\phi(s) = E\left[s^\xi\right] = \sum_{k=0}^{\infty} s^k \frac{\lambda^k e^{-\lambda}}{k!}$$

$$= e^{-\lambda} \sum_{k=0}^{\infty} \frac{(\lambda s)^k}{k!}$$

$$= e^{-\lambda} e^{\lambda s} = e^{-\lambda(1-s)} \quad \text{for } |s| < 1.$$

Then,

$$\frac{d\phi(s)}{ds} = \lambda e^{-\lambda(1-s)}; \qquad \frac{d\phi(s)}{ds}\bigg|_{s=1} = \lambda;$$

$$\frac{d^2\phi(s)}{ds^2} = \lambda^2 e^{-\lambda(1-s)}; \qquad \frac{d^2\phi(s)}{ds^2}\bigg|_{s=1} = \lambda^2.$$

From (3.106) and (3.107), we verify that

$$E[\xi] = \lambda,$$
$$\text{Var}[\xi] = \lambda^2 + \lambda - (\lambda)^2 = \lambda.$$

3.9.1 Generating Functions and Extinction Probabilities

Consider a branching process whose population size at stage n is denoted by X_n. Assume that the offspring distribution $p_k = \Pr\{\xi = k\}$ has the generating function

$\phi(s) = E\left[s^{\xi}\right] = \sum_k s^k p_k$. If $u_n = \Pr\{X_n = 0\}$ is the probability of extinction by stage n, then the recursion (3.101) in terms of generating functions becomes

$$u_n = \sum_{k=0}^{\infty} p_k (u_{n-1})^k = \phi(u_{n-1}).$$

That is, knowing the generating function $\phi(s)$, we may successively compute the extinction probabilities u_n beginning with $u_0 = 0$ and then $u_1 = \phi(u_0), u_2 = \phi(u_1)$, and so on.

Example The extinction probabilities when there are no offspring with probability $p_0 = \frac{1}{4}$ and two offspring with probability $p_2 = \frac{3}{4}$ were computed in the example in Section 3.8.3. We now reexamine this example using the offspring generating function $\phi(s) = \frac{1}{4} + \frac{3}{4}s^2$. This generating function is plotted as Figure 3.7. From the figure, it is clear that the extinction probabilities converge upward to the smallest solution of the equation $u = \phi(u)$. This, in fact, occurs in the most general case. If u_∞ denotes this smallest solution to $u = \phi(u)$, then u_∞ gives the probability that the population eventually becomes extinct at some indefinite, but finite, time. The alternative is that the population grows infinitely large, and this occurs with probability $1 - u_\infty$.

For the example at hand, $\phi(s) = \frac{1}{4} + \frac{3}{4}s^2$, and the equation $u = \phi(u)$ is the simple quadratic $u = \frac{1}{4} + \frac{3}{4}u^2$, which gives

$$u = \frac{4 \pm \sqrt{16 - 12}}{6} = 1, \frac{1}{3}.$$

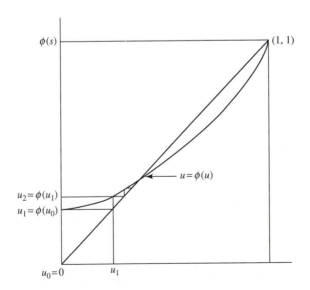

Figure 3.7 The generating function corresponding to the offspring distribution $p_0 = \frac{1}{4}$ and $p_2 = \frac{3}{4}$. Here $u_k = \Pr\{X_k = 0\}$ is the probability of extinction by generation k.

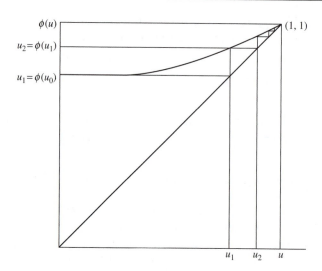

Figure 3.8 The generating function corresponding to the offspring distribution $p_0 = \frac{3}{4}$ and $p_2 = \frac{1}{4}$.

The smaller solution is $u_\infty = \frac{1}{3}$, which is to be compared with the apparent limit of the sequence u_n computed in the example in Section 3.8.3.

It may happen that $u_\infty = 1$, i.e., the population is sure to die out at some time. An example is depicted in Figure 3.8: The offspring distribution is $p_0 = \frac{3}{4}$ and $p_2 = \frac{1}{4}$. We solve $u = \phi(u) = \frac{3}{4} + \frac{1}{4}u^2$ to obtain

$$u = \frac{4 \pm \sqrt{16-12}}{2} = 1, 3.$$

The smaller solution is $u_\infty = 1$, the probability of eventual extinction.

In general, the key is whether or not the generating function $\phi(s)$ crosses the $45°$ line $\phi(s) = s$, and this, in turn, can be determined from the slope

$$\phi'(1) = \frac{d\phi(s)}{ds}\bigg|_{s=1}$$

of the generating function at $s = 1$. If this slope is less than or equal to one, then no crossing takes place, and the probability of eventual extinction is $u_\infty = 1$. On the other hand, if the slope $\phi'(1)$ exceeds one, then the equation $u = \phi(u)$ has a smaller solution that is less than one, and extinction is not a certain event.

But the slope $\phi'(1)$ of a generating function at $s = 1$ is the mean $E[\xi]$ of the corresponding distribution. We have thus arrived at the following important conclusion: If the mean offspring size $E[\xi] \leq 1$, then $u_\infty = 1$ and extinction is certain. If $E[\xi] > 1$, then $u_\infty < 1$ and the population may grow unboundedly with positive probability.

The borderline case $E[\xi] = 1$ merits some special attention. Here, $E[X_n|X_0=1]=1$ for all n, so the mean population size is constant. Yet the population is sure to die out eventually! This is a simple example in which the mean population size alone does not adequately describe the population behavior.

3.9.2 Probability Generating Functions and Sums of Independent Random Variables

Let ξ and η be independent nonnegative integer-valued random variables having the probability generating functions (p.g.f.s)

$$\phi(s) = E[s^\xi] \quad \text{and} \quad \psi(s) = E[s^\eta] \quad \text{for } |s| < 1.$$

The probability generating function of the sum $\xi + \eta$ is simply the product $\phi(s)\psi(s)$ because

$$\begin{aligned} E[s^{\xi+\eta}] &= E[s^\xi s^\eta] \\ &= E[s^\xi]E[s^\eta] \quad \text{(because ξ and η are independent)} \\ &= \phi(s)\psi(s). \end{aligned} \tag{3.108}$$

The converse is also true. Specifically, if the product of the p.g.f.s of two independent random variables is a p.g.f. of a third random variable, then the third random variable equals (in distribution) the sum of the other two.

Let $\xi_1, \xi_2, \ldots$ be independent and identically distributed nonnegative integer-valued random variables with p.g.f. $\phi(s) = E[s^\xi]$. Direct induction of (3.108) implies that the sum $\xi_1 + \cdots + \xi_m$ has p.g.f.

$$E[s^{\xi_1+\cdots+\xi_m}] = [\phi(s)]^m. \tag{3.109}$$

We extend this result to determine the p.g.f. of a sum of a random number of independent summands. Accordingly, let N be a nonnegative integer-valued random variable, independent of $\xi_1, \xi_2, \ldots$, with p.g.f. $g_N(s) = E[s^N]$, and consider the random sum (see Chapter 2, Section 2.3).

$$X = \xi_1 + \cdots + \xi_N.$$

Let $h_X(s) = E[s^X]$ be the p.g.f. of X. We claim that $h_X(s)$ takes the simple form

$$h_X(s) = g_N[\phi(s)]. \tag{3.110}$$

To establish (3.110), consider

$$
\begin{aligned}
h_X(s) &= \sum_{k=0}^{\infty} \Pr\{X = k\} s^k \\
&= \sum_{k=0}^{\infty} \left(\sum_{n=0}^{\infty} \Pr\{X = k | N = n\} \Pr\{N = n\} \right) s^k \\
&= \sum_{k=0}^{\infty} \left(\sum_{n=0}^{\infty} \Pr\{\xi_1 + \cdots + \xi_n = k | N = n\} \Pr\{N = n\} \right) s^k \\
&= \sum_{k=0}^{\infty} \sum_{n=0}^{\infty} \Pr\{\xi_1 + \cdots + \xi_n = k\} \Pr\{N = n\} s^k
\end{aligned}
$$

[because N is independent of $\xi_1, \xi_2, \ldots$]

$$
\begin{aligned}
&= \sum_{n=0}^{\infty} \left(\sum_{k=0}^{\infty} \Pr\{\xi_1 + \cdots + \xi_n = k\} s^k \right) \Pr\{N = n\} \\
&= \sum_{n=0}^{\infty} \phi(s)^n \Pr\{N = n\} \quad \text{[using (3.109)]} \\
&= g_n[\phi(s)] \quad \text{[by the definition of } g_n(s)].
\end{aligned}
$$

With the aid of (3.110), the basic branching process equation

$$
X_{n+1} = \xi_1^{(n)} + \cdots + \xi_{X_n}^{(n)} \tag{3.111}
$$

can be expressed equivalently and succinctly by means of generating functions. To this end, let $\phi_n(s) = E[s^{X_n}]$ be the p.g.f. of the population size X_n at generation n, assuming that $X_0 = 1$. Then easily, $\phi_0(s) = E[s^1] = s$, and $\phi_1(s) = \phi(s) = E[s^\xi]$. To obtain the general expression, we apply (3.110) to (3.111) to yield

$$
\phi_{n+1}(s) = \phi_n[\phi(s)]. \tag{3.112}
$$

This expression may be iterated in the manner

$$
\begin{aligned}
\phi_{n+1}(s) &= \phi_{n-1}\{\phi[\phi(s)]\} \\
&= \underbrace{\phi\{\cdots \phi[\phi(s)]\}}_{(n+1) \text{ iterations}} \\
&= \phi[\phi_n(s)].
\end{aligned} \tag{3.113}
$$

That is, we obtain the generating function for the population size X_n at generation n, given that $X_0 = 1$, by repeated substitution in the probability generating function of the offspring distribution.

For general initial population sizes $X_0 = k$, the p.g.f. is

$$\sum_{j=0}^{\infty} \Pr\{X_n = j | X_0 = k\}s^j = [\phi_n(s)]^k, \tag{3.114}$$

exactly that of a sum of k independent lines of descents. From this perspective, the branching process evolves as the sum of k independent branching processes, one for each initial parent.

Example Let $\phi(s) = q + ps$, where $0 < p < 1$ and $p + q = 1$. The associated branching process is a pure death process. In each period, each individual dies with probability q and survives with probability p. The iterates $\phi_n(s)$ in this case are readily determined, e.g., $\phi_2(s) = q + p, (q + ps) = 1 - p^2 + p^2 s$, and generally, $\phi_n(s) = 1 - p^n + p^n s$. If we follow (3.114), the nth generation p.g.f. starting from an initial population size of k is $[\phi_n(s)]^k = [1 - p^n + p^n s]^k$.

The probability distribution of the time T to extinction may be determined from the p.g.f. as follows:

$$\begin{aligned}
\Pr\{T = n | X(0) = k\} &= \Pr\{X_n = 0 | X_0 = k\} - \Pr\{X_{n-1} = 0 | X_0 = k\} \\
&= [\phi_n(0)]^k - [\phi_{n-1}(0)]^k \\
&= \left(1 - p^n\right)^k - \left(1 - p^{n-1}\right)^k.
\end{aligned}$$

3.9.3 Multiple Branching Processes

Population growth processes often involve several life history phases (e.g., juvenile, reproductive adult, senescence) with different viability and behavioral patterns. We consider a number of examples of branching processes that take account of this characteristic.

For the first example, suppose that a mature individual produces offspring according to the p.g.f. $\phi(s)$. Consider a population of immature individuals, each of which grows to maturity with probability p and then reproduces independently of the status of the remaining members of the population. With probability $1 - p$, an immature individual will not attain maturity and thus will leave no descendants. With probability p, an individual will reach maturity and reproduce a number of offspring determined according to the p.g.f. $\phi(s)$. Therefore, the progeny size distribution (or equivalently the p.g.f.) of a typical immature individual taking account of both contingencies is

$$(1 - p) + p\phi(s). \tag{3.115}$$

If a census is taken of individuals at the adult (mature) stage, the aggregate number of mature individuals contributed by a mature individual will now have p.g.f.

$$\phi(1 - p + ps). \tag{3.116}$$

(The student should verify this finding.)

It is worth emphasis that the p.g.f.s (3.115) and (3.116) have the same mean $p\phi'(1)$ but generally not the same variance, the first being

$$p\left[\phi''(1)+\phi'(1)-(\phi'(1))^2\right]$$

as compared with

$$p^2\phi''(1)+p\phi'(1)-p^2(\phi'(1))^2.$$

Example A second example leading to (3.116), as opposed to (3.115), concerns the different forms of mortality that affect a population. We appraise the strength (*stability*) of a population as the probability of indefinite survivorship $= 1-$ probability of eventual extinction.

In the absence of mortality, the offspring number X of a single individual has the p.g.f. $\phi(s)$. Assume, consistent with the postulates of a branching process, that all offspring in the population behave independently governed by the same probability laws. Assume also an adult population of size $X = k$. We consider three types of mortality:

(a) *Mortality of Individuals* Let p be the probability of an offspring surviving to reproduce, independently of what happens to others. Thus, the contribution of each litter (family) to the adult population of the next generation has a binomial distribution with parameters (N,p), where N is the progeny size of the parent with p.g.f. $\phi(s)$. The p.g.f. of the adult numbers contributed by a single parent is, therefore, $\phi(q+ps), q = 1-p$, and for the population as a whole is

$$\psi_1(s) = [\phi(q+ps)]^k. \tag{3.117}$$

This type of mortality might reflect predation on adults.

(b) *Mortality of Litters* Independently of what happens to other litters, each litter survives with probability p and is wiped out with probability $q = 1-p$. That is, given an actual litter size ξ, the effective litter size is ξ with probability p, and 0 with probability q. The p.g.f. of adults in the following generation is accordingly

$$\psi_2(s) = [q+p\phi(s)]^k. \tag{3.118}$$

This type of mortality might reflect predation on juveniles or on nests and eggs in the case of birds.

(c) *Mortality of Generations* An entire generation survives with probability p and is wiped out with probability q. This type of mortality might represent environmental catastrophes (e.g., forest fire, flood). The p.g.f. of population size in the next generation in this case is

$$\psi_3(s) = q+p[\phi(s)]^k. \tag{3.119}$$

All the p.g.f.s (3.117) through (3.119) have the same mean but usually different variances.

It is interesting to assess the relative stability of these three models. That is, we need to compare the smallest positive roots of $\psi_i(s) = s, i = 1, 2, 3$, which we will denote by $s_i^*, i = 1, 2, 3$, respectively.

We will show by convexity analysis that

$$\psi_1(s) \leq \psi_2(s) \leq \psi_3(s).$$

A function $f(x)$ is convex in x if for every x_1 and x_2 and $0 < \lambda < 1$, then $f[\lambda x_1 + (1 - \lambda)x_2] \leq \lambda f(x_1) + (1 - \lambda)f(x_2)$. In particular, the function $\phi(s) = \sum_{k=0}^{\infty} p_k s^k$ for $0 < s < 1$ is convex in s, since for each positive integer k, $[(\lambda s_1) + (1 - \lambda)s_2]^k \leq \lambda s_1^k + (1 - \lambda)s_2^k$ for $0 < \lambda, s_1, s_2 < 1$. Now, $\psi_1(s) = [\phi(q + ps)]^k < [q\phi(1) + p\phi(s)]^k = [q + p\phi(s)]^k = \psi_2(s)$, and then $s_1^* < s_2^*$. Thus, the first model is more stable than the second model.

Observe further that due to the convexity of $f(x) = x^k, x > 0, \psi_2(s) = [p\phi(s) + q]^k < p[\phi(s)]^k + q \times 1^k = \psi_3(s)$, and thus $s_2^* < s_3^*$, implying that the second model is more stable than the third model. In conjunction we get the ordering $s_1^* < s_2^* < s_3^*$.

Exercises

3.9.1 Suppose that the offspring distribution is Poisson with mean $\lambda = 1.1$. Compute the extinction probabilities $u_n = \Pr\{X_n = 0 | X_0 = 1\}$ for $n = 0, 1, \ldots, 5$. What is u_∞, the probability of ultimate extinction?

3.9.2 Determine the probability generating function for the offspring distribution in which an individual either dies, with probability p_0, or is replaced by two progeny, with probability p_2, where $p_0 + p_2 = 1$.

3.9.3 Determine the probability generating function corresponding to the offspring distribution in which each individual produces 0 or N direct descendants, with probabilities p and q, respectively.

3.9.4 Let $\phi(s)$ be the generating function of an offspring random variable ξ. Let Z be a random variable whose distribution is that of ξ, but conditional on $\xi > 0$. That is,

$$\Pr\{Z = k\} = \Pr\{\xi = k | \xi > 0\} \quad \text{for } k = 1, 2, \ldots.$$

Express the generating function for Z in terms of ϕ.

Problems

3.9.1 One-fourth of the married couples in a far-off society have no children at all. The other three-fourths of couples have exactly three children, with each child equally likely to be a boy or a girl. What is the probability that the male line of descent of a particular husband will eventually die out?

3.9.2 One-fourth of the married couples in a far-off society have exactly three children. The other three-fourths of couples continue to have children until the first boy and then cease childbearing. Assume that each child is equally likely to be a boy or girl. What is the probability that the male line of descent of a particular husband will eventually die out?

3.9.3 Consider a large region consisting of many subareas. Each subarea contains a branching process that is characterized by a Poisson distribution with parameter λ. Assume, furthermore, that the value of λ varies with the subarea, and its distribution over the whole region is that of a gamma distribution. Formally, suppose that the offspring distribution is given by

$$\pi(k|\lambda) = \frac{e^{-\lambda}\lambda^k}{k!} \quad \text{for } k = 0, 1, \ldots,$$

where λ itself is a random variable having the density function

$$f(\lambda) = \frac{\theta^\alpha \lambda^{\alpha-1} e^{-\theta\lambda}}{\Gamma(\alpha)} \quad \text{for } \lambda > 0,$$

where θ and α are positive constants. Determine the marginal offspring distribution $p_k = \int \pi(k|\lambda)f(\lambda)d\lambda$.

Hint: Refer to the last example of Chapter 2, Section 2.4.

3.9.4 Let $\phi(s) = 1 - p(1-s)^\beta$, where p and β are constants with $0 < p, \beta < 1$. Prove that $\phi(s)$ is a probability generating function and that its iterates are

$$\phi_n(s) = 1 - p^{1+\beta+\cdots+\beta^{n-1}}(1-s)^{\beta^n} \quad \text{for } n = 1, 2, \ldots.$$

3.9.5 At time 0, a blood culture starts with one red cell. At the end of 1 min, the red cell dies and is replaced by one of the following combinations with the probabilities as indicated:

Two red cells	$\frac{1}{4}$
One red, One white	$\frac{2}{3}$
Two white	$\frac{1}{12}$

Each red cell lives for 1 min and gives birth to offspring in the same way as the parent cell. Each white cell lives for 1 min and dies without reproducing. Assume that individual cells behave independently.

(a) At time $n + \frac{1}{2}$ min after the culture begins, what is the probability that no white cells have yet appeared?

(b) What is the probability that the entire culture eventually dies out entirely?

3.9.6 Let $\phi(s) = as^2 + bs + c$, where a, b, c are positive and $\phi(1) = 1$. Assume that the probability of extinction is u_∞, where $0 < u_\infty < 1$. Prove that $u_\infty = c/a$.

3.9.7 Families in a certain society choose the number of children that they will have according to the following rule: If the first child is a girl, they have exactly one more child. If the first child is a boy, they continue to have children until the first girl and then cease childbearing. Let ξ be the number of male children in a particular family. What is the generating function of ξ? Determine the mean of ξ directly and by differentiating the generating function.

3.9.8 Consider a branching process whose offspring follow the geometric distribution $p_k = (1 - c)c^k$ for $k = 0, 1, \ldots$, where $0 < c < 1$. Determine the probability of eventual extinction.

3.9.9 One-fourth of the married couples in a distant society have no children at all. The other three-fourths of couples continue to have children until the first girl and then cease childbearing. Assume that each child is equally likely to be a boy or girl.

  (a) For $k = 0, 1, 2, \ldots$, what is the probability that a particular husband will have k male offspring?

  (b) What is the probability that the husband's male line of descent will cease to exist by the fifth generation?

3.9.10 Suppose that in a branching process the number of offspring of an initial particle has a distribution whose generating function is $f(s)$. Each member of the first generation has a number of offspring whose distribution has generating function $g(s)$. The next generation has generating function f, the next has g, and the distributions continue to alternate in this way from generation to generation.

  (a) Determine the extinction probability of the process in terms of $f(s)$ and $g(s)$.

  (b) Determine the mean population size at generation n.

  (c) Would any of these quantities change if the process started with the $g(s)$ process and then continued to alternate?

4 The Long Run Behavior of Markov Chains

4.1 Regular Transition Probability Matrices

Suppose that a transition probability matrix $\mathbf{P} = \|P_{ij}\|$ on a finite number of states labeled $0, 1, \ldots, N$ has the property that when raised to some power k, the matrix $\mathbf{P}^k$ has all of its elements strictly positive. Such a transition probability matrix, or the corresponding Markov chain, is called *regular*. The most important fact concerning a regular Markov chain is the existence of a *limiting probability distribution* $\pi = (\pi_0, \pi_1, \ldots, \pi_N)$, where $\pi_j > 0$ for $j = 0, 1, \ldots, N$ and $\Sigma_j \pi_j = 1$, and this distribution is independent of the initial state. Formally, for a regular transition probability matrix $\mathbf{P} = \|P_{ij}\|$, we have the convergence

$$\lim_{n \to \infty} P_{ij}^{(n)} = \pi_j > 0 \quad \text{for } j = 0, 1, \ldots, N,$$

or, in terms of the Markov chain $\{X_n\}$,

$$\lim_{n \to \infty} \Pr\{X_n = j | X_0 = i\} = \pi_j > 0 \quad \text{for } j = 0, 1, \ldots, N.$$

This convergence means that, in the long run ($n \to \infty$), the probability of finding the Markov chain in state j is approximately π_j no matter in which state the chain began at time 0.

Example The Markov chain whose transition probability matrix is

$$\mathbf{P} = \begin{matrix} & \begin{matrix} 0 & \quad 1 \end{matrix} \\ \begin{matrix} 0 \\ 1 \end{matrix} & \left\| \begin{matrix} 1-a & a \\ b & 1-b \end{matrix} \right\| \end{matrix} \tag{4.1}$$

is regular when $0 < a, b < 1$, and in this case, the limiting distribution is $\pi = (b/(a+b), a/(a+b))$. To give a numerical example, we will suppose that

$$\mathbf{P} = \left\| \begin{matrix} 0.33 & 0.67 \\ 0.75 & 0.25 \end{matrix} \right\|.$$

An Introduction to Stochastic Modeling
© 2011 Elsevier Inc. All rights reserved.

The first several powers of P are given as follows:

$$\mathbf{P}^2 = \begin{Vmatrix} 0.6114 & 0.3886 \\ 0.4350 & 0.5650 \end{Vmatrix}, \quad \mathbf{P}^3 = \begin{Vmatrix} 0.4932 & 0.5068 \\ 0.5673 & 0.4327 \end{Vmatrix},$$

$$\mathbf{P}^4 = \begin{Vmatrix} 0.5428 & 0.4572 \\ 0.5117 & 0.4883 \end{Vmatrix}, \quad \mathbf{P}^5 = \begin{Vmatrix} 0.5220 & 0.4780 \\ 0.5350 & 0.4560 \end{Vmatrix},$$

$$\mathbf{P}^6 = \begin{Vmatrix} 0.5307 & 0.4693 \\ 0.5253 & 0.4747 \end{Vmatrix}, \quad \mathbf{P}^7 = \begin{Vmatrix} 0.5271 & 0.4729 \\ 0.5294 & 0.4706 \end{Vmatrix}.$$

By $n = 7$, the entries agree row-to-row to two decimal places. The limiting probabilities are $b/(a + b) = 0.5282$ and $a/(a + b) = 0.4718$.

Example Sociologists often assume that the social classes of successive generations in a family can be regarded as a Markov chain. Thus, the occupation of a son is assumed to depend only on his father's occupation and not on his grandfather's. Suppose that such a model is appropriate and that the transition probability matrix is given by

		Son's class		
		Lower	Middle	Upper
	Lower	0.40	0.50	0.10
Father's	Middle	0.05	0.70	0.25
class	Upper	0.05	0.50	0.45

For such a population, what fraction of people are middle class in the long run?

For the time being, we will answer the question by computing sufficiently high powers of $\mathbf{P}^n$. A better method for determining the limiting distribution will be presented later in this section.

We compute

$$\mathbf{P}^2 = \mathbf{P} \times \mathbf{P} = \begin{Vmatrix} 0.1900 & 0.6000 & 0.2100 \\ 0.0675 & 0.6400 & 0.2925 \\ 0.0675 & 0.6000 & 0.3325 \end{Vmatrix},$$

$$\mathbf{P}^4 = \mathbf{P}^2 \times \mathbf{P}^2 = \begin{Vmatrix} 0.0908 & 0.6240 & 0.2852 \\ 0.0758 & 0.6256 & 0.2986 \\ 0.0758 & 0.6240 & 0.3002 \end{Vmatrix},$$

$$\mathbf{P}^8 = \mathbf{P}^4 \times \mathbf{P}^4 = \begin{Vmatrix} 0.0772 & 0.6250 & 0.2978 \\ 0.0769 & 0.6250 & 0.2981 \\ 0.0769 & 0.6250 & 0.2981 \end{Vmatrix}.$$

Note that we have not computed $\mathbf{P}^n$ for consecutive values of n but have speeded up the calculations by evaluating the successive squares $\mathbf{P}^2, \mathbf{P}^4, \mathbf{P}^8$.

In the long run, approximately 62.5% of the population are middle class under the assumptions of the model.

Computing the limiting distribution by raising the transition probability matrix to a high power suffers from being inexact, since $n = \infty$ is never attained, and it also requires more computational effort than is necessary. Theorem 4.1 provides an alternative computational approach by asserting that the limiting distribution is the unique solution to a set of linear equations. For this social class example, the exact limiting distribution, computed using the method of Theorem 4.1, is $\pi_0 = \frac{1}{13} = 0.0769$, $\pi_1 = \frac{5}{8} = 0.6250$, and $\pi_2 = \frac{31}{104} = 0.2981$.

If a transition probability matrix $\mathbf{P}$ on N states is regular, then $\mathbf{P}^{N^2}$ will have no zero elements. Equivalently, if $\mathbf{P}^{N^2}$ is not strictly positive, then the Markov chain is not regular. Furthermore, once it happens that $\mathbf{P}^k$ has no zero entries, then every higher power $\mathbf{P}^{k+n}, n = 1, 2, \ldots$, will have no zero entries. Thus, it suffices to check the successive squares $\mathbf{P}, \mathbf{P}^2, \mathbf{P}^4, \mathbf{P}^8, \ldots$.

Finally, to determine whether or not the square of a transition probability matrix has only strictly positive entries, it is not necessary to perform the actual multiplication, but only to record whether or not the product is nonzero.

Example Consider the transition probability matrix

$$\mathbf{P} = \begin{Vmatrix} 0.9 & 0.1 & 0 & 0 & 0 & 0 & 0 \\ 0.9 & 0 & 0.1 & 0 & 0 & 0 & 0 \\ 0.9 & 0 & 0 & 0.1 & 0 & 0 & 0 \\ 0.9 & 0 & 0 & 0 & 0.1 & 0 & 0 \\ 0.9 & 0 & 0 & 0 & 0 & 0.1 & 0 \\ 0.9 & 0 & 0 & 0 & 0 & 0 & 0.1 \\ 0.9 & 0 & 0 & 0 & 0 & 0 & 0.1 \end{Vmatrix}.$$

We recognize this as a success runs Markov chain. We record the nonzero entries as $+$ and write $\mathbf{P} \times \mathbf{P}$ in the form

$$\mathbf{P} \times \mathbf{P} = \begin{Vmatrix} + & + & 0 & 0 & 0 & 0 & 0 \\ + & 0 & + & 0 & 0 & 0 & 0 \\ + & 0 & 0 & + & 0 & 0 & 0 \\ + & 0 & 0 & 0 & + & 0 & 0 \\ + & 0 & 0 & 0 & 0 & + & 0 \\ + & 0 & 0 & 0 & 0 & 0 & + \\ + & 0 & 0 & 0 & 0 & 0 & + \end{Vmatrix}$$

$$\times \begin{Vmatrix} + & + & 0 & 0 & 0 & 0 & 0 \\ + & 0 & + & 0 & 0 & 0 & 0 \\ + & 0 & 0 & + & 0 & 0 & 0 \\ + & 0 & 0 & 0 & + & 0 & 0 \\ + & 0 & 0 & 0 & 0 & + & 0 \\ + & 0 & 0 & 0 & 0 & 0 & + \\ + & 0 & 0 & 0 & 0 & 0 & + \end{Vmatrix}$$

$$
= \begin{Vmatrix}
+ & + & + & 0 & 0 & 0 & 0 \\
+ & + & 0 & + & 0 & 0 & 0 \\
+ & + & 0 & 0 & + & 0 & 0 \\
+ & + & 0 & 0 & 0 & + & 0 \\
+ & + & 0 & 0 & 0 & 0 & + \\
+ & + & 0 & 0 & 0 & 0 & + \\
+ & + & 0 & 0 & 0 & 0 & +
\end{Vmatrix} = \mathbf{P}^2,
$$

$$
\mathbf{P}^4 = \begin{Vmatrix}
+ & + & + & + & + & 0 & 0 \\
+ & + & + & + & 0 & + & 0 \\
+ & + & + & + & 0 & 0 & + \\
+ & + & + & + & 0 & 0 & + \\
+ & + & + & + & 0 & 0 & + \\
+ & + & + & + & 0 & 0 & + \\
+ & + & + & + & 0 & 0 & +
\end{Vmatrix},
$$

$$
\mathbf{P}^8 = \begin{Vmatrix}
+ & + & + & + & + & + & + \\
+ & + & + & + & + & + & + \\
+ & + & + & + & + & + & + \\
+ & + & + & + & + & + & + \\
+ & + & + & + & + & + & + \\
+ & + & + & + & + & + & + \\
+ & + & + & + & + & + & +
\end{Vmatrix}.
$$

We see that $\mathbf{P}^8$ has all strictly positive entries, and therefore, $\mathbf{P}$ is regular. The limiting distribution for a similar matrix is computed in Section 4.2.2.

Every transition probability matrix on the states $0, 1, \ldots, N$ that satisfies the following two conditions is regular:

1. For every pair of states i, j there is a path $k_1, \ldots, k_r$ for which $P_{ik_1} P_{k_1 k_2} \cdots P_{k_r j} > 0$.
2. There is at least one state i for which $P_{ii} > 0$.

Theorem 4.1. *Let $\mathbf{P}$ be a regular transition probability matrix on the states $0, 1, \ldots, N$. Then the limiting distribution $\boldsymbol{\pi} = (\pi_0, \pi_1, \ldots, \pi_N)$ is the unique nonnegative solution of the equations*

$$
\pi_j = \sum_{k=0}^{N} \pi_k P_{kj}, \quad j = 0, 1, \ldots, N, \tag{4.2}
$$

$$
\sum_{k=0}^{N} \pi_k = 1. \tag{4.3}
$$

Proof. Because the Markov chain is regular, we have a limiting distribution, $\lim_{n \to \infty} P_{ij}^{(n)} = \pi_j$, for which $\sum_{k=0}^{N} \pi_k = 1$. Write $\mathbf{P}^n$ as the matrix product $\mathbf{P}^{n-1} \mathbf{P}$

in the form

$$P_{ij}^{(n)} = \sum_{k=0}^{N} P_{ik}^{(n-1)} P_{kj}, \quad j = 0, \ldots, N, \tag{4.4}$$

and now let $n \to \infty$. Then, $P_{ij}^{(n)} \to \pi_j$, while $P_{ik}^{(n-1)} \to \pi_k$, and (4.4) passes into $\pi_j = \sum_{k=0}^{N} \pi_k P_{kj}$ as claimed.

It remains to show that the solution is unique. Suppose that $x_0, x_1, \ldots, x_N$ solves

$$x_j = \sum_{k=0}^{N} x_k P_{kj} \quad \text{for } j = 0, \ldots, N \tag{4.5}$$

and

$$\sum_{k=0}^{N} x_k = 1. \tag{4.6}$$

We wish to show that $x_j = \pi_j$, the limiting probability. Begin by multiplying (4.5) on the right by P_{jl} and then sum over j to get

$$\sum_{j=0}^{N} x_j P_{jl} = \sum_{j=0}^{N} \sum_{k=0}^{N} x_k P_{kj} P_{jl} = \sum_{k=0}^{N} x_k P_{kl}^{(2)}. \tag{4.7}$$

But by (4.5), we have $x_l = \sum_{j=0}^{N} x_j P_{jl}$, whence (4.7) becomes

$$x_l = \sum_{k=0}^{N} x_k P_{kl}^{(2)} \quad \text{for } l = 0, \ldots, N.$$

Repeating this argument n times we deduce that

$$x_l = \sum_{k=0}^{N} x_k P_{kl}^{(n)} \quad \text{for } l = 0, \ldots, N,$$

and then passing to the limit in n and using that $P_{kl}^{(n)} \to \pi_l$, we see that

$$x_l = \sum_{k=0}^{N} x_k \pi_l, \quad l = 0, \ldots, N.$$

But by (4.6), we have $\Sigma_k x_k = 1$, whence $x = \pi_l$ as claimed. ∎

Example For the social class matrix

$$
\begin{array}{c c c c}
 & 0 & 1 & 2 \\
0 & 0.40 & 0.50 & 0.10 \\
\mathbf{P} = 1 & 0.05 & 0.70 & 0.25 \\
2 & 0.05 & 0.50 & 0.45
\end{array},
$$

the equations determining the limiting distribution (π_0, π_1, π_2) are

$$0.40\pi_0 + 0.05\pi_1 + 0.05\pi_2 = \pi_0, \tag{4.8}$$

$$0.50\pi_0 + 0.70\pi_1 + 0.50\pi_2 = \pi_1, \tag{4.9}$$

$$0.10\pi_0 + 0.25\pi_1 + 0.45\pi_2 = \pi_2, \tag{4.10}$$

$$\pi_0 + \quad \pi_1 + \quad \pi_2 = 1. \tag{4.11}$$

One of the equations (4.8), (4.9), and (4.10) is redundant because of the linear constraint $\Sigma_k P_{ik} = 1$. We arbitrarily strike out (4.10) and simplify the remaining equations to get

$$-60\pi_0 + 5\pi_1 + 5\pi_2 = 0, \tag{4.12}$$

$$5\pi_0 - 3\pi_1 + 5\pi_2 = 0, \tag{4.13}$$

$$\pi_0 + \pi_1 + \pi_2 = 1. \tag{4.14}$$

We eliminate π_2 by subtracting (4.12) from (4.13) and five times (4.14) to reduce the system to

$$65\pi_0 - 8\pi_1 = 0,$$

$$65\pi_0 \qquad = 5.$$

Then, $\pi_0 = \frac{5}{65} = \frac{1}{13}, \pi_1 = \frac{5}{8}$, and then $\pi_2 = 1 - \pi_0 - \pi_1 = \frac{31}{104}$, as given earlier.

4.1.1 Doubly Stochastic Matrices

A transition probability matrix is called *doubly stochastic* if the columns sum to one as well as the rows. Formally, $\mathbf{P} = \|P_{ij}\|$ is doubly stochastic if

$$P_{ij} \geq 0 \quad \text{and} \quad \sum_k P_{ik} = \sum_k P_{kj} = 1 \quad \text{for all } i, j.$$

Consider a doubly stochastic transition probability matrix on the N states 0, $1, \ldots, N-1$. If the matrix is regular, then the unique limiting distribution is the uniform distribution $\pi = (1/N, \ldots, 1/N)$. Because there is only one solution to $\pi_j = \Sigma_k \pi_k P_{kj}$ and $\Sigma_k \pi_k = 1$ when P is regular, we need only to check that $\pi = (1/N, \ldots, 1/N)$ is a solution where $\mathbf{P}$ is doubly stochastic in order to establish the claim. By using the doubly stochastic feature $\Sigma_j P_{jk} = 1$, we verify that

$$\frac{1}{N} = \sum_j \frac{1}{N} P_{jk} = \frac{1}{N}.$$

As an example, let Y_n be the sum of n independent rolls of a fair die and consider the problem of determining with what probability Y_n is a multiple of 7 in the long run. Let X_n be the remainder when Y_n is divided by 7. Then, X_n is a Markov chain on the states $0, 1, \ldots, 6$ with transition probability matrix

$$\mathbf{P} = \begin{array}{c} \\ 0 \\ 1 \\ 2 \\ 3 \\ 4 \\ 5 \\ 6 \end{array} \begin{array}{cccccccc} 0 & 1 & 2 & 3 & 4 & 5 & 6 \\ \left\| \begin{array}{ccccccc} 0 & \frac{1}{6} & \frac{1}{6} & \frac{1}{6} & \frac{1}{6} & \frac{1}{6} & \frac{1}{6} \\ \frac{1}{6} & 0 & \frac{1}{6} & \frac{1}{6} & \frac{1}{6} & \frac{1}{6} & \frac{1}{6} \\ \frac{1}{6} & \frac{1}{6} & 0 & \frac{1}{6} & \frac{1}{6} & \frac{1}{6} & \frac{1}{6} \\ \frac{1}{6} & \frac{1}{6} & \frac{1}{6} & 0 & \frac{1}{6} & \frac{1}{6} & \frac{1}{6} \\ \frac{1}{6} & \frac{1}{6} & \frac{1}{6} & \frac{1}{6} & 0 & \frac{1}{6} & \frac{1}{6} \\ \frac{1}{6} & \frac{1}{6} & \frac{1}{6} & \frac{1}{6} & \frac{1}{6} & 0 & \frac{1}{6} \\ \frac{1}{6} & \frac{1}{6} & \frac{1}{6} & \frac{1}{6} & \frac{1}{6} & \frac{1}{6} & 0 \end{array} \right\| \end{array}.$$

The matrix is doubly stochastic, and it is regular ($\mathbf{P}^2$ has only strictly positive entries), hence the limiting distribution is $\pi = \left(\frac{1}{7}, \ldots, \frac{1}{7}\right)$. Furthermore, Y_n is a multiple of 7 if and only if $X_n = 0$. Thus, the limiting probability that Y_n is a multiple of 7 is $\frac{1}{7}$.

4.1.2 Interpretation of the Limiting Distribution

Given a regular transition matrix $\mathbf{P}$ for a Markov process $\{X_n\}$ on the $N+1$ states $0, 1, \ldots, N$, we solve the linear equations

$$\pi_i = \sum_{k=0}^{N} \pi_k P_{ki} \quad \text{for } i = 0, 1, \ldots, N$$

and

$$\pi_0 + \pi_1 + \cdots + \pi_N = 1.$$

The primary interpretation of the solution $(\pi_0, \ldots, \pi_N)$ is as the limiting distribution

$$\pi_j = \lim_{n \to \infty} P_{ij}^{(n)} = \lim_{n \to \infty} \Pr\{X_n = j | X_0 = i\}.$$

In words, after the process has been in operation for a long duration, the probability of finding the process in state j is π_j, irrespective of the starting state.

There is a second interpretation of the limiting distribution $\pi = (\pi_0, \pi_1, \ldots, \pi_N)$ that plays a major role in many models. We claim that π_j also gives the long *run* mean fraction of time that the process $\{X_n\}$ is in state j. Thus, if each visit to state j incurs a "cost" of c_j, then the long run mean cost per unit time associated with this Markov chain is

$$\text{Long run mean cost per unit time} = \sum_{j=0}^{N} \pi_j c_j.$$

To verify this interpretation, recall that if a sequence $a_0, a_1, \ldots$ of real numbers converges to a limit a, then the averages of these numbers also converge in the manner

$$\lim_{m \to \infty} \frac{1}{m} \sum_{k=0}^{m-1} a_k = a.$$

We apply this result to the convergence $\lim_{n \to \infty} P_{ij}^{(n)} = \pi_j$ to conclude that

$$\lim_{m \to \infty} \frac{1}{m} \sum_{k=0}^{m-1} P_{ij}^{(k)} = \pi_j.$$

Now, $(1/m) \sum_{k=0}^{m-1} P_{ij}^{(k)}$ is exactly the mean fraction of time during steps $0, 1, \ldots, m-1$ that the process spends in state j. Indeed, the actual (random) fraction of time in state j is

$$\frac{1}{m} \sum_{k=0}^{m-1} \mathbf{1}\{X_k = j\},$$

where

$$\mathbf{1}\{X_k = j\} = \begin{cases} 1 & \text{if } X_k = j, \\ 0 & \text{if } X_k \neq j. \end{cases}$$

Therefore, the *mean* fraction of visits is obtained by taking expected values according to

$$E\left[\frac{1}{m} \sum_{k=0}^{m-1} \mathbf{1}\{X_k = i\} | X_0 = j \right] = \frac{1}{m} \sum_{k=0}^{m-1} E[\mathbf{1}\{X_k = j\} | X_0 = i]$$

$$= \frac{1}{m} \sum_{k=0}^{m-1} \Pr\{X_k = j | X_0 = i\}$$

$$= \frac{1}{m} \sum_{k=0}^{m-1} P_{ij}^{(k)}.$$

Because $\lim_{n\to\infty} P_{ij}^{(n)} = \pi_j$, the long run mean fraction of time that the process spends in state j is

$$\lim_{m\to\infty} E\left[\frac{1}{m}\sum_{k=0}^{m-1} \mathbf{1}\{X_k = j\}|X_0 = i\right] = \lim_{m\to\infty}\frac{1}{m}\sum_{k=0}^{m-1} P_{ij}^{(k)} = \pi_j,$$

independent of the starting state i.

Exercises

4.1.1 A Markov chain $X_0, X_1, X_2, \ldots$ has the transition probability matrix

$$\mathbf{P} = \begin{array}{c} \\ 0 \\ 1 \\ 2 \end{array} \begin{array}{ccc} 0 & 1 & 2 \\ \left\| \begin{array}{ccc} 0.7 & 0.2 & 0.1 \\ 0 & 0.6 & 0.4 \\ 0.5 & 0 & 0.5 \end{array} \right\| \end{array}.$$

Determine the limiting distribution.

4.1.2 A Markov chain $X_0, X_1, X_2, \ldots$ has the transition probability matrix

$$\mathbf{P} = \begin{array}{c} \\ 0 \\ 1 \\ 2 \end{array} \begin{array}{ccc} 0 & 1 & 2 \\ \left\| \begin{array}{ccc} 0.6 & 0.3 & 0.1 \\ 0.3 & 0.3 & 0.4 \\ 0.4 & 0.1 & 0.5 \end{array} \right\| \end{array}.$$

Determine the limiting distribution.

4.1.3 A Markov chain $X_0, X_1, X_2, \ldots$ has the transition probability matrix

$$\mathbf{P} = \begin{array}{c} \\ 0 \\ 1 \\ 2 \end{array} \begin{array}{ccc} 0 & 1 & 2 \\ \left\| \begin{array}{ccc} 0.1 & 0.1 & 0.8 \\ 0.2 & 0.2 & 0.6 \\ 0.3 & 0.3 & 0.4 \end{array} \right\| \end{array}.$$

What fraction of time, in the long run, does the process spend in state 1?

4.1.4 A Markov chain $X_0, X_1, X_2, \ldots$ has the transition probability matrix

$$\mathbf{P} = \begin{array}{c} \\ 0 \\ 1 \\ 2 \end{array} \begin{array}{ccc} 0 & 1 & 2 \\ \left\| \begin{array}{ccc} 0.3 & 0.2 & 0.5 \\ 0.5 & 0.1 & 0.4 \\ 0.5 & 0.2 & 0.3 \end{array} \right\| \end{array}.$$

Every period that the process spends in state 0 incurs a cost of \$2. Every period that the process spends in state 1 incurs a cost of \$5. Every period that the process spends in state 2 incurs a cost of \$3. What is the long run cost per period associated with this Markov chain?

4.1.5 Consider the Markov chain whose transition probability matrix is given by

$$
\mathbf{P} = \begin{array}{c} \\ 0 \\ 1 \\ 2 \\ 3 \end{array}
\begin{array}{cccc}
0 & 1 & 2 & 3 \\
\left\|\begin{array}{cccc}
0.1 & 0.5 & 0 & 0.4 \\
0 & 0 & 1 & 0 \\
0 & 0 & 0 & 1 \\
1 & 0 & 0 & 0
\end{array}\right\|
\end{array}.
$$

Determine the limiting distribution for the process.

4.1.6 Compute the limiting distribution for the transition probability matrix

$$
\mathbf{P} = \begin{array}{c} \\ 0 \\ \\ 1 \\ \\ 2 \end{array}
\begin{array}{ccc}
0 & 1 & 2 \\
\left\|\begin{array}{ccc}
\dfrac{1}{2} & \dfrac{1}{2} & 0 \\[2mm]
\dfrac{1}{3} & \dfrac{1}{3} & \dfrac{1}{3} \\[2mm]
\dfrac{1}{6} & \dfrac{1}{2} & \dfrac{1}{3}
\end{array}\right\|
\end{array}.
$$

4.1.7 A Markov chain on the states $0, 1, 2, 3$ has the transition probability matrix

$$
\mathbf{P} = \begin{array}{c} \\ 0 \\ 1 \\ 2 \\ 3 \end{array}
\begin{array}{cccc}
0 & 1 & 2 & 3 \\
\left\|\begin{array}{cccc}
0.1 & 0.2 & 0.3 & 0.4 \\
0 & 0.3 & 0.3 & 0.4 \\
0 & 0 & 0.6 & 0.4 \\
1 & 0 & 0 & 0
\end{array}\right\|
\end{array}.
$$

Determine the corresponding limiting distribution.

4.1.8 Suppose that the social classes of successive generations in a family follow a Markov chain with transition probability matrix given by

		Son's class		
		Lower	Middle	Upper
Father's class	Lower	0.7	0.2	0.1
	Middle	0.2	0.6	0.2
	Upper	0.1	0.4	0.5

What fraction of families are upper class in the long run?

4.1.9 Determine the limiting distribution for the Markov chain whose transition probability matrix is

$$\mathbf{P} = \begin{matrix} & \begin{matrix} 0 & 1 & 2 \end{matrix} \\ \begin{matrix} 0 \\ 1 \\ 2 \end{matrix} & \left\| \begin{matrix} \frac{1}{2} & \frac{1}{2} & 0 \\ \frac{1}{3} & \frac{1}{2} & \frac{1}{6} \\ 0 & \frac{1}{4} & \frac{3}{4} \end{matrix} \right\| \end{matrix}.$$

4.1.10 A bus in a mass transit system is operating on a continuous route with intermediate stops. The arrival of the bus at a stop is classified into one of three states, namely

1. Early arrival;
2. On-time arrival;
3. Late arrival.

Suppose that the successive states form a Markov chain with transition probability matrix

$$\mathbf{P} = \begin{matrix} & \begin{matrix} 1 & 2 & 3 \end{matrix} \\ \begin{matrix} 1 \\ 2 \\ 3 \end{matrix} & \left\| \begin{matrix} 0.5 & 0.4 & 0.1 \\ 0.2 & 0.5 & 0.3 \\ 0.1 & 0.2 & 0.7 \end{matrix} \right\| \end{matrix}.$$

Over a long period of time, what fraction of stops can be expected to be late?

Problems

4.1.1 Five balls are distributed between two urns, labeled A and B. Each period, an urn is selected at random, and if it is not empty, a ball from that urn is removed and placed into the other urn. In the long run what fraction of time is urn A empty?

4.1.2 Five balls are distributed between two urns, labeled A and B. Each period, one of the five balls is selected at random, and whichever urn it's in, it is moved to the other urn. In the long run, what fraction of time is urn A empty?

4.1.3 A Markov chain has the transition probability matrix

$$\mathbf{P} = \begin{matrix} & \begin{matrix} 0 & 1 & 2 & 3 & 4 & 5 \end{matrix} \\ \begin{matrix} 0 \\ 1 \\ 2 \\ 3 \\ 4 \\ 5 \end{matrix} & \left\| \begin{matrix} \alpha_1 & \alpha_2 & \alpha_3 & \alpha_4 & \alpha_5 & \alpha_6 \\ 1 & 0 & 0 & 0 & 0 & 0 \\ 0 & 1 & 0 & 0 & 0 & 0 \\ 0 & 0 & 1 & 0 & 0 & 0 \\ 0 & 0 & 0 & 1 & 0 & 0 \\ 0 & 0 & 0 & 0 & 1 & 0 \end{matrix} \right\| \end{matrix},$$

where $\alpha_i \geq 0, i = 1, \ldots, 6$, and $\alpha_1 + \cdots + \alpha_6 = 1$. Determine the limiting probability of being in state 0.

4.1.4 A finite-state regular Markov chain has transition probability matrix $\mathbf{P} = \|P_{ij}\|$ and limiting distribution $\pi = \|\pi_i\|$. In the long run, what fraction of the *transitions* are from a prescribed state k to a prescribed state m?

4.1.5 The four towns A, B, C, and D are connected by railroad lines as shown in the following diagram:

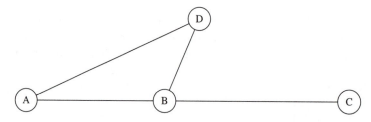

Figure 4.1 A graph whose nodes represent towns and whose arcs represent railroad lines.

Each day, in whichever town it is in, a train chooses one of the lines out of that town at random and traverses it to the next town, where the process repeats the next day. In the long run, what is the probability of finding the train in town D?

4.1.6 Determine the following limits in terms of the transition probability matrix $\mathbf{P} = \|P_{ij}\|$ and limiting distribution $\pi = \|\pi_j\|$ of a finite-state regular Markov chain $\{X_n\}$:

(a) $\lim_{n \to \infty} \Pr\{X_{n+1} = j | X_0 = i\}$.

(b) $\lim_{n \to \infty} \Pr\{X_n = k, X_{n+1} = j | X_0 = i\}$.

(c) $\lim_{n \to \infty} \Pr\{X_{n-1} = k, X_n = j | X_0 = i\}$.

4.1.7 Determine the limiting distribution for the Markov chain whose transition probability matrix is

$$
\mathbf{P} = \begin{array}{c} \\ 0 \\ 1 \\ 2 \\ 3 \end{array}
\begin{array}{cccc}
0 & 1 & 2 & 3 \\
\left\| \begin{array}{cccc}
\frac{1}{2} & 0 & 0 & \frac{1}{2} \\
1 & 0 & 0 & 0 \\
0 & \frac{1}{2} & \frac{1}{3} & \frac{1}{6} \\
0 & 0 & 1 & 0
\end{array} \right\|
\end{array}.
$$

4.1.8 Show that the transition probability matrix

$$
\begin{array}{c c c c c c}
 & 0 & 1 & 2 & 3 & 4 \\
\mathbf{P} = \begin{array}{c} 0 \\ 1 \\ 2 \\ 3 \\ 4 \end{array} &
\left\|\begin{array}{ccccc}
0 & \frac{1}{2} & \frac{1}{2} & 0 & 0 \\
\frac{1}{2} & 0 & \frac{1}{2} & 0 & 0 \\
\frac{1}{3} & \frac{1}{3} & 0 & \frac{1}{3} & 0 \\
0 & 0 & \frac{1}{2} & 0 & \frac{1}{2} \\
\frac{1}{2} & 0 & 0 & \frac{1}{2} & 0
\end{array}\right\|
\end{array}
$$

is regular and compute the limiting distribution.

4.1.9 Determine the long run, or limiting, distribution for the Markov chain whose transition probability matrix is

$$
\begin{array}{c c c c c}
 & 0 & 1 & 2 & 3 \\
\mathbf{P} = \begin{array}{c} 0 \\ 1 \\ 2 \\ 3 \end{array} &
\left\|\begin{array}{cccc}
0 & 0 & 1 & 0 \\
0 & 0 & 0 & 1 \\
\frac{1}{4} & \frac{1}{4} & \frac{1}{4} & \frac{1}{4} \\
0 & 0 & \frac{1}{2} & \frac{1}{2}
\end{array}\right\|.
\end{array}
$$

4.1.10 Consider a Markov chain with transition probability matrix

$$
\mathbf{P} = \left\|\begin{array}{ccccc}
p_0 & p_1 & p_2 & \cdots & p_N \\
p_N & p_0 & p_1 & \cdots & p_{N-1} \\
p_{N-1} & p_N & p_0 & \cdots & p_{N-2} \\
\vdots & \vdots & \vdots & & \vdots \\
p_1 & p_2 & p_3 & \cdots & p_0
\end{array}\right\|,
$$

where $0 < p_0 < 1$ and $p_0 + p_1 + \cdots + p_N = 1$. Determine the limiting distribution.

4.1.11 Suppose that a production process changes state according to a Markov process whose transition probability matrix is given by

$$
\begin{array}{c c c c c}
 & 0 & 1 & 2 & 3 \\
\mathbf{P} = \begin{array}{c} 0 \\ 1 \\ 2 \\ 3 \end{array} &
\left\|\begin{array}{cccc}
0.3 & 0.5 & 0 & 0.2 \\
0.5 & 0.2 & 0.2 & 0.1 \\
0.2 & 0.3 & 0.4 & 0.1 \\
0.1 & 0.2 & 0.4 & 0.3
\end{array}\right\|.
\end{array}
$$

It is known that $\pi_1 = \frac{119}{379} = 0.3140$ and $\pi_2 = \frac{81}{379} = 0.2137$.

(a) Determine the limiting probabilities π_0 and π_3.

(b) Suppose that states 0 and 1 are "In-Control" while states 2 and 3 are deemed "Out-of-Control." In the long run, what fraction of time is the process Out-of-Control?

(c) In the long run, what *fraction of transitions* are from an In-Control state to an Out-of-Control state?

4.1.12 Let **P** be the transition probability matrix of a finite-state regular Markov chain, and let Π be the matrix whose rows are the stationary distribution π. Define $\mathbf{Q} = \mathbf{P} - \boldsymbol{\Pi}$.

(a) Show that $\mathbf{P}^n = \boldsymbol{\Pi} + \mathbf{Q}^n$.

(b) When

$$\mathbf{P} = \left\| \begin{array}{ccc} \dfrac{1}{2} & \dfrac{1}{2} & 0 \\[2mm] \dfrac{1}{4} & \dfrac{1}{2} & \dfrac{1}{4} \\[2mm] 0 & \dfrac{1}{2} & \dfrac{1}{2} \end{array} \right\|$$

obtain an explicit expression for $\mathbf{Q}^n$ and then for $\mathbf{P}^n$.

4.1.13 A Markov chain has the transition probability matrix

$$\mathbf{P} = \begin{array}{c} \\ 0 \\ 1 \\ 2 \end{array} \left\| \begin{array}{ccc} 0 & 1 & 2 \\ 0.4 & 0.4 & 0.2 \\ 0.6 & 0.2 & 0.2 \\ 0.4 & 0.2 & 0.4 \end{array} \right\|.$$

After a long period of time, you observe the chain and see that it is in state 1. What is the conditional probability that the previous state was state 2? That is, find

$$\lim_{n \to \infty} \Pr\{X_{n-1} = 2 | X_n = 1\}.$$

4.2 Examples

Markov chains arising in meteorology, reliability, statistical quality control, and management science are presented next, and the long run behavior of each Markov chain is developed and interpreted in terms of the phenomenon under study.

4.2.1 Including History in the State Description

Often a phenomenon that is not naturally a Markov process can be modeled as a Markov process by including part of the history in the state description. To illustrate this technique, we suppose that the weather on any day depends on the weather conditions for the previous 2 days. To be exact, we suppose that if it was sunny today and yesterday, then it will be sunny tomorrow with probability 0.8; if it was sunny today but cloudy yesterday, then it will be sunny tomorrow with probability 0.6; if it was cloudy today but sunny yesterday, then it will be sunny tomorrow with probability 0.4; if it was cloudy for the last 2 days, then it will be sunny tomorrow with probability 0.1.

Such a model can be transformed into a Markov chain, provided we say that the state at any time is determined by the weather conditions during both that day and the previous day. We say the process is in

State (S, S) if it was sunny both today and yesterday,

State (S, C) if it was sunny yesterday but cloudy today,

State (C, S) if it was cloudy yesterday but sunny today,

State (C, C) if it was cloudy both today and yesterday.

Then, the transition probability matrix is

		\multicolumn{4}{c}{Today's state}
		(S,S)	(S,C)	(C,S)	(C,C)
Yesterday's state	(S,S)	0.8	0.2		
	(S,C)			0.4	0.6
	(C,S)	0.6	0.4		
	(C,C)			0.1	0.9

The equations determining the limiting distribution are

$$
\begin{aligned}
0.8\pi_0 & + 0.6\pi_2 & = \pi_0, \\
0.2\pi_0 & + 0.4\pi_2 & = \pi_1, \\
0.4\pi_1 & + 0.1\pi_3 & = \pi_2, \\
0.6\pi_1 & \quad 0.9\pi_3 & = \pi_3, \\
\pi_0 + \pi_1 + \pi_2 + \pi_3 & = 1.
\end{aligned}
$$

Again, one of the top four equations is redundant. Striking out the first equation and solving the remaining four equations gives $\pi_0 = \frac{3}{11}, \pi_1 = \frac{1}{11}, \pi_2 = \frac{1}{11}$, and $\pi_3 = \frac{6}{11}$.

We recover the fraction of days, in the long run, on which it is sunny by summing the appropriate terms in the limiting distribution. It can be sunny today in conjunction with either being sunny or cloudy tomorrow. Therefore, the long run fraction of days in which it is sunny is $\pi_0 + \pi_1 = \pi(S, S) + \pi(S, C) = \frac{4}{11}$. Formally, $\lim_{n\to\infty} \Pr\{X_n = S\} = \lim_{n\to\infty}[\Pr\{X_n = S, X_{n+1} = S\} + \Pr\{X_n = S, X_{n+1} = C\}] = \pi_0 + \pi_1$.

4.2.2 Reliability and Redundancy

An airline reservation system has two computers, only one of which is in operation at any given time. A computer may break down on any given day with probability p.

There is a single repair facility that takes 2 days to restore a computer to normal. The facilities are such that only one computer at a time can be dealt with. Form a Markov chain by taking as states the pairs (x, y), where x is the number of machines in operating condition at the end of a day and y is 1 if a day's labor has been expended on a machine not yet repaired and 0 otherwise. The transition matrix is

$$
\begin{array}{c}
\text{To state} \\
\hline
\end{array} \rightarrow \quad (2,0) \quad (1,0) \quad (1,1) \quad (0,1)
$$

$$
\begin{array}{c}
\text{From state} \\
\downarrow
\end{array}
$$

$$
\mathbf{P} =
\begin{array}{c}
(2,0) \\
(1,0) \\
(1,1) \\
(0,1)
\end{array}
\left\|
\begin{array}{cccc}
q & p & 0 & 0 \\
0 & 0 & q & p \\
q & p & 0 & 0 \\
0 & 1 & 0 & 0
\end{array}
\right\|,
$$

where $p + q = 1$.

We are interested in the long run probability that both machines are inoperative. Let $(\pi_0, \pi_1, \pi_2, \pi_3)$ be the limiting distribution of the Markov chain. Then, the long run probability that neither computer is operating is π_3, and the availability, the probability that at least one computer is operating, is $1 - \pi_3 = \pi_0 + \pi_1 + \pi_2$.

The equations for the limiting distributions are

$$
\begin{aligned}
q\pi_0 \quad + q\pi_2 \quad &= \pi_0, \\
p\pi_0 \quad + p\pi_2 + \pi_3 &= \pi_1, \\
q\pi_1 \quad &= \pi_2, \\
p\pi_1 \quad &= \pi_3
\end{aligned}
$$

and

$$
\pi_0 + \pi_1 + \pi_2 + \pi_3 = 1.
$$

The solution is

$$
\pi_0 = \frac{q^2}{1+p^2}, \quad \pi_2 = \frac{qp}{1+p^2},
$$

$$
\pi_1 = \frac{p}{1+p^2}, \quad \pi_3 = \frac{p^2}{1+p^2}.
$$

The availability is $R_1 = 1 - \pi_3 = 1/(1+p^2)$.

In order to increase the system availability, it is proposed to add a duplicate repair facility so that both computers can be repaired simultaneously. The corresponding transition matrix is now

$$
\begin{array}{c}
\text{To state} \\
\hline
\end{array} \rightarrow (2,0) \quad (1,0) \quad (1,1) \quad (0,1)
$$

$$
\mathbf{P} = \begin{array}{c} \text{From state} \\ \downarrow \\ \begin{array}{c} (2,0) \\ (1,0) \\ (1,1) \\ (0,1) \end{array} \end{array}
\begin{Vmatrix}
q & p & 0 & 0 \\
0 & 0 & q & p \\
q & p & 0 & 0 \\
0 & 0 & 1 & 0
\end{Vmatrix},
$$

and the limiting distribution is

$$
\pi_0 = \frac{q}{1+p+p^2}, \quad \pi_2 = \frac{p}{1+p+p^2},
$$

$$
\pi_1 = \frac{p}{1+p+p^2}, \quad \pi_3 = \frac{p^2}{1+p+p^2}.
$$

Thus, availability has increased to $R_2 = 1 - \pi_3 = (1+p)/(1+p+p^2)$.

4.2.3 A Continuous Sampling Plan

Consider a production line where each item has probability p of being defective. Assume that the condition of a particular item (defective or nondefective) does not depend on the conditions of other items. The following sampling plan is used.

Initially every item is sampled as it is produced; this procedure continues until i consecutive nondefective items are found. Then, the sampling plan calls for sampling only one out of every r items at random until a defective one is found. When this happens the plan calls for reverting to 100% sampling until i consecutive nondefective items are found. The process continues in the same way.

State $E_k (k = 0, 1, \ldots, i-1)$ denotes that k consecutive nondefective items have been found in the 100% sampling portion of the plan, while state E_i denotes that the plan is in the second stage (sampling one out of r). Time m is considered to follow the mth item, whether sampled or not. Then, the sequence of states is a Markov chain with

$$
P_{jk} = \Pr\{\text{in state } E_k \text{ after } m+1 \text{ items} | \text{in state } E_j \text{ after } m \text{ items}\}
$$

$$
= \begin{cases}
p & \text{for } k = 0, 0 \le j < i, \\
1 - p & \text{for } k = j+1 \le i, \\
\dfrac{p}{r} & \text{for } k = 0, j = i, \\
1 - \dfrac{p}{r} & \text{for } k = j = i, \\
0 & \text{otherwise.}
\end{cases}
$$

Let π_k be the limiting probability that the system is in state E_k for $k = 0, 1, \ldots, i$. The equations determining these limiting probabilities are

$$(0) \qquad p\pi_0 + \qquad p\pi_1 + \cdots + \qquad p\pi_{i-1} + \quad (p/r)\pi_i = \pi_0,$$

$$(1)\ (1-p)\pi_0 \hspace{7.5cm} = \pi_1,$$

$$(2) \hspace{2.3cm} (1-p)\pi_1 \hspace{5.1cm} = \pi_2,$$

$$\vdots$$

$$(i) \hspace{4.3cm} (1-p)\pi_{i-1} + (1 - p/r)\pi_i = \pi_i$$

together with

$$(*) \hspace{3cm} \pi_0 + \pi_1 + \cdots + \pi_i = 1.$$

From equations (1) through (i), we deduce that $\pi_k = (1-p)\pi_{k-1}$ so that $\pi_k = (1-p)^k \pi_0$ for $k = 0, \ldots, i-1$, while equation (i) yields $\pi_i = (r/p)(1-p)\pi_{i-1}$ or $\pi_i = (r/p)(1-p)^i \pi_0$. Having determined π_k in terms of π_0 for $k = 0, \ldots, i$, we place these values in (*) to obtain

$$\left\{ [1 + \cdots + (1-p)^{i-1}] + \frac{r}{p}(1-p)^i \right\} \pi_0 = 1.$$

The geometric series simplifies, and after elementary algebra, the solution is

$$\pi_0 = \frac{p}{1 + (r-1)(1-p)^i},$$

whence

$$\pi_k = \frac{p(1-p)^k}{1 + (r-1)(1-p)^i} \qquad \text{for } k = 0, \ldots, i-1,$$

while

$$\pi_i = \frac{r(1-p)^i}{1 + (r-1)(1-p)^i}.$$

Let AFI (Average Fraction Inspected) denote the long run fraction of items that are inspected. Since each item is inspected while in states $E_0, \ldots, E_{i-1}$ but only one out of r is inspected in state E_i, we have

$$\begin{aligned} \text{AFI} &= (\pi_0 + \cdots + \pi_{i-1}) + (1/r)\pi_i \\ &= (1 - \pi_i) + (1/r)\pi_i \\ &= \frac{1}{1 + (r-1)(1-p)^i}. \end{aligned}$$

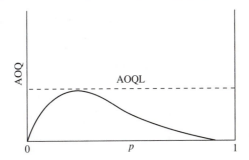

Figure 4.2 The average outgoing quality (AOQ) as a function of the input quality level p.

Figure 4.3 Black-box picture of a continuous inspection scheme as a method of guaranteeing outgoing quality.

Let us assume that each item found to be defective is replaced by an item known to be good. The average outgoing quality (AOQ) is defined to be the fraction of defectives in the output of such an inspection scheme. The average fraction not inspected is

$$1 - \text{AFI} = \frac{(r-1)(1-p)^i}{1 + (r-1)(1-p)^i},$$

and of these on the average p are defective. Hence

$$\text{AOQ} = \frac{(r-1)(1-p)^i p}{1 + (r-1)(1-p)^i}.$$

This average outgoing quality is zero if $p = 0$ or $p = 1$, and rises to a maximum at some intermediate value, as shown in Figure 4.2. The maximum AOQ is called the *average outgoing quality limit* (AOQL), and it has been determined numerically and tabulated as a function of i and r.

This quality control scheme guarantees an output quality better than the AOQL regardless of the input fraction defective, as shown in Figure 4.3.

4.2.4 Age Replacement Policies

A component of a computer has an active life, measured in discrete units, that is a random variable T, where $\Pr[T = k] = a_k$ for $k = 1, 2, \ldots$. Suppose one starts with a fresh component, and each component is replaced by a new component upon failure. Let X_n be the age of the component in service at time n. Then, (X_n) is a success runs Markov chain. (See Chapter 3, Section 3.5.4.)

In an attempt to replace components before they fail in service, an *age replacement* policy is instituted. This policy calls for replacing the component upon its failure or upon its reaching age N, whichever occurs first. Under this age replacement policy, the Markov chain $\{X_n\}$ has the transition probability matrix

$$
\mathbf{P} =
\begin{array}{c}
\\ 0 \\ 1 \\ 2 \\ \vdots \\ N-1
\end{array}
\begin{array}{c}
\begin{array}{cccccc}
0 & 1 & 2 & 3 & & N-1
\end{array} \\
\left\|
\begin{array}{cccccc}
p_0 & 1-p_0 & 0 & 0 & \cdots & 0 \\
p_1 & 0 & 1-p_1 & 0 & & 0 \\
p_2 & 0 & 0 & 1-p_2 & \cdots & 0 \\
\vdots & \vdots & \vdots & \vdots & & \vdots \\
1 & 0 & 0 & 0 & & 0
\end{array}
\right\|
\end{array},
$$

where

$$
p_k = \frac{a_{k+1}}{a_{k+1} + a_{k+2} + \cdots} \quad \text{for } k = 0, 1, \ldots, N-2.
$$

State 0 corresponds to a new component, and therefore, the limiting probability π_0 corresponds to the long run probability of replacement during any single time unit, or the long run replacement per unit time. Some of these replacements are planned or age replacements, and some correspond to failures in service. A planned replacement occurs in each period for which $X_n = N - 1$, and therefore, the long run planned replacements per unit time is the limiting probability π_{N-1}. The difference $\pi_0 - \pi_{N-1}$ is the long run rate of failures in service. The equations for the limiting distribution $\pi = (\pi_0, \pi_1, \ldots, \pi_{N-1})$ are

$$
\begin{aligned}
p_0\pi_0 + \quad p_1\pi_1 + \cdots + \quad p_{N-2}\pi_{N-2} + \quad \pi_{N-1} &= \pi_0, \\
(1-p_0)\pi_0 \quad\quad\quad\quad\quad\quad\quad\quad\quad\quad &= \pi_1, \\
(1-p_1)\pi_1 \quad\quad\quad\quad\quad &= \pi_2, \\
&\;\;\vdots \\
(1-p_{N-2})\pi_{N-2} \quad\quad &= \pi_{N-1}, \\
\pi_0 + \quad\quad \pi_1 + \cdots \quad\quad\quad\quad +\pi_{N-1} &= 1.
\end{aligned}
$$

Solving in terms of π_0, we obtain

$$
\begin{aligned}
\pi_0 &= \pi_0, \\
\pi_1 &= (1-p_0)\pi_0, \\
\pi_2 &= (1-p_1)\pi_1 = (1-p_1)(1-p_0)\pi_0, \\
&\;\;\vdots \\
\pi_k &= (1-p_{k-1})\pi_{k-1} = (1-p_{k-1})(1-p_{k-2})\cdots(1-p_0)\pi_0, \\
&\;\;\vdots \\
\pi_{N-1} &= (1-p_{N-2})\pi_{N-2} = (1-p_{N-2})(1-p_{N-3})\cdots(1-p_0)\pi_0,
\end{aligned}
$$

and since $\pi_0 + \pi_1 + \cdots + \pi_{N-1} = 1$, we have

$$1 = [1 + (1 - p_0) + (1 - p_0)(1 - p_1) + \cdots$$
$$+ (1 - p_0)(1 - p_1) \cdots (1 - p_{N-2})]\pi_0,$$

or

$$\pi_0 = \frac{1}{1 + (1 - p_0) + (1 - p_0)(1 - p_1) + \cdots + (1 - p_0)(1 - p_1) \cdots (1 - p_{N-2})}.$$

If $A_j = a_j + a_{j+1} + \cdots$ for $j = 1, 2, \ldots,$ where $A_1 = 1$, then $p_k = a_{k+1}/A_{k+1}$ and $1 - p_k = A_{k+2}/A_{k+1}$, which simplifies the expression for π_0 to

$$\pi_0 = \frac{1}{A_1 + A_2 + \cdots + A_N},$$

and then

$$\pi_{N-1} = A_N \pi_0 = \frac{A_N}{A_1 + A_2 + \cdots + A_N}.$$

In practice, one determines the cost C of a replacement and the additional cost K that is incurred when a failure in service occurs. Then, the long run total cost per unit time is $C\pi_0 + K(\pi_0 - \pi_{N-1})$, and the replacement age N is chosen so as to minimize this total cost per unit time.

Observe that

$$\frac{1}{\pi_0} = A_1 + A_2 + \cdots + A_N = \sum_{j=1}^{N} \Pr\{T \geq j\} = \sum_{k=0}^{N-1} \Pr\{T > k\}$$

$$= \sum_{k=0}^{\infty} \Pr\{\min\{T, N\} > k\} = E[\min\{T, N\}].$$

In words, the reciprocal of the mean time between replacements $E[\min\{T, N\}]$ yields the long run replacements per unit time π_0. This relation will be further explored in the chapter on renewal processes.

4.2.5 Optimal Replacement Rules

A common industrial activity is the periodic inspection of some system as part of a procedure for keeping it operative. After each inspection, a decision must be made whether or not to alter the system at that time. If the inspection procedure and the ways of modifying the system are fixed, an important problem is that of determining, according to some cost criterion, the optimal rule for making the appropriate decision. Here, we consider the case in which the only possible act is to replace the system with a new one.

Suppose that the system is inspected at equally spaced points in time and that after each inspection it is classified into one of the $L+1$ possible states $0, 1, \ldots, L$. A system is in state 0 if and only if it is new and is in state L if and only if it is inoperative. Let the inspection times be $n = 0, 1, \ldots$, and let X_n denote the observed state of the system at time n. In the absence of replacement, we assume that $\{X_n\}$ is a Markov chain with transition probabilities $p_{ij} = \Pr\{X_{n+1} = j | X_n = i\}$ for all i, j, and n.

It is possible to replace the system at any time before failure. The motivation for doing so may be to avoid the possible consequences of further deterioration or of failure of the system. A replacement rule, denoted by R, is a specification of those states at which the system will be replaced. Replacement takes place at the next inspection time. A replacement rule R modifies the behavior of the system and results in a modified Markov chain $\{X_n(R); n = 0, 1, \ldots\}$. The corresponding modified transition probabilities $p_{ij}(R)$ are given by

$$p_{ij}(R) = p_{ij} \quad \text{if the system is not replaced at state } i,$$
$$p_{i0}(R) = 1,$$

and

$$p_{ij}(R) = 0, \quad j \neq 0 \quad \text{if the system is replaced at state } i.$$

It is assumed that each time the equipment is replaced, a replacement cost of K units is incurred. Further it is assumed that each unit of time the system is in state j incurs an operating cost of a_j. Note that a_L may be interpreted as failure (inoperative) cost. This interpretation leads to the one period cost function $c_i(R)$ given for $i = 0, \ldots, L$ by

$$c_i(R) = \begin{cases} a_i & \text{if } p_{i0}(R) = 0, \\ K + a_i & \text{if } p_{i0}(R) = 1. \end{cases}$$

We are interested in replacement rules that minimize the expected long run time average cost. This cost is given by the expected cost under the limiting distribution for the Markov chain $\{X_n(R)\}$. Denoting this average cost by $\phi(R)$, we have

$$\phi(R) = \sum_{i=0}^{L} \pi_i(R) c_i(R),$$

where $\pi_i(R) = \lim_{n \to \infty} \Pr\{X_n(R) = i\}$. The limiting distribution $\pi_i(R)$ is determined by the equations

$$\pi_i(R) = \sum_{k=0}^{L} \pi_k(R) p_{ki}(R), \quad i = 0, \ldots, L,$$

and

$$\pi_0(R) + \pi_1(R) + \cdots + \pi_L(R) = 1.$$

We define a control limit rule to be a replacement rule of the form

"Replace the system if and only if $X_n \geq k$,"

where k, called the control limit, is some fixed state between 0 and L. We let R_k denote the control limit rule with control limit equal to k. Then, R_0 is the rule "Replace the system at every step," and R_L is the rule "Replace only upon failure (State L)."

Control limit rules seem reasonable provided that the states are labeled monotonically from best (0) to worst (L) in some sense. Indeed, it can be shown that a control limit rule is optimal whenever the following two conditions hold:

1. $a_0 \leq a_1 \leq \cdots \leq a_L$.

2. If $i \leq j$, then $\sum_{m=k}^{L} p_{im} \leq \sum_{m=k}^{L} p_{jm}$ for every $k = 0, \ldots, L$.

Condition (1) asserts that the one-stage costs are higher in the "worse" states. Condition (2) asserts that further deterioration is more likely in the "worse" states.

Let us suppose that conditions (1) and (2) prevail. Then, we need only check the $L+1$ control limit rules $R_0, \ldots, R_L$ in order to find an optimal rule. Furthermore, it can be shown that a control limit k^* satisfying $\phi(R_{k^*-1}) \geq \phi(R_{k^*}) \leq \phi(R_{k^*+1})$ is optimal so that not always do all $L+1$ control limit rules need to be checked.

Under control limit k, we have the cost vector

$$c(R_k) = (a_0, \ldots, a_{k-1}, K + a_k, \ldots, K + a_L)$$

and the transition probabilities

$$\mathbf{P}(R_k) = \begin{array}{c} \\ 0 \\ 1 \\ k-1 \\ k \\ \\ L \end{array} \begin{array}{c} 0 \quad 1 \quad\quad\quad k-1 \quad\quad k \quad\quad\quad L \\ \left\| \begin{array}{ccccccc} 0 & p_{01} & \cdots & p_{0,k-1} & p_{0k} & \cdots & p_{0L} \\ 0 & p_{11} & \cdots & p_{1,k-1} & p_{1k} & \cdots & p_{1L} \\ 0 & p_{k-1,1} & \cdots & p_{k-1,k-1} & p_{k-1,k} & \cdots & p_{k-1,L} \\ 1 & 0 & \cdots & 0 & 0 & \cdots & 0 \\ \vdots & & & & & & \\ 1 & 0 & \cdots & 0 & 0 & \cdots & 0 \end{array} \right\| \end{array}.$$

To look at a numerical example, we will find the optimal control limit k^* for the following data: $L = 5$ and

$$\mathbf{P} = \begin{array}{c} \\ 0 \\ 1 \\ 2 \\ 3 \\ 4 \\ 5 \end{array} \begin{array}{c} 0 \quad 1 \quad\; 2 \quad\; 3 \quad\; 4 \quad\; 5 \\ \left\| \begin{array}{cccccc} 0 & 0.2 & 0.2 & 0.2 & 0.2 & 0.2 \\ 0 & 0.1 & 0.2 & 0.2 & 0.2 & 0.3 \\ 0 & 0 & 0.1 & 0.2 & 0.3 & 0.4 \\ 0 & 0 & 0 & 0.1 & 0.4 & 0.5 \\ 0 & 0 & 0 & 0 & 0.4 & 0.6 \\ 0 & 0 & 0 & 0 & 0 & 1 \end{array} \right\| \end{array},$$

$a_0 = \cdots = a_{L-1} = 0, a_L = 5$, and $K = 3$. When $k = 1$, the transition matrix is

$$
\mathbf{P}(R_1) =
\begin{array}{c@{\;}c}
 & \begin{array}{cccccc} 0 & 1 & 2 & 3 & 4 & 5 \end{array} \\
\begin{array}{c} 0 \\ 1 \\ 2 \\ 3 \\ 4 \\ 5 \end{array} &
\left\|
\begin{array}{cccccc}
0 & 0.2 & 0.2 & 0.2 & 0.2 & 0.2 \\
1 & 0 & 0 & 0 & 0 & 0 \\
1 & 0 & 0 & 0 & 0 & 0 \\
1 & 0 & 0 & 0 & 0 & 0 \\
1 & 0 & 0 & 0 & 0 & 0 \\
1 & 0 & 0 & 0 & 0 & 0
\end{array}
\right\|
\end{array},
$$

which implies the following equations for the stationary distribution:

$$
\begin{aligned}
\pi_1 + \pi_2 + \pi_3 + \pi_4 + \pi_5 &= \pi_0, \\
0.2\pi_0 &= \pi_1, \\
0.2\pi_0 &= \pi_2, \\
0.2\pi_0 &= \pi_3, \\
0.2\pi_0 &= \pi_4, \\
0.2\pi_0 &= \pi_5, \\
\pi_0 + \pi_1 + \pi_2 + \pi_3 + \pi_4 + \pi_5 &= 1.
\end{aligned}
$$

The solution is $\pi_0 = 0.5$ and $\pi_1 = \pi_2 = \pi_3 = \pi_4 = \pi_5 = 0.1$. The average cost associated with $k = 1$ is

$$
\begin{aligned}
\phi_1 &= 0.5(0) + 0.1(3) + 0.1(3) + 0.1(3) + 0.1(3) + 0.1(3 + 5) \\
&= 2.0.
\end{aligned}
$$

When $k = 2$, the transition matrix is

$$
\mathbf{P}(R_2) =
\begin{array}{c@{\;}c}
 & \begin{array}{cccccc} 0 & 1 & 2 & 3 & 4 & 5 \end{array} \\
\begin{array}{c} 0 \\ 1 \\ 2 \\ 3 \\ 4 \\ 5 \end{array} &
\left\|
\begin{array}{cccccc}
0 & 0.2 & 0.2 & 0.2 & 0.2 & 0.2 \\
0 & 0.1 & 0.2 & 0.2 & 0.2 & 0.3 \\
1 & 0 & 0 & 0 & 0 & 0 \\
1 & 0 & 0 & 0 & 0 & 0 \\
1 & 0 & 0 & 0 & 0 & 0 \\
1 & 0 & 0 & 0 & 0 & 0
\end{array}
\right\|
\end{array},
$$

and the associated stationary distribution is $\pi_0 = 0.450, \pi_1 = 0.100, \pi_2 = \pi_3 = \pi_4 = 0.110, \pi_5 = 0.120$. We evaluate the average cost to be

$$
\begin{aligned}
\phi_2 &= 0.45(0) + 0.10(0) + 0.11(3) + 0.11(3) + 0.11(3) + 0.12(8) \\
&= 1.95.
\end{aligned}
$$

Continuing in this manner, we obtain the following table:

Control Limit	Stationary Distribution						Average Cost
k	π_0	π_1	π_2	π_3	π_4	π_5	ϕ_k
1	0.5000	0.1000	0.1000	0.1000	0.1000	0.1000	2.0000
2	0.4500	0.1000	0.1100	0.1100	0.1100	0.1200	1.9500
3	0.4010	0.0891	0.1089	0.1198	0.1307	0.1505	1.9555
4	0.3539	0.9786	0.0961	0.1175	0.1623	0.1916	2.0197
5	0.2785	0.06189	0.0756	0.0925	0.2139	0.2785	2.2280

The optimal control limit is $k^* = 2$, and the corresponding minimum average cost per unit time is $\phi_2 = 1.95$.

Exercises

4.2.1 On a southern Pacific island, a sunny day is followed by another sunny day with probability 0.9, whereas a rainy day is followed by another rainy day with probability 0.2. Supposing that there are only sunny or rainy days, in the long run on what fraction of days is it sunny?

4.2.2 In the reliability example of Section 4.2.2, what fraction of time is the repair facility idle? When a second repair facility is added, what fraction of time is each facility idle?

4.2.3 Determine the average fraction inspected, AFI, and the average outgoing quality, AOQ, of Section 4.2.3 for $p = 0, 0.05, 0.10, 0.15, \ldots, 0.50$ when
(a) $r = 10$ and $i = 5$.
(b) $r = 5$ and $i = 10$.

4.2.4 Section 4.2.2 determined the availability R of a certain computer system to be

$$R_1 = \frac{1}{1+p^2} \qquad \text{for one repair facility,}$$

$$R_2 = \frac{1+p}{1+p+p^2} \qquad \text{for two repair facilities,}$$

where p is the computer failure probability on a single day. Compute and compare R_1 and R_2 for $p = 0.01, 0.02, 0.05,$ and 0.10.

4.2.5 From purchase to purchase, a particular customer switches brands among products $A, B,$ and C according to a Markov chain whose transition probability matrix is

$$\mathbf{P} = \begin{array}{c} A \\ B \\ C \end{array} \begin{array}{ccc} A & B & C \\ \left\| \begin{array}{ccc} 0.6 & 0.2 & 0.2 \\ 0.1 & 0.7 & 0.2 \\ 0.1 & 0.1 & 0.8 \end{array} \right\| \end{array}.$$

In the long run, what fraction of time does this customer purchase brand A?

4.2.6 A component of a computer has an active life, measured in discrete units, that is a random variable T where

$$\Pr\{T = 1\} = 0.1, \quad \Pr\{T = 3\} = 0.3,$$
$$\Pr\{T = 2\} = 0.2, \quad \Pr\{T = 4\} = 0.4.$$

Suppose one starts with a fresh component, and each component is replaced by a new component upon failure. Determine the long run probability that a failure occurs in a given period.

4.2.7 Consider a machine whose condition at any time can be observed and classified as being in one of the following three states:

State 1: Good operating order
State 2: Deteriorated operating order
State 3: In repair

We observe the condition of the machine at the end of each period in a sequence of periods. Let X_n denote the condition of the machine at the end of period n for $n = 1, 2, \ldots$. Let X_0 be the condition of the machine at the start. We assume that the sequence of machine conditions is a Markov chain with transition probabilities

$$P_{11} = 0.9, \quad P_{12} = 0.1, \quad P_{13} = 0,$$
$$P_{21} = 0, \quad\;\; P_{22} = 0.9, \quad P_{23} = 0.1,$$
$$P_{31} = 1, \quad\;\; P_{32} = 0, \quad\;\; P_{33} = 0,$$

and that the process starts in state $X_0 = 1$.
(a) Find $\Pr\{X_4 = 1\}$.
(b) Calculate the limiting distribution.
(c) What is the long run rate of repairs per unit time?

4.2.8 At the end of a month, a large retail store classifies each receivable account according to
0: Current
1: 30–60 days overdue
2: 60–90 days overdue
3: Over 90 days
Each such account moves from state to state according to a Markov chain with transition probability matrix

$$
\mathbf{P} = \begin{array}{c|cccc}
 & 0 & 1 & 2 & 3 \\
\hline
0 & 0.95 & 0.05 & 0 & 0 \\
1 & 0.50 & 0 & 0.50 & 0 \\
2 & 0.20 & 0 & 0 & 0.80 \\
3 & 0.10 & 0 & 0 & 0.90
\end{array}.
$$

In the long run, what fraction of accounts are over 90 days overdue?

Problems

4.2.1 Consider a discrete-time periodic review inventory model (see Chapter 3, Section 3.3.1), and let ξ_n be the total demand in period n. Let X_n be the inventory quantity on hand at the end of period n. Instead of following an (s, S) policy, a (q, Q) policy will be used: If the stock level at the end of a period is less than or equal to $q = 2$ units, then $Q = 2$ additional units will be ordered and will be available at the beginning of the next period. Otherwise, no ordering will take place. This is a (q, Q) policy with $q = 2$ and $Q = 2$. Assume that demand that is not filled in a period is lost (no back ordering).

(a) Suppose that $X_0 = 4$ and that the period demands turn out to be $\xi_1 = 3$, $\xi_2 = 4, \xi_3 = 0, \xi_4 = 2$. What are the end-of-period stock levels for periods $n = 1, 2, 3, 4$?

(b) Suppose that $\xi_1, \xi_2, \ldots$ are independent random variables, each having the probability distribution where

k =	0	1	2	3	4
$\Pr\{\xi = k\}$ =	0.1	0.3	0.3	0.2	0.1

Then, $X_0, X_1, \ldots$ is a Markov chain. Determine the transition probability distribution and the limiting distribution.

(c) In the long run, during what fraction of periods are orders placed?

4.2.2 A system consists of two components operating *in parallel*: The system functions if at least one of the components is operating. In any single period, if both components are operating at the beginning of the period, then each will fail, independently, during the period with probability α. When one component has failed, the remaining component fails during a period with a higher probability β. There is a single repair facility, and it takes two periods to repair a component.

(a) Define an appropriate set of states for the system in the manner of the *Reliability and Redundancy* example and specify the transition probabilities in terms of α and β.

(b) When $\alpha = 0.1$ and $\beta = 0.2$, in the long run what fraction of time is the system operating?

4.2.3 Suppose that a production process changes state according to a Markov process whose transition probability matrix is given by

$$
\mathbf{P} = \begin{array}{c} 0 \\ 1 \\ 2 \\ 3 \end{array}
\begin{array}{cccc}
0 & 1 & 2 & 3 \\
\left\|\begin{array}{cccc}
0.2 & 0.2 & 0.4 & 0.2 \\
0.5 & 0.2 & 0.2 & 0.1 \\
0.2 & 0.3 & 0.4 & 0.1 \\
0.1 & 0.2 & 0.4 & 0.3
\end{array}\right\|
\end{array}.
$$

(a) Determine the limiting distribution for the process.

(b) Suppose that states 0 and 1 are "In-Control," while states 2 and 3 are deemed "Out-of-Control." In the long run, what fraction of time is the process Out-of-Control?

(c) In the long run, what fraction of transitions are from an In-Control state to an Out-of-Control state?

4.2.4 A component of a computer has an active life, measured in discrete units, that is a random variable ξ, where

k =	1	2	3	4
$\Pr\{\xi = k\}$ =	0.1	0.3	0.2	0.4

Suppose that one starts with a fresh component, and each component is replaced by a new component upon failure. Let X_n be the *remaining life* of the component in service at the *end* of period n. When $X_n = 0$, a new item is placed into service at the *start* of the next period.

(a) Set up the transition probability matrix for $\{X_n\}$.

(b) By showing that the chain is regular and solving for the limiting distribution, determine the long run probability that the item in service at the end of a period has no remaining life and therefore will be replaced.

(c) Relate this to the mean life of a component.

4.2.5 Suppose that the weather on any day depends on the weather conditions during the previous 2 days. We form a Markov chain with the following states:

State (S, S) if it was sunny both today and yesterday,

State (S, C) if it was sunny yesterday but cloudy today,

State (C, S) if it was cloudy yesterday but sunny today,

State (C, C) if it was cloudy both today and yesterday,

and transition probability matrix

Today's state

$$
\mathbf{P} =
\begin{array}{c}
 \\
(S,S) \\
(S,C) \\
(C,S) \\
(C,C)
\end{array}
\begin{array}{c}
\begin{array}{cccc}
(S,S) & (S,C) & (C,S) & (C,C)
\end{array} \\
\left\|
\begin{array}{cccc}
0.7 & 0.3 & 0 & 0 \\
0 & 0 & 0.4 & 0.6 \\
0.5 & 0.5 & 0 & 0 \\
0 & 0 & 0.2 & 0.8
\end{array}
\right\|
\end{array}.
$$

(a) Given that it is sunny on days 0 and 1, what is the probability it is sunny on day 5?

(b) In the long run, what fraction of days are sunny?

4.2.6 Consider a computer system that fails on a given day with probability p and remains "up" with probability $q = 1 - p$. Suppose the repair time is a random variable N having the probability mass function $p(k) = \beta(1 - \beta)^{k-1}$ for $k = 1, 2, \ldots$, where $0 < \beta < 1$. Let $X_n = 1$ if the computer is operating on day n and

$X_1 = 0$ if not. Show that $\{X_n\}$ is a Markov chain with transition matrix

$$
\begin{array}{c c c}
 & 0 & 1 \\
0 & \alpha & \beta \\
1 & p & q
\end{array}
$$

and $\alpha = 1 - \beta$. Determine the long run probability that the computer is operating in terms of α, β, p, and q.

4.2.7 Customers arrive for service and take their place in a waiting line. There is a single service facility, and a customer undergoing service at the beginning of a period will complete service and depart at the end of the period with probability β and will continue service into the next period with probability $\alpha = 1 - \beta$, and then the process repeats. This description implies that the service time η of an individual is a random variable with the geometric distribution,

$$\Pr\{\eta = k\} = \beta\alpha^{k-1} \quad \text{for } k = 1, 2, \ldots,$$

and the service times of distinct customers are independent random variables.

At most a single customer can arrive during a period. We suppose that the actual number of arrivals during the nth period is a random variable ξ_n taking on the values 0 or 1 according to

$$\Pr\{\xi_n = 0\} = p$$

and

$$\Pr\{\xi_n = 1\} = q = 1 - p \quad \text{for } n = 0, 1, \ldots.$$

The state X_n of the system at the start of period n is defined to be the number of customers in the system, either waiting or being served. Then, $\{X_n\}$ is a Markov chain. Specify the following transition probabilities in terms of α, β, p, and q: $P_{00}, P_{01}, P_{02}, P_{10}, P_{11}$, and P_{12}. State any additional assumptions that you make.

4.2.8 An airline reservation system has a single computer, which breaks down on any given day with probability p. It takes 2 days to restore a failed computer to normal service. Form a Markov chain by taking as states the pairs (x, y), where x is the number of machines in operating condition at the end of a day and y is 1 if a day's labor has been expended on a machine, and 0 otherwise. The transition probability matrix is

To state From state ↓	→ $(1,0)$	$(0,0)$	$(0,1)$
$(1,0)$	q	p	0
$\mathbf{P} = (0,0)$	0	0	1
$(0,1)$	1	0	0

Compute the system availability $\pi_{(1,0)}$ for $p = 0.01, 0.02, 0.05$, and 0.10.

4.3   The Classification of States

Not all Markov chains are regular. We consider some examples.

The Markov chain whose transition probability matrix is the identity matrix

$$\mathbf{P} = \begin{array}{cc} & \begin{array}{cc} 0 & 1 \end{array} \\ \begin{array}{c} 0 \\ 1 \end{array} & \left\| \begin{array}{cc} 1 & 0 \\ 0 & 1 \end{array} \right\| \end{array}$$

remains always in the state in which it starts. State trivially $\mathbf{P}^n = \mathbf{P}$ for all n, the Markov chain X_n has a limiting distribution, but it obviously depends on the initial state.

The Markov chain whose transition probability matrix is

$$\mathbf{P} = \begin{array}{cc} & \begin{array}{cc} 0 & 1 \end{array} \\ \begin{array}{c} 0 \\ 1 \end{array} & \left\| \begin{array}{cc} 1 & 0 \\ 0 & 1 \end{array} \right\| \end{array}$$

oscillates deterministically between the two states. The Markov chain is *periodic*, and no limiting distribution exists. When n is an odd number, then $\mathbf{P}^n = \mathbf{P}$, but when n is even, then $\mathbf{P}^n$ is the 2×2 identity matrix.

When $\mathbf{P}$ is the matrix

$$\mathbf{P} = \begin{array}{cc} & \begin{array}{cc} 0 & 1 \end{array} \\ \begin{array}{c} 0 \\ 1 \end{array} & \left\| \begin{array}{cc} \dfrac{1}{2} & \dfrac{1}{2} \\ 0 & 1 \end{array} \right\| \end{array},$$

$\mathbf{P}^n$ is given by

$$\mathbf{P}^n = \begin{array}{cc} & \begin{array}{cc} 0 & 1 \end{array} \\ \begin{array}{c} 0 \\ 1 \end{array} & \left\| \begin{array}{cc} \left(\dfrac{1}{2}\right)^n & 1 - \left(\dfrac{1}{2}\right)^n \\ 0 & 1 \end{array} \right\| \end{array},$$

and the limit is

$$\lim_{n \to \infty} \mathbf{P}^n = \begin{array}{cc} & \begin{array}{cc} 0 & 1 \end{array} \\ \begin{array}{c} 0 \\ 1 \end{array} & \left\| \begin{array}{cc} 0 & 1 \\ 0 & 1 \end{array} \right\| \end{array}.$$

Here, state 0 is *transient*; after the process starts from state 0, there is a positive probability that it will never return to that state.

The three matrices just presented illustrated three distinct types of behavior in addition to the convergence exemplified by a regular Markov chain. Various and more elaborate combinations of these behaviors are also possible. Some definitions and classifications of states and matrices are needed in order to sort out the variety of possibilities.

4.3.1 Irreducible Markov Chains

j is said to be *accessible* from state i if $P_{ij}^{(n)} > 0$ for some integer $n \geq 0$; i.e., state j is accessible from state i if there is positive probability that state j can be reached starting from state i in some finite number of transitions. Two states i and j, each accessible to the other, are said to *communicate*, and we write $i \leftrightarrow j$. If two states i and j do not communicate, then either

$$P_{ij}^{(n)} = 0 \quad \text{for all } n \geq 0$$

or

$$P_{ji}^{(n)} = 0 \quad \text{for all } n \geq 0$$

or both relations are true. The concept of communication is an equivalence relation:

1. $i \leftrightarrow i$ (reflexivity), a consequence of the definition of

$$P_{ij}^{(0)} = \delta_{ij} = \begin{cases} 1 & i = j, \\ 0 & i \neq j. \end{cases}$$

2. If $i \leftrightarrow j$, then $j \leftrightarrow i$ (symmetry), from the definition of communication.
3. If $i \leftrightarrow j$ and $j \leftrightarrow k$, then $i \leftrightarrow k$ (transitivity).

The proof of transitivity proceeds as follows: $i \leftrightarrow j$ and $j \leftrightarrow k$ imply that there exist integers n and m such that $P_{ij}^{(n)} > 0$ and $P_{jk}^{(m)} > 0$. Consequently, by the nonnegativity of each $P_{rs}^{(t)}$, we conclude that

$$P_{ik}^{(n+m)} = \sum_{r=0}^{\infty} P_{ir}^{(n)} P_{rk}^{(m)} \geq P_{ij}^{(n)} P_{jk}^{(m)} > 0.$$

A similar argument shows the existence of an integer ν such that $P_{ki}^{(\nu)} > 0$, as desired.

We can now partition the totality of states into equivalence classes. The states in an equivalence class are those that communicate with each other. It may be possible starting in one class to enter some other class with positive probability; if so, however, it is clearly not possible to return to the initial class, or else the two classes would together form a single class. We say that the Markov chain is irreducible if the equivalence relation induces only one class. In other words, a process is irreducible if all states communicate with each other.

To illustrate this concept, we consider the transition probability matrix

$$
\mathbf{P} =
\begin{Vmatrix}
\frac{1}{2} & \frac{1}{2} & \vdots & 0 & 0 & 0 \\
\frac{1}{4} & \frac{3}{4} & \vdots & 0 & 0 & 0 \\
\cdots & \cdots & \cdots & \cdots & \cdots & \cdots \\
0 & 0 & \vdots & 0 & 1 & 0 \\
0 & 0 & \vdots & \frac{1}{2} & 0 & \frac{1}{2} \\
0 & 0 & \vdots & 0 & 1 & 0
\end{Vmatrix}
=
\begin{Vmatrix}
\mathbf{P}_1 & 0 \\
0 & \mathbf{P}_2
\end{Vmatrix},
$$

where $\mathbf{P}_1$ is an abbreviation for the matrix formed from the initial two rows and columns of $\mathbf{P}$, and similarly for $\mathbf{P}_2$. This Markov chain clearly divides into the two classes composed of states $\{1, 2\}$ and states $\{3, 4, 5\}$.

If the state of X_0 lies in the first class, then the state of the system thereafter remains in this class, and for all purposes the relevant transition matrix is $\mathbf{P}_1$. Similarly, if the initial state belongs to the second class, then the relevant transition matrix is P_2. This is a situation where we have two completely unrelated processes labeled together.

In the random walk model with transition matrix

$$
\begin{array}{c}
\phantom{\mathbf{P} =} \quad \text{states} \\
\mathbf{P} =
\begin{Vmatrix}
1 & 0 & 0 & 0 & \cdots & 0 & 0 & 0 \\
q & 0 & p & 0 & \cdots & 0 & 0 & 0 \\
0 & q & 0 & p & \cdots & 0 & 0 & 0 \\
\vdots & & & & & \vdots & \vdots & \vdots \\
0 & & & \cdots & & q & 0 & p \\
0 & & & \cdots & & 0 & 0 & 1
\end{Vmatrix}
\begin{matrix}
0 \\ 1 \\ 2 \\ \vdots \\ a-1 \\ a
\end{matrix}
\end{array}
\tag{4.15}
$$

we have the three classes $\{0\}$, $\{1, 2, \ldots, a-1\}$, and $\{a\}$. In this example, it is possible to reach the first class or third class from the second class, but it is not possible to return to the second class from either the first or the third class.

4.3.2 Periodicity of a Markov Chain

We define the *period* of state i, written $d(i)$, to be the greatest common divisor (g.c.d.) of all integers $n \geq 1$ for which $P_{ii}^{(n)} > 0$. (If $P_{ii}^{(n)} = 0$ for all $n \geq 1$, define $d(i) = 0$.) In a random walk (4.15), every transient state $1, 2, \ldots, \alpha - 1$ has period 2. If $P_{ii} > 0$ for some single state i, then that state now has period 1, since the system can remain in this state any length of time.

In a finite Markov chain of n states with transition matrix

$$
\mathbf{P} =
\overbrace{
\begin{Vmatrix}
0 & 1 & 0 & 0 & \cdots & 0 \\
0 & 0 & 1 & 0 & \cdots & 0 \\
\vdots & & & & & \vdots \\
0 & 0 & & & \cdots & 1 \\
1 & 0 & 0 & & \cdots & 0
\end{Vmatrix}
}^{n},
$$

each state has period n.

Consider the Markov chain whose transition probability matrix is

$$
\mathbf{P} = \begin{array}{c@{}c}
 & \begin{array}{cccc} 0 & 1 & 2 & 3 \end{array} \\
\begin{array}{c} 0 \\ 1 \\ 2 \\ 3 \end{array} &
\left\|\begin{array}{cccc}
0 & 1 & 0 & 0 \\
0 & 0 & 1 & 0 \\
0 & 0 & 0 & 1 \\
\frac{1}{2} & 0 & \frac{1}{2} & 0
\end{array}\right\|.
\end{array}
$$

We evaluate $P_{00} = 0$, $P_{00}^{(2)} = 0$, $P_{00}^{(3)} = 0$, $P_{00}^{(4)} = \frac{1}{2}$, $P_{00}^{(5)} = 0$, $P_{00}^{(6)} = \frac{1}{4}$. The set of integers $n \geq 1$ for which $P_{00}^{(n)} > 0$ is $\{4, 6, 8, \ldots\}$. The period of state 0 is $d(0) = 2$, the greatest common divisor of this set.

Example Suppose that the precipitation in a certain locale depends on the season (Wet or Dry) as well as on the precipitation level (High or Low) during the preceding season. We model the process as a Markov chain whose states are of the form (x, y), where x denotes the season ($W = $ Wet, $D = $ Dry) and y denotes the precipitation level ($H = $ High, $L = $ Low). Suppose the transition probability matrix is

$$
\mathbf{P} = \begin{array}{c@{}c}
 & \begin{array}{cccc} (W,H) & (W,L) & (D,H) & (D,L) \end{array} \\
\begin{array}{c} (W,H) \\ (W,L) \\ (D,H) \\ (D,L) \end{array} &
\left\|\begin{array}{cccc}
0 & 0 & 0.8 & 0.2 \\
0 & 0 & 0.4 & 0.6 \\
0.7 & 0.3 & 0 & 0 \\
0.2 & 0.8 & 0 & 0
\end{array}\right\|.
\end{array}
$$

All states are periodic with period $d = 2$.

A situation in which the demand for an inventory item depends on the month of the year as well as on the demand during the previous month would lead to a Markov chain whose states had period $d = 12$.

The random walk on the states $0, \pm 1, \pm 2, \ldots$ with probabilities $P_{i,i+1} = p$, $P_{i,i-1} = q = 1 - p$ is periodic with period $d = 2$.

We state, without proof, three basic properties of the period of a state:

1. If $i \leftrightarrow j$, then $d(i) = d(j)$.

   This assertion shows that the period is a constant in each class of communicating states.

2. If state i has period $d(i)$, then there exists an integer N depending on i such that for all integers $n \geq N$,

$$
P_{ii}^{(nd(i))} > 0.
$$

   This asserts that a return to state i can occur at all sufficiently large multiples of the period $d(i)$.

3. If $P_{ji}^{(m)} > 0$, then $P_{ji}^{(m+nd(i))} > 0$ for all n (a positive integer) sufficiently large.

A Markov chain in which each state has period 1 is called *aperiodic*. The vast majority of Markov chain processes we deal with are aperiodic. Results will be developed for the aperiodic case, and the modified conclusions for the general case will be stated, usually without proof.

4.3.3   Recurrent and Transient States

Consider an arbitrary, but fixed, state i. We define, for each integer $n \geq 1$,

$$f_{ii}^{(n)} = \Pr\{X_n = i, X_\nu \neq i, \nu = 1, 2, \ldots, n-1 | X_0 = i\}.$$

In other words, $P_{ii}^{(n)}$ is the probability that starting from state i, the first return to state i occurs at the nth transition. Clearly, $f_{ii}^{(1)} = P_{ii}$, and $f_{ii}^{(n)}$ may be calculated recursively according to

$$P_{ii}^{(n)} = \sum_{k=0}^{n} f_{ii}^{(k)} P_{ii}^{(n-k)}, \quad n \geq 1, \tag{4.16}$$

where we define $f_{ii}^{(0)} = 0$ for all i. Equation (4.16) is derived by decomposing the event from which $P_{ii}^{(n)}$ is computed according to the time of the first return to state i. Indeed, consider all the possible realizations of the process for which $X_0 = i, X_n = i$, and the first return to state i occurs at the kth transition. Call this event E_k. The events E_k ($k = 1, 2, \ldots, n$) are clearly mutually exclusive. The probability of the event that the first return is at the kth transition is by definition $f_{ii}^{(k)}$. In the remaining $n - k$ transitions, we are dealing only with those realizations for which $X_n = i$. Using the Markov property, we have

$$\Pr\{E_k\} = \Pr\{\text{first return is at } k\text{th transition} | X_0 = i\} \Pr\{X_n = i | X_k = i\}$$
$$= f_{ii}^{(k)} P_{ii}^{(n-k)}, \quad 1 \leq k \leq n$$

(recall that $P_{ii}^0 = 1$). Hence,

$$\Pr\{X_n = i | X_0 = i\} = \sum_{k=1}^{n} \Pr\{E_k\} = \sum_{k=1}^{n} f_{ii}^{(k)} P_{ii}^{(n-k)} = \sum_{k=0}^{n} f_{ii}^{(k)} P_{ii}^{(n-k)},$$

since by definition $f_{ii}^{(0)} = 0$. The verification of (4.16) is now complete.

When the process starts from state i, the probability that it returns to state i at some time is

$$f_{ii} = \sum_{n=0}^{\infty} f_{ii}^{(n)} = \lim_{N \to \infty} \sum_{n=0}^{N} f_{ii}^{(n)}. \tag{4.17}$$

We say that a state i is *recurrent* if $f_{ii} = 1$. This definition says that a state i is recurrent if and only if, after the process starts from state i, the probability of its returning to state i after some finite length of time is one. A nonrecurrent state is said to be *transient*.

Consider a transient state i. Then, the probability that a process starting from state i returns to state i at least once is $f_{ii} < 1$. Because of the Markov property, the probability that the process returns to state i at least twice is $(f_{ii})^2$, and repeating the argument, we see that the probability that the process returns to i at least k times is $(f_{ii})^k$ for $k = 1, 2, \ldots$. Let M be the random variable that counts the number of times that the process returns to i. Then, we have shown that M has the geometric distribution in which

$$\Pr\{M \geq k | X_0 = i\} = (f_{ii})^k \quad \text{for } k = 1, 2, \ldots \tag{4.18}$$

and

$$E[M | X_0 = i] = \frac{f_{ii}}{1 - f_{ii}}. \tag{4.19}$$

Theorem 4.2 establishes a criterion for the recurrence of a state i in terms of the transition probabilities $P_{ii}^{(n)}$.

Theorem 4.2. *A state i is recurrent if and only if*

$$\sum_{n=1}^{\infty} P_{ii}^{(n)} = \infty.$$

Equivalently, state i is transient if and only if $\sum_{n=1}^{\infty} P_{ii}^{(n)} < \infty$.

Proof. Suppose first that state i is transient so that, by definition, $f_{ii} < 1$, and let M count the total number of returns to state i. We write M in terms of indicator random variables as

$$M = \sum_{n=1}^{\infty} 1\{X_n = i\},$$

where

$$1\{X_n = i\} = \begin{cases} 1 & \text{if } X_n = i, \\ 0 & \text{if } X_n \neq i. \end{cases}$$

Now, equation (3.5) shows that $E[M | X_0 = i] < \infty$ when i is transient. But then

$$\infty > E[M | X_0 = i] = \sum_{n=1}^{\infty} E[1\{X_n = i\} | X_0 = i]$$

$$= \sum_{n=1}^{\infty} P_{ii}^{(n)},$$

as claimed.   ∎

Conversely, suppose $\sum_{n=1}^{\infty} P_{ii}^{(n)} < \infty$. Then, M is a random variable whose mean is finite, and thus, M must be finite. That is, starting from state i, the process returns to state i only a finite number of times. Then, there must be a positive probability that, starting from state i, the process never returns to that state. In other words, $1 - f_{ii} > 0$ or $f_{ii} < 1$, as claimed.

Corollary 4.1. *If $i \leftrightarrow j$ and if i is recurrent, then j is recurrent.*

Proof. Since $i \leftrightarrow j$, there exists $m, n \geq 1$ such that

$$P_{ij}^{(n)} > 0 \quad \text{and} \quad P_{ji}^{(m)} > 0.$$

Let $v > 0$. We obtain, by the usual argument (see Section 4.3.1), $P_{jj}^{(m+n+v)} \geq P_{ji}^{(m)} P_{ii}^{(v)} P_{ij}^{(n)}$ and, on summing,

$$\sum_{v=0}^{\infty} P_{jj}^{(m+n+v)} \geq \sum_{v=0}^{\infty} P_{ji}^{(m)} P_{ii}^{(v)} P_{ij}^{(n)} = P_{ji}^{(m)} P_{ij}^{(n)} \sum_{v=0}^{\infty} P_{ii}^{(v)}.$$

Hence, if $\sum_{v=0}^{\infty} P_{ii}^{(v)}$ diverges, then $\sum_{v=0}^{\infty} P_{jj}^{(v)}$ also diverges. ∎

This corollary proves that recurrence, like periodicity, is a class property; that is, all states in an equivalence class are either recurrent or nonrecurrent.

Example Consider the one-dimensional random walk on the positive and negative integers, where at each transition the particle moves with probability p one unit to the right and with probability q one unit to the left ($p + q = 1$). Hence,

$$P_{00}^{(2n+1)} = 0, \quad n = 0, 1, 2, \ldots,$$

and

$$P_{00}^{(2n)} = \binom{2n}{n} p^n q^n = \frac{(2n)!}{n!n!} p^n q^n. \tag{4.20}$$

We appeal now to Stirling's formula (see Chapter 1, (1.60)),

$$n! \sim n^{n+1/2} e^{-n} \sqrt{2\pi}. \tag{4.21}$$

Applying (4.21) to (4.20), we obtain

$$P_{00}^{(2n)} \sim \frac{(pq)^n 2^{2n}}{\sqrt{\pi n}} = \frac{(4pq)^n}{\sqrt{\pi n}}.$$

It is readily verified that $p(1-p) = pq \le \frac{1}{4}$, with equality holding if and only if $p = q = \frac{1}{2}$. Hence, $\sum_{n=0}^{\infty} P_{00}^{(n)} = \infty$ if and only if $p = \frac{1}{2}$. Therefore, from Theorem 4.2, the one-dimensional random walk is recurrent if and only if $p = q = \frac{1}{2}$. Remember that recurrence is a class property. Intuitively, if $p \ne q$, there is positive probability that a particle initially at the origin will drift to $+\infty$ if $p > q$ (to $-\infty$ if $p < q$) without ever returning to the origin.

Exercises

4.3.1 A Markov chain has a transition probability matrix

$$
\begin{array}{c|cccccccc}
 & 0 & 1 & 2 & 3 & 4 & 5 & 6 & 7 \\
\hline
0 & 0 & 1 & 0 & 0 & 0 & 0 & 0 & 0 \\
1 & 0 & 0 & 1 & 0 & 0 & 0 & 0 & 0 \\
2 & 0 & 0 & 0 & 1 & 0 & 0 & 0 & 0 \\
3 & 0 & 0 & 0 & 0 & 1 & 0 & 0 & 0 \\
4 & 0.5 & 0 & 0 & 0 & 0 & 0.5 & 0 & 0 \\
5 & 0 & 0 & 0 & 0 & 0 & 0 & 1 & 0 \\
6 & 0 & 0 & 0 & 0 & 0 & 0 & 0 & 1 \\
7 & 1 & 0 & 0 & 0 & 0 & 0 & 0 & 0 \\
\end{array}
$$

Find the equivalence classes. For which integers $n = 1, 2, \ldots, 20$, is it true that

$$P_{00}^{(n)} > 0?$$

What is the period of the Markov chain?

Hint: One need not compute the actual probabilities. See Section 4.1.1.

4.3.2 Which states are transient and which are recurrent in the Markov chain whose transition probability matrix is

$$
\begin{array}{c|cccccc}
 & 0 & 1 & 2 & 3 & 4 & 5 \\
\hline
0 & \frac{1}{3} & 0 & \frac{1}{3} & 0 & 0 & \frac{1}{3} \\
1 & \frac{1}{2} & \frac{1}{4} & \frac{1}{4} & 0 & 0 & 0 \\
2 & 0 & 0 & 0 & 0 & 1 & 0 \\
3 & \frac{1}{4} & \frac{1}{4} & \frac{1}{4} & 0 & 0 & \frac{1}{4} \\
4 & 0 & 0 & 1 & 0 & 0 & 0 \\
5 & 0 & 0 & 0 & 0 & 0 & 1 \\
\end{array}
$$?

4.3.3 A Markov chain on states $\{0, 1, 2, 3, 4, 5\}$ has transition probability matrix

(a)
$$
\begin{Vmatrix}
\frac{1}{3} & 0 & \frac{2}{3} & 0 & 0 & 0 \\
0 & \frac{1}{4} & 0 & \frac{3}{4} & 0 & 0 \\
\frac{2}{3} & 0 & \frac{1}{3} & 0 & 0 & 0 \\
0 & \frac{1}{5} & 0 & \frac{4}{5} & 0 & 0 \\
\frac{1}{4} & \frac{1}{4} & 0 & 0 & \frac{1}{4} & \frac{1}{4} \\
\frac{1}{6} & \frac{1}{6} & \frac{1}{6} & \frac{1}{6} & \frac{1}{6} & \frac{1}{6}
\end{Vmatrix}.
$$

(b)
$$
\begin{Vmatrix}
1 & 0 & 0 & 0 & 0 & 0 \\
0 & \frac{3}{4} & \frac{1}{4} & 0 & 0 & 0 \\
0 & \frac{1}{8} & \frac{7}{8} & 0 & 0 & 0 \\
\frac{1}{4} & \frac{1}{4} & 0 & \frac{1}{8} & \frac{3}{8} & 0 \\
\frac{1}{3} & 0 & \frac{1}{6} & \frac{1}{4} & \frac{1}{4} & 0 \\
0 & 0 & 0 & 0 & 0 & 1
\end{Vmatrix}.
$$

Find all communicating classes; which classes are transient and which are recurrent?

4.3.4 Determine the communicating classes and period for each state of the Markov chain whose transition probability matrix is

$$
\begin{array}{c}
 & \begin{array}{cccccc} 0 & 1 & 2 & 3 & 4 & 5 \end{array} \\
\begin{array}{c} 0 \\ 1 \\ 2 \\ 3 \\ 4 \\ 5 \end{array} &
\begin{Vmatrix}
\frac{1}{2} & 0 & 0 & 0 & \frac{1}{2} & 0 \\
0 & 0 & 1 & 0 & 0 & 0 \\
0 & 0 & 0 & 1 & 0 & 0 \\
0 & 0 & 0 & 0 & 1 & 0 \\
0 & 0 & 0 & 0 & 0 & 1 \\
0 & 0 & \frac{1}{3} & \frac{1}{3} & 0 & \frac{1}{3}
\end{Vmatrix}.
\end{array}
$$

Problems

4.3.1 A two-state Markov chain has the transition probability matrix

$$\mathbf{P} = \begin{array}{c} \\ 0 \\ 1 \end{array} \begin{array}{cc} 0 & 1 \\ \left\| \begin{array}{cc} 1-a & a \\ b & 1-b \end{array} \right\| \end{array}.$$

(a) Determine the first return distribution

$$f_{00}^{(n)} = \Pr\{X_1 \neq 0, \ldots, X_{n-1} \neq 0, X_n = 0 | X_0 = 0\}.$$

(b) Verify equation (4.16) when $i = 0$. (Refer to Chapter 3, (4.40).)

4.3.2 Show that a finite-state aperiodic irreducible Markov chain is regular and recurrent.

4.3.3 Recall the first return distribution (Section 4.3.3),

$$f_{ii}^{(n)} = \Pr\{X_1 \neq i, X_2 \neq j \ldots, X_{n-1} \neq i, X_n = i | X_0 = i\} \quad \text{for } n = 1, 2, \ldots,$$

with $f_{ii}^{(0)} = 0$ by convention. Using equation (4.16), determine $f_{00}^{(n)}, n = 1, 2, 3, 4, 5$, for the Markov chain whose transition probability matrix is

$$\begin{array}{c} \\ 0 \\ 1 \\ 2 \\ 3 \end{array} \begin{array}{cccc} 0 & 1 & 2 & 3 \\ \left\| \begin{array}{cccc} 0 & \frac{1}{2} & 0 & \frac{1}{2} \\ 0 & 0 & 1 & 0 \\ 0 & 0 & 0 & 1 \\ \frac{1}{2} & 0 & 0 & \frac{1}{2} \end{array} \right\| \end{array}.$$

4.4 The Basic Limit Theorem of Markov Chains

Consider a recurrent state i. Then,

$$f_{ii}^{(n)} = \Pr\{X_n = i, X_v \neq i \quad \text{for} \quad v = 1, \ldots, n-1 | X_0 = i\} \tag{4.22}$$

is the probability distribution of the *first return time*

$$R_i = \min\{n \geq 1; X_n = i\}. \tag{4.23}$$

This is

$$f_{ii}^{(n)} = \Pr\{R_i = n | X_0 = i\} \quad \text{for } n = 1, 2, \ldots. \tag{4.24}$$

Since state i is recurrent by assumption, then $f_{ii} = \sum_{n=1}^{\infty} = f_{ii}^{(n)} = 1$, and R_i is a finite-valued random variable. The mean duration between visits to state i is

$$m_i = E[R_i | X_0 = i] = \sum_{n=1}^{\infty} n f_{ii}^{(n)}. \tag{4.25}$$

After starting in i, then, on the average, the process is in state i once every $m_i = E[R_i | X_0 = i]$ units of time. The basic limit theorem of Markov chains states this result in a sharpened form.

Theorem 4.3. *The basic limit theorem of Markov chains*

(a) Consider a recurrent irreducible aperiodic Markov chain. Let $P_{ii}^{(n)}$ be the probability of entering state i at the nth transition, $n = 0, 1, 2, \ldots$, given that $X_0 = i$ (the initial state is i). By our earlier convention $P_{ii}^{(0)} = 1$. Let $f_{ii}^{(n)}$ be the probability of first returning to state i at the nth transition, $n = 0, 1, 2, \ldots$, where $f_{ii}^{(0)} = 0$. Then,

$$\lim_{n \to \infty} P_{ii}^{(n)} = \frac{1}{\sum_{n=0}^{\infty} n f_{ii}^{(n)}} = \frac{1}{m_i}. \tag{4.26}$$

(b) under the same conditions as in (a), $\lim_{n \to \infty} P_{ji}^{(n)} = \lim_{n \to \infty} P_{ii}^{(n)}$ for all states j.

Remark Let C be a recurrent class. Then, $P_{ij}^{(n)} = 0$ for $i \in C, j \notin C$, and every n. Hence, once in C, it is not possible to leave C. It follows that the submatrix $\|P_{ij}\|$, i, $j \in C$, is a transition probability matrix and the associated Markov chain is irreducible and recurrent. The limit theorem, therefore, applies verbatim to any aperiodic recurrent class.

If $\lim_{n \to \infty} P_{ii}^{(n)} > 0$ for one i in an aperiodic recurrent class, then $\pi_j > 0$ for all j in the class of i. In this case, we call the class *positive recurrent* or strongly ergodic. If each $\pi_i = 0$ and the class is recurrent, we speak of the class as *null recurrent* or weakly ergodic. In terms of the first return time $R_i = \min\{n \geq 1; X_n = i\}$, state i is positive recurrent if $m_i = E[R_i | X_0 = i] < \infty$ and null recurrent if $m_i = \infty$. This statement is immediate from the equality $\lim_{n \to \infty} P_{ii}^{(n)} = \pi_i = 1/m_i$. An alternative method for determining the limiting distribution π_i for a positive recurrent aperiodic class is given in Theorem 4.4.

Theorem 4.4. *In a positive recurrent aperiodic class with states $j = 0, 1, 2, \ldots$,*

$$\lim_{n \to \infty} P_{jj}^{(n)} = \pi_j = \sum_{i=0}^{\infty} \pi_i P_{ij}, \quad \sum_{i=0}^{\infty} \pi_i = 1,$$

and the π's are uniquely determined by the set of equations

$$\pi_i \geq 0, \quad \sum_{i=0}^{\infty} \pi_i = 1, \quad and \quad \pi_j = \sum_{i=0}^{\infty} \pi_i P_{ij} \quad for\ j = 0, 1, \ldots. \tag{4.27}$$

Any set $(\pi_i)_{i=0}^{\infty}$ satisfying (4.27) is called a *stationary probability distribution* of the Markov chain. The term "stationary" derives from the property that a Markov chain started according to a stationary distribution will follow this distribution at all points of time. Formally, if $\Pr\{X_0 = i\} = \pi_i$, then $\Pr\{X_n = i\} = \pi_i$ for all $n = 1, 2, \ldots$. We check this for the case $n = 1$; the general case follows by induction. We write

$$\Pr\{X_1 = i\} = \sum_{k=0}^{\infty} \Pr\{X_0 = k\} \Pr\{X_1 = i | X_0 = k\}$$

$$= \sum_{k=0}^{\infty} \pi_k P_{ki} = \pi_i,$$

where the last equality follows because $\pi = (\pi_0, \pi_1, \ldots)$ is a stationary distribution. When the initial state X_0 is selected according to the stationary distribution, then the joint probability distribution of (X_n, X_{n+1}) is given by

$$\Pr\{X_n = i, X_{n+1} = j\} = \pi_i P_{ij}.$$

The reader should supply the proof.

A limiting distribution, when it exists, is always a stationary distribution, but the converse is not true. There may exist a stationary distribution but no limiting distribution. For example, there is no limiting distribution for the periodic Markov chain whose transition probability matrix is

$$\mathbf{P} = \begin{Vmatrix} 0 & 1 \\ 1 & 0 \end{Vmatrix},$$

but $\pi = (\frac{1}{2}, \frac{1}{2})$ is a stationary distribution, since

$$\left(\frac{1}{2}, \frac{1}{2}\right) \begin{Vmatrix} 0 & 1 \\ 1 & 0 \end{Vmatrix} = \left(\frac{1}{2}, \frac{1}{2}\right).$$

Example Consider the class of random walks whose transition matrices are given by

$$\mathbf{P} = \|P_{ij}\| = \begin{Vmatrix} 0 & 1 & 0 & & \cdots \\ q_1 & 0 & p_1 & & \cdots \\ 0 & q_2 & 0 & p_2 & \cdots \\ \vdots & & & & \end{Vmatrix}.$$

This Markov chain has period 2. Nevertheless, we investigate the existence of a stationary probability distribution; that is, we wish to determine the positive solutions of

$$x_i = \sum_{j=0}^{\infty} x_j P_{ji} = p_{i-1} x_{i-1} + q_{i+1} x_{i+1}, \quad i = 0, 1, \ldots, \tag{4.28}$$

under the normalization

$$\sum_{i=0}^{\infty} x_i = 1,$$

where $p_{-1} = 0$ and $p_0 = 1$, and thus, $x_0 = q_1 x_l$. Using equation (4.28) for $i = 1$, we could determine x_2 in terms of x_0. Equation (4.28) for $i = 2$ determines x_3 in terms of x_0, and so forth. It is immediately verified that

$$x_i = \frac{p_{i-1} p_{i-2} \cdots p_1}{q_i q_{i-1} \cdots q_1} x_0 = x_0 \prod_{k=0}^{i-1} \frac{p_k}{q_{k+1}}, \quad i \geq 1,$$

is a solution of (4.28), with x_0 still to be determined. Now, since

$$1 = x_0 + \sum_{i=1}^{\infty} x_0 \prod_{k=0}^{i-1} \frac{p_k}{q_{k+1}},$$

we have

$$x_0 = \frac{1}{1 + \sum_{i=1}^{\infty} \prod_{k=0}^{i-1} \frac{p_k}{q_{k+1}}},$$

and so

$$x_0 > 0 \text{ if and only if } \sum_{i=1}^{\infty} \prod_{k=0}^{i-1} \frac{p_k}{q_{k+1}} < \infty.$$

In particular, if $p_k = p$ and $q_k = q = 1 - p$ for $k \geq 1$, the series

$$\sum_{i=1}^{\infty} \prod_{k=0}^{i-1} \frac{p_k}{q_{k+1}} = \frac{1}{p} \sum_{i=1}^{\infty} \left(\frac{p}{q}\right)^i$$

converges only when $p < q$, and then

$$\frac{1}{p} \sum_{i=1}^{\infty} \left(\frac{p}{q}\right)^i = \frac{1}{p} \frac{p/q}{1 - p/q} = \frac{1}{q - p},$$

and

$$x_0 = \frac{1}{1 + 1/(q - p)} = \frac{q - p}{1 + q - p} = \frac{1}{2}\left(1 - \frac{p}{q}\right),$$

$$x_k = \frac{1}{p}\left(\frac{p}{q}\right)^k x_0 = \frac{1}{2p}\left(1 - \frac{p}{q}\right)\left(\frac{p}{q}\right)^k \quad \text{for } k = 1, 2, \ldots.$$

Example Consider now the Markov chain that represents the success runs of binomial trials. The transition probability matrix is

$$\begin{Vmatrix} p_0 & 1-p_0 & 0 & 0 & \cdots \\ p_1 & 0 & 1-p_1 & 0 & \cdots \\ p_1 & 0 & 0 & 1-p_2 & \cdots \\ \vdots & \vdots & \vdots & \vdots & \end{Vmatrix} \qquad (0 < p_k < 1).$$

The states of this Markov chain all belong to the same equivalence class (any state can be reached from any other state). Since recurrence is a class property (see Corollary 4.1), we will investigate recurrence for the zeroth state.

Let $R_0 = \min\{n \geq 1; X_n = 0\}$ be the time of first return to state 0. It is easy to evaluate

$$\Pr\{R_0 > 1 | X_0 = 0\} = (1 - p_0),$$
$$\Pr\{R_0 > 2 | X_0 = 0\} = (1 - p_0)(1 - p_1),$$
$$\Pr\{R_0 > 3 | X_0 = 0\} = (1 - p_0)(1 - p_1)(1 - p_2),$$

$$\vdots$$

$$\Pr\{R_0 > k | X_0 = 0\} = (1 - p_0)(1 - p_1) \cdots (1 - p_{k-1}) = \prod_{i=0}^{k-1}(1 - p_i).$$

In terms of the first return distribution

$$f_{00}^{(n)} = \Pr\{R_0 = n | X_0 = 0\},$$

we have

$$\Pr\{R_0 > k | X_0 = 0\} = 1 - \sum_{n=1}^{k} f_{00}^{(n)},$$

or

$$\sum_{n=1}^{k} f_{00}^{(n)} = 1 - \Pr\{R_0 > k | X_0 = 0\} = 1 - \prod_{i=0}^{k-1}(1 - p_i).$$

By definition, state 0 is recurrent provided $\sum_{n=1}^{\infty} f_{00}^{(n)} = 1$. In terms of $p_0, p_1, \ldots$ then, state 0 is recurrent whenever $\lim_{k \to \infty} \Pi_{i=0}^{k-1}(1 - p_i) = \Pi_{i=0}^{\infty}(1 - p_i) = 0$. Lemma 4.1 shows that $\Pi_{i=0}^{\infty}(1 - p_i) = 0$ is equivalent, in this case, to the condition $\sum_{i=0}^{\infty} p_i = \infty$.

Lemma 4.1. *If $0 < p_i < 1, i = 0, 1, 2, \ldots$, then $u_m = \Pi_{i=0}^{m}(1 - p_l) \to 0$ as $m \to \infty$ if and only if $\sum_{i=0}^{\infty} p_i = \infty$.*

Proof. Assume $\sum_{i=0}^{\infty} p_i = \infty$. Since the series expansion for $\exp(-p_i)$ is an alternating series with terms decreasing in absolute value, we can write

$$1 - p_i < 1 - p_i + \frac{p_i^2}{2!} - \frac{p_i^3}{3!} + \cdots = \exp(-p_i), \quad i = 0, 1, 2, \ldots. \tag{4.29}$$

Since (4.29) holds for all i, we obtain $\Pi_{i=0}^{m}(1 - p_i) < \exp\left(-\sum_{i=0}^{m} p_i\right)$. But by assumption,

$$\lim_{m \to \infty} \sum_{i=0}^{m} p_i = \infty;$$

hence,

$$\lim_{m \to \infty} \prod_{i=0}^{m}(1 - p_i) = 0.$$

To prove necessity, observe that from a straightforward induction,

$$\prod_{i=j}^{m}(1 - p_i) > (1 - p_j - p_{j+1} - \cdots - p_m)$$

for any j and all $m = j+1, j+2, \ldots$. Assume now that $\sum_{i=1}^{\infty} p_i < \infty$; then, $0 < \sum_{i=j}^{\infty} p_i < 1$ for some $j > 1$. Thus,

$$\lim_{m \to \infty} \prod_{i=j}^{m}(1 - p_i) > \lim_{m \to \infty} \left(1 - \sum_{i=j}^{m} p_i\right) > 0,$$

which contradicts $u_m \to 0$. ■

State 0 is recurrent when $\Pi_{i=0}^{\infty}(1 - p_i) = 0$, or equivalently, when $\sum_{i=0}^{\infty} p_i = \infty$. The state is positive recurrent when $m_0 = E[R_0 | X_0 = 0] < \infty$. But

$$m_0 = \sum_{k=0}^{\infty} \Pr\{R_0 > k | X_0 = 0\}$$

$$= 1 + \sum_{k=1}^{\infty} \prod_{i=0}^{k-1}(1 - p_i).$$

Thus, positive recurrence requires the stronger condition that $\sum_{k=1}^{\infty} \Pi_{i=0}^{k-1}(1 - p_i) < \infty$, and in this case, the stationary probability π_0 is given by

$$\pi_0 = \frac{1}{m_0} = \frac{1}{1 + \sum_{k=1}^{\infty} \Pi_{i=0}^{k-1}(1 - p_i)}.$$

From the equations for the stationary distribution, we have

$$(1 - p_0)\pi_0 = \pi_1,$$
$$(1 - p_1)\pi_1 = \pi_2,$$
$$(1 - p_2)\pi_2 = \pi_3,$$
$$\vdots$$

or

$$\pi_1 = (1 - p_0)\pi_0,$$
$$\pi_2 = (1 - p_1)\pi_1 = (1 - p_1)(1 - p_0)\pi_0,$$
$$\pi_3 = (1 - p_2)\pi_2 = (1 - p_2)(1 - p_1)(1 - p_0)\pi_0,$$

and, in general,

$$\pi_k = \pi_0 \prod_{i=0}^{k-1} (1 - p_i) \quad \text{for } k \geq 1.$$

In the special case where $p_i = p = 1 - q$ for $i = 0, 1, \ldots$, then $\Pi_{i=0}^{k-1}(1 - p_i) = q^k$,

$$m_0 = 1 + \sum_{k=1}^{\infty} q^k = \frac{1}{p}$$

so that $\pi_k = pq^k$ for $k = 0, 1, \ldots$.

Remark Suppose $a_0, a_1, a_2, \ldots$ is a convergent sequence of real numbers where $a_n \to a$ as $n \to \infty$. Then, it can be proved by elementary methods that the partial averages of the sequence also converge in the form

$$\lim_{n \to \infty} \frac{1}{n} \sum_{k=0}^{n-1} a_k = a. \tag{4.30}$$

Applying (4.30) with $a_n = P_{ii}^{(n)}$, where i is a member of a positive recurrent aperiodic class, we obtain

$$\lim_{n \to \infty} \frac{1}{n} \sum_{m=0}^{n-1} P_{ii}^{(m)} = \pi_i = \frac{1}{m_i} > 0, \tag{4.31}$$

where $\pi = (\pi_0, \pi_1, \ldots)$ is the stationary distribution and where m_i is the mean return time for state i. Let $M_i^{(n)}$ be the random variable that counts the total number of visits to state i during time periods $0, 1, \ldots, n - 1$. We may write

$$M_i^{(n)} = \sum_{k=0}^{n-1} 1\{X_k = i\}, \tag{4.32}$$

where

$$1\{X_k = i\} = \begin{cases} 1 & \text{if } X_k = i \\ 0 & \text{if } X_k \neq i \end{cases} \tag{4.33}$$

and then see that

$$E\left[M_i^{(n)} | X_0 = i\right] = \sum_{k=0}^{n-1} E[1\{X_k = i\} | X_0 = i] = \sum_{k=0}^{n-1} P_{ii}^{(k)}. \tag{4.34}$$

Then, referring to (4.31), we have

$$\lim_{n \to \infty} \frac{1}{n} E\left[M_i^{(n)} | X_0 = i\right] = \frac{1}{m_i}. \tag{4.35}$$

In words, the long run ($n \to \infty$) mean visits to state i per unit time equals π_i, the probability of state i under the stationary distribution.

Next, let $r(i)$ define a cost or rate to be accumulated upon each visit to state i. The total cost accumulated during the first n stages is

$$\begin{aligned} R^{(n-1)} &= \sum_{k=0}^{n-1} r(X_k) = \sum_{k=0}^{n-1} \sum_i 1\{X_k = i\} r(i) \\ &= \sum_{i=0}^{\infty} M_i^{(n)} r(i). \end{aligned} \tag{4.36}$$

This leads to the following derivation showing that the long run mean cost per unit time equals the mean cost evaluated over the stationary distribution:

$$\begin{aligned} \lim_{n \to \infty} \frac{1}{n} E[R^{(n-1)} | X_0 = i] &= \lim_{n \to \infty} \sum_{i=0}^{\infty} \frac{1}{n} E\left[M_i^{(n)} | X_0 = i\right] r(i) \\ &= \sum_{i=0}^{\infty} \pi_i r(i). \end{aligned} \tag{4.37}$$

(When the Markov chain has an infinite number of states, then the derivation requires that a limit and infinite sum be interchanged. A sufficient condition to justify this interchange is that $r(i)$ be a bounded function of i.)

Remark *The Periodic Case* If i is a member of a recurrent periodic irreducible Markov chain with period d, one can show that $P_{ii}^m = 0$ if m is not a multiple of d (i.e., if $m \neq nd$ for any n), and that

$$\lim_{n \to \infty} P_{ii}^{nd} = \frac{d}{m_i}.$$

These last two results are easily combined with (4.30) to show that (4.31) also holds in the periodic case. If $m_i < \infty$, then the chain is positive recurrent and

$$\lim_{n\to\infty} \frac{1}{n} \sum_{m=0}^{n-1} P_{ii}^{(m)} = \pi_i = \frac{1}{m_i}, \tag{4.38}$$

where $\pi = (\pi_0, \pi_1, \ldots)$ is given as the unique nonnegative solution to

$$\pi_j = \sum_{k=0}^{\infty} \pi_k P_{kj}, \quad j = 0, 1, \ldots,$$

and

$$\sum_{j=0}^{\infty} \pi_j = 1.$$

That is, a unique stationary distribution $\pi = (\pi_0, \pi_1, \ldots)$ exists for a positive recurrent periodic irreducible Markov chain, and the mean fraction of time in state i converges to π_i as the number of stages n grows to infinity.

The convergence of (4.38) does not require the chain to start in state i. Under the same conditions,

$$\lim_{n\to\infty} \frac{1}{n} \sum_{m=0}^{n-1} P_{ki}^{(m)} = \pi_i = \frac{1}{m_i}$$

holds for all states $k = 0, 1, \ldots$ as well.

Exercises

4.4.1 Determine the limiting distribution for the Markov chain whose transition probability matrix is

$$\mathbf{P} = \begin{array}{c c} & \begin{array}{c c c c c} 0 & 1 & 2 & 3 & 4 \end{array} \\ \begin{array}{c} 0 \\ 1 \\ 2 \\ 3 \\ 4 \end{array} & \left\| \begin{array}{c c c c c} q & p & 0 & 0 & 0 \\ q & 0 & p & 0 & 0 \\ q & 0 & 0 & p & 0 \\ q & 0 & 0 & 0 & p \\ 1 & 0 & 0 & 0 & 0 \end{array} \right\| \end{array},$$

where $p > 0$, $q > 0$, and $p + q = 1$.

4.4.2 Consider the Markov chain whose transition probability *matrix* is given by

$$
\mathbf{P} =
\begin{array}{c c}
 & \begin{array}{cccc} 0 & 1 & 2 & 3 \end{array} \\
\begin{array}{c} 0 \\ 1 \\ 2 \\ 3 \end{array} &
\left\| \begin{array}{cccc}
0 & 1 & 0 & 0 \\
0.1 & 0.4 & 0.2 & 0.3 \\
0.2 & 0.2 & 0.5 & 0.1 \\
0.3 & 0.3 & 0.4 & 0
\end{array} \right\|
\end{array}
$$

(a) Determine the limiting probability π_0 that the process is in state 0.

(b) By pretending that state 0 is absorbing, use a first step analysis (Chapter 3, Section 3.4) and calculate the mean time m_{10} for the process to go from state 1 to state 0.

(c) Because the process always goes directly to state 1 from state 0, the mean *return* time to state 0 is $m_0 = 1 + m_{10}$. Verify equation (4.26), $\pi_0 = 1/m_0$.

4.4.3 Determine the stationary distribution for the periodic Markov chain whose transition probability matrix is

$$
\mathbf{P} =
\begin{array}{c c}
 & \begin{array}{cccc} 0 & 1 & 2 & 3 \end{array} \\
\begin{array}{c} 0 \\ 1 \\ 2 \\ 3 \end{array} &
\left\| \begin{array}{cccc}
0 & \frac{1}{2} & 0 & \frac{1}{2} \\
\frac{1}{4} & 0 & \frac{3}{4} & 0 \\
0 & \frac{1}{3} & 0 & \frac{2}{3} \\
\frac{1}{2} & 0 & \frac{1}{2} & 0
\end{array} \right\|
\end{array}.
$$

Problems

4.4.1 Consider the Markov chain on $\{0, 1\}$ whose transition probability matrix is

$$
\begin{array}{c c}
 & \begin{array}{cc} 0 & 1 \end{array} \\
\begin{array}{c} 0 \\ 1 \end{array} &
\left\| \begin{array}{cc}
1 - \alpha & \alpha \\
\beta & 1 - \beta
\end{array} \right\|
\end{array}, \quad 0 < \alpha, \beta < 1.
$$

(a) Verify that $(\pi_0, \pi_1) = (\beta/(\alpha + \beta), \alpha/(\alpha + \beta))$ is a stationary distribution.

(b) Show that the first return distribution to state 0 is given by $f_{00}^{(1)} = (1 - \alpha)$ and $f_{00}^{(n)} = \alpha\beta(1 - \beta)^{n-2}$ for $n = 2, 3, \ldots$.

(c) Calculate the mean return time $m_0 = \sum_{n=1}^{\infty} n f_{00}^{(n)}$ and verify that $\pi_0 = 1/m_0$.

4.4.2 Determine the stationary distribution for the Markov chain whose transition probability matrix is

$$
\mathbf{P} =
\begin{array}{c}
\\
0 \\
1 \\
2 \\
3
\end{array}
\begin{array}{cccc}
0 & 1 & 2 & 3 \\
\left\|\begin{array}{cccc}
0 & 0 & \dfrac{1}{2} & \dfrac{1}{2} \\[6pt]
0 & 0 & \dfrac{1}{3} & \dfrac{2}{3} \\[6pt]
\dfrac{1}{4} & \dfrac{3}{4} & 0 & 0 \\[6pt]
\dfrac{1}{3} & \dfrac{2}{3} & 0 & 0
\end{array}\right\|
\end{array}.
$$

4.4.3 Consider a random walk Markov chain on state $0, 1, \ldots, N$ with transition probability matrix

$$
\begin{array}{c}
\\
0 \\
1 \\
2 \\
3 \\
\vdots \\
N-1 \\
N
\end{array}
\begin{array}{c}
\begin{array}{ccccccccc}
0 & 1 & 2 & 3 & 4 & 5 & & N-1 & N
\end{array} \\
\left\|\begin{array}{ccccccccc}
0 & 1 & 0 & 0 & 0 & 0 & \cdots & 0 & 0 \\
q_1 & 0 & p_1 & 0 & 0 & 0 & \cdots & 0 & 0 \\
0 & q_2 & 0 & p_2 & 0 & 0 & \cdots & 0 & 0 \\
0 & 0 & q_3 & 0 & p_3 & 0 & \cdots & 0 & 0 \\
\vdots & \vdots & \vdots & \vdots & \vdots & \vdots & \vdots & \vdots & \vdots \\
0 & 0 & 0 & 0 & 0 & 0 & \cdots & 0 & p_{N-1} \\
0 & 0 & 0 & 0 & 0 & 0 & \cdots & 1 & 0
\end{array}\right\|
\end{array},
$$

where $p_i + q_i = 1, p_i > 0, q_i > 0$ for all i.

The transition probabilities from state 0 and N "reflect" the process back into state $1, 2, \ldots, N-1$. Determine the limiting distribution.

4.4.4 Let $\{\alpha_i : i = 1, 2, \ldots\}$ be a probability distribution, and consider the Markov chain whose transition probability matrix is

$$
\begin{array}{c}
\\
0 \\
1 \\
2 \\
3 \\
4 \\
\vdots
\end{array}
\begin{array}{c}
\begin{array}{ccccccc}
0 & 1 & 2 & 3 & 4 & 5 & \cdots
\end{array} \\
\left\|\begin{array}{ccccccc}
\alpha_1 & \alpha_2 & \alpha_3 & \alpha_4 & \alpha_5 & \alpha_6 & \cdots \\
1 & 0 & 0 & 0 & 0 & 0 & \cdots \\
0 & 1 & 0 & 0 & 0 & 0 & \cdots \\
0 & 0 & 1 & 0 & 0 & 0 & \cdots \\
0 & 0 & 0 & 1 & 0 & 0 & \cdots \\
\vdots & \vdots & \vdots & \vdots & \vdots & \vdots & \vdots
\end{array}\right\|
\end{array}.
$$

What condition on the probability distribution $\{\alpha_i : i = 1, 2, \ldots\}$ is necessary and sufficient in order that a limiting distribution exist, and what is this limiting distribution? Assume $\alpha_1 > 0$ and $\alpha_2 > 0$ so that the chain is aperiodic.

4.4.5 Let P be the transition probability matrix of a finite-state regular Markov chain. Let $\mathbf{M} = \|m_{ij}\|$ be the matrix of mean *return* times.

(a) Use a first step argument to establish that

$$m_{ij} = 1 + \sum_{k \neq j} P_{ik} m_{kj}.$$

(b) Multiply both sides of the preceding by π_i and sum to obtain

$$\sum_i \pi_i m_{ij} = \sum_i \pi_i + \sum_{k \neq j} \sum_i \pi_i P_{ik} m_{kj}.$$

Simplify this to show (see equation (4.26))

$$\pi_j m_{jj} = 1, \quad \text{or } \pi_j = 1/m_{jj}.$$

4.4.6 Determine the period of state 0 in the Markov chain whose transition probability matrix is

$$
\mathbf{P} =
\begin{array}{c|cccccccc}
 & 3 & 2 & 1 & 0 & -1 & -2 & -3 & -4 \\
\hline
3 & 0 & 0 & 0 & 1 & 0 & 0 & 0 & 0 \\
2 & 1 & 0 & 0 & 0 & 0 & 0 & 0 & 0 \\
1 & 0 & 1 & 0 & 0 & 0 & 0 & 0 & 0 \\
0 & 0 & 0 & \frac{1}{2} & 0 & \frac{1}{2} & 0 & 0 & 0 \\
-1 & 0 & 0 & 0 & 0 & 0 & 1 & 0 & 0 \\
-2 & 0 & 0 & 0 & 0 & 0 & 0 & 1 & 0 \\
-3 & 0 & 0 & 0 & 0 & 0 & 0 & 0 & 1 \\
-4 & 0 & 0 & 0 & 1 & 0 & 0 & 0 & 0 \\
\end{array}.
$$

4.4.7 An individual either drives his car or walks in going from his home to his office in the morning, and from his office to his home in the afternoon. He uses the following strategy: If it is raining in the morning, then he drives the car, provided it is at home to be taken. Similarly, if it is raining in the afternoon and his car is at the office, then he drives the car home. He walks on any morning or afternoon that it is not raining or the car is not where he is. Assume that, independent of the past, it rains during successive mornings and afternoons with constant probability p. In the long run, on what fraction of *days* does our man walk in the rain? What if he owns two cars?

4.4.8 A Markov chain on states $0, 1, \ldots$ has transition probabilities

$$P_{ij} = \frac{1}{i+2} \quad \text{for } j = 0, 1, \ldots, i, i+1.$$

Find the stationary distribution.

4.5 Reducible Markov Chains*

Recall that states i and j *communicate* if it is possible to reach state j starting from state i, and vice versa, and a Markov chain is *irreducible* if all pairs of states communicate. In this section, we show, mostly by example, how to analyze more general Markov chains.

Consider first the Markov chain whose transition probability matrix is

$$
\mathbf{P} = \left\|
\begin{array}{cccc}
\dfrac{1}{2} & \dfrac{1}{2} & 0 & 0 \\[2mm]
\dfrac{1}{4} & \dfrac{3}{4} & 0 & 0 \\[2mm]
0 & 0 & \dfrac{1}{3} & \dfrac{2}{3} \\[2mm]
0 & 0 & \dfrac{2}{3} & \dfrac{1}{3}
\end{array}
\right\| .
$$

Which we write in the form

$$
\mathbf{P} = \left\|
\begin{array}{cc}
\mathbf{P}_1 & \mathbf{0} \\
\mathbf{0} & \mathbf{P}_2
\end{array}
\right\| ,
$$

where

$$
\mathbf{P} = \left\|
\begin{array}{cc}
\dfrac{1}{2} & \dfrac{1}{2} \\[2mm]
\dfrac{1}{4} & \dfrac{3}{4}
\end{array}
\right\|
\quad \text{and} \quad
\mathbf{P} = \left\|
\begin{array}{cc}
\dfrac{1}{3} & \dfrac{2}{3} \\[2mm]
\dfrac{2}{3} & \dfrac{1}{3}
\end{array}
\right\| .
$$

The chain has two communicating classes, the first two states forming one class and the last two states forming the other. Then,

$$
\mathbf{P}^2 = \left\|
\begin{array}{cccc}
\dfrac{1}{2} & \dfrac{1}{2} & 0 & 0 \\[2mm]
\dfrac{1}{4} & \dfrac{3}{4} & 0 & 0 \\[2mm]
0 & 0 & \dfrac{1}{3} & \dfrac{2}{3} \\[2mm]
0 & 0 & \dfrac{2}{3} & \dfrac{1}{3}
\end{array}
\right\|
\times
\left\|
\begin{array}{cccc}
\dfrac{1}{2} & \dfrac{1}{2} & 0 & 0 \\[2mm]
\dfrac{1}{4} & \dfrac{3}{4} & 0 & 0 \\[2mm]
0 & 0 & \dfrac{1}{3} & \dfrac{2}{3} \\[2mm]
0 & 0 & \dfrac{2}{3} & \dfrac{1}{3}
\end{array}
\right\|
$$

---

* The overwhelming majority of Markov chains encountered in stochastic modeling are irreducible. Reducible Markov chains form a specialized topic.

$$= \begin{Vmatrix} \dfrac{3}{8} & \dfrac{5}{8} & 0 & 0 \\[2mm] \dfrac{5}{16} & \dfrac{11}{16} & 0 & 0 \\[2mm] 0 & 0 & \dfrac{5}{9} & \dfrac{4}{9} \\[2mm] 0 & 0 & \dfrac{4}{9} & \dfrac{5}{9} \end{Vmatrix} = \begin{Vmatrix} \mathbf{P}_1^2 & \mathbf{0} \\ \mathbf{0} & \mathbf{P}_2^2 \end{Vmatrix},$$

and, in general,

$$\mathbf{P}^n = \begin{Vmatrix} \mathbf{P}_1^n & \mathbf{0} \\ \mathbf{0} & \mathbf{P}_2^n \end{Vmatrix}, \quad n \geq 1. \tag{4.39}$$

Equation (4.39) is the mathematical expression of the property that it is not possible to communicate back and forth between distinct communicating classes; once in the first class, the process remains there thereafter; and similarly, once in the second class, the process remains there. In effect, two completely unrelated processes have been labeled together. The transition probability matrix $\mathbf{P}$ is reducible to the irreducible matrices $\mathbf{P}_1$ and $\mathbf{P}_2$. It follows from (4.39) that

$$\lim_{n \to \infty} \mathbf{P}^n = \begin{Vmatrix} \pi_0^{(1)} & \pi_1^{(1)} & 0 & 0 \\ \pi_0^{(1)} & \pi_1^{(1)} & 0 & 0 \\ 0 & 0 & \pi_0^{(2)} & \pi_1^{(2)} \\ 0 & 0 & \pi_0^{(2)} & \pi_1^{(2)} \end{Vmatrix},$$

where

$$\lim_{n \to \infty} \mathbf{P}_1^n = \begin{Vmatrix} \pi_0^{(1)} & \pi_1^{(1)} \\ \pi_0^{(1)} & \pi_1^{(1)} \end{Vmatrix} \quad \text{and} \quad \lim_{n \to \infty} \mathbf{P}_2^n = \begin{Vmatrix} \pi_0^{(2)} & \pi_1^{(2)} \\ \pi_0^{(2)} & \pi_1^{(2)} \end{Vmatrix}.$$

We solve for $\pi^{(1)} = \left(\pi_0^{(1)}, \pi_1^{(1)} \right)$ and $\pi^{(2)} = \left(\pi_0^{(2)}, \pi_1^{(2)} \right)$ in the usual way:

$$\frac{1}{2}\pi_0^{(1)} + \frac{1}{4}\pi_1^{(1)} = \pi_0^{(1)},$$

$$\frac{1}{2}\pi_0^{(1)} + \frac{3}{4}\pi_1^{(1)} = \pi_1^{(1)},$$

$$\pi_0^{(1)} + \pi_1^{(1)} = 1,$$

or

$$\pi_0^{(1)} = \frac{1}{3}, \quad \pi_1^{(1)} = \frac{2}{3}, \tag{4.40}$$

and because $\mathbf{P}_2$ is doubly stochastic (see Section 4.1.1), it follows that $\pi_0^{(2)} = \frac{1}{2}$, $\pi_1^{(2)} = \frac{1}{2}$.

The basic limit theorem of Markov chains, Theorem 4.3, referred to an irreducible Markov chain. The limit theorem applies verbatim to any aperiodic recurrent class in a reducible Markov chain. If i, j are in the same aperiodic recurrent class, then $P_{ij}^{(n)} \to 1/m_j \geq 0$ as $n \to \infty$. If i, j are in the same periodic recurrent class, then $n^{-1} \sum_{m=0}^{n-1} P_{ij}^{(m)} \to 1/m_j \geq 0$ as $n \to \infty$.

If j is a transient state, then $P_{jj}^{(n)} \to 0$ as $n \to \infty$, and, more generally, $P_{ij}^{(n)} \to 0$ as $n \to \infty$ for all initial states i.

In order to complete the discussion of the limiting behavior of $P_{ij}^{(n)}$, we still must consider the case where i is transient and j is recurrent. Consider the transition probability matrix

$$\mathbf{P} = \begin{array}{c} \\ 0 \\ 1 \\ 2 \\ 3 \end{array} \begin{array}{cccc} 0 & 1 & 2 & 3 \\ \left\| \begin{array}{cccc} \frac{1}{2} & \frac{1}{2} & 0 & 0 \\ \frac{1}{4} & \frac{3}{4} & 0 & 0 \\ \frac{1}{4} & \frac{1}{4} & \frac{1}{4} & \frac{1}{4} \\ 0 & 0 & 0 & 1 \end{array} \right\| \end{array}.$$

There are three classes: $\{0, 1\}$, $\{2\}$, and $\{3\}$, and of these, $\{0, 1\}$ and $\{3\}$ are recurrent, while $\{2\}$ is transient. Starting from state 2, the process ultimately gets absorbed in one of the other classes. The question is, Which one? or more precisely, What are the probabilities of absorption in the two recurrent classes starting from state 2?

A first step analysis answers the question. Let u denote the probability of absorption in class $\{0, 1\}$ starting from state 2. Then, $1 - u$ is the probability of absorption in class $\{3\}$. Conditioning on the first step, we have

$$u = \left(\frac{1}{4} + \frac{1}{4} \right) 1 + \frac{1}{4} u + \frac{1}{4} (0) = \frac{1}{2} + \frac{1}{4} u,$$

or $u = \frac{2}{3}$. With probability $\frac{2}{3}$, the process enters $\{0, 1\}$ and remains there ever after. The stationary distribution for the recurrent class $\{0, 1\}$, computed in (5.2), is $\pi_0 = \frac{1}{3}, \pi_1 = \frac{2}{3}$. Therefore, $\lim_{n \to \infty} P_{20}^{(n)} = \frac{2}{3} \times \frac{1}{3} = \frac{2}{9}, \lim_{n \to \infty} P_{21}^{(n)} = \frac{2}{3} \times \frac{2}{3} = \frac{4}{9}$. That is, we multiply the probability of entering the class $\{0, 1\}$ by the appropriate probabilities under the stationary distribution for the various states in the class. In matrix form, the limiting behavior of $\mathbf{P}^n$ is given by

$$\lim_{n \to \infty} \mathbf{P}^n = \begin{Vmatrix} \dfrac{1}{3} & \dfrac{2}{3} & 0 & 0 \\ \dfrac{1}{3} & \dfrac{2}{3} & 0 & 0 \\ \dfrac{2}{9} & \dfrac{4}{9} & 0 & \dfrac{1}{3} \\ 0 & 0 & 0 & 1 \end{Vmatrix}.$$

To firm up the principles, consider one last example:

$$\mathbf{P}^n = \begin{array}{c|cc|cc|cc} & 0 & 1 & 2 & 3 & 4 & 5 \\ \hline 0 & \dfrac{1}{2} & \dfrac{1}{2} & 0 & 0 & 0 & 0 \\ 1 & \dfrac{1}{3} & \dfrac{2}{3} & 0 & 0 & 0 & 0 \\ \hline 2 & \dfrac{1}{3} & 0 & 0 & \dfrac{1}{3} & \dfrac{1}{6} & \dfrac{1}{6} \\ 3 & \dfrac{1}{6} & \dfrac{1}{6} & \dfrac{1}{6} & 0 & \dfrac{1}{3} & \dfrac{1}{6} \\ \hline 4 & 0 & 0 & 0 & 0 & 0 & 1 \\ 5 & 0 & 0 & 0 & 0 & 1 & 0 \end{array}.$$

There are three classes: $C_1 = \{0, 1\}$, $C_2 = \{2, 3\}$, and $C_3 = \{4, 5\}$. The stationary distribution in C_1 is (π_0, π_1), where

$$\frac{1}{2}\pi_0 + \frac{1}{3}\pi_1 = \pi_0,$$

$$\frac{1}{2}\pi_0 + \frac{2}{3}\pi_1 = \pi_1,$$

$$\pi_0 + \pi_1 = 1.$$

Then, $\pi_0 = \frac{2}{5}$ and $\pi_1 = \frac{3}{5}$.

Class C_3 is periodic, and $P_{ij}^{(n)}$ does not converge for i, j in $C_3 = \{4, 5\}$. The time averages do converge, however; and $\lim_{n \to \infty} n^{-1} \sum_{m=0}^{n-1} P_{ij}^{(m)} = \frac{1}{2}$ for $i = 3, 4$ and $j = 3, 4$.

For the transient class $C_2 = \{2, 3\}$, let u_i be the probability of ultimate absorption in class $C_1 = \{0, 1\}$ starting from state i for $i = 2, 3$. From a first step analysis, then

$$u_2 = \frac{1}{3}(1) + 0(1) + 0u_2 + \frac{1}{3}u_3 + \frac{1}{6}(0) + \frac{1}{6}(0),$$
$$u_3 = \frac{1}{6}(1) + \frac{1}{6}(1) + \frac{1}{6}u_2 + 0u_3 + \frac{1}{3}(0) + \frac{1}{6}(0),$$

or

$$u_2 = \frac{1}{3}\frac{1}{3}u_3; \quad u_3 = \frac{1}{3} + \frac{1}{6}u_2.$$

The solution is $u_2 = \frac{8}{17}$ and $u_3 = \frac{7}{17}$. Combining these partial answers in matrix form, we have

$$\lim_{n\to\infty} \mathbf{P} = \begin{array}{c} \\ 0 \\ 1 \\ 2 \\ 3 \\ 4 \\ 5 \end{array}\begin{array}{cccccc} 0 & 1 & 2 & 3 & 4 & 5 \end{array}\left\| \begin{array}{cccccc} \frac{2}{5} & \frac{3}{5} & 0 & 0 & 0 & 0 \\ \frac{2}{5} & \frac{3}{5} & 0 & 0 & 0 & 0 \\ \left(\frac{8}{17}\right)\left(\frac{2}{5}\right) & \left(\frac{8}{17}\right)\left(\frac{3}{5}\right) & 0 & 0 & X & X \\ \left(\frac{7}{17}\right)\left(\frac{2}{5}\right) & \left(\frac{7}{17}\right)\left(\frac{3}{5}\right) & 0 & 0 & X & X \\ 0 & 0 & 0 & 0 & X & X \\ 0 & 0 & 0 & 0 & X & X \end{array} \right\|,$$

where X denotes that the limit does not exist. For the time average, we have

$$\lim_{n\to\infty} \frac{1}{n}\sum_{m=0}^{n-1} \mathbf{P}^m = \begin{array}{c} \\ 0 \\ 1 \\ 2 \\ 3 \\ 4 \\ 5 \end{array}\begin{array}{cccccc} 0 & 1 & 2 & 3 & 4 & 5 \end{array}\left\| \begin{array}{cccccc} \frac{2}{5} & \frac{3}{5} & 0 & 0 & 0 & 0 \\ \frac{2}{5} & \frac{3}{5} & 0 & 0 & 0 & 0 \\ \left(\frac{8}{17}\right)\left(\frac{2}{5}\right) & \left(\frac{8}{17}\right)\left(\frac{3}{5}\right) & 0 & 0 & \left(\frac{9}{17}\right)\left(\frac{1}{2}\right) & \left(\frac{9}{17}\right)\left(\frac{1}{2}\right) \\ \left(\frac{7}{17}\right)\left(\frac{2}{5}\right) & \left(\frac{7}{17}\right)\left(\frac{3}{5}\right) & 0 & 0 & \left(\frac{10}{17}\right)\left(\frac{1}{2}\right) & \left(\frac{10}{17}\right)\left(\frac{1}{2}\right) \\ 0 & 0 & 0 & 0 & \frac{1}{2} & \frac{1}{2} \\ 0 & 0 & 0 & 0 & \frac{1}{2} & \frac{1}{2} \end{array} \right\|.$$

One possible behavior remains to be illustrated. It can occur only when there are an infinite number of states. In this case, it is possible that all states are transient or null recurrent and $\lim_{n\to\infty} P_{ij}^{(n)} = 0$ for all states i, j. For example, consider the deterministic Markov chain described by $X_n = X_0 + n$. The transition probability matrix is

$$
\mathbf{P}^n =
\begin{array}{c}
\begin{array}{ccccc} 0 & 1 & 2 & 3 \end{array} \\
\begin{array}{c} 0 \\ 1 \\ 2 \\ 3 \\ \vdots \end{array}
\left\|
\begin{array}{ccccc}
0 & 1 & 0 & 0 & \cdots \\
0 & 0 & 1 & 0 & \cdots \\
0 & 0 & 0 & 1 & \cdots \\
0 & 0 & 0 & 0 & \cdots \\
\vdots & \vdots & \vdots & \vdots &
\end{array}
\right\|
\end{array}.
$$

Then, all states are transient, and $\lim_{n\to\infty} P_{ij}^{(n)} = \lim_{n\to\infty} \Pr\{X_n = j | X_0 = i\} = 0$ for all states i, j.

If there is only a finite number M of states, then there are no null recurrent states and not all states can be transient. In fact, since $\sum_{j=0}^{M-1} P_{ij}^{(n)} = 1$ for all n, it cannot happen that $\lim_{n\to\infty} P_{ij}^{(n)} = 0$ for all j.

Exercises

4.5.1 Given the transition matrix

$$
\mathbf{P} =
\begin{array}{c}
\begin{array}{ccccc} 0 & 1 & 2 & 3 & 4 \end{array} \\
\begin{array}{c} 0 \\ 1 \\ 2 \\ 3 \\ 4 \end{array}
\left\|
\begin{array}{ccccc}
\dfrac{1}{4} & \dfrac{3}{4} & 0 & 0 & 0 \\
\dfrac{1}{2} & \dfrac{1}{2} & 0 & 0 & 0 \\
0 & 0 & 1 & 0 & 0 \\
0 & 0 & \dfrac{1}{3} & \dfrac{2}{3} & 0 \\
1 & 0 & 0 & 0 & 0
\end{array}
\right\|
\end{array},
$$

determine the limits, as $n \to \infty$, of $P_{i0}^{(n)}$ for $i = 0, 1, \ldots, 4$.

4.5.2 Given the transition matrix

$$
P = \begin{array}{c|ccccccc}
 & 1 & 2 & 3 & 4 & 5 & 6 & 7 \\
\hline
1 & \frac{1}{3} & \frac{2}{3} & 0 & 0 & 0 & 0 & 0 \\
2 & \frac{1}{4} & \frac{3}{4} & 0 & 0 & 0 & 0 & 0 \\
3 & 0 & 0 & 0 & \frac{2}{3} & \frac{1}{3} & 0 & 0 \\
4 & 0 & 0 & 1 & 0 & 0 & 0 & 0 \\
5 & 0 & 0 & 1 & 0 & 0 & 0 & 0 \\
6 & \frac{1}{6} & 0 & \frac{1}{6} & \frac{1}{6} & 0 & \frac{1}{4} & \frac{1}{4} \\
7 & 0 & 0 & 0 & 0 & 0 & 0 & 1
\end{array},
$$

derive the following limits, where they exist:

(a) $\lim_{n\to\infty} P_{11}^{(n)}$ (e) $\lim_{n\to\infty} P_{21}^{(n)}$

(b) $\lim_{n\to\infty} P_{31}^{(n)}$ (f) $\lim_{n\to\infty} P_{33}^{(n)}$

(c) $\lim_{n\to\infty} P_{61}^{(n)}$ (g) $\lim_{n\to\infty} P_{67}^{(n)}$

(d) $\lim_{n\to\infty} P_{63}^{(n)}$ (h) $\lim_{n\to\infty} P_{64}^{(n)}$

Problems

4.5.1 Describe the limiting behavior of the Markov chain whose transition probability matrix is

	0	1	2	3	4	5	6	7
0	0.1	0.1	0.2	0.2	0.1	0.1	0.1	0.1
1	0	0.1	0.1	0.1	0	0.3	0.2	0.2
2	0.6	0	0	0.1	0.1	0.1	0.1	0.0
3	0	0	0	0.3	0.7	0	0	0
4	0	0	0	0.7	0.3	0	0	0
5	0	0	0	0	0	0.3	0.4	0.3
6	0	0	0	0	0	0.1	0	0.9
7	0	0	0	0	0	0.8	0.2	0

Hint: First consider the matrices

$$
\mathbf{P}_A =
\begin{array}{c}
\\
0 \\
1 \\
2 \\
3\text{-}4 \\
5\text{-}7
\end{array}
\begin{array}{c}
\begin{array}{ccccc}
0 & 1 & 2 & 3\text{-}4 & 5\text{-}7
\end{array} \\
\left\|
\begin{array}{ccccc}
0.1 & 0.1 & 0.2 & 0.3 & 0.3 \\
0 & 0.1 & 0.1 & 0.1 & 0.7 \\
0.6 & 0 & 0 & 0.2 & 0.2 \\
0 & 0 & 0 & 1 & 0 \\
0 & 0 & 0 & 0 & 1
\end{array}
\right\|
\end{array}
$$

and

$$
\mathbf{P}_B =
\begin{array}{c}
\\
3 \\
4
\end{array}
\begin{array}{c}
\begin{array}{cc}
3 & 4
\end{array} \\
\left\|
\begin{array}{cc}
0.3 & 0.7 \\
0.7 & 0.3
\end{array}
\right\|
\end{array},
\qquad
\mathbf{P}_C =
\begin{array}{c}
\begin{array}{ccc}
5 & 6 & 7
\end{array} \\
\left\|
\begin{array}{ccc}
0.3 & 0.4 & 0.3 \\
0.1 & 0 & 0.9 \\
0.8 & 0.2 & 0
\end{array}
\right\|
\end{array}.
$$

4.5.2 Determine the limiting behavior of the Markov chain whose transition probability matrix is

$$
\mathbf{P} =
\begin{array}{c}
\\
0 \\
1 \\
2 \\
3 \\
4 \\
5 \\
6 \\
7
\end{array}
\begin{array}{c}
\begin{array}{cccccccc}
0 & 1 & 2 & 3 & 4 & 5 & 6 & 7
\end{array} \\
\left\|
\begin{array}{cccccccc}
0.1 & 0.2 & 0.1 & 0.1 & 0.2 & 0.1 & 0.1 & 0.1 \\
0 & 0.1 & 0.2 & 0.1 & 0 & 0.3 & 0.1 & 0.2 \\
0.5 & 0 & 0 & 0.2 & 0.1 & 0.1 & 0.1 & 0 \\
0 & 0 & 0.3 & 0.7 & 0 & 0 & 0 & 0 \\
0 & 0 & 0.6 & 0.4 & 0 & 0 & 0 & 0 \\
0 & 0 & 0 & 0 & 0 & 0.3 & 0.4 & 0.3 \\
0 & 0 & 0 & 0 & 0 & 0.2 & 0.2 & 0.6 \\
0 & 0 & 0 & 0 & 0 & 0.9 & 0.1 & 0
\end{array}
\right\|
\end{array}.
$$

5 Poisson Processes

5.1 The Poisson Distribution and the Poisson Process

Poisson behavior is so pervasive in natural phenomena and the Poisson distribution is so amenable to extensive and elaborate analysis as to make the Poisson process a cornerstone of stochastic modeling.

5.1.1 The Poisson Distribution

The Poisson distribution with parameter $\mu > 0$ is given by

$$p_k = \frac{e^{-\mu}\mu^k}{k!} \quad \text{for } k = 0, 1, \ldots. \tag{5.1}$$

Let X be a random variable having the Poisson distribution in (5.1). We evaluate the mean, or first moment, via

$$
\begin{aligned}
E[X] &= \sum_{k=0}^{\infty} k p_k = \sum_{k=1}^{\infty} \frac{k e^{-\mu}\mu^k}{k!} \\
&= \mu e^{-\mu} \sum_{k=1}^{\infty} \frac{\mu^{(k-1)}}{(k-1)!} \\
&= \mu.
\end{aligned}
$$

To evaluate the variance, it is easier first to determine

$$
\begin{aligned}
E[X(X-1)] &= \sum_{k=2}^{\infty} k(k-1) p_k \\
&= \mu^2 e^{-\mu} \sum_{k=2}^{\infty} \frac{\mu^{(k-2)}}{(k-2)!} \\
&= \mu^2.
\end{aligned}
$$

Then

$$
\begin{aligned}
E[X^2] &= E[X(X-1)] + E[X] \\
&= \mu^2 + \mu,
\end{aligned}
$$

while

$$\sigma_X^2 = \text{Var}[X] = E[X^2] - \{E[X]\}^2$$
$$= \mu^2 + \mu - \mu^2 = \mu.$$

Thus, the Poisson distribution has the unusual characteristic that both the mean and the variance are given by the same value μ.

Two fundamental properties of the Poisson distribution, which will arise later in a variety of forms, concern the sum of independent Poisson random variables and certain random decompositions of Poisson phenomena. We state these properties formally as Theorems 5.1 and 5.2.

Theorem 5.1. *Let X and Y be independent random variables having Poisson distributions with parameters μ and ν, respectively. Then the sum $X + Y$ has a Poisson distribution with parameter $\mu + \nu$.*

Proof. By the law of total probability,

$$\Pr\{X + Y = n\} = \sum_{k=0}^{n} \Pr\{X = k, Y = n - k\}$$

$$= \sum_{k=0}^{n} \Pr\{X = k\} \Pr\{Y = n - k\}$$

$$\text{(X and Y are independent)}$$

$$= \sum_{k=0}^{n} \left\{ \frac{\mu^k e^{-\mu}}{k!} \right\} \left\{ \frac{\nu^{n-k} e^{-\nu}}{(n-k)!} \right\}$$

$$= \frac{e^{-(\mu+\nu)}}{n!} \sum_{k=0}^{n} \frac{n!}{k!\,(n-k)!} \mu^k \nu^{n-k}. \tag{5.2}$$

The binomial expansion of $(\mu + \nu)^n$ is, of course,

$$(\mu + \nu)^n = \sum_{k=0}^{n} \frac{n!}{k!\,(n-k)!} \mu^k \nu^{n-k},$$

and so (5.2) simplifies to

$$\Pr\{X + Y = n\} = \frac{e^{-(\mu+\nu)}(\mu + \nu)^n}{n!}, \quad n = 0, 1, \ldots,$$

the desired Poisson distribution.

To describe the second result, we consider first a Poisson random variable N where the parameter is $\mu > 0$. Write N as a sum of ones in the form

$$N = \underbrace{1 + 1 + \cdots + 1}_{N \text{ ones}},$$

and next, considering each one separately and independently, erase it with probability $1 - p$ and keep it with probability p. What is the distribution of the resulting sum M, of the form $M = 1 + 0 + 0 + 1 + \cdots + 1$?

The next theorem states and answers the question in a more precise wording.   ■

Theorem 5.2. *Let N be a Poisson random variable with parameter μ, and conditional on N, let M have a binomial distribution with parameters N and p. Then the unconditional distribution of M is Poisson with parameter μp.*

Proof. The verification proceeds via a direct application of the law of total probability. Then

$$\Pr\{M = k\} = \sum_{n=0}^{\infty} \Pr\{M = k | N = n\} \Pr\{N = n\}$$

$$= \sum_{n=k}^{\infty} \left\{ \frac{n!}{k!\,(n-k)!} p^k (1-p)^{n-k} \right\} \left\{ \frac{\mu^n e^{-\mu}}{n!} \right\}$$

$$= \frac{e^{-\mu}(\mu p)^k}{k!} \sum_{n=k}^{\infty} \frac{[\mu(1-p)]^{n-k}}{(n-k)!}$$

$$= \frac{e^{-\mu}(\mu p)^k}{k!} e^{\mu(1-p)}$$

$$= \frac{e^{-\mu p}(\mu p)^k}{k!} \quad \text{for } k = 0, 1, \ldots,$$

which is the claimed Poisson distribution.   ■

5.1.2 The Poisson Process

The Poisson process entails notions of both independence and the Poisson distribution.

Definition A Poisson process of intensity, or rate, $\lambda > 0$ is an integer-valued stochastic process $\{X(t); t \geq 0\}$ for which

1. for any time points $t_0 = 0 < t_1 < t_2 < \cdots < t_n$, the process increments

$$X(t_1) - X(t_0),\ X(t_2) - X(t_1), \ldots, X(t_n) - X(t_{n-1})$$

are independent random variables;

2. for $s \geq 0$ and $t > 0$, the random variable $X(s+t) - X(s)$ has the Poisson distribution

$$\Pr\{X(s+t) - X(s) = k\} = \frac{(\lambda t)^k e^{-\lambda t}}{k!} \quad \text{for } k = 0, 1, \ldots;$$

3. $X(0) = 0$.

In particular, observe that if $X(t)$ is a Poisson process of rate $\lambda > 0$, then the moments are

$$E[X(t)] = \lambda t \quad \text{and} \quad \text{Var}[X(t)] = \sigma^2_{X(t)} = \lambda t.$$

Example Defects occur along an undersea cable according to a Poisson process of rate $\lambda = 0.1$ per mile. (a) What is the probability that no defects appear in the first two miles of cable? (b) Given that there are no defects in the first two miles of cable, what is the conditional probability of no defects between mile points two and three? To answer (a) we observe that $X(2)$ has a Poisson distribution whose parameter is $(0.1)(2) = 0.2$. Thus, $\Pr\{X(2) = 0\} = e^{-0.2} = 0.8187$. In part (b), we use the independence of $X(3) - X(2)$ and $X(2) - X(0) = X(2)$. Thus, the conditional probability is the same as the unconditional probability, and

$$\Pr\{X(3) - X(2) = 0\} = \Pr\{X(1) = 0\} = e^{-0.1} = 0.9048.$$

Example Customers arrive in a certain store according to a Poisson process of rate $\lambda = 4$ per hour. Given that the store opens at 9:00 A.M., what is the probability that exactly one customer has arrived by 9:30 and a total of five have arrived by 11:30 A.M.?

Measuring time t in hours from 9:00 A.M., we are asked to determine $\Pr\{X(\frac{1}{2}) = 1,$ $X(\frac{5}{2}) = 5\}$. We use the independence of $X(\frac{5}{2}) - X(\frac{1}{2})$ and $X(\frac{1}{2})$ to reformulate the question thus:

$$\Pr\left\{X\left(\frac{1}{2}\right) = 1, X\left(\frac{5}{2}\right) = 5\right\} = \Pr\left\{X\left(\frac{1}{2}\right) = 1, X\left(\frac{5}{2}\right) - X\left(\frac{1}{2}\right) = 4\right\}$$

$$= \left\{\frac{e^{-4(1/2)}4\left(\frac{1}{2}\right)}{1!}\right\}\left\{\frac{e^{-4(2)}[4(2)]^4}{4!}\right\}$$

$$= \left(2e^{-2}\right)\left(\frac{512}{3}e^{-8}\right) = 0.0154965.$$

5.1.3 Nonhomogeneous Processes

The rate λ in a Poisson process $X(t)$ is the proportionality constant in the probability of an event occurring during an arbitrarily small interval. To explain this more precisely,

$$\Pr\{X(t+h) - X(t) = 1\} = \frac{(\lambda h)e^{-\lambda h}}{1!}$$

$$= (\lambda h)\left(1 - \lambda h + \frac{1}{2}\lambda^2 h^2 - \cdots\right)$$

$$= \lambda h + o(h),$$

where $o(h)$ denotes a general and unspecified remainder term of smaller order than h.

It is pertinent in many applications to consider rates $\lambda = \lambda(t)$ that vary with time. Such a process is termed a *nonhomogeneous* or *nonstationary* Poisson process to distinguish it from the stationary, or homogeneous, process that we primarily consider. If $X(t)$ is a nonhomogeneous Poisson process with rate $\lambda(t)$, then an increment $X(t) - X(s)$, giving the number of events in an interval $(s, t]$, has a Poisson distribution with parameter $\int_s^t \lambda(u)\,du$, and increments over disjoint intervals are independent random variables.

Example Demands on a first aid facility in a certain location occur according to a nonhomogeneous Poisson process having the rate function

$$\lambda(t) = \begin{cases} 2t & \text{for } 0 \leq t < 1, \\ 2 & \text{for } 1 \leq t < 2, \\ 4 - t & \text{for } 2 \leq t \leq 4, \end{cases}$$

where t is measured in hours from the opening time of the facility. What is the probability that two demands occur in the first 2 h of operation and two in the second 2 h? Since demands during disjoint intervals are independent random variables, we can answer the two questions separately. The mean for the first 2 h is $\mu = \int_0^1 2t\,dt + \int_1^2 2\,dt = 3$, and thus

$$\Pr\{X(2) = 2\} = \frac{e^{-3}(3)^2}{2!} = 0.2240.$$

For the second 2 h, $\mu = \int_2^4 (4 - t)\,dt = 2$, and

$$\Pr\{X(4) - X(2) = 2\} = \frac{e^{-2}(2)^2}{2!} = 0.2707.$$

Let $X(t)$ be a nonhomogeneous Poisson process of rate $\lambda(t) > 0$ and define $\Lambda(t) = \int_0^t \lambda(u)\,du$. Make a deterministic change in the time scale and define a new process $Y(s) = X(t)$, where $s = \Lambda(t)$. Observe that $\Delta s = \lambda(t)\Delta t + o(\Delta t)$. Then

$$\Pr\{Y(s + \Delta s) - Y(s) = 1\} = \Pr\{X(t + \Delta t) - X(t) = 1\}$$
$$= \lambda(t)\Delta t + o(\Delta t)$$
$$= \Delta s + o(\Delta s),$$

so that $Y(s)$ is a homogeneous Poisson process of unit rate. By this means, questions about nonhomogeneous Poisson processes can be transformed into corresponding questions about homogeneous processes. For this reason, we concentrate our exposition on the latter.

5.1.4 Cox Processes

Suppose that $X(t)$ is a nonhomogeneous Poisson process, but where the rate function $\{\lambda(t), t \geq 0\}$ is itself a stochastic process. Such processes were introduced in 1955 as models for fibrous threads by Sir David Cox, who called them *doubly stochastic Poisson processes*. Now they are most often referred to as *Cox processes* in honor of their discoverer. Since their introduction, Cox processes have been used to model a myriad of phenomena, e.g., bursts of rainfall, where the likelihood of rain may vary with the season; inputs to a queueing system, where the rate of input varies over time, depending on changing and unmeasured factors; and defects along a fiber, where the rate and type of defect may change due to variations in material or manufacture. As these applications suggest, the process increments over disjoint intervals are, in general, statistically dependent in a Cox process, as contrasted with their postulated independence in a Poisson process.

Let $\{X(t); t \geq 0\}$ be a Poisson process of constant rate $\lambda = 1$. The very simplest Cox process, sometimes called a *mixed Poisson process*, involves choosing a single random variable Θ, and then observing the process $X'(t) = X(\Theta t)$. Given Θ, then X' is, conditionally, a Poisson process of constant rate $\lambda = \Theta$, but Θ is random, and typically, unobservable. If Θ is a continuous random variable with probability density function $f(\theta)$, then, upon removing the condition via the law of total probability, we obtain the marginal distribution

$$\Pr\{X'(t) = k\} = \int_0^\infty \frac{(\theta t)^k e^{-\theta t}}{k!} f(\theta) d\theta. \tag{5.3}$$

Problem 5.1.12 calls for carrying out the integration in (5.3) in the particular instance in which Θ has an exponential density.

Chapter 6, Section 6.7 develops a model for defects along a fiber in which a Markov chain in continuous time is the random intensity function for a Poisson process. A variety of functionals are evaluated for the resulting Cox process.

Exercises

5.1.1 Defects occur along the length of a filament at a rate of $\lambda = 2$ per foot.
   (a) Calculate the probability that there are no defects in the first foot of the filament.
   (b) Calculate the conditional probability that there are no defects in the second foot of the filament, given that the first foot contained a single defect.

5.1.2 Let $p_k = \Pr\{X = k\}$ be the probability mass function corresponding to a Poisson distribution with parameter λ. Verify that $p_0 = \exp\{-\lambda\}$, and that p_k may be computed recursively by $p_k = (\lambda/k)p_{k-1}$.

5.1.3 Let X and Y be independent Poisson distributed random variables with parameters α and β, respectively. Determine the conditional distribution of X, given that $N = X + Y = n$.

5.1.4 Customers arrive at a service facility according to a Poisson process of rate λ customer/hour. Let $X(t)$ be the number of customers that have arrived up to time t.
   (a) What is $\Pr\{X(t) = k\}$ for $k = 0, 1, \ldots$?
   (b) Consider fixed times $0 < s < t$. Determine the conditional probability $\Pr\{X(t) = n + k | X(s) = n\}$ and the expected value $E[X(t)X(s)]$.

5.1.5 Suppose that a random variable X is distributed according to a Poisson distribution with parameter λ. The parameter λ is itself a random variable, exponentially distributed with density $f(x) = \theta e^{-\theta x}$ for $x \geq 0$. Find the probability mass function for X.

5.1.6 Messages arrive at a telegraph office as a Poisson process with mean rate of 3 messages per hour.

   (a) What is the probability that no messages arrive during the morning hours 8:00 A.M. to noon?

   (b) What is the distribution of the time at which the first afternoon message arrives?

5.1.7 Suppose that customers arrive at a facility according to a Poisson process having rate $\lambda = 2$. Let $X(t)$ be the number of customers that have arrived up to time t. Determine the following probabilities and conditional probabilities:

   (a) $\Pr\{X(1) = 2\}$.

   (b) $\Pr\{X(1) = 2 \text{ and } X(3) = 6\}$.

   (c) $\Pr\{X(1) = 2 | X(3) = 6\}$.

   (d) $\Pr\{X(3) = 6 | X(1) = 2\}$.

5.1.8 Let $\{X(t); t \geq 0\}$ be a Poisson process having rate parameter $\lambda = 2$. Determine the numerical values to two decimal places for the following probabilities:

   (a) $\Pr\{X(1) \leq 2\}$.

   (b) $\Pr\{X(1) = 1 \text{ and } X(2) = 3\}$.

   (c) $\Pr\{X(1) \geq 2 | X(1) \geq 1\}$.

5.1.9 Let $\{X(t); t \geq 0\}$ be a Poisson process having rate parameter $\lambda = 2$. Determine the following expectations:

   (a) $E[X(2)]$.

   (b) $E\left[\{X(1)\}^2\right]$.

   (c) $E[X(1)X(2)]$.

Problems

5.1.1 Let $\xi_1, \xi_2, \ldots$ be independent random variables, each having an exponential distribution with parameter λ. Define a new random variable X as follows: If $\xi_1 > 1$, then $X = 0$; if $\xi_1 \leq 1$ but $\xi_1 + \xi_2 > 1$, then set $X = 1$; in general, set $X = k$ if

$$\xi_1 + \cdots + \xi_k \leq 1 < \xi_1 + \cdots + \xi_k + \xi_{k+1}.$$

Show that X has a Poisson distribution with parameter λ. (Thus, the method outlined can be used to simulate a Poisson distribution.)

Hint: $\xi_1 + \cdots + \xi_k$ has a gamma density

$$f_k(x) = \frac{\lambda^k x^{k-1}}{(k-1)!} e^{-\lambda x} \quad \text{for } x > 0.$$

Condition on $\xi_1 + \cdots + \xi_k$ and use the law of total probability to show

$$\Pr\{X = k\} = \int_0^1 [1 - F(1 - x)] f_k(x) \, dx,$$

where $F(x)$ is the exponential distribution function.

5.1.2 Suppose that minor defects are distributed over the length of a cable as a Poisson process with rate α, and that, independently, major defects are distributed over the cable according to a Poisson process of rate β. Let $X(t)$ be the number of defects, either major or minor, in the cable up to length t. Argue that $X(t)$ must be a Poisson process of rate $\alpha + \beta$.

5.1.3 The *generating function* of a probability mass function $p_k = \Pr\{X = k\}$, for $k = 0, 1, \ldots$, is defined by

$$g_X(s) = E\left[s^X\right] = \sum_{k=0}^{\infty} p_k s^k \quad \text{for } |s| < 1.$$

Show that the generating function for a Poisson random variable X with mean μ is given by

$$g_X(s) = e^{-\mu(1-s)}.$$

5.1.4 (Continuation) Let X and Y be independent random variables, Poisson distributed with parameters α and β, respectively. Show that the generating function of their sum $N = X + Y$ is given by

$$g_N(s) = e^{-(\alpha+\beta)(1-s)}.$$

Hint: Verify and use the fact that the generating function of a sum of independent random variables is the product of their respective generating functions. See Chapter 3, Section 3.9.2.

5.1.5 For each value of $h > 0$, let $X(h)$ have a Poisson distribution with parameter λh. Let $p_k(h) = \Pr\{X(h) = k\}$ for $k = 0, 1, \ldots$. Verify that

$$\lim_{h \to 0} \frac{1 - p_0(h)}{h} = \lambda, \quad \text{or } p_0(h) = 1 - \lambda h + o(h);$$

$$\lim_{h \to 0} \frac{p_1(h)}{h} = \lambda, \quad \text{or } p_1(h) = \lambda h + o(h);$$

$$\lim_{h \to 0} \frac{p_2(h)}{h} = 0, \quad \text{or } p_2(h) = o(h).$$

Here $o(h)$ stands for any remainder term of order less than h as $h \to 0$.

5.1.6 Let $\{X(t); t \geq 0\}$ be a Poisson process of rate λ. For $s, t > 0$, determine the conditional distribution of $X(t)$, given that $X(t + s) = n$.

5.1.7 Shocks occur to a system according to a Poisson process of rate λ. Suppose that the system survives each shock with probability α, independently of other shocks, so that its probability of surviving k shocks is α^k. What is the probability that the system is surviving at time t?

5.1.8 Find the probability $\Pr\{X(t) = 1, 3, 5, \ldots\}$ that a Poisson process having rate λ is odd.

5.1.9 Arrivals of passengers at a bus stop form a Poisson process $X(t)$ with rate $\lambda = 2$ per unit time. Assume that a bus departed at time $t = 0$ leaving no customers behind. Let T denote the arrival time of the next bus. Then, the number of passengers present when it arrives is $X(T)$. Suppose that the bus arrival time T is independent of the Poisson process and that T has the uniform probability density function

$$f_T(t) = \begin{cases} 1 & \text{for } 0 \leq t \leq 1, \\ 0 & \text{elsewhere.} \end{cases}$$

(a) Determine the conditional moments $E[X(T)|T = t]$ and $E\left[\{X(T)\}^2 | T = t\right]$.
(b) Determine the mean $E[X(T)]$ and variance $\text{Var}[X(T)]$.

5.1.10 Customers arrive at a facility at random according to a Poisson process of rate λ. There is a waiting time cost of c per customer per unit time. The customers gather at the facility and are processed or dispatched in groups at fixed times $T, 2T, 3T, \ldots$. There is a dispatch cost of K. The process is depicted in the following graph.

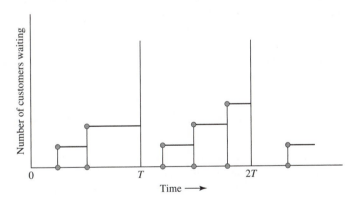

Figure 5.1 The number of customers in a dispatching system as a function of time.

(a) What is the total dispatch cost during the first cycle from time 0 to time T?
(b) What is the mean total customer waiting cost during the first cycle?
(c) What is the mean total customer waiting + dispatch cost per unit time during the first cycle?
(d) What value of T minimizes this mean cost per unit time?

5.1.11 Assume that a device fails when a cumulative effect of k shocks occurs. If the shocks happen according to a Poisson process of parameter λ, what is the density function for the life T of the device?

5.1.12 Consider the mixed Poisson process of Section 5.1.4, and suppose that the mixing parameter Θ has the exponential density $f(\theta) = e^{-\theta}$ for $\theta > 0$.

(a) Show that equation (5.3) becomes

$$\Pr\{X'(t) = j\} = \left(\frac{t}{1+t}\right)^j \left(\frac{1}{1+t}\right), \quad \text{for } j = 0, 1, \dots.$$

(b) Show that

$$\Pr\{X'(t) = j, X'(t+s) = j+k\} = \binom{j+k}{j} t^j s^k \left(\frac{1}{1+s+t}\right)^{j+k+1},$$

so that $X'(t)$ and the increment $X'(t+s) - X'(t)$ are not independent random variables, in contrast to the simple Poisson process as defined in Section 5.1.2.

5.2 The Law of Rare Events

The common occurrence of the Poisson distribution in nature is explained by the *law of rare events*. Informally, this law asserts that where a certain event may occur in any of a large number of possibilities, but where the probability that the event does occur in any given possibility is small, then the total number of events that do happen should follow, approximately, the Poisson distribution.

A more formal statement in a particular instance follows. Consider a large number N of independent Bernoulli trials where the probability p of success on each trial is small and constant from trial to trial. Let $X_{N,p}$ denote the total number of successes in the N trials, where $X_{N,p}$ follows the binomial distribution

$$\Pr\{X_{N,p} = k\} = \frac{N!}{k!\,(N-k)!} p^k (1-p)^{N-k} \quad \text{for } k = 0, \dots, N. \tag{5.4}$$

Now let us consider the limiting case in which $N \to \infty$ and $p \to 0$ in such a way that $Np = \mu > 0$ where μ is constant. It is a familiar fact (see Chapter 1, Section 1.3) that the distribution for $X_{N,p}$ becomes, in the limit, the Poisson distribution

$$\Pr\{X_\mu = k\} = \frac{e^{-\mu}\mu^k}{k!} \quad \text{for } k = 0, 1, \dots. \tag{5.5}$$

This form of the law of rare events is stated as a limit. In stochastic modeling, the law is used to suggest circumstances under which one might expect the Poisson distribution to prevail, at least approximately. For example, a large number of cars may pass through a given stretch of highway on any particular day. The probability that any specified car is in an accident is, we hope, small. Therefore, one might expect that the actual number of accidents on a given day along that stretch of highway would be, at least approximately, Poisson distributed.

While we have formally stated this form of the law of rare events as a mathematical limit, in older texts, (5.5) is often called "the Poisson approximation" to the binomial, the idea being that when N is large and p is small, the binomial probability (5.4) may be approximately evaluated by the Poisson probability (5.5) with $\mu = Np$. With today's computing power, exact binomial probabilities are not difficult to obtain, so there is little need to approximate them with Poisson probabilities. Such is not the case if the problem is altered slightly by allowing the probability of success to vary from trial to trial. To examine this proposal in detail, let $\epsilon_1, \epsilon_2, \ldots$ be independent Bernoulli random variables, where

$$\Pr\{\epsilon_i = 1\} = p_i \quad \text{and} \quad \Pr\{\epsilon_i = 0\} = 1 - p_i,$$

and let $S_n = \epsilon_1 + \cdots + \epsilon_n$. When $p_1 = p_2 = \cdots = p$, then S_n has a binomial distribution, and the probability $\Pr\{S_n = k\}$ for some $k = 0, 1, \ldots$ is easily computed. It is not so easily evaluated when the p's are unequal, with the binomial formula generalizing to

$$\Pr\{S_n = k\} = \Sigma^{(k)} \prod_{i=1}^{n} p_i^{x_i} (1 - p_i)^{1-x_i}, \tag{5.6}$$

where $\Sigma^{(k)}$ denotes the sum over all 0, 1 valued x_i's such that $x_1 + \cdots + x_n = k$.

Fortunately, Poisson approximation may still prove accurate and allow the computational challenges presented by equation (5.6) to be avoided.

Theorem 5.3. *Let $\epsilon_1, \epsilon_2, \ldots$ be independent Bernoulli random variables, where*

$$\Pr\{\epsilon_i = 1\} = p_i \quad \text{and} \quad \Pr\{\epsilon_i = 0\} = 1 - p_i,$$

and let $S_n = \epsilon_1 + \cdots + \epsilon_n$. The exact probabilities for S_n, determined using (5.6), and Poisson probabilities with $\mu = p_1 + \cdots + p_n$ differ by at most

$$\left| \Pr\{S_n = k\} - \frac{\mu^k e^{-\mu}}{k!} \right| \leq \sum_{i=1}^{n} p_i^2. \tag{5.7}$$

Not only does the inequality of Theorem 5.3 extend the law of rare events to the case of unequal p's, it also directly confronts the approximation issue by providing a numerical measure of the approximation error. Thus, the Poisson distribution provides a good approximation to the exact probabilities whenever the p_i's are uniformly small as measured by the right side of (5.7). For instance, when $p_1 = p_2 = \cdots = \mu/n$, then the right side of (5.7) reduces to μ^2/n, which is small when n is large, and thus (5.7) provides another means of obtaining the Poisson distribution (5.5) as a limit of the binomial probabilities (5.4).

We defer the proof of Theorem 5.3 to the end of this section, choosing to concentrate now on its implications. As an immediate consequence, e.g., in the context of the earlier car accident vignette, we see that the individual cars need not all have the same accident probabilities in order for the Poisson approximation to apply.

5.2.1 The Law of Rare Events and the Poisson Process

Consider events occurring along the positive axis $[0, \infty)$ in the manner shown in Figure 5.2. Concrete examples of such processes are the time points of the X-ray emissions of a substance undergoing radioactive decay, the instances of telephone calls originating in a given locality, the occurrence of accidents at a certain intersection, the location of faults or defects along the length of a fiber or filament, and the successive arrival times of customers for service.

Let $N((a, b])$ denote the number of events that occur during the interval $(a, b]$. That is, if $t_1 < t_2 < t_3 < \cdots$ denote the times (or locations, etc.) of successive events, then $N((a, b])$ is the number of values t_i for which $a < t_i \le b$.

We make the following postulates:

1. The numbers of events happening in disjoint intervals are independent random variables. That is, for every integer $m = 2, 3, \ldots$ and time points $t_0 = 0 < t_1 < t_2 < \cdots < t_m$, the random variables

   $$N((t_0, t_1]), N((t_1, t_2]), \ldots, N((t_{m-1}, t_m])$$

   are independent.

2. For any time t and positive number h, the probability distribution of $N((t, t + h])$, the number of events occurring between time t and $t + h$, depends only on the interval length h and not on the time t.

3. There is a positive constant λ for which the probability of at least one event happening in a time interval of length h is

   $$\Pr\{N((t, t + h]) \ge 1\} = \lambda h + o(h) \quad \text{as } h \downarrow 0.$$

(Conforming to a common notation, here $o(h)$ as $h \downarrow 0$ stands for a general and unspecified remainder term for which $o(h)/h \to 0$ as $h \downarrow 0$. That is, a remainder term of smaller order than h as h vanishes.)

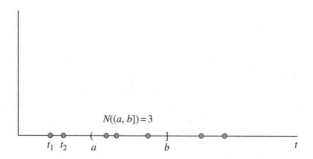

$$N((a, b]) = 3$$

Figure 5.2 A Poisson point process.

4. The probability of two or more events occurring in an interval of length h is $o(h)$, or

$$\Pr\{N((t, t+h]) \geq 2\} = o(h), \quad h \downarrow 0.$$

Postulate 3 is a specific formulation of the notion that events are rare. Postulate 4 is tantamount to excluding the possibility of the simultaneous occurrence of two or more events. In the presence of Postulates 1 and 2, Postulates 3 and 4 are equivalent to the apparently weaker assumption that events occur singly and discretely, with only a finite number in any finite interval. In the concrete illustrations cited earlier, this requirement is usually satisfied.

Disjoint intervals are independent by 1, and 2 asserts that the distribution of $N((s, t])$ is the same as that of $N((0, t-s])$. Therefore, to describe the probability law of the system, it suffices to determine the probability distribution of $N((0, t])$ for an arbitrary value of t. Let

$$P_k(t) = \Pr\{N((0, t]) = k\}.$$

We will show that Postulates 1 through 4 require that $P_k(t)$ be the Poisson distribution

$$P_k(t) = \frac{(\lambda t)^k e^{-\lambda t}}{k!} \quad \text{for } k = 0, 1, \ldots. \tag{5.8}$$

To establish (5.8), divide the interval $(0, t]$ into n subintervals of equal length $h = t/n$, and let

$$\epsilon_i = \begin{cases} 1 & \text{if there is at least one event in the interval } ((i-1)t/n, it/n], \\ 0 & \text{otherwise.} \end{cases}$$

Then, $S_n = \epsilon_1 + \cdots + \epsilon_n$ counts the total number of subintervals that contain at least one event, and

$$p_i = \Pr\{\epsilon_i = 1\} = \lambda t/n + o(t/n) \tag{5.9}$$

according to Postulate 3. Upon applying (5.7), we see that

$$\left| \Pr\{S_n = k\} - \frac{\mu^k e^{-\mu}}{k!} \right| \leq n[\lambda t/n + o(t/n)]^2$$

$$= \frac{(\lambda t)^2}{n} + 2\lambda t o\left(\frac{t}{n}\right) + n o\left(\frac{t}{n}\right)^2,$$

where

$$\mu = \sum_{i=1}^{n} p_i = \lambda t + n o(t/n). \tag{5.10}$$

Because $o(h) = o(t/n)$ is a term of order smaller than $h = t/n$ for large n, it follows that

$$no(t/n) = t\frac{o(t/n)}{t/n} = t\frac{o(h)}{h}.$$

vanishes for arbitrarily large n. Passing to the limit as $n \to \infty$, then, we deduce that

$$\lim_{n \to \infty} \Pr\{S_n = k\} = \frac{\mu^k e^{-\mu}}{k!}, \quad \text{with } \mu = \lambda t.$$

To complete the demonstration, we need only show that

$$\lim_{n \to \infty} \Pr\{S_n = k\} = \Pr\{N((0, t]) = k\} = P_k(t).$$

But S_n differs from $N((0, t])$ only if at least one of the subintervals contains two or more events, and Postulate 4 precludes this because

$$|P_k(t) - \Pr\{S_n = k\}| \leq \Pr\{N((0, t]) \neq S_n\}$$

$$\leq \sum_{i=1}^{n} \Pr\left\{N\left(\left(\frac{(i-1)t}{n}, \frac{it}{n}\right]\right) \geq 2\right\}$$

$$\leq no(t/n) \quad \text{(by Postulate 4)}$$

$$\to 0 \quad \text{as } n \to \infty.$$

By making n arbitrarily large, or equivalently, by dividing the interval $(0, t]$ into arbitrarily small subintervals, we see that it must be the case that

$$\Pr\{N((0, t]) = k\} = \frac{(\lambda t)^k e^{-\lambda t}}{k!} \quad \text{for } k \geq 0,$$

and Postulates 1 through 4 imply the Poisson distribution.

Postulates 1 through 4 arise as physically plausible assumptions in many circumstances of stochastic modeling. The postulates seem rather weak. Surprisingly, they are sufficiently strong to force the Poisson behavior just derived. This motivates the following definition.

Definition Let $N((s, t])$ be a random variable counting the number of events occurring in an interval $(s, t]$. Then, $N((s, t])$ is a Poisson point process of intensity $\lambda > 0$ if

1. for every $m = 2, 3, \ldots$ and distinct time points $t_0 = 0 < t_1 < t_2 < \cdots < t_m$, the random variables

$$N((t_0, t_1]), N((t_1, t_2]), \ldots, N((t_{m-1}, t_m])$$

are independent; and

2. for any times $s < t$ the random variable $N((s, t])$ has the Poisson distribution

$$\Pr\{N((s, t]) = k\} = \frac{[\lambda(t - s)]^k e^{-\lambda(t-s)}}{k!}, \quad k = 0, 1, \dots.$$

Poisson point processes often arise in a form where the time parameter is replaced by a suitable spatial parameter. The following formal example illustrates this vein of ideas. Consider an array of points distributed in a space E (E is a Euclidean space of dimension $d \geq 1$). Let $N(A)$ denote the number of points (finite or infinite) contained in the region A of E. We postulate that $N(A)$ is a random variable. The collection $\{N(A)\}$ of random variables, where A varies over all possible subsets of E, is said to be a homogeneous Poisson process if the following assumptions are fulfilled:

1. The numbers of points in nonoverlapping regions are independent random variables.
2. For any region A of finite volume, $N(A)$ is Poisson distributed with mean $\lambda |A|$, where $|A|$ is the volume of A. The parameter λ is fixed and measures in a sense the intensity component of the distribution, which is independent of the size or shape. Spatial Poisson processes arise in considering such phenomena as the distribution of stars or galaxies in space, the spatial distribution of plants and animals, and the spatial distribution of bacteria on a slide. These ideas and concepts will be further studied in Section 5.5.

5.2.2 Proof of Theorem 5.3

First, some notation. Let $\epsilon(p)$ denote a Bernoulli random variable with success probability p, and let $X(\theta)$ be a Poisson distributed random variable with parameter θ. We are given probabilities $p_1, \dots, p_n$ and let $\mu = p_1 + \cdots + p_n$. With $\epsilon(p_1), \dots, \epsilon(p_n)$ assumed to be independent, we have $S_n = \epsilon(p_1) + \cdots + \epsilon(p_n)$, and according to Theorem 5.1, we may write $X(\mu)$ as the sum of independent Poisson distributed random variables in the form $X(\mu) = X(p_1) + \cdots + X(p_n)$. We are asked to compare $\Pr\{S_n = k\}$ with $\Pr\{X(\mu) = k\}$, and, as a first step, we observe that if S_n and $X(\mu)$ are unequal, then at least one of the pairs $\epsilon(p_k)$ and $X(p_k)$ must differ, whence

$$|\Pr\{S_n = k\} - \Pr\{X(\mu) = k\}| \leq \sum_{k=1}^{n} \Pr\{\epsilon(p_k) \neq X(p_k)\}. \tag{5.11}$$

As the second step, observe that the quantities that are compared on the left of (5.11) are the marginal distributions of S_n and $X(\mu)$, while the bound on the right is a joint probability. This leaves us free to choose the joint distribution that makes our task the easiest. That is, we are free to specify the joint distribution of each $\epsilon(p_k)$ and $X(p_k)$, as we please, provided only that the marginal distributions are Bernoulli and Poisson, respectively.

To complete the proof, we need to show that $\Pr\{\epsilon(p) \neq X(p)\} \leq p^2$ for some Bernoulli random variable $\epsilon(p)$ and Poisson random variable $X(p)$, since this reduces the right side of (5.11) to that of (5.7). Equivalently, we want to show that $1 - p^2 \leq \Pr\{\epsilon(p) = X(p)\} = \Pr\{\epsilon(p) = X(p) = 0\} + \Pr\{\epsilon(p) = X(p) = 1\}$, and we are free to choose the joint distribution, provided that the marginal distributions are correct.

Let U be a random variable that is uniformly distributed over the interval $(0, 1]$. Define

$$\epsilon(p) = \begin{cases} 1 & \text{if } 0 < U \le p, \\ 0 & \text{if } p < U \le 1, \end{cases}$$

and for $k = 0, 1, \ldots$, set

$$X(p) = k \quad \text{when} \quad \sum_{i=0}^{k-1} \frac{p^i e^{-p}}{i!} < U \le \sum_{i=0}^{k} \frac{p^i e^{-p}}{i!}.$$

It is elementary to verify that $\epsilon(p)$ and $X(p)$ have the correct marginal distributions. Furthermore, because $1 - p \le e^{-p}$, we have $\epsilon(p) = X(p) = 0$ only for $U \le 1 - p$, whence $\Pr\{\epsilon(p) = X(p) = 0\} = 1 - p$. Similarly, $\epsilon(p) = X(p) = 1$ only when $e^{-p} < U \le (1+p)e^{-p}$, whence $\Pr\{\epsilon(p) = X(p) = 1\} = pe^{-p}$. Upon summing these two evaluations, we obtain

$$\Pr\{\epsilon(p) = X(p)\} = 1 - p + pe^{-p} = 1 - p^2 + p^3/2 \cdots \ge 1 - p^2$$

as was to be shown. This completes the proof of (5.7).

Problem 2.10 calls for the reader to review the proof and to discover the single line that needs to be changed in order to establish the stronger result

$$|\Pr\{S_n \text{ in } I\} - \Pr\{X(\mu) \text{ in } I\}| \le \sum_{k=1}^{n} p_i^2$$

for any set of nonnegative integers I.

Exercises

5.2.1 Determine numerical values to three decimal places for $\Pr\{X = k\}$, $k = 0, 1, 2$, when
   (a) X has a binomial distribution with parameters $n = 20$ and $p = 0.06$.
   (b) X has a binomial distribution with parameters $n = 40$ and $p = 0.03$.
   (c) X has a Poisson distribution with parameter $\lambda = 1.2$.

5.2.2 Explain in general terms why it might be plausible to assume that the following random variables follow a Poisson distribution:
   (a) The number of customers that enter a store in a fixed time period.
   (b) The number of customers that enter a store and buy something in a fixed time period.
   (c) The number of atomic particles in a radioactive mass that disintegrate in a fixed time period.

5.2.3 A large number of distinct pairs of socks are in a drawer, all mixed up. A small number of individual socks are removed. Explain in general terms why it might be plausible to assume that the number of pairs among the socks removed might follow a Poisson distribution.

5.2.4 Suppose that a book of 600 pages contains a total of 240 typographical errors. Develop a Poisson approximation for the probability that three particular successive pages are error-free.

Problems

5.2.1 Let $X(n,p)$ have a binomial distribution with parameters n and p. Let $n \to \infty$ and $p \to 0$ in such a way that $np = \lambda$. Show that

$$\lim_{n \to \infty} \Pr\{X(n,p) = 0\} = e^{-\lambda}$$

and

$$\lim_{n \to \infty} \frac{\Pr\{X(n,p) = k+1\}}{\Pr\{X(n,p) = k\}} = \frac{\lambda}{k+1} \quad \text{for } k = 0, 1, \ldots.$$

5.2.2 Suppose that 100 tags, numbered $1, 2, \ldots, 100$, are placed into an urn, and 10 tags are drawn successively, *with replacement*. Let A be the event that no tag is drawn twice. Show that

$$\Pr\{A\} = \left(1 - \frac{1}{100}\right)\left(1 - \frac{2}{100}\right) \cdots \left(1 - \frac{9}{100}\right) = 0.6282.$$

Use the approximation

$$1 - x \approx e^{-x} \quad \text{for } x \approx 0$$

to get

$$\Pr\{A\} \approx \exp\left\{-\frac{1}{100}(1 + 2 + \cdots + 9)\right\} = e^{-0.45} = 0.6376.$$

Interpret this in terms of the law of rare events.

5.2.3 Suppose that N pairs of socks are sent to a laundry, where they are washed and thoroughly mixed up to create a mass of unmatched socks. Then, n socks are drawn at random without replacement from the pile. Let A be the event that no pair is among the n socks so selected. Show that

$$\Pr\{A\} = \frac{2^n \binom{N}{n}}{\binom{2N}{n}} = \prod_{i=1}^{n-1}\left(1 - \frac{i}{2N - i}\right).$$

Use the approximation

$$1 - x \approx e^{-x} \quad \text{for } x \approx 0$$

to get

$$\Pr\{A\} \approx \exp\left\{-\sum_{i=1}^{n-1} \frac{i}{2N-i}\right\} \approx \exp\left\{-\frac{n(n-1)}{4N}\right\},$$

the approximations holding when n is small relative to N, which is large. Evaluate the exact expression and each approximation when $N = 100$ and $n = 10$. Is the approximation here consistent with the actual number of pairs of socks among the n socks drawn having a Poisson distribution?

Answer: Exact 0.7895; Approximate 0.7985.

5.2.4 Suppose that N points are uniformly distributed over the interval $[0, N)$. Determine the probability distribution for the number of points in the interval $[0, 1)$ as $N \to \infty$.

5.2.5 Suppose that N points are uniformly distributed over the surface of a circular disk of radius r. Determine the probability distribution for the number of points within a distance of one of the origin as $N \to \infty$, $r \to \infty$, $N/(\pi r^2) = \lambda$.

5.2.6 Certain computer coding systems use randomization to assign memory storage locations to account numbers. Suppose that $N = M\lambda$ different accounts are to be randomly located among M storage locations. Let X_i be the number of accounts assigned to the ith location. If the accounts are distributed independently and each location is equally likely to be chosen, show that $\Pr\{X_i = k\} \to e^{-\lambda}\lambda^k/k!$ as $N \to \infty$. Show that X_i and X_j are independent random variables in the limit, for distinct locations $i \neq j$. In the limit, what fraction of storage locations have two or more accounts assigned to them?

5.2.7 N bacteria are spread independently with uniform distribution on a microscope slide of area A. An arbitrary region having area a is selected for observation. Determine the probability of k bacteria within the region of area a. Show that as $N \to \infty$ and $a \to 0$ such that $(a/A)N \to c(0 < c < \infty)$, then $p(k) \to e^{-c}c^k/k!$.

5.2.8 Using (5.6), evaluate the exact probabilities for S_n and the Poisson approximation and error bound in (5.7) when $n = 4$ and $p_1 = 0.1, p_2 = 0.2, p_3 = 0.3$, and $p_4 = 0.4$.

5.2.9 Using (5.6), evaluate the exact probabilities for S_n and the Poisson approximation and error bound in (5.7) when $n = 4$ and $p_1 = 0.1, p_2 = 0.1, p_3 = 0.1$, and $p_4 = 0.2$.

5.2.10 Review the proof of Theorem 5.3 in Section 5.2.2 and establish the stronger result

$$|\Pr\{S_n \text{ in } I\} - \Pr\{X(\mu) \text{ in } I\}| \leq \sum_{k=1}^{n} p_i^2$$

for any set of nonnegative integers I.

5.2.11 Let X and Y be jointly distributed random variables and B an arbitrary set. Fill in the details that justify the inequality $|\Pr\{X \text{ in } B\} - \Pr\{Y \text{ in } B\}| \leq \Pr\{X \neq Y\}$.

Hint: Begin with

$$\{X \text{ in } B\} = \{X \text{ in } B \text{ and } Y \text{ in } B\} \quad \text{or} \quad \{X \text{ in } B \text{ and } Y \text{ not in } B\}$$
$$\subset \{Y \text{ in } B\} \quad \text{or} \quad \{X \neq Y\}.$$

5.2.12 *Computer Challenge* Most computers have available a routine for simulating a sequence $U_0, U_1, \ldots$ of independent random variables, each uniformly distributed on the interval $(0, 1)$. Plot, say, 10,000 pairs (U_{2n}, U_{2n+1}) on the unit square. Does the plot look like what you would expect? Repeat the experiment several times. Do the points in a fixed number of disjoint squares of area 1/10,000 look like independent unit Poisson random variables?

5.3 Distributions Associated with the Poisson Process

A *Poisson point process* $N((s, t])$ counts the number of events occurring in an interval $(s, t]$. A *Poisson counting process*, or more simply a *Poisson process* $X(t)$, counts the number of events occurring up to time t. Formally, $X(t) = N((0, t])$.

Poisson events occurring in space can best be modeled as a point process. For Poisson events occurring on the positive time axis, whether we view them as a Poisson point process or Poisson counting process is largely a matter of convenience, and we will freely do both. The two descriptions are equivalent for Poisson events occurring along a line. The Poisson process is the more common and traditional description in this case because it allows a pictorial representation as an increasing integer-valued random function taking unit steps.

Figure 5.3 shows a typical sample path of a Poisson process where W_n is the time of occurrence of the nth event, the so-called *waiting time*. It is often convenient to set $W_0 = 0$. The differences $S_n = W_{n+1} - W_n$ are called *sojourn times*; S_n measures the duration that the Poisson process sojourns in state n.

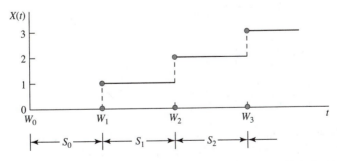

Figure 5.3 A typical sample path of a Poisson process showing the waiting times W_n and the sojourn times S_n.

In this section, we will determine a number of probability distributions associated with the Poisson process $X(t)$, the waiting times W_n, and the sojourn times S_n.

Theorem 5.4. *The waiting time W_n has the gamma distribution whose probability density function is*

$$f_{W_n}(t) = \frac{\lambda^n t^{n-1}}{(n-1)!} e^{-\lambda t}, \quad n = 1, 2, \ldots, t \geq 0. \tag{5.12}$$

In particular, W_1, the time to the first event, is exponentially distributed:

$$f_{W_1}(t) = \lambda e^{-\lambda t}, \quad t \geq 0. \tag{5.13}$$

Proof. The event $W_n \leq t$ occurs if and only if there are at least n events in the interval $(0, t]$, and since the number of events in $(0, t]$ has a Poisson distribution with mean λt we obtain the cumulative distribution function of W_n via

$$F_{W_n}(t) = \Pr\{W_n \leq t\} = \Pr\{X(t) \geq n\}$$

$$= \sum_{k=n}^{\infty} \frac{(\lambda t)^k e^{-\lambda t}}{k!}$$

$$= 1 - \sum_{k=0}^{n-1} \frac{(\lambda t)^k e^{-\lambda t}}{k!}, \quad n = 1, 2, \ldots, t \geq 0.$$

We obtain the probability density function $f_{W_n}(t)$ by differentiating the cumulative distribution function. Then

$$f_{W_n}(t) = \frac{d}{dt} F_{W_n}(t)$$

$$= \frac{d}{dt} \left\{ 1 - e^{-\lambda t} \left[1 - \frac{\lambda t}{1!} + \frac{(\lambda t)^2}{2!} + \cdots + \frac{(\lambda t)^{n-1}}{(n-1)!} \right] \right\}$$

$$= -e^{-\lambda t} \left[\lambda + \frac{\lambda(\lambda t)}{1!} + \lambda \frac{(\lambda t)^2}{2!} + \cdots + \lambda \frac{(\lambda t)^{n-2}}{(n-2)!} \right]$$

$$+ \lambda e^{-\lambda t} \left[1 + \frac{\lambda t}{1!} + \frac{(\lambda t)^2}{2!} + \cdots + \frac{(\lambda t)^{n-1}}{(n-1)!} \right]$$

$$= \frac{\lambda^n t^{n-1}}{(n-1)!} e^{-\lambda t}, \quad n = 1, 2, \ldots, t \geq 0.$$

There is an alternative derivation of the density in (5.12) that uses the Poisson point process $N((s, t])$ and proceeds directly without differentiation. The event $t < W_n \leq t + \Delta t$ corresponds exactly to $n - 1$ occurrences in $(0, t]$ and one in $(t, t + \Delta t]$, as depicted in Figure 5.4.

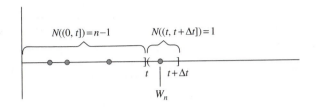

Figure 5.4

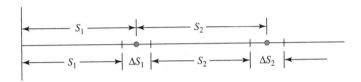

Figure 5.5

Then

$$f_{W_n}(t)\Delta t \approx \Pr\{t < W_n \leq t + \Delta t\} + o(\Delta t) \quad \text{[see Chapter 1, equation (1.5)]}$$

$$= \Pr\{N((0,t]) = n - 1\}\Pr\{N((t, t + \Delta t]) = 1\} + o(\Delta t)$$

$$= \frac{(\lambda t)^{n-1}e^{-\lambda t}}{(n-1)!}\lambda(\Delta t) + o(\Delta t).$$

Dividing by Δt and passing to the limit as $\Delta t \to 0$ we obtain (5.13).

Observe that $\Pr\{N((t, t + \Delta t]) \geq 1\} = \Pr\{N((t, t + \Delta t]) = 1\} + o(\Delta t) = \lambda(\Delta t) + o(\Delta t)$. ∎

Theorem 5.5. *The sojourn times $S_0, S_1, \ldots, S_{n-1}$ are independent random variables, each having the exponential probability density function*

$$f_{S_k}(s) = \lambda e^{-\lambda s}, \quad s \geq 0. \tag{5.14}$$

Proof. We are being asked to show that the joint probability density function of $S_0, S_1, \ldots, S_{n-1}$ is the product of the exponential densities given by

$$f_{S_0, S_1, \ldots, S_{n-1}}(s_0, s_1, \ldots, s_{n-1}) = \left(\lambda e^{-\lambda s_0}\right)\left(\lambda e^{-\lambda s_1}\right)\cdots\left(\lambda e^{-\lambda s_{n-1}}\right). \tag{5.15}$$

We give the proof only in the case $n = 2$, the general case being entirely similar. Referring to Figure 5.5 we see that the joint occurrence of

$$s_1 < S_1 < s_1 + \Delta s_1 \quad \text{and} \quad s_2 < S_2 < s_2 + \Delta s_2$$

corresponds to no events in the intervals $(0, s_1]$ and $(s_1 + \Delta s_1, s_1 + \Delta s_1 + s_2]$ and exactly one event in each of the intervals $(s_1, s_1 + \Delta s_1]$ and $(s_1 + \Delta s_1 + s_2, s_1 + \Delta s_1 + s_2 + \Delta s_2]$. Thus

$$
\begin{aligned}
f_{S_1, S_2}(s_1, s_2) \Delta s_1 \Delta s_2 &= \Pr\{s_1 < S_1 < s_1 + \Delta s_1, s_2 < S_2 < s_2 + \Delta s_2\} \\
&\quad + o(\Delta s_1 \Delta s_2) \\
&= \Pr\{N((0, s_1]) = 0\} \\
&\quad \times \Pr\{N((s_1 + \Delta s_1, s_1 + \Delta s_1 + s_2]) = 0\} \\
&\quad \times \Pr\{N((s_1, s_1 + \Delta s_1]) = 1\} \\
&\quad \times \Pr\{N((s_1 + \Delta s_1 + s_2, s_1 + \Delta s_1 + s_2 + \Delta s_2]) = 1\} \\
&\quad + o(\Delta s_1 \Delta s_2) \\
&= e^{-\lambda s_1} e^{-\lambda s_2} e^{-\lambda \Delta s_1} e^{-\lambda \Delta s_2} \lambda(\Delta s_1) \lambda(\Delta s_2) + o(\Delta s_1 \Delta s_2) \\
&= (\lambda e^{-\lambda s_1})(\lambda e^{-\lambda s_2})(\Delta s_1)(\Delta s_2) + o(\Delta s_1 \Delta s_2).
\end{aligned}
$$

Upon dividing both sides by $(\Delta s_1)(\Delta s_2)$ and passing to the limit as $\Delta s_1 \to 0$ and $\Delta s_2 \to 0$, we obtain (5.15) in the case $n = 2$. ∎

The binomial distribution also arises in the context of Poisson processes.

Theorem 5.6. *Let $\{X(t)\}$ be a Poisson process of rate $\lambda > 0$. Then for $0 < u < t$ and $0 \le k \le n$,*

$$
\Pr\{X(u) = k | X(t) = n\} = \frac{n!}{k!(n-k)!} \left(\frac{u}{t}\right)^k \left(1 - \frac{u}{t}\right)^{n-k}. \tag{5.16}
$$

Proof. Straightforward computations give

$$
\begin{aligned}
\Pr\{X(u) = k | X(t) = n\} &= \frac{\Pr\{X(u) = k \text{ and } X(t) = n\}}{\Pr\{X(t) = n\}} \\
&= \frac{\Pr\{X(u) = k \text{ and } X(t) - X(u) = n - k\}}{\Pr\{X(t) = n\}} \\
&= \frac{\{e^{-\lambda u}(\lambda u)^k / k!\} \{e^{-\lambda(t-u)}[\lambda(t-u)]^{n-k} / (n-k)!\}}{e^{-\lambda t}(\lambda t)^n / n!} \\
&= \frac{n!}{k!(n-k)!} \frac{u^k (t-u)^{n-k}}{t^n},
\end{aligned}
$$

which establishes (5.16). ∎

Exercises

5.3.1 A radioactive source emits particles according to a Poisson process of rate $\lambda = 2$ particles per minute. What is the probability that the first particle appears after 3 min?

5.3.2 A radioactive source emits particles according to a Poisson process of rate $\lambda = 2$ particles per minute.
  (a) What is the probability that the first particle appears some time after 3 min but before 5 min?
  (b) What is the probability that exactly one particle is emitted in the interval from 3 to 5 min?

5.3.3 Customers enter a store according to a Poisson process of rate $\lambda = 6$ per hour. Suppose it is known that only a single customer entered during the first hour. What is the conditional probability that this person entered during the first 15 min?

5.3.4 Let $X(t)$ be a Poisson process of rate $\xi = 3$ per hour. Find the conditional probability that there were two events in the first hour, given that there were five events in the first 3 h.

5.3.5 Let $X(t)$ be a Poisson process of rate θ per hour. Find the conditional probability that there were m events in the first t hours, given that there were n events in the first T hours. Assume $0 \le m \le n$ and $0 < t < T$.

5.3.6 For $i = 1, \ldots, n$, let $\{X_i(t); t \ge 0\}$ be independent Poisson processes, each with the same parameter λ. Find the distribution of the first time that at least one event has occurred in every process.

5.3.7 Customers arrive at a service facility according to a Poisson process of rate λ customers/hour. Let $X(t)$ be the number of customers that have arrived up to time t. Let $W_1, W_2, \ldots$ be the successive arrival times of the customers. Determine the conditional mean $E[W_5 | X(t) = 3]$.

5.3.8 Customers arrive at a service facility according to a Poisson process of rate $\lambda = 5$ per hour. Given that 12 customers arrived during the first two hours of service, what is the conditional probability that 5 customers arrived during the first hour?

5.3.9 Let $X(t)$ be a Poisson process of rate λ. Determine the cumulative distribution function of the gamma density as a sum of Poisson probabilities by first verifying and then using the identity $W_r \le t$ if and only if $X(t) \ge r$.

Problems

5.3.1 Let $X(t)$ be a Poisson process of rate λ. Validate the identity

$$\{W_1 > w_1, W_2 > w_2\}$$

if and only if

$$\{X(w_1) = 0, X(w_2) - X(w_1) = 0 \text{ or } 1\}.$$

Use this to determine the joint upper tail probability

$$\Pr\{W_1 > w_1, W_2 > w_2\} = \Pr\{X(w_1) = 0, X(w_2) - X(w_1) = 0 \text{ or } 1\}$$
$$= e^{-\lambda w_1}[1 + \lambda(w_2 - w_1)]e^{-\lambda(w_2 - w_1)}.$$

Finally, differentiate twice to obtain the joint density function

$$f(w_1, w_2) = \lambda^2 \exp\{-\lambda w_2\} \quad \text{for } 0 < w_1 < w_2.$$

5.3.2 The joint probability density function for the waiting times W_1 and W_2 is given by

$$f(w_1, w_2) = \lambda^2 \exp\{-\lambda w_2\} \quad \text{for } 0 < w_1 < w_2.$$

Determine the conditional probability density function for W_1, given that $W_2 = w_2$. How does this result differ from that in Theorem 5.6 when $n = 2$ and $k = 1$?

5.3.3 The joint probability density function for the waiting times W_1 and W_2 is given by

$$f(w_1, w_2) = \lambda^2 \exp\{-\lambda w_2\} \quad \text{for } 0 < w_1 < w_2.$$

Change variables according to

$$S_0 = W_1 \quad \text{and} \quad S_1 = W_2 - W_1$$

and determine the joint distribution of the first two sojourn times. Compare with Theorem 5.5.

5.3.4 The joint probability density function for the waiting times W_1 and W_2 is given by

$$f(w_1, w_2) = \lambda^2 \exp\{-\lambda w_2\} \quad \text{for } 0 < w_1 < w_2.$$

Determine the marginal density functions for W_1 and W_2, and check your work by comparison with Theorem 5.4.

5.3.5 Let $X(t)$ be a Poisson process with parameter λ. Independently, let T be a random variable with the exponential density

$$f_T(t) = \theta e^{-\theta t} \quad \text{for } t > 0.$$

Determine the probability mass function for $X(T)$.

Hint: Use the law of total probability and Chapter 1, (1.54). Alternatively, use the results of Chapter 1, Section 1.5.2.

5.3.6 Customers arrive at a holding facility at random according to a Poisson process having rate λ. The facility processes in batches of size Q. That is, the first $Q - 1$ customers wait until the arrival of the Qth customer. Then, all are passed simultaneously, and the process repeats. Service times are instantaneous. Let $N(t)$ be the number of customers in the holding facility at time t. Assume that $N(0) = 0$ and let $T = \min\{t \geq 0 : N(t) = Q\}$ be the first dispatch time. Show that $E[T] = Q/\lambda$ and $E\left[\int_0^T N(t)dt\right] = [1 + 2 + \cdots + (Q - 1)]/\lambda = Q(Q - 1)/2\lambda$.

5.3.7 A critical component on a submarine has an operating lifetime that is exponentially distributed with mean 0.50 years. As soon as a component fails, it is replaced by a new one having statistically identical properties. What is the smallest number of *spare* components that the submarine should stock if it is leaving for a one-year tour and wishes the probability of having an inoperable unit caused by failures exceeding the spare inventory to be less than 0.02?

5.3.8 Consider a Poisson process with parameter λ. Given that $X(t) = n$ events occur in time t, find the density function for W_r, the time of occurrence of the rth event. Assume that $r \leq n$.

5.3.9 The following calculations arise in certain highly simplified models of learning processes. Let $X_1(t)$ and $X_2(t)$ be independent Poisson processes having parameters λ_1 and λ_2, respectively.
(a) What is the probability that $X_1(t) = 1$ before $X_2(t) = 1$?
(b) What is the probability that $X_1(t) = 2$ before $X_2(t) = 2$?

5.3.10 Let $\{W_n\}$ be the sequence of waiting times in a Poisson process of intensity $\lambda = 1$. Show that $X_n = 2^n \exp\{-W_n\}$ defines a nonnegative martingale.

5.4 The Uniform Distribution and Poisson Processes

The major result of this section, Theorem 5.7, provides an important tool for computing certain functionals on a Poisson process. It asserts that, conditioned on a fixed total number of events in an interval, the locations of those events are uniformly distributed in a certain way.

After a complete discussion of the theorem and its proof, its application in a wide range of problems will be given.

In order to completely understand the theorem, consider first the following experiment. We begin with a line segment t units long and a fixed number n of darts and throw darts at the line segment in such a way that each dart's position upon landing is uniformly distributed along the segment, independent of the location of the other darts. Let U_1 be the position of the first dart thrown, U_2 the position of the second, and so on up to U_n. The probability density function is the uniform density

$$
f_U(u) = \begin{cases} \dfrac{1}{t} & \text{for } 0 \leq u \leq t, \\ 0 & \text{elsewhere.} \end{cases}
$$

Now let $W_1 \leq W_2 \leq \cdots \leq W_n$ denote these same positions, not in the order in which the darts were thrown, but instead in the order in which they appear along the line. Figure 5.6 depicts a typical relation between $U_1, U_2, \ldots, U_n$ and $W_1, W_2, \ldots, W_n$.

The joint probability density function for $W_1, W_2, \ldots, W_n$ is

$$
f_{W_1,\ldots,W_n}(w_1,\ldots,w_n) = n!\, t^{-n} \quad \text{for } 0 < w_1 < w_2 < \cdots < w_n \leq t. \tag{5.17}
$$

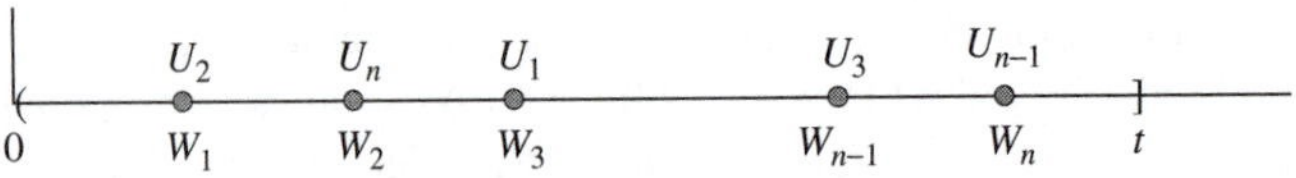

Figure 5.6 $W_1, W_2, \ldots, W_n$ are the values $U_1, U_2, \ldots, U_n$ arranged in increasing order.

For example, to establish (5.17) in the case $n = 2$ we have

$$f_{W_1,W_2}(w_1, w_2)\Delta w_1 \Delta w_2$$

$$= \Pr\{w_1 < W_1 \le w_1 + \Delta w_1, w_2 < W_2 \le w_2 + \Delta w_2\}$$

$$= \Pr\{w_1 < U_1 \le w_1 + \Delta w_1, w_2 < U_2 < w_2 + \Delta w_2\}$$

$$+ \Pr\{w_1 < U_2 \le w_1 + \Delta w_1, w_2 < U_1 \le w_2 + \Delta w_2\}$$

$$= 2\left(\frac{\Delta w_1}{t}\right)\left(\frac{\Delta w_2}{t}\right) = 2t^{-2}\Delta w_1 \Delta w_2.$$

Dividing by $\Delta w_1 \Delta w_2$ and passing to the limit gives (5.17). When $n = 2$, there are two ways that U_1 and U_2 can be ordered; either U_1 is less than U_2, or U_2 is less than U_1. In general, there are $n!$ arrangements of $U_1, \ldots, U_n$ that lead to the same ordered values $W_1 \le \cdots \le W_n$, thus giving (5.17).

Theorem 5.7. *Let $W_1, W_2, \ldots$ be the occurrence times in a Poisson process of rate $\lambda > 0$. Conditioned on $N(t) = n$, the random variables $W_1, W_2, \ldots, W_n$ have the joint probability density function*

$$f_{W_1,\ldots,W_n|X(t)=n}(w_1, \ldots, w_n) = n! t^{-n} \quad \text{for } 0 < w_1 < \cdots < w_n \le t. \tag{5.18}$$

Proof. The event $w_i < W_i \le w_i + \Delta w_i$ for $i = 1, \ldots, n$ and $N(t) = n$ corresponds to no events occurring in any of the intervals $(0, w_1], (w_1 + \Delta w_1, w_2], \ldots, (w_{n-1} + \Delta w_{n-1}, w_n], (w_n + \Delta w_n, t]$, and exactly one event in each of the intervals $(w_1, w_1 + \Delta w_1], (w_2, w_2 + \Delta w_2], \ldots, (w_n, w_n + \Delta w_n]$. These intervals are disjoint, and

$$\Pr\{N((0, w_1]) = 0, \ldots, N((w_n + \Delta w_n, t]) = 0\}$$

$$= e^{-\lambda w_1} e^{-\lambda(w_2 - w_1 - \Delta w_1)} \cdots e^{-\lambda(w_n - w_{n-1} - \Delta w_{n-1})} e^{-\lambda(t - w_n - \Delta w_n)}$$

$$= e^{-\lambda t}\left[e^{\lambda(\Delta w_1 + \cdots + \Delta w_n)}\right]$$

$$= e^{-\lambda t}[1 + o(\max\{\Delta w_i\})],$$

while

$$\Pr\{N((w_1, w_1 + \Delta w_1]) = 1, \ldots, N((w_n, w_n + \Delta w_n]) = 1\}$$

$$= \lambda(\Delta w_1) \cdots \lambda(\Delta w_n)[1 + o(\max\{\Delta w_i\})].$$

Thus

$$f_{W_1,\ldots,W_n|X(t)=n}(w_1,\ldots,w_n)\Delta w_1\cdots\Delta w_n$$

$$= \Pr\{w_1 < W_1 \le w_1 + \Delta w_1,\ldots,w_n < W_n \le w_n + \Delta w_n|N(t) = n\}$$
$$+ o(\Delta w_1\cdots\Delta w_n)$$

$$= \frac{\Pr\{w_i < W_i \le w_i + \Delta w_i, i = 1,\ldots,n, N(t) = n\}}{\Pr\{N(t) = n\}}$$
$$+ o(\Delta w_1\cdots\Delta w_1)$$

$$= \frac{e^{-\lambda t}\lambda(\Delta w_1)\cdots\lambda(\Delta w_n)}{e^{-\lambda t}(\lambda t)^n/n!}[1 + o(\max\{\Delta w_i\})]$$

$$= n!t^{-n}(\Delta w_1)\cdots(\Delta w_n)[1 + o(\max\{\Delta w_i\})].$$

Dividing both sides by $(\Delta w_1)\cdots(\Delta w_n)$ and letting $\Delta w_1 \to 0,\ldots,\Delta w_n \to 0$ establishes (5.18). ∎

Theorem 5.7 has important applications in evaluating certain symmetric functionals on Poisson processes. Some sample instances follow.

Example Customers arrive at a facility according to a Poisson process of rate λ. Each customer pays \$1 on arrival, and it is desired to evaluate the expected value of the total sum collected during the interval $(0, t]$ discounted back to time 0. This quantity is given by

$$M = E\left[\sum_{k=1}^{X(t)} e^{-\beta W_k}\right],$$

where β is the discount rate, $W_1, W_2,\ldots$ are the arrival times, and $X(t)$ is the total number of arrivals in $(0, t]$. The process is shown in Figure 5.7.

We evaluate the mean total discounted sum M by conditioning on $X(t) = n$. Then

$$M = \sum_{n=1}^{\infty} E\left[\sum_{k=1}^{n} e^{-\beta W_k}\Big|X(t) = n\right]\Pr\{X(t) = n\}. \tag{5.19}$$

Let $U_1,\ldots,U_n$ denote independent random variables that are uniformly distributed in $(0, t]$. Because of the symmetry of the functional $\sum_{k=1}^{n}\exp\{-\beta W_k\}$ and Theorem 5.7,

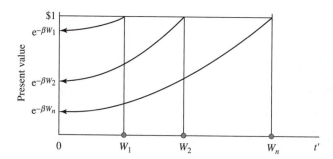

Figure 5.7 A dollar received at time W_k is discounted to a present value at time 0 of $\exp\{-\beta W_k\}$.

we have

$$E\left[\sum_{k=1}^{n} e^{-\beta W_k} \Big| X(t) = n\right] = E\left[\sum_{k=1}^{n} e^{-\beta U_k}\right]$$

$$= nE\left[e^{-\beta U_1}\right]$$

$$= nt^{-1}\int_{0}^{t} e^{-\beta u}\,du$$

$$= \frac{n}{\beta t}\left[1 - e^{-\beta t}\right].$$

Substitution into (5.19) then gives

$$M = \frac{1}{\beta t}\left[1 - e^{-\beta t}\right]\sum_{n=1}^{\infty} n\Pr\{X(t) = n\}$$

$$= \frac{1}{\beta t}\left[1 - e^{-\beta t}\right]E[X(t)]$$

$$= \frac{\lambda}{\beta}\left[1 - e^{-\beta t}\right].$$

Example Viewing a fixed mass of a certain radioactive material, suppose that *alpha* particles appear in time according to a Poisson process of intensity λ. Each particle exists for a random duration and is then annihilated. Suppose that the successive lifetimes $Y_1, Y_2, \ldots$ of distinct particles are independent random variables having the common distribution function $G(y) = \Pr\{Y_k \leq y\}$. Let $M(t)$ count the number of alpha particles existing at time t. The process is depicted in Figure 5.8.

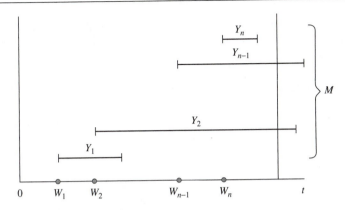

Figure 5.8 A particle created at time $W_k \leq t$ still exists at time t if $W_k + Y_k \geq t$.

We will use Theorem 5.7 to evaluate the probability distribution of $M(t)$ under the condition that $M(0) = 0$.

Let $X(t)$ be the number of particles created up to time t, by assumption a Poisson process of intensity λ. Observe that $M(t) \leq X(t)$; the number of existing particles cannot exceed the number of particles created. Condition on $X(t) = n$ and let $W_1, \ldots, W_n \leq t$ be the times of particle creation. Then, particle k exists at time t if and only if $W_k + Y_k \geq t$. Let

$$\mathbf{1}\{W_k + Y_k \geq t\} = \begin{cases} 1 & \text{if } W_k + Y_k \geq t, \\ 0 & \text{if } W_k + Y_k < t. \end{cases}$$

Then, $\mathbf{1}\{W_k + Y_k \geq t\} = 1$ if and only if the kth particle is alive at time t. Thus

$$\Pr\{M(t) = m | X(t) = n\} = \Pr\left\{\sum_{k=1}^{n} \mathbf{1}\{W_k + Y_k \geq t\} = m | X(t) = n\right\}.$$

Invoking Theorem 5.7 and the symmetry among particles, we have

$$\Pr\left\{\sum_{k=1}^{n} \mathbf{1}\{W_k + Y_k \geq t\} = m | X(t) = n\right\}$$

$$= \Pr\left\{\sum_{k=1}^{n} \mathbf{1}\{U_k + Y_k \geq t\} = m\right\}, \tag{5.20}$$

where $U_1, U_2, \ldots, U_n$ are independent and uniformly distributed on $(0, t]$. The right-hand side of (5.20) is readily recognized as the binomial distribution in which

$$p = \Pr\{U_k + Y_k \geq t\} = \frac{1}{t} \int_0^t \Pr\{Y_k \geq t - u\} du$$

$$= \frac{1}{t} \int_0^t [1 - G(t - u)] du \tag{5.21}$$

$$= \frac{1}{t} \int_0^t [1 - G(z)] dz.$$

Thus, explicitly writing the binomial distribution, we have

$$\Pr\{M(t) = m | X(t) = n\} = \frac{n!}{m!\,(n - m)!} p^m (1 - p)^{n-m},$$

with p^n given by (5.21). Finally,

$$\Pr\{M(t) = m\} = \sum_{n=m}^{\infty} \Pr\{M(t) = m | X(t) = n\} \Pr\{X(t) = n\}$$

$$= \sum_{n=m}^{\infty} \frac{n!}{m!\,(n - m)!} p^m (1 - p)^{n-m} \frac{(\lambda t)^n e^{-\lambda t}}{n!} \tag{5.22}$$

$$= e^{-\lambda t} \frac{(\lambda p t)^m}{m!} \sum_{n=m}^{\infty} \frac{(1 - p)^{n-m} (\lambda t)^{n-m}}{(n - m)!}.$$

The infinite sum is an exponential series and reduces according to

$$\sum_{n=m}^{\infty} \frac{(1 - p)^{n-m} (\lambda t)^{n-m}}{(n - m)!} = \sum_{j=0}^{\infty} \frac{[\lambda t(1 - p)]^j}{j!} = e^{\lambda t(1 - p)},$$

and this simplifies (5.22) to

$$\Pr\{M(t) = m\} = \frac{e^{-\lambda p t} (\lambda p t)^m}{m!} \quad \text{for } m = 0, 1, \ldots.$$

In words, the number of particles existing at time t has a Poisson distribution with mean

$$\lambda p t = \lambda \int_0^t [1 - G(y)] dy. \tag{5.23}$$

It is often relevant to let $t \to \infty$ in (5.23) and determine the corresponding long run distribution. Let $\mu = E[Y_k] = \int_0^\infty [1 - G(y)]dy$ be the mean lifetime of an alpha particle. It is immediate from (5.23) that as $t \to \infty$, the distribution of $M(t)$ converges to the Poisson distribution with parameter $\lambda\mu$. A great simplification has taken place. In the long run, the probability distribution for existing particles depends only on the mean lifetime μ, and not otherwise on the lifetime distribution $G(y)$. In practical terms, this statement implies that in order to apply this model, only the mean lifetime μ need be known.

5.4.1 Shot Noise

The shot noise process is a model for fluctuations in electrical currents that are due to chance arrivals of electrons to an anode. Variants of the phenomenon arise in physics and communication engineering. Assume:

1. Electrons arrive at an anode according to a Poisson process $\{X(t); t \geq 0\}$ of constant rate λ;
2. An arriving electron produces a current whose intensity x time units after arrival is given by the *impulse response function* $h(x)$.

The intensity of the current at time t is, then, the shot noise

$$I(t) = \sum_{k=1}^{X(t)} h(t - W_k), \tag{5.24}$$

where W_1, W_2 are the arrival times of the electrons.

Common impulse response functions include triangles, rectangles, decaying exponentials of the form

$$h(x) = e^{-\theta x}, \quad x > 0,$$

where $\theta > 0$ is a parameter, and *power law* shot noise for which

$$h(x) = x^{-\theta}, \quad \text{for } x > 0.$$

We will show that *for a fixed time point t*, the shot noise $I(t)$ has the same probability distribution as a certain random sum that we now describe. Independent of the Poisson process $X(t)$, let $U_1, U_2, \ldots$ be independent random variables, each uniformly distributed over the interval $(0, t]$, and define $\epsilon_k = h(U_k)$ for $k = 1, 2, \ldots$. The claim is that $I(t)$ has the same probability distribution as the random sum

$$S(t) = \epsilon_1 + \cdots + \epsilon_{X(t)}. \tag{5.25}$$

With this result in hand, the mean, variance, and distribution of the shot noise $I(t)$ may be readily obtained using the results on random sums developed in Chapter 2,

Section 2.3. For example, Chapter 2, equation (2.30) immediately gives us

$$E[I(t)] = E[S(t)] = \lambda t E[h(U_1)] = \lambda \int_0^t h(u)du$$

and

$$\text{Var}[I(t)] = \lambda t \left\{ \text{Var}[h(U_1)] + E[h(U_1)]^2 \right\}$$

$$= \lambda t E\left[h(U_1)^2\right] = \lambda \int_0^t h(u)^2 du.$$

In order to establish that the shot noise $I(t)$ and the random sum $S(t)$ share the same probability distribution, we need to show that $\Pr\{I(t) \leq x\} = \Pr\{S(t) \leq x\}$ for a fixed $t > 0$. Begin with

$$\Pr\{I(t) \leq x\} = \Pr\left\{ \sum_{k=1}^{X(t)} h(t - W_k) \leq x \right\}$$

$$= \sum_{n=0}^{\infty} \Pr\left\{ \sum_{k=1}^{X(t)} h(t - W_k) \leq x | X(t) = n \right\} \Pr\{X(t) = n\}$$

$$= \sum_{n=0}^{\infty} \Pr\left\{ \sum_{k=1}^{n} h(t - W_k) \leq x | X(t) = n \right\} \Pr\{X(t) = n\},$$

and now invoking Theorem 5.7,

$$= \sum_{n=0}^{\infty} \Pr\left\{ \sum_{k=1}^{n} h(t - U_k) \leq x \right\} \Pr\{X(t) = n\}$$

$$= \sum_{n=0}^{\infty} \Pr\left\{ \sum_{k=1}^{n} h(U_k) \leq x \right\} \Pr\{X(t) = n\}$$

(because U_k and $t - U_k$ have the same distribution)

$$= \sum_{n=0}^{\infty} \Pr\{\epsilon_1 + \cdots + \epsilon_n \leq x\} \Pr\{X(t) = n\}$$

$$= \Pr\{\epsilon_1 + \cdots + \epsilon_{X(t)} \leq x\}$$

$$= \Pr\{S(t) \leq x\},$$

which completes the claim.

5.4.2 Sum Quota Sampling

A common procedure in statistical inference is to observe a fixed number n of independent and identically distributed random variables $X_1, \ldots, X_n$ and use their sample mean

$$\overline{X}_n = \frac{X_1 + \cdots + X_n}{n}$$

as an estimate of the population mean or expected value $E[X_1]$. But suppose we are asked to advise an airline that wishes to estimate the failure rate *in service* of a particular component, or, what is nearly the same thing, to estimate the mean service life of the part. The airline monitors a new plane for two years and observes that the original component lasted 7 months before failing. Its replacement lasted 5 months, and the third component lasted 9 months. No further failures were observed during the remaining 3 months of the observation period. Is it correct to estimate the mean life in service as the observed average $(7 + 5 + 9)/3 = 7$ months?

This airline scenario provides a realistic example of a situation in which the sample size is not fixed in advance but is determined by a preassigned quota $t > 0$. In *sum quota sampling*, a sequence of independent and identically distributed nonnegative random variables $X_1, X_2, \ldots$ is observed sequentially, with the sampling continuing as long as the sum of the observations is less than the quota t. Let this random sample size be denoted by $N(t)$. Formally,

$$N(t) = \max\{n \geq 0; X_1 + \cdots + X_n < t\}.$$

The sample mean is

$$\overline{X}_{N(t)} = \frac{W_{N(t)}}{N(t)} = \frac{X_1 + \cdots + X_{N(t)}}{N(t)}.$$

Of course it is possible that $X_1 \geq t$, and then $N(t) = 0$, and the sample mean is undefined. Thus, we must assume, or condition on, the event that $N(t) \geq 1$. An important question in statistical theory is whether or not this sample mean is unbiased. That is, how does the expected value of this sample mean relate to the expected value of, say, X_1?

In general, the determination of the expected value of the sample mean under sum quota sampling is very difficult. It can be carried out, however, in the special case in which the individual X summands are exponentially distributed with common parameter λ, so that $N(t)$ is a Poisson process. One hopes that the results in the special case will shed some light on the behavior of the sample mean under other distributions.

The key is the use of Theorem 5.7 to evaluate the conditional expectation

$$E[W_{N(t)}|N(t) = n] = E[\max\{U_1, \ldots, U_n\}]$$

$$= t\left(\frac{n}{n+1}\right),$$

where $U_1, \ldots, U_n$ are independent and uniformly distributed over the interval $(0, t]$. Note also that

$$\Pr\{N(t) = n | N(t) > 0\} = \frac{(\lambda t)^n e^{-\lambda t}}{n!\,(1 - e^{-\lambda t})}.$$

Then

$$E\left[\frac{W_{N(t)}}{N(t)} \middle| N(t) > 0\right] = \sum_{n=1}^{\infty} E\left[\frac{W_n}{n} \middle| N(t) = n\right] \Pr\{N(t) = n | N(t) > 0\}$$

$$= \sum_{n=1}^{\infty} t\left(\frac{n}{n+1}\right)\left(\frac{1}{n}\right)\left\{\frac{(\lambda t)^n e^{-\lambda t}}{n!\,(1 - e^{-\lambda t})}\right\}$$

$$= \frac{1}{\lambda}\left(\frac{1}{e^{\lambda t} - 1}\right)\sum_{n=1}^{\infty} \frac{(\lambda t)^{n+1}}{(n+1)!}$$

$$= \frac{1}{\lambda}\left(\frac{1}{e^{\lambda t} - 1}\right)\left(e^{\lambda t} - 1 - \lambda t\right)$$

$$= \frac{1}{\lambda}\left(1 - \frac{\lambda t}{e^{\lambda t} - 1}\right).$$

We can perhaps more clearly see the effect of the sum quota sampling if we express the preceding calculation in terms of the ratio of the bias to the true mean $E[X_1] = 1/\lambda$. We then have

$$\frac{E[X_1] - E[\overline{X}_{N(t)}]}{E[X_1]} = \frac{\lambda t}{e^{\lambda t} - 1} = \frac{E[N(t)]}{e^{E[N(t)]} - 1}.$$

The left side is the fraction of bias, and the right side expresses this fraction bias as a function of the expected sample size under sum quota sampling. The following table relates some values:

Fraction Bias	$E[N(t)]$
0.58	1
0.31	2
0.16	3
0.17	4
0.03	5
0.015	6
0.0005	10

In the airline example, we observed $N(t) = 3$ failures in the two-year period, and upon consulting the above table, we might estimate the fraction bias to be something on the order of -16%. Since we observed $\overline{X}_{N(t)} = 7$, a more accurate estimate of the mean time between failures (MTBF $= E[X_1]$) might be $7/.84 = 8.33$, an estimate that attempts to correct, at least on average, for the bias due to the sampling method.

Looking once again at the table, we may conclude in general, that the bias due to sum quota sampling can be made acceptably small by choosing the quota t sufficiently large so that, on average, the sample size so selected is reasonably large. If the individual observations are exponentially distributed, the bias can be kept within 0.05% of the true value, provided that the quota t is large enough to give an average sample size of 10 or more.

Exercises

5.4.1 Let $\{X(t); t \geq 0\}$ be a Poisson process of rate λ. Suppose it is known that $X(1) = n$. For $n = 1, 2, \ldots$, determine the mean of the first arrival time W_1.

5.4.2 Let $\{X(t); t \geq 0\}$ be a Poisson process of rate λ. Suppose it is known that $X(1) = 2$. Determine the mean of $W_1 W_2$, the product of the first two arrival times.

5.4.3 Customers arrive at a certain facility according to a Poisson process of rate λ. Suppose that it is known that five customers arrived in the first hour. Determine the mean total waiting time $E[W_1 + W_2 + \cdots + W_5]$.

5.4.4 Customers arrive at a service facility according to a Poisson process of intensity λ. The service times $Y_1, Y_2, \ldots$ of the arriving customers are independent random variables having the common probability distribution function $G(y) = \Pr\{Y_k \leq y\}$. Assume that there is no limit to the number of customers that can be serviced simultaneously; i.e., there is an infinite number of servers available. Let $M(t)$ count the number of customers in the system at time t. Argue that $M(t)$ has a Poisson distribution with mean $\lambda p t$, where

$$p = t^{-1} \int_0^t [1 - G(y)] dy.$$

5.4.5 Customers arrive at a certain facility according to a Poisson process of rate λ. Suppose that it is known that five customers arrived in the first hour. Each customer spends a time in the store that is a random variable, exponentially distributed with parameter α and independent of the other customer times, and then departs. What is the probability that the store is empty at the end of this first hour?

Problems

5.4.1 Let $W_1, W_2, \ldots$ be the event times in a Poisson process $\{X(t); t \geq 0\}$ of rate λ. Suppose it is known that $X(1) = n$. For $k < n$, what is the conditional density function of $W_1, \ldots, W_{k-1}, W_{k+1}, \ldots, W_n$, given that $W_k = w$?

5.4.2 Let $\{N(t); t \geq 0\}$ be a Poisson process of rate λ, representing the arrival process of customers entering a store. Each customer spends a duration in the store that is a random variable with cumulative distribution function G. The customer durations are independent of each other and of the arrival process. Let $X(t)$ denote the number of customers remaining in the store at time t, and let $Y(t)$ be the number of customers who have arrived and departed by time t. Determine the joint distribution of $X(t)$ and $Y(t)$.

5.4.3 Let $W_1, W_2, \ldots$ be the waiting times in a Poisson process $\{X(t); t \geq 0\}$ of rate λ. Under the condition that $X(1) = 3$, determine the joint distribution of $U = W_1/W_2$ and $V = (1 - W_3)/(1 - W_2)$.

5.4.4 Let $W_1, W_2, \ldots$ be the waiting times in a Poisson process $\{X(t); t \geq 0\}$ of rate λ. Independent of the process, let $Z_1, Z_2, \ldots$ be independent and identically distributed random variables with common probability density function $f(x), 0 < x < \infty$. Determine $\Pr\{Z > z\}$, where

$$Z = \min\{W_1 + Z_1, W_2 + Z_2, \ldots\}.$$

5.4.5 Let $W_1, W_2, \ldots$ be the waiting times in a Poisson process $\{N(t); t \geq 0\}$ of rate λ. Determine the limiting distribution of W_1, under the condition that $N(t) = n$ as $n \to \infty$ and $t \to \infty$ in such a way that $n/t = \beta > 0$.

5.4.6 Customers arrive at a service facility according to a Poisson process of rate λ customers/hour. Let $X(t)$ be the number of customers that have arrived up to time t. Let $W_1, W_2, \ldots$ be the successive arrival times of the customers.
   (a) Determine the conditional mean $E[W_1 | X(t) = 2]$.
   (b) Determine the conditional mean $E[W_3 | X(t) = 5]$.
   (c) Determine the conditional probability density function for W_2, given that $X(t) = 5$.

5.4.7 Let $W_1, W_2, \ldots$ be the event times in a Poisson process $\{X(t); t \geq 0\}$ of rate λ, and let $f(w)$ be an arbitrary function. Verify that

$$E\left[\sum_{i=1}^{X(t)} f(W_i)\right] = \lambda \int_0^t f(w) dw.$$

5.4.8 Electrical pulses with independent and identically distributed random amplitudes $\xi_1, \xi_2, \ldots$ arrive at a detector at random times $W_1, W_2, \ldots$ according to a Poisson process of rate λ. The detector output $\theta_k(t)$ for the kth pulse at time t is

$$\theta_k(t) = \begin{cases} 0 & \text{for } t < W_k, \\ \xi_k \exp\{-\alpha(t - W_k)\} & \text{for } t \geq W_k. \end{cases}$$

That is, the amplitude impressed on the detector when the pulse arrives is ξ_k, and its effect thereafter decays exponentially at rate α. Assume that the detector is additive, so that if $N(t)$ pulses arrive during the time interval $[0, t]$, then the output at time t is

$$Z(t) = \sum_{k=1}^{N(t)} \theta_k(t).$$

Determine the mean output $E[Z(t)]$ assuming $N(0) = 0$. Assume that the amplitudes $\xi_1, \xi_2, \dots$ are independent of the arrival times $W_1, W_2, \dots$.

5.4.9 Customers arrive at a service facility according to a Poisson process of rate λ customers per hour. Let $N(t)$ be the number of customers that have arrived up to time t, and let $W_1, W_2, \dots$ be the successive arrival times of the customers. Determine the expected value of the product of the waiting times up to time t. (Assume that $W_1 W_2 \cdots W_{N(t)} = 1$ when $N(t) = 0$.)

5.4.10 Compare and contrast the example immediately following Theorem 5.7, the shot noise process of Section 5.4.1, and the model of Problem 4.8. Can you formulate a general process of which these three examples are special cases?

5.4.11 *Computer Challenge* Let $U_0, U_1, \dots$ be independent random variables, each uniformly distributed on the interval $(0, 1)$. Define a stochastic process $\{S_n\}$ recursively by setting

$$S_0 = 0 \quad \text{and} \quad S_{n+1} = U_n(1 + S_n) \quad \text{for } n > 0.$$

(This is an example of a discrete-time, continuous-state, Markov process.) When n becomes large, the distribution of S_n approaches that of a random variable $S = S_\infty$, and S must have the same probability distribution as $U(1 + S)$, where U and S are independent. We write this in the form

$$S \overset{\mathscr{D}}{=} U(1 + S),$$

from which it is easy to determine that $E[S] = 1$, $\text{Var}[S] = \frac{1}{2}$, and even (the Laplace transform)

$$E\left[e^{-\theta S}\right] = \exp\left\{ -\int_{0+}^{\theta} \frac{1 - e^{-u}}{u} du \right\}, \quad \theta > 0.$$

The probability density function $f(s)$ satisfies

$$f(s) = 0 \quad \text{for} \quad s \leq 0, \quad \text{and}$$
$$\frac{df}{ds} = \frac{1}{s} f(s - 1), \quad \text{for } s > 0.$$

What is the 99th percentile of the distribution of S? (Note: Consider the shot noise process of Section 5.4.1. When the Poisson process has rate $\lambda = 1$ and the impulse response function is the exponential $h(x) = \exp\{-x\}$, then the shot noise $I(t)$ has, in the limit for large t, the same distribution as S.)

5.5 Spatial Poisson Processes

In this section, we define some versions of multidimensional Poisson processes and describe some examples and applications.

Let S be a set in n-dimensional space and let $\mathscr{A}$ be a family of subsets of S. A *point process* in S is a stochastic process $N(A)$ indexed by the sets A in $\mathscr{A}$ and having the set of nonnegative integers $\{0, 1, 2, \ldots\}$ as its possible values. We think of "points" being scattered over S in some random manner and of $N(A)$ as counting the number of points in the set A. Because $N(A)$ is a counting function, there are certain obvious requirements that it must satisfy. For example, if A and B are disjoint sets in $\mathscr{A}$ whose union $A \cup B$ is also in $\mathscr{A}$, then it must be that $N(A \cup B) = N(A) + N(B)$. In words, the number of points in A or B equals the number of points in A plus the number of points in B when A and B are disjoint.

The one-dimensional case, in which S is the positive half line and $\mathscr{A}$ comprises all intervals of the form $A = (s, t]$, for $0 \leq s < t$, was introduced in Section 5.3. The straightforward generalization to the plane and three-dimensional space that is now being discussed has relevance when we consider the spatial distribution of stars or galaxies in astronomy, of plants or animals in ecology, of bacteria on a slide in medicine, and of defects on a surface or in a volume in reliability engineering.

Let S be a subset of the real line, two-dimensional plane, or three-dimensional space; let $\mathscr{A}$ be the family of subsets of S and for any set A in $\mathscr{A}$; let $|A|$ denote the size (length, area, or volume, respectively) of A. Then, $\{N(A); A \text{ in } \mathscr{A}\}$ is a *homogeneous Poisson point process* of intensity $\lambda > 0$ if

1. for each A in $\mathscr{A}$, the random variable $N(A)$ has a Poisson distribution with parameter $\lambda|A|$;
2. for every finite collection $\{A_1, \ldots, A_n\}$ of disjoint subsets of S, the random variables $N(A_1)$, $\ldots, N(A_n)$ are independent.

In Section 5.2, the law of rare events was invoked to derive the Poisson process as a consequence of certain physically plausible postulates. This implication serves to justify the Poisson process as a model in those situations where the postulates may be expected to hold. An analogous result is available in the multidimensional case at hand. Given an arbitrary point process $\{N(A); A \text{ in } \mathscr{A}\}$, the required postulates are as follows:

1. The possible values for $N(A)$ are the nonnegative integers $\{0, 1, 2, \ldots\}$ and $0 < \Pr\{N(A) = 0\} < 1$ if $0 < |A| < \infty$.
2. The probability distribution of $N(A)$ depends on the set A only through its size (length, area, or volume) $|A|$, with the further property that $\Pr\{N(A) \geq 1\} = \lambda|A| + o(|A|)$ as $|A| \downarrow 0$.
3. For $m = 2, 3, \ldots$, if $A_1, A_2, \ldots, A_m$ are disjoint regions, then $N(A_1), N(A_2), \ldots, N(A_m)$ are independent random variables and $N(A_1 \cup A_2 \cup \cdots \cup A_m) = N(A_1) + N(A_2) + \cdots + N(A_m)$.
4. $$\lim_{|A| \to 0} \frac{\Pr\{N(A) \geq 1\}}{\Pr\{N(A) = 1\}} = 1.$$

The motivation and interpretation of these postulates is quite evident. Postulate 2 asserts that the probability distribution of $N(A)$ does not depend on the shape or location of A, but only on its size. Postulate 3 requires that the outcome in one region not influence or be influenced by the outcome in a second region that does not overlap the first. Postulate 4 precludes the possibility of two points occupying the same location.

If a random point process $N(A)$ defined with respect to subsets A of Euclidean n-space satisfies Postulates 1 through 4, then $N(A)$ is a homogeneous Poisson point process of intensity $\lambda > 0$, and

$$\Pr\{N(A) = k\} = \frac{e^{-\lambda|A|}(\lambda|A|)^k}{k!} \quad \text{for } k = 0, 1, \ldots. \tag{5.26}$$

As in the one-dimensional case, homogeneous Poisson point processes in n-dimensions are highly amenable to analysis, and many results are known for them. We elaborate a few of these consequences next, beginning with the uniform distribution of a single point. Consider a region A of positive size $|A| > 0$, and suppose it is known that A contains exactly one point; i.e., $N(A) = 1$. Where in A is this point located? We claim that the point is uniformly distributed in the sense that

$$\Pr\{N(B) = 1|N(A) = 1\} = \frac{|B|}{|A|} \quad \text{for any set } B \subset A. \tag{5.27}$$

In words, the probability of the point being in any subset B of A is proportional to the size of B; i.e., the point is uniformly distributed in A. The uniform distribution expressed in (5.27) is an immediate consequence of elementary conditional probability manipulations. We write $A = B \cup C$, where B is an arbitrary subset of A and C is the portion of A not included in B. Then, B and C are disjoint, so that $N(B)$ and $N(C)$ are independent Poisson random variables with respective means $\lambda|B|$ and $\lambda|C|$. Then

$$\begin{aligned}
\Pr\{N(B) = 1|N(A) = 1\} &= \frac{\Pr\{N(B) = 1, N(C) = 0\}}{\Pr\{N(A) = 1\}} \\
&= \frac{\lambda|B|e^{-\lambda|B|}e^{-\lambda|C|}}{\lambda|A|e^{-\lambda|A|}} \\
&= \frac{|B|}{|A|} \quad \text{(because } |B| + |C| = |A|\text{)},
\end{aligned}$$

and the proof is complete.

The generalization to n points in a region A is stated as follows. Consider a set A of positive size $|A| > 0$ and containing $N(A) = n \geq 1$ points. Then, these n points are independent and uniformly distributed in A in the sense that for any disjoint partition $A_1, \ldots, A_m$ of A, where $A_1 \cup \cdots \cup A_m = A$, and any positive integers $k_1, \ldots, k_m$, where $k_1 + \cdots + k_m = n$, we have

$$\Pr\{N(A_1) = k_1, \ldots, N(A_m) = k_m|N(A) = n\} \tag{5.28}$$

$$= \frac{n!}{k_1! \cdots k_m!} \left(\frac{|A_1|}{|A|}\right)^{k_1} \cdots \left(\frac{|A_m|}{|A|}\right)^{k_m}.$$

Equation (5.28) expresses the multinomial distribution for the conditional distribution of $N(A_1), \ldots, N(A_m)$ given that $N(A) = n$.

Example *An Application in Astronomy* Consider stars distributed in space in accordance with a three-dimensional Poisson point process of intensity $\lambda > 0$.

Let $\mathbf{x}$ and $\mathbf{y}$ designate general three-dimensional vectors, and assume that the light intensity exerted at $\mathbf{x}$ by a star located at $\mathbf{y}$ is $f(\mathbf{x}, \mathbf{y}, \alpha) = \alpha/\|\mathbf{x} - \mathbf{y}\|^2 = \alpha/\left[(x_1 - y_1)^2 + (x_2 - y_2)^2 + (x_3 - y_3)^2\right]$, where α is a random parameter depending on the intensity of the star at $\mathbf{y}$. We assume that the intensities α associated with different stars are independent, identically distributed random variables possessing a common mean μ_α and variance σ_α^2. We also assume that the combined intensity exerted at the point $\mathbf{x}$ due to light created by different stars accumulates additively. Let $Z(\mathbf{x}, A)$ denote the total light intensity at the point $\mathbf{x}$ due to signals emanating from all sources located in region A. Then

$$Z(\mathbf{x}, A) = \sum_{r=1}^{N(A)} f(\mathbf{x}, \mathbf{y}_r, \alpha_r)$$
$$= \sum_{r=1}^{N(A)} \frac{\alpha_r}{\|\mathbf{x} - \mathbf{y}_r\|^2}, \tag{5.29}$$

where $\mathbf{y}_r$ is the location of the rth star in A. We recognize (5.29) as a random sum, as discussed in Chapter 2, Section 2.3.2. Accordingly, we have the mean intensity at $\mathbf{x}$ given by

$$E[Z(\mathbf{x}, A)] = (E[N(A)])(E[f(\mathbf{x}, \mathbf{y}, \alpha)]). \tag{5.30}$$

Note that $E[N(A)] = \lambda |A|$, while because we have assumed α and $\mathbf{y}$ to be independent,

$$E[f(\mathbf{x}, \mathbf{y}, \alpha)] = E[\alpha] E\left[\|\mathbf{x} - \mathbf{y}\|^{-2}\right].$$

But as a consequence of the Poisson distribution of stars in space, we may take $\mathbf{y}$ to be uniformly distributed in A. Thus

$$E\left[\|\mathbf{x} - \mathbf{y}\|^{-2}\right] = \frac{1}{|A|} \int_A \frac{d\mathbf{y}}{\|\mathbf{x} - \mathbf{y}\|^2}.$$

With $\mu_\alpha = E[\alpha]$, then (5.30) reduces to

$$E[Z(\mathbf{x}, A)] = \lambda \mu_\alpha \int_A \frac{d\mathbf{y}}{\|\mathbf{x} - \mathbf{y}\|2}.$$

Exercises

5.5.1 Bacteria are distributed throughout a volume of liquid according to a Poisson process of intensity $\theta = 0.6$ organisms per mm^3. A measuring device counts the number of bacteria in a 10 mm^3 volume of the liquid. What is the probability that more than two bacteria are in this measured volume?

5.5.2 Customer arrivals at a certain service facility follow a Poisson process of unknown rate. Suppose it is known that 12 customers have arrived during the first 3 h. Let N_i be the number of customers who arrive during the ith hour, $i = 1, 2, 3$. Determine the probability that $N_1 = 3, N_2 = 4$, and $N_3 = 5$.

5.5.3 Defects (air bubbles, contaminants, chips) occur over the surface of a varnished tabletop according to a Poisson process at a mean rate of one defect per top. If two inspectors each check separate halves of a given table, what is the probability that both inspectors find defects?

Problems

5.5.1 A piece of a fibrous composite material is sliced across its circular cross section of radius R, revealing fiber ends distributed across the circular area according to a Poisson process of rate 100 fibers per cross section. The locations of the fibers are measured, and the radial distance of each fiber from the center of the circle is computed. What is the probability density function of this radial distance X for a randomly chosen fiber?

5.5.2 Points are placed on the surface of a circular disk of radius one according to the following scheme. First, a Poisson distributed random variable N is observed. If $N = n$, then n random variables $\theta_1, \ldots, \theta_n$ are independently generated, each uniformly distributed over the interval $[0, 2\pi)$, and n random variables $R_1, \ldots, R_n$ are independently generated, each with the triangular density $f(r) = 2r, 0 < r < 1$. Finally, the points are located at the positions with polar coordinates $(R_i, \theta_i), i = 1, \ldots, n$. What is the distribution of the resulting point process on the disk?

5.5.3 Let $\{N(A); A \in \mathbf{R}^2\}$ be a homogeneous Poisson point process in the plane, where the intensity is λ. Divide the $(0, t] \times (0, t]$ square into n^2 boxes of side length $d = t/n$. Suppose there is a reaction between two or more points whenever they are located within the same box. Determine the distribution for the number of reactions, valid in the limit as $t \to \infty$ and $d \to 0$ in such a way that $td \to \mu > 0$.

5.5.4 Consider spheres in three-dimensional space with centers distributed according to a Poisson distribution with parameter $\lambda|A|$, where $|A|$ now represents the volume of the set A. If the radii of all spheres are distributed according to $F(r)$ with density $f(r)$ and finite third moment, show that the number of spheres that cover a point $\mathbf{t}$ is a Poisson random variable with parameter $\frac{4}{3}\lambda\pi \int_0^\infty r^3 f(r) dr$.

5.5.5 Consider a two-dimensional Poisson process of particles in the plane with intensity parameter ν. Determine the distribution $F_D(x)$ of the distance between a particle and its nearest neighbor. Compute the mean distance.

5.5.6 Suppose that stars are distributed in space following a Poisson point process of intensity λ. Fix a star *alpha* and let R be the distance from alpha to its nearest neighbor. Show that R has the probability density function

$$f_R(x) = (4\lambda\pi x^2)\exp\left\{\frac{-4\lambda\pi x^3}{3}\right\}, \quad x > 0.$$

5.5.7 Consider a collection of circles in the plane whose centers are distributed according to a spatial Poisson process with parameter $\lambda|A|$, where $|A|$ denotes the area of the set A. (In particular, the number of centers $\xi(A)$ in the set A follows the distribution law $\Pr\{\xi(A) = k\} = e^{-\lambda|A|}\left[(\lambda|A|)^k/k!\right]$.) The radius of each circle is assumed to be a random variable independent of the location of the center of the circle, with density function $f(r)$ and finite second moment.

    (a) Show that $C(r)$, defined to be the number of circles that cover the origin and have centers at a distance less than r from the origin, determines a variable-time Poisson process, where the time variable is now taken to be the distance r.

    Hint: Prove that an event occurring between r and $r + dr$ (i.e., there is a circle that covers the origin and whose center is in the ring of radius r to $r + dr$) has probability $\lambda 2\pi r\, dr \int_r^\infty f(\rho)\, d\rho + o(dr)$, and events occurring over disjoint intervals constitute independent random variables. Show that $C(r)$ is a variable-time (nonhomogeneous) Poisson process with parameter

$$\lambda(r) = 2\pi \lambda r \int_r^\infty f(\rho)\mathrm{d}\rho.$$

    (b) Show that the number of circles that cover the origin is a Poisson random variable with parameter $\lambda \int_0^\infty \pi r^2 f(r)\mathrm{d}r$.

5.6 Compound and Marked Poisson Processes

Given a Poisson process $X(t)$ of rate $\lambda > 0$, suppose that each event has associated with it a random variable, possibly representing a value or a cost. Examples will appear shortly. The successive values $Y_1, Y_2, \ldots$ are assumed to be independent, independent of the Poisson process, and random variables sharing the common distribution function

$$G(y) = \Pr\{Y_k \le y\}.$$

A *compound Poisson process* is the cumulative value process defined by

$$Z(t) = \sum_{k=1}^{X(t)} Y_k \quad \text{for } t \ge 0. \tag{5.31}$$

A *marked Poisson process* is the sequence of pairs $(W_i, Y_1), (W_2, Y_2), \ldots$, where $W_1, W_2, \ldots$ are the waiting times or event times in the Poisson process $X(t)$.

Both compound Poisson and marked Poisson processes appear often as models of physical phenomena.

5.6.1 *Compound Poisson Processes*

Consider the compound Poisson process $Z(t) = \sum_{k=1}^{X(t)} Y_k$. If $\lambda > 0$ is the rate for the process $X(t)$ and $\mu = E[Y_1]$ and $\nu^2 = \text{Var}[Y_1]$ are the common mean and variance for $Y_1, Y_2, \ldots$ then the moments of $Z(t)$ can be determined from the random sums formulas of Chapter 2, Section 2.3.2 and are

$$E[Z(t)] = \lambda \mu t; \quad \text{Var}[Z(t)] = \lambda \left(v^2 + \mu^2\right) t. \tag{5.32}$$

Examples

(a) *Risk Theory* Suppose claims arrive at an insurance company in accordance with a Poisson process having rate λ. Let Y_k be the magnitude of the kth claim. Then, $Z(t) = \sum_{k=1}^{X(t)} Y_k$ represents the cumulative amount claimed up to time t.

(b) *Stock Prices* Suppose that transactions in a certain stock take place according to a Poisson process of rate λ. Let Y_k denote the change in market price of the stock between the kth and $(k-1)$th transaction.

The *random walk hypothesis* asserts that $Y_1, Y_2, \ldots$ are independent random variables. The random walk hypothesis, which has a history dating back to 1900, can be deduced formally from certain assumptions describing a "perfect market."

Then, $Z(t) = \sum_{k=1}^{X(t)} Y_k$ represents the total price change up to time t.

This stock price model has been proposed as an explanation for why stock price changes do not follow a Gaussian (normal) distribution.

The distribution function for the compound Poisson process $Z(t) = \sum_{k=1}^{X(t)} Y_k$ can be represented explicitly after conditioning on the values of $X(t)$. Recall the convolution notation

$$G^{(n)}(y) = \Pr\{Y_1 + \cdots + Y_n \leq y\}$$

$$= \int\limits_{-\infty}^{+\infty} G^{(n-1)}(y-z) dG(z) \tag{5.33}$$

with

$$G^{(0)}(y) = \begin{cases} 1 & \text{for } y \geq 0, \\ 0 & \text{for } y < 0. \end{cases}$$

Then

$$\Pr\{Z(t) \leq z\} = \Pr\left\{ \sum_{k=1}^{X(t)} Y_k \leq z \right\}$$

$$= \sum_{n=0}^{\infty} \Pr\left\{ \sum_{k=1}^{X(t)} Y_k \leq z | X(t) = n \right\} \frac{(\lambda t)^n e^{-\lambda t}}{n!} \tag{5.34}$$

$$= \sum_{n=0}^{\infty} \frac{(\lambda t)^n e^{-\lambda t}}{n!} G^{(n)}(z) \quad \begin{array}{l} \text{(since } X(t) \text{ is independent} \\ \text{of } Y_1, Y_2, \ldots). \end{array}$$

Example *A Shock Model* Let $X(t)$ be the number of shocks to a system up to time t and let Y_k be the damage or wear incurred by the kth shock. We assume that damage is positive, i.e., that $\Pr\{Y_k \geq 0\} = 1$, and that the damage accumulates additively, so that $Z(t) = \Sigma_{k=1}^{X(t)} Y_k$ represents the total damage sustained up to time t. Suppose that the system continues to operate as long as this total damage is less than some critical value a and fails in the contrary circumstance. Let T be the time of system failure. Then

$$\{T > t\} \quad \text{if and only if} \quad \{Z(t) < a\}. \quad \text{(Why?)} \tag{5.35}$$

In view of (5.34) and (5.35), we have

$$\Pr\{T > t\} = \sum_{n=0}^{\infty} \frac{(\lambda t)^n e^{-\lambda t}}{n!} G^{(n)}(a).$$

All summands are nonnegative, so we may interchange integration and summation to get the mean system failure time

$$E[T] = \int_0^{\infty} \Pr\{T > t\}\, dt$$

$$= \sum_{n=0}^{\infty} \left(\int_0^{\infty} \frac{(\lambda t)^n e^{-\lambda t}}{n!}\, dt \right) G^{(n)}(a)$$

$$= \lambda^{-1} \sum_{n=0}^{\infty} G^{(n)}(a).$$

This expression simplifies greatly in the special case in which $Y_1, Y_2, \ldots$ are each exponentially distributed according to the density $g_Y(y) = \mu e^{-\mu y}$ for $y \geq 0$. Then, the sum $Y_1 + \cdots + Y_n$ has the gamma distribution

$$G^{(n)}(z) = 1 - \sum_{k=0}^{n-1} \frac{(\mu z)^k e^{-\mu z}}{k!} = \sum_{k=n}^{\infty} \frac{(\mu z)^k e^{-\mu z}}{k!},$$

and

$$\sum_{n=0}^{\infty} G^{(n)}(a) = \sum_{n=0}^{\infty} \sum_{k=n}^{\infty} \frac{(\mu a)^k e^{-\mu a}}{k!}$$

$$= \sum_{k=0}^{\infty} \sum_{n=0}^{k} \frac{(\mu a)^k e^{-\mu a}}{k!}$$

$$= \sum_{k=0}^{\infty} (1+k) \frac{(\mu a)^k e^{-\mu a}}{k!}$$

$$= 1 + \mu a.$$

When $Y_1, Y_2, \ldots$ are exponentially distributed, then

$$E[T] = \frac{1 + \mu a}{\lambda}.$$

5.6.2 Marked Poisson Processes

Again suppose that a random variable Y_k is associated with the kth event in a Poisson process of rate λ. We stipulate that $Y_1, Y_2, \ldots$ are independent, independent of the Poisson process, and share the common distribution function

$$G(y) = \Pr\{Y_k \leq y\}.$$

The sequence of pairs $(W_1, Y_1), (W_2, Y_2), \ldots$ is called a *marked Poisson process*.

We begin the analysis of marked Poisson processes with one of the simplest cases. For a fixed value $p(0 < p < 1)$, suppose

$$\Pr\{Y_k = 1\} = p, \quad \Pr\{Y_k = 0\} = q = 1 - p.$$

Now consider separately the processes of points marked with ones and of points marked with zeros. In this case, we can define the relevant Poisson processes explicitly by

$$X_1(t) = \sum_{k=1}^{X(t)} Y_k \quad \text{and} \quad X_0(t) = X(t) - X_1(t).$$

Then, nonoverlapping increments in $X_1(t)$ are independent random variables, $X_1(0) = 0$, and finally, Theorem 5.2 applies to assert that $X_1(t)$ has a Poisson distribution with mean $\lambda p t$. In summary, $X_1(t)$ is a Poisson process with rate λp, and the parallel argument shows that $X_0(t)$ is a Poisson process with rate $\lambda(1 - p)$. *What is even more interesting and surprising is that $X_0(t)$ and $X_1(t)$ are independent processes!* The relevant property to check is that $\Pr\{X_0(t) = j$ and $X_1(t) = k\} = \Pr\{X_0(t) = j\} \times \Pr\{X_1(t) = k\}$ for $j, k = 0, 1, \ldots$. We establish this independence by writing

$$
\begin{aligned}
\Pr\{X_0(t) = j, X_1(t) = k\} &= \Pr\{X_0(t) = j + k, X_1(t) = k\} \\
&= \Pr\{X_1(t) = k | X(t) = j + k\} \Pr\{X(t) = j + k\} \\
&= \frac{(j+k)!}{j! k!} p^k (1 - p)^j \frac{(\lambda t)^{j+k} e^{-\lambda t}}{(j+k)!} \\
&= \left[\frac{e^{-\lambda p t} (\lambda p t)^k}{k!} \right] \left[\frac{e^{-\lambda(1-p)t} (\lambda(1-p)t)^j}{j!} \right] \\
&= \Pr\{X_1(t) = k\} \Pr\{X_0(t) = j\} \\
&\qquad\qquad\qquad\qquad\qquad \text{for } j, k = 0, 1, \ldots.
\end{aligned}
$$

Example Customers enter a store according to a Poisson process of rate $\lambda = 10$ per hour. Independently, each customer buys something with probability $p = 0.3$ and leaves without making a purchase with probability $q = 1 - p = 0.7$. What is the probability that during the first hour 9 people enter the store and that 3 of these people make a purchase and 6 do not?

Let $X_1 = X_1(1)$ be the number of customers who make a purchase during the first hour and $X_0 = X_0(1)$ be the number of people who do not. Then, X_1 and X_0 are independent Poisson random variables having respective rates $0.3(10) = 3$ and $0.7(10) = 7$. According to the Poisson distribution,

$$\Pr\{X_1 = 3\} = \frac{3^3 e^{-3}}{3!} = 0.2240,$$

$$\Pr\{X_0 = 6\} = \frac{7^6 e^{-7}}{6!} = 0.1490,$$

and

$$\Pr\{X_1 = 3, X_0 = 6\} = \Pr\{X_1 = 3\}\Pr\{X_0 = 6\} = (0.2240)(0.1490) = 0.0334.$$

In our study of marked Poisson processes, let us next consider the case where the value random variables $Y_1, Y_2, \ldots$ are discrete, with possible values $0, 1, 2, \ldots$ and

$$\Pr\{Y_n = k\} = a_k > 0 \quad \text{for } k = 0, 1, \ldots, \text{ with } \sum_k a_k = 1.$$

In Figure 5.9, the original Poisson event times $W_1, W_2, \ldots$ are shown on the bottom axis. Then, a point is placed in the (t, y) plane at (W_n, Y_n) for every n. For every integer $k = 0, 1, 2, \ldots$, one obtains a point process that corresponds to the times W_n for which $Y_n = k$. The same reasoning as in the zero–one case applies to imply that each of these

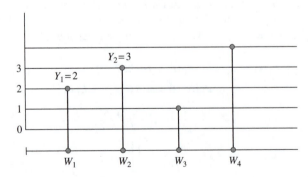

Figure 5.9 A marked Poisson process. $W_1, W_2, \ldots$ are the event times in a Poisson process of rate λ. The random variables $Y_1, Y_2, \ldots$ are the markings, assumed to be independent and identically distributed, and independent of the Poisson process.

processes is Poisson, the rate for the kth process being λa_k, and that processes for distinct values of k are independent.

To state the corresponding decomposition result when the values $Y_1, Y_2, \ldots$ are continuous random variables requires a higher level of sophistication, although the underlying ideas are basically the same. To set the stage for the formal statement, we first define what we mean by a nonhomogeneous Poisson point process in the plane, thus extending the homogeneous processes of the previous section. Let $\theta = \theta(x, y)$ be a nonnegative function defined on a region S in the (x, y) plane. For each subset A of S, let $\mu(A) = \iint_A \theta(x, y) \, dx \, dy$ be the volume under $\theta(x, y)$ enclosed by A. A nonhomogeneous Poisson point process of intensity function $\theta(x, y)$ is a point process $\{N(A); A \subset S\}$ for which

1. for each subset A of S, the random variable $N(A)$ has a Poisson distribution with mean $\mu(A)$;
2. for disjoint subsets $A_1, \ldots, A_m$ of S, the random variables $N(A_1), \ldots, N(A_m)$ are independent.

It is easily seen that the homogeneous Poisson point process of intensity λ corresponds to the function $\theta(x, y)$ being constant, and $\theta(x, y) = \lambda$ for all x, y.

With this definition in hand, we state the appropriate decomposition result for general marked Poisson processes.

Theorem 5.8. *Let $(W_1, Y_1), (W_2, Y_2), \ldots$ be a marked Poisson process where W_1, $W_2, \ldots$ are the waiting times in a Poisson process of rate λ and $Y_1, Y_2, \ldots$ are independent identically distributed continuous random variables having probability density function $g(y)$. Then $(W_1, Y_1), (W_2, Y_2), \ldots$ form a two-dimensional nonhomogeneous Poisson point process in the (t, y) plane, where the mean number of points in a region A is given by*

$$\mu(A) = \iint_A \lambda g(y) \, dy \, dt. \tag{5.36}$$

Figure 5.10 diagrams the scene.

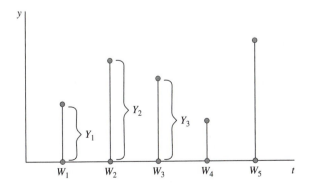

Figure 5.10 A marked Poisson process.

Theorem 5.8 asserts that the numbers of points in disjoint intervals are independent random variables. For example, the waiting times corresponding to positive values $Y_1, Y_2, \ldots$ form a Poisson process, as do the times associated with negative values, and these two processes are independent.

Example *Crack Failure* The following model is proposed to describe the failure time of a sheet or volume of material subjected to a constant stress σ. The failure time is viewed in two parts, crack initiation and crack propagation.

Crack initiation occurs according to a Poisson process whose rate per unit time and unit volume is a constant $\lambda_\sigma > 0$ depending on the stress level σ. Then, crack initiation per unit time is a Poisson process of rate $\lambda_\sigma |V|$, where $|V|$ is the volume of material under consideration. We let $W_1, W_2, \ldots$ be the times of crack initiation.

Once begun, a crack grows at a random rate until it reaches a critical size, at which instant structural failure occurs. Let Y_k be the time to reach critical size for the kth crack. The cumulative distribution function $G_\sigma(y) = \Pr\{Y_k \le y\}$ depends on the constant stress level σ.

We assume that crack initiations are sufficiently sparse as to make $Y_1, Y_2, \ldots$ independent random variables. That is, we do not allow two small cracks to join and form a larger one.

The structural failure time Z is the smallest of $W_1 + Y_1, W_2 + Y_2, \ldots$. It is not necessarily the case that the first crack to appear will cause system failure. A later crack may grow to critical size faster.

In the (t, y) plane, the event $\{\min\{W_k + Y_k\} > z\}$ corresponds to no points falling in the triangle $\triangle = \{(t, y) : t + y \le z, t \ge 0, y \ge 0\}$, as shown in Figure 5.11.

The number of points $N(\triangle)$ falling in the triangle $\triangle$ has a Poisson distribution with mean $\mu(\triangle)$ given, according to (5.36), by

$$\mu(\triangle) = \iint_\triangle \lambda_\sigma |V| ds g_\sigma(u) du$$

$$= \int_0^z \lambda_\sigma |V| \left\{ \int_0^{z-s} g_\sigma(u) du \right\} ds$$

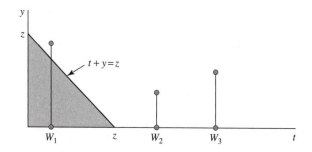

Figure 5.11 A crack failure model.

$$= \lambda_\sigma |V| \int_0^z G_\sigma(z-s)ds$$

$$= \lambda_\sigma |V| \int_0^z G_\sigma(v)dv.$$

From this we obtain the cumulative distribution function for structural failure time,

$$\Pr\{Z \le z\} = 1 - \Pr\{Z > z\} = 1 - \Pr\{N(\triangle) = 0\}$$

$$= 1 - \exp\left\{-\lambda_\sigma |V| \int_0^z G_\sigma(v)dv\right\}.$$

Observe the appearance of the so-called *size effect* in the model, wherein the structure volume $|V|$ affects the structural failure time even at constant stress level σ. The parameter λ_σ and distribution function $G_\sigma(y)$ would require experimental determination.

Example *The Strength Versus Length of Filaments* It was noted that the logarithm of mean tensile strength of brittle fibers, such as boron filaments, in general varies linearly with the logarithm of the filament length, but that this relation did not hold for short filaments. It was suspected that the breakdown in the log linear relation might be due to testing or measurement problems, rather than being an inherent property of short filaments. Evidence supporting this idea was the observation that short filaments would break in the test clamps, rather than between them as desired, more often than would long filaments. Some means of correcting observed mean strengths to account for filaments breaking in, rather than between, the clamps was desired. It was decided to compute the ratio between the actual mean strength and an ideal mean strength, obtained under the assumption that there was no stress in the clamps, as a correction factor.

Since the molecular bonding strength is several orders of magnitude higher than generally observed strengths, it was felt that failure typically was caused by flaws. There are a number of different types of flaws, both internal flaws such as voids, inclusions, and weak grain boundaries, and external, or surface, flaws such as notches and cracks that cause stress concentrations. Let us suppose that flaws occur independently in a Poisson manner along the length of the filament. We let Y_k be the strength of the filament at the kth flaw and suppose Y_k has the cumulative distribution function $G(y), y > 0$. We have plotted this information in Figure 5.12. The flaws reduce the strength. Opposing the strength is the stress in the filament. Ideally, the stress should be constant along the filament between the clamp faces and zero within the clamp. In practice, the stress tapers off to zero over some positive length in the clamp. As a first approximation it is reasonable to assume that the stress decreases linearly. Let l be the length of the clamp and t the distance between the clamps, called the *gauge length*, as illustrated in Figure 5.12 on the next page.

The filament holds as long as the stress has not exceeded the strength as determined by the weakest flaw. That is, the filament will support a stress of y as long as no flaw points fall in the stress trapezoid of Figure 5.12. The number of points in this trapezoid has a Poisson distribution with mean $\mu(B) + 2\mu(A)$. In particular, no points fall there with probability $e^{-[\mu(B)+2\mu(A)]}$. If we let S be the strength of the filament, then

$$\Pr\{S > y\} = e^{-2\mu(A)-\mu(B)}.$$

We compute

$$\mu(A) = \int_0^l G\left(\frac{xy}{l}\right)\lambda\,dx$$

and

$$\mu(B) = \lambda t G(y).$$

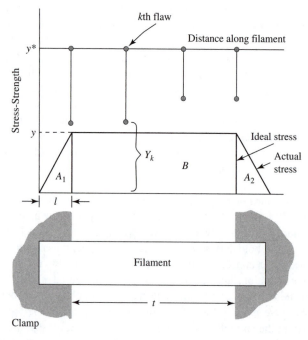

Figure 5.12 The stress versus strength of a filament under tension. Flaws reducing the strength of a filament below its theoretical maximum y^* are distributed randomly along its length. The stress in the filament is constant at the level y between clamps and tapers off to zero within the clamps. The filament fails if at any point along its length a flaw reduces its strength below its stress.

Finally, the mean strength of the filament is

$$E[S] = \int_0^\infty \Pr\{S > y\}\, dy = \int_0^\infty \exp\left\{-\lambda\left[tG(y) + 2\int_0^l c\left(\frac{xy}{l}\right)dx\right]\right\} dy.$$

For an ideal filament, we use the same expression but with $l = 0$.

Exercises

5.6.1 Customers demanding service at a central processing facility arrive according to a Poisson process of intensity $\theta = 8$ per unit time. Independently, each customer is classified as *high priority* with probability $\alpha = 0.2$, or *low priority* with probability $1 - \alpha = 0.8$. What is the probability that three high priority and five low priority customers arrive during the first unit of time?

5.6.2 Shocks occur to a system according to a Poisson process of intensity λ. Each shock causes some damage to the system, and these damages accumulate. Let $N(t)$ be the number of shocks up to time t, and let Y_i be the damage caused by the ith shock. Then

$$X(t) = Y_1 + \cdots + Y_{N(t)}$$

is the total damage up to time t. Determine the mean and variance of the total damage at time t when the individual shock damages are exponentially distributed with parameter θ.

5.6.3 Let $\{N(t); t \geq 0\}$ be a Poisson process of intensity λ, and let $Y_1, Y_2, \ldots$ be independent and identically distributed nonnegative random variables with cumulative distribution function $G(y) = \Pr\{Y \leq y\}$. Determine $\Pr\{Z(t) > z | N(t) > 0\}$, where

$$Z(t) = \min\{Y_1, Y_2, \ldots, Y_{N(t)}\}.$$

5.6.4 Men and women enter a supermarket according to independent Poisson processes having respective rates of two and four per minute.

    (a) Starting at an arbitrary time, what is the probability that at least two men arrive before the first woman arrives?

    (b) What is the probability that at least two men arrive before the third woman arrives?

5.6.5 Alpha particles are emitted from a fixed mass of material according to a Poisson process of rate λ. Each particle exists for a random duration and is then annihilated. Suppose that the successive lifetimes $Y_1, Y_2, \ldots$ of distinct particles are independent random variables having the common distribution function $G(y) = \Pr\{Y_k \leq y\}$. Let $M(t)$ be the number of particles existing at time t. By considering the lifetimes as markings, identify the region in the lifetime, arrival-time space that corresponds to $M(t)$, and thereby deduce the probability distribution of $M(t)$.

Problems

5.6.1 Suppose that points are distributed over the half line $[0, \infty)$ according to a Poisson process of rate λ. A sequence of independent and identically distributed nonnegative random variables $Y_1, Y_2, \ldots$ is used to reposition the points so that a point formerly at location W_k is moved to the location $W_k + Y_k$. Completely describe the distribution of the relocated points.

5.6.2 Suppose that particles are distributed on the surface of a circular region according to a spatial Poisson process of intensity λ particles per unit area. The polar coordinates of each point are determined, and each angular coordinate is shifted by a random amount, with the amounts shifted for distinct points being independent random variables following a fixed probability distribution. Show that at the end of the point movement process, the points are still Poisson distributed over the region.

5.6.3 Shocks occur to a system according to a Poisson process of intensity λ. Each shock causes some damage to the system, and these damages accumulate. Let $N(t)$ be the number of shocks up to time t, and let Y_i be the damage caused by the ith shock. Then

$$X(t) = Y_1 + \cdots + Y_{N(t)}$$

is the total damage up to time t. Suppose that the system continues to operate as long as the total damage is strictly less than some critical value a, and fails in the contrary circumstance. Determine the mean time to system failure when the individual damages Y_i have a geometric distribution with $\Pr\{Y = k\} = p(1 - p)^k, k = 0, 1, \ldots$.

5.6.4 Let $\{X(t); t \geq 0\}$ and $\{Y(t); t \geq 0\}$ be independent Poisson processes with respective parameters λ and μ. For a fixed integer a, let $T_a = \min\{t \geq 0; Y(t) = a\}$ be the random time that the Y process first reaches the value a. Determine $\Pr\{X(T_a) = k\}$ for $k = 0, 1, \ldots$.

Hint: First consider $\xi = X(T_1)$ in the case in which $a = 1$. Then, ξ has a geometric distribution. Then, argue that $X(T_a)$ for general a has the same distribution as the sum of a independent ξs and hence has a negative binomial distribution.

5.6.5. Let $\{X(t); t \geq 0\}$ and $\{Y(t); t \geq 0\}$ be independent Poisson processes with respective parameters λ and μ. Let $T = \min\{t \geq 0; Y(t) = 1\}$ be the random time of the first event in the Y process. Determine $\Pr\{X(T/2) = k\}$ for $k = 0, 1, \ldots$.

5.6.6 Let $W_1, W_2, \ldots$ be the event times in a Poisson process $\{X(t); t \geq 0\}$ of rate λ. A new point process is created as follows: Each point W_k is replaced by two new points located at $W_k + X_k$ and $W_k + Y_k$, where $X_1, Y_1, X_2, Y_2, \ldots$ are independent and identically distributed nonnegative random variables, independent of the Poisson process. Describe the distribution of the resulting point process.

5.6.7 Let $\{N(t); t \geq 0\}$ be a Poisson process of intensity λ, and let $Y_1, Y_2, \ldots$ be independent and identically distributed nonnegative random variables with cumulative distribution function

$$G(y) = y^\alpha \quad \text{for } 0 < y < 1.$$

Determine $\Pr\{Z(t) > z | N(t) > 0\}$, where

$$Z(t) = \min\{Y_1, Y_2, \ldots, Y_{N(t)}\}.$$

Describe the behavior for large t.

5.6.8 Let $\{N(t); t \geq 0\}$ be a nonhomogeneous Poisson process of intensity $\lambda(t)$, $t > 0$, and let $Y_1, Y_2, \ldots$ be independent and identically distributed nonnegative random variables with cumulative distribution function

$$G(y) = y^\alpha \quad \text{for } 0 < y < 1.$$

Suppose that the intensity process averages out in the sense that

$$\lim_{t \to \infty} \frac{1}{t} \int_0^t \lambda(u) du = \theta.$$

Let

$$Z(t) = \min\{Y_1, Y_2, \ldots, Y_{N(t)}\}.$$

Determine

$$\lim_{t \to \infty} \Pr\left\{t^{1/\alpha} Z(t) > z\right\}.$$

5.6.9 Let $W_1, W_2, \ldots$ be the event times in a Poisson process of rate λ, and let $N(t) = N((0, t])$ be the number of points in the interval $(0, t]$. Evaluate

$$E\left[\sum_{k=1}^{N(t)} (W_k)^2\right].$$

Note: $\Sigma_{k=1}^0 (W_k)^2 = 0$.

5.6.10 *A Bidding Model* Let $U_1, U_2, \ldots$ be independent random variables, each uniformly distributed over the interval $(0, 1]$. These random variables represent successive bids on an asset that you are trying to sell, and that you must sell by time $t = 1$, when the asset becomes worthless. As a strategy, you adopt a secret number θ, and you will accept the first offer that is greater than θ. For example, you accept the second offer if $U_1 \leq \theta$ while $U_2 > \theta$. Suppose that the offers arrive according to a unit rate Poisson process ($\lambda = 1$).

(a) What is the probability that you sell the asset by time $t = 1$?

(b) What is the value for θ that maximizes your expected return? (You get nothing if you don't sell the asset by time $t = 1$.)

(c) To improve your return, you adopt a new strategy, which is to accept an offer at time t if it exceeds $\theta(t) = (1 - t)/(3 - t)$. What are your new chances of selling the asset, and what is your new expected return?

6 Continuous Time Markov Chains

6.1 Pure Birth Processes

In this chapter, we present several important examples of continuous time, discrete state, and Markov processes. Specifically, we deal here with a family of random variables $\{X(t); 0 \le t < \infty\}$ where the possible values of $X(t)$ are the nonnegative integers. We shall restrict attention to the case where $\{X(t)\}$ is a Markov process with stationary transition probabilities. Thus, the transition probability function for $t > 0$,

$$P_{ij}(t) = \Pr\{X(t+u) = j | X(u) = i\}, \qquad i, j = 0, 1, 2, \ldots,$$

is independent of $u \ge 0$.

It is usually more natural in investigating particular stochastic models based on physical phenomena to prescribe the so-called infinitesimal probabilities relating to the process and then derive from them an explicit expression for the transition probability function. For the case at hand, we will postulate the form of $P_{ij}(h)$ for h small, and, using the Markov property, we will derive a system of differential equations satisfied by $P_{ij}(t)$ for all $t > 0$. The solution of these equations under suitable boundary conditions gives $P_{ij}(t)$.

By way of introduction to the general pure birth process, we review briefly the axioms characterizing the Poisson process.

6.1.1 Postulates for the Poisson Process

The Poisson process is the prototypical pure birth process. Let us point out the relevant properties. The Poisson process is a Markov process on the nonnegative integers for which

(i) $\Pr\{X(t+h) - X(t) = 1 | X(t) = x\} = \lambda h + o(h)$ as $h \downarrow 0$
$$(x = 0, 1, 2, \ldots).$$
(ii) $\Pr\{X(t+h) - X(t) = 0 | X(t) = x\} = 1 - \lambda h + o(h)$ as $h \downarrow 0$.
(iii) $X(0) = 0$.

The precise interpretation of (i) is the relationship

$$\lim_{h \to 0+} \frac{\Pr\{X(t+h) - X(t) = 1 | X(t) = x\}}{h} = \lambda.$$

The $o(h)$ symbol represents a negligible remainder term in the sense that if we divide the term by h, then the resulting value tends to zero as h tends to zero. Notice that the right side of (i) is independent of x.

These properties are easily verified by direct computation, since the explicit formulas for all the relevant properties are available. Problem 6.1.13 calls for showing that these properties, in fact, define the Poisson process.

6.1.2 Pure Birth Process

A natural generalization of the Poisson process is to permit the chance of an event occurring at a given instant of time to depend upon the number of events that have already occurred. An example of this phenomenon is the reproduction of living organisms (and hence the name of the process), in which under certain conditions—e.g., sufficient food, no mortality, no migration—the infinitesimal probability of a birth at a given instant is proportional (directly) to the population size at that time. This example is known as the *Yule process* and will be considered in detail later.

Consider a sequence of positive numbers, $\{\lambda_k\}$. We define a pure birth process as a Markov process satisfying the following postulates:

1. $\Pr\{X(t+h) - X(t) = 1 | X(t) = k\} = \lambda_k h + o_{1,k}(h)(h \to 0+)$.
2. $\Pr\{X(t+h) - X(t) = 0 | X(t) = k\} = 1 - \lambda_k h + o_{2,k}(h)$.   (6.1)
3. $\Pr\{X(t+h) - X(t) < 0 | X(t) = k\} = 0 \ (k \geq 0)$.

   As a matter of convenience, we often add the postulate
4. $X(0) = 0$.

With this postulate, $X(t)$ does not denote the population size but, rather, the number of births in the time interval $(0, t]$.

Note that the left sides of Postulates (1) and (2) are just $P_{k,k+1}(h)$ and $P_{k,k}(h)$, respectively (owing to stationarity), so that $o_{1,k}(h)$ and $o_{2,k}(h)$ do not depend upon t.

We define $P_n(t) = \Pr\{X(t) = n\}$, assuming $X(0) = 0$.

By analyzing the possibilities at time t just prior to time $t+h$ (h small), we will derive a system of differential equations satisfied by $P_n(t)$ for $t \geq 0$, namely

$$P'_0(t) = -\lambda_0 P_0(t),$$
$$P'_n(t) = -\lambda_n P_n(t) + \lambda_{n-1} P_{n-1}(t) \qquad \text{for } n \geq 1,$$ (6.2)

with initial conditions

$$P_0(0) = 1, \qquad P_n(0) = 0, \qquad n > 0.$$

Indeed, if $h > 0, n \geq 1$, then by invoking the law of total probability, the Markov property, and Postulate (3), we obtain

$$P_n(t+h) = \sum_{k=0}^{\infty} P_k(t) \Pr\{X(t+h) = n | X(t) = k\}$$

$$= \sum_{k=0}^{\infty} P_k(t) \Pr\{X(t+h) - X(t) = n - k | X(t) = k\}$$

$$= \sum_{k=0}^{n} P_k(t) \Pr\{X(t+h) - X(t) = n - k | X(t) = k\}.$$

Now for $k = 0, 1, \ldots, n-2$, we have

$$
\begin{aligned}
&\Pr\{X(t+h) - X(t) = n - k | X(t) = k\} \\
&\quad \le \Pr\{X(t+h) - X(t) \ge 2 | X(t) = k\} \\
&\quad = o_{1,k}(h) + o_{2,k}(h),
\end{aligned}
$$

or

$$
\Pr\{X(t+h) - X(t) = n - k | X(t) = k\} = o_{3,n,k}(h), \qquad k = 0, \ldots, n-2.
$$

Thus,

$$
\begin{aligned}
P_n(t+h) = {}& P_n(t)\big[1 - \lambda_n h + o_{2,n}(h)\big] + P_{n-1}(t)\big[\lambda_{n-1} h + o_{1,n-1}(h)\big] \\
& + \sum_{k=0}^{n-2} P_k(t) o_{3,n,k}(h) k,
\end{aligned}
$$

or

$$
\begin{aligned}
& P_n(t+h) - P_n(t) \\
& \quad = P_n(t)\big[-\lambda_n h + o_{2,n}(h)\big] + P_{n-1}(t)\big[\lambda_{n-1} h + o_{1,n-1}(h)\big] + o_n(h),
\end{aligned}
\tag{6.3}
$$

where, clearly, $\lim_{h\downarrow 0} o_n(h)/h = 0$ uniformly in $t \ge 0$, since $o_n(h)$ is bounded by the finite sum $\Sigma_{k=0}^{n-2} o_{3,n,k}(h)$, which does not depend on t.

Dividing by h and passing to the limit $h \downarrow 0$, we validate the relations (6.2), where on the left side we should, to be precise, write the derivative from the right. With a little more care, however, we can derive the same relation involving the derivative from the left. In fact, from (6.3), we see at once that the $P_n(t)$ are continuous functions of t. Replacing t by $t - h$ in (6.3), dividing by h, and passing to the limit $h \downarrow 0$, we find that each $P_n(t)$ has a left derivative that also satisfies equation (6.2).

The first equation of (6.2) can be solved immediately and yields

$$
P_0(t) = \exp\{-\lambda_0 t\} \qquad \text{for } t > 0.
\tag{6.4}
$$

Define S_k as the time between the kth and the $(k+1)$st birth, so that

$$
P_n(t) = \Pr\left\{ \sum_{i=0}^{n-1} S_i \le t < \sum_{i=0}^{n} S_i \right\}.
$$

The random variables S_k are called the "sojourn times" between births, and

$$
W_k = \sum_{i=0}^{k-1} S_i = \text{the time at which the } k\text{th birth occurs.}
$$

We have already seen that $P_0(t) = \exp\{-\lambda_0 t\}$. Therefore,

$$\Pr\{S_0 \leq t\} = 1 - \Pr\{X(t) = 0\} = 1 - \exp\{-\lambda_0 t\};$$

that is S_0 has an exponential distribution with parameter λ_0. It may be deduced from Postulates (1) through (4) that $S_k, k > 0$, also has an exponential distribution with parameter λ_k and that the S_i's are mutually independent.

This description characterizes the pure birth process in terms of its sojourn times, in contrast to the infinitesimal description corresponding to (6.1).

To solve the differential equations of (6.2) recursively, introduce $Q_n(t) = e^{\lambda_n t} P_n(t)$ for $n = 0, 1, \ldots$. Then,

$$\begin{aligned}
Q_n'(t) &= \lambda_n e^{\lambda_n t} P_n(t) + e^{\lambda_n t} P_n'(t) \\
&= e^{\lambda_n t} \left[\lambda_n P_n(t) + P_n'(t) \right] \\
&= e^{\lambda_n t} \lambda_{n-1} P_{n-1}(t) \qquad \text{[using (6.2)]}.
\end{aligned}$$

Integrating both sides of these equations and using the boundary condition $Q_n(0) = 0$ for $n \geq 1$ gives

$$Q_n(t) = \int_0^t e^{\lambda_n x} \lambda_{n-1} P_{n-1}(x) \, dx,$$

or

$$P_n(t) = \lambda_{n-1} e^{-\lambda_n t} \int_0^t e^{\lambda_n x} P_{n-1}(x) \, dx, \qquad n = 1, 2, \ldots. \tag{6.5}$$

It is now clear that all $P_k(t) \geq 0$, but there is still a possibility that

$$\sum_{n=0}^{\infty} P_n(t) < 1.$$

To secure the validity of the process, i.e., to assure that $\sum_{n=0}^{\infty} P_n(t) = 1$ for all t, we must restrict the λ_k according to the following:

$$\sum_{n=0}^{\infty} P_n(t) = 1 \quad \text{if and only if} \quad \sum_{n=0}^{\infty} \frac{1}{\lambda_n} = \infty. \tag{6.6}$$

The intuitive argument for this result is as follows: The time S_k between consecutive births is exponentially distributed with a corresponding parameter λ_k. Therefore, the quantity $\sum_n 1/\lambda_n$ equals the expected time before the population becomes infinite. By comparison, $1 - \sum_{n=0}^{\infty} P_n(t)$ is the probability that $X(t) = \infty$.

If $\Sigma_n \lambda_n^{-1} < \infty$, then the expected time for the population to become infinite is finite. It is then plausible that for all $t > 0$, the probability that $X(t) = \infty$ is positive.

When no two of the birth parameters $\lambda_0, \lambda_1, \ldots$ are equal, the integral equation (6.5) may be solved to give the explicit formula

$$P_0(t) = e^{-\lambda_0 t}, \tag{6.7}$$

$$P_1(t) = \lambda_0 \left(\frac{1}{\lambda_1 - \lambda_0} e^{-\lambda_0 t} + \frac{1}{\lambda_0 - \lambda_1} e^{-\lambda_1 t} \right)$$

and

$$P_n(t) = \Pr\{X(t) = n | X(0) = 0\} \tag{6.8}$$
$$= \lambda_0 \cdots \lambda_{n-1} \left[B_{0,n} e^{-\lambda_0 t} + \cdots + B_{n,n} e^{-\lambda_n t} \right] \qquad \text{for } n > 1,$$

where

$$B_{0,n} = \frac{1}{(\lambda_1 - \lambda_0) \cdots (\lambda_n - \lambda_0)}, \tag{6.9}$$

$$B_{k,n} = \frac{1}{(\lambda_0 - \lambda_k) \cdots (\lambda_{k-1} - \lambda_k)(\lambda_{k+1} - \lambda_k) \cdots (\lambda_n - \lambda_k)}$$

$$\text{for } 0 < k < n$$

and

$$B_{n,n} = \frac{1}{(\lambda_0 - \lambda_n) \cdots (\lambda_{n-1} - \lambda_n)}.$$

Because $\lambda_j \neq \lambda_k$ when $j \neq k$ by assumption, the denominator in (6.9) does not vanish, and $B_{k,n}$ is well defined.

We will verify that $P_1(t)$, as given by (6.7), satisfies (6.5). Equation (6.4) gives $P_0(t) = e^{-\lambda_0 t}$. We next substitute this in (6.5) when $n = 1$, thereby obtaining

$$P_1(t) = \lambda_0 e^{-\lambda_1 t} \int_0^t e^{\lambda_1 x} e^{-\lambda_0 x} dx$$

$$= \lambda_0 e^{-\lambda_1 t} (\lambda_0 - \lambda_1)^{-1} \left[1 - e^{-(\lambda_0 - \lambda_1)t} \right]$$

$$= \lambda_0 \left(\frac{1}{\lambda_1 - \lambda_0} e^{-\lambda_0 t} + \frac{1}{\lambda_0 - \lambda_1} e^{-\lambda_1 t} \right),$$

in agreement with (6.7).

The induction proof for general n involves tedious and difficult algebra. The case $n = 2$ is suggested as a problem.

6.1.3 The Yule Process

The Yule process arises in physics and biology and describes the growth of a population in which each member has a probability $\beta h + o(h)$ of giving birth to a new member during an interval of time of length $h(\beta > 0)$. Assuming independence and no interaction among members of the population, the binomial theorem gives

$$\Pr\{X(t+h) - X(t) = 1 \mid X(t) = n\} = \binom{n}{1} [\beta h + o(h)][1 - \beta h + o(h)]^{n-1}$$

$$= n\beta h + o_n(h);$$

for the Yule process the infinitesimal parameters are $\lambda_n = n\beta$. In words, the total population birth rate is directly proportional to the population size, the proportionality constant being the individual birth rate β. As such, the Yule process forms a stochastic analog of the deterministic population growth model represented by the differential equation $dy/dt = \alpha y$. In the deterministic model, the rate dy/dt of population growth is directly proportional to population size y. In the stochastic model, the infinitesimal deterministic increase dy is replaced by the probability of a unit increase during the infinitesimal time interval dt. Similar connections between deterministic rates and birth (and death) parameters arise frequently in stochastic modeling. Examples abound in this chapter.

The system of equations (6.2) in the case that $X(0) = 1$ becomes

$$P'_n(t) = -\beta \left[nP_n(t) - (n-1)P_{n-1}(t) \right], \qquad n = 1, 2, \ldots,$$

under the initial conditions

$$P_1(0) = 1, \qquad P_n(0) = 0, \qquad n = 2, 3, \ldots.$$

Its solution is

$$P_n(t) = e^{-\beta t} \left(1 - e^{-\beta t} \right)^{n-1}, \qquad n \geq 1, \tag{6.10}$$

as may be verified directly. We recognize (6.10) as the geometric distribution in Chapter 1, (1.26) with $p = e^{-\beta t}$.

The general solution analogous to (6.8) but for pure birth processes starting from $X(0) = 1$ is

$$P_n(t) = \lambda_1 \cdots \lambda_{n-1} \left[B_{1,n} e^{-\lambda_1 t} + \cdots + B_{n,n} e^{-\lambda_n t} \right], \qquad n > 1. \tag{6.11}$$

When $\lambda_n = \beta n$, we will show that (6.11) reduces to the solution given in (6.10) for a Yule process with parameter β. Then,

$$B_{1,n} = \frac{1}{(\lambda_2 - \lambda_1)(\lambda_3 - \lambda_1) \cdots (\lambda_n - \lambda_1)}$$

$$= \frac{1}{\beta^{n-1}(1)(2) \cdots (n-1)}$$

$$= \frac{1}{\beta^{n-1}(n-1)!},$$

$$B_{2,n} = \frac{1}{(\lambda_1 - \lambda_2)(\lambda_3 - \lambda_2) \cdots (\lambda_n - \lambda_2)}$$

$$= \frac{1}{\beta^{n-1}(-1)(1)(2) \cdots (n-2)}$$

$$= \frac{-1}{\beta^{n-1}(n-2)!},$$

and

$$\beta_{k,n} = \frac{1}{(\lambda_1 - \lambda_k) \cdots (\lambda_{k-1} - \lambda_k)(\lambda_{k+1} - \lambda_k) \cdots (\lambda_n - \lambda_k)}$$

$$= \frac{(-1)^{k-1}}{\beta^{n-1}(k-1)!(n-k)!}.$$

Thus, according to (6.11),

$$P_n(t) = \beta^{n-1}(n-1)! \left(B_{1,n} e^{-\beta t} + \cdots + B_{n,n} e^{-n\beta t} \right)$$

$$= \sum_{k=1}^{n} \frac{(n-1)!}{(k-1)!(n-k)!} (-1)^{k-1} e^{-k\beta t}$$

$$= e^{-\beta t} \sum_{j=0}^{n-1} \frac{(n-1)!}{j!(n-1-j)!} \left(-e^{-\beta t} \right)^j$$

$$= e^{-\beta t} \left(1 - e^{-\beta t} \right)^{n-1} \qquad \text{[see Chapter 1, (1.67)],}$$

which establishes (6.10).

Exercises

6.1.1 A pure birth process starting from $X(0) = 0$ has birth parameters $\lambda_0 = 1$, $\lambda_1 = 3, \lambda_2 = 2$, and $\lambda_3 = 5$. Determine $P_n(t)$ for $n = 0, 1, 2, 3$.

6.1.2 A pure birth process starting from $X(0) = 0$ has birth parameters $\lambda_0 = 1$, $\lambda_1 = 3, \lambda_2 = 2$, and $\lambda_3 = 5$. Let W_3 be the random time that it takes the process to reach state 3.
   (a) Write W_3 as a sum of sojourn times and thereby deduce that the mean time is $E[W_3] = \frac{11}{6}$.
   (b) Determine the mean of $W_1 + W_2 + W_3$.
   (c) What is the variance of W_3?

6.1.3 A population of organisms evolves as follows. Each organism exists, independent of the other organisms, for an exponentially distributed length of time with parameter θ, and then splits into two new organisms, each of which exists, independent of the other organisms, for an exponentially distributed length of time with parameter θ, and then splits into two new organisms, and so on. Let $X(t)$ denote the number of organisms existing at time t. Show that $X(t)$ is a Yule process.

6.1.4 Consider an experiment in which a certain event will occur with probability αh and will not occur with probability $1 - \alpha h$, where α is a fixed positive parameter and h is a small ($h < 1/\alpha$) positive variable. Suppose that n independent trials of the experiment are carried out, and the total number of times that the event occurs is noted. Show that

(a) The probability that the event never occurs during the n trials is $1 - n\alpha h + o(h)$;

(b) The probability that the event occurs exactly once is $n\alpha h + o(h)$;

(c) The probability that the event occurs twice or more is $o(h)$.

Hint: Use the binomial expansion

$$(1 - \alpha h)^n = 1 - n\alpha h + \frac{n(n-1)}{2}(\alpha h)^2 - \cdots.$$

6.1.5 Using equation (6.10), calculate the mean and variance for the Yule process where $X(0) = 1$.

6.1.6 Operations 1, 2, and 3 are to be performed in succession on a major piece of equipment. Operation k takes a random duration S_k that is exponentially distributed with parameter λ_k for $k = 1, 2, 3$, and all operation times are independent. Let $X(t)$ denote the operation being performed at time t, with time $t = 0$ marking the start of the first operation. Suppose that $\lambda_1 = 5, \lambda_2 = 3$, and $\lambda_3 = 13$. Determine

(a) $P_1(t) = \Pr\{X(t) = 1\}$.

(b) $P_2(t) = \Pr\{X(t) = 2\}$.

(c) $P_3(t) = \Pr\{X(t) = 3\}$.

Problems

6.1.1 Let $X(t)$ be a Yule process that is observed at a random time U, where U is uniformly distributed over $[0, 1)$. Show that $\Pr\{X(U) = k\} = p^k/(\beta k)$ for $k = 1, 2, \ldots$, with $p = 1 - e^{-\beta}$.

Hint: Integrate (6.10) over t between 0 and 1.

6.1.2 A Yule process with immigration has birth parameters $\lambda_k = \alpha + k\beta$ for $k = 0, 1, 2, \ldots$. Here, α represents the rate of immigration into the population, and β represents the individual birth rate. Supposing that $X(0) = 0$, determine $P_n(t)$ for $n = 0, 1, 2, \ldots$.

6.1.3 Consider a population comprising a fixed number N of individuals. Suppose that at time $t = 0$, there is exactly one *infected* individual and $N - 1$ *susceptible* individuals in the population. Once infected, an individual remains in that state forever. In any short time interval of length h, *any given infected person* will transmit the disease to *any given susceptible person* with probability $\alpha h + o(h)$. (The parameter α is the *individual infection rate*.) Let $X(t)$ denote the number of infected individuals in the population at time $t \geq 0$. Then, $X(t)$ is a pure birth process on the states $0, 1, \ldots, N$. Specify the birth parameters.

6.1.4 A new product (a "Home Helicopter" to solve the commuting problem) is being introduced. The sales are expected to be determined by both media (newspaper and television) advertising and word-of-mouth advertising, wherein satisfied customers tell others about the product. Assume that media advertising creates new customers according to a Poisson process of rate $\alpha = 1$ customer per month. For the word-of-mouth advertising, assume that each purchaser of a Home Helicopter will generate sales to new customers at a rate of $\theta = 2$ customers per month. Let $X(t)$ be the total number of Home Helicopter customers up to time t.

(a) Model $X(t)$ as a pure birth process by specifying the birth parameters λ_k, for $k = 0, 1, \ldots$.

(b) What is the probability that exactly two Home Helicopters are sold during the first month?

6.1.5 Let W_k be the time to the kth birth in a pure birth process starting from $X(0) = 0$. Establish the equivalence

$$\Pr\{W_1 > t, W_2 > t + s\} = P_0(t)[P_0(s) + P_1(s)].$$

From this relation together with equation (6.7), determine the joint density for W_1 and W_2, and then the joint density of $S_0 = W_1$ and $S_1 = W_2 - W_1$.

6.1.6 A fatigue model for the growth of a crack in a discrete lattice proposes that the size of the crack evolves as a pure birth process with parameters

$$\lambda_k = (1 + k)^\rho \qquad \text{for } k = 1, 2, \ldots.$$

The theory behind the model postulates that the growth rate of the crack is proportional to some power of the stress concentration at its ends and that this stress concentration is itself proportional to some power of $1 + k$, where k is the crack length. Use the sojourn time description to deduce that the mean time for the crack to grow to infinite length is finite when $\rho > 1$ and that, therefore, the failure time of the system is a well-defined and finite-valued random variable.

6.1.7 Let λ_0, λ_1, and λ_2 be the parameters of the independent exponentially distributed random variables S_0, S_1, and S_2. Assume that no two of the parameters are equal.

(a) Verify that

$$\Pr\{S_0 > t\} = e^{-\lambda_0 t},$$

$$\Pr\{S_0 + S_1 > t\} = \frac{\lambda_1}{\lambda_1 - \lambda_0} e^{-\lambda_0 t} + \frac{\lambda_0}{\lambda_0 - \lambda_1} e^{-\lambda_1 t},$$

and evaluate in similar terms

$$\Pr\{S_0 + S_1 + S_2 > t\}.$$

(b) Verify equation (6.8) in the case that $n = 2$ by evaluating

$$P_2(t) = \Pr\{X(t) = 2\} = \Pr\{S_0 + S_1 + S_2 > t\} - \Pr\{S_0 + S_1 > t\}.$$

6.1.8 Let $N(t)$ be a pure birth process for which

$$\Pr\{\text{an event happens in}(t, t+h)|N(t) \text{ is odd}\} = \alpha h + o(h),$$
$$\Pr\{\text{an event happens in}(t, t+h)|N(t) \text{ is even}\} = \beta h + o(h),$$

where $o(h)/h \to 0$ as $h \downarrow 0$. Take $N(0) = 0$. Find the following probabilities:

$$P_0(t) = \Pr\{N(t) \text{ is even}\}; \qquad P_1(t) = \Pr\{N(t) \text{ is odd}\}.$$

Hint: Derive the differential equations

$$P_0'(t) = \alpha P_1(t) - \beta P_0(t) \quad \text{and} \quad P_1'(t) = -\alpha P_1(t) + \beta P_0(t)$$

and solve them by using $P_0(t) + P_1(t) = 1$.

6.1.9 Under the conditions of Problem 6.8, determine $E[N(t)]$.

6.1.10 Consider a pure birth process on the states $0, 1, \ldots, N$ for which $\lambda_k = (N-k)\lambda$ for $k = 0, 1, \ldots, N$. Suppose that $X(0) = 0$. Determine $P_n(t) = \Pr\{X(t) = n\}$ for $n = 0, 1$, and 2.

6.1.11 Beginning with $P_0(t) = e^{-\lambda_0 t}$ and using equation (6.5), calculate $P_1(t), P_2(t)$, and $P_3(t)$ and verify that these probabilities conform with equation (6.7), assuming distinct birth parameters.

6.1.12 Verify that $P_2(t)$, as given by (6.8), satisfies (6.5) by following the calculations in the text that showed that $P_1(t)$ satisfies (6.5).

6.1.13 Using (6.5), derive $P_n(t)$ when all birth parameters are the same constant λ and show that

$$P_n(t) = \frac{(\lambda t)^n e^{-\lambda t}}{n!}, \qquad n = 0, 1, \ldots.$$

Thus, the postulates of Section 6.1.1 serve to define the Poisson processes.

6.2 Pure Death Processes

Complementing the increasing pure birth process is the decreasing pure death process. It moves successively through states $N, N-1, \ldots, 2, 1$ and ultimately is absorbed in state 0 (extinction). The process is specified by the death parameters $\mu_k > 0$ for $k = 1, 2, \ldots, N$, where the sojourn time in state k is exponentially distributed with parameter μ_k, all sojourn times being independent. A typical sample path is depicted in Figure 6.1.

Alternatively, we have the infinitesimal description of a pure death process as a Markov process $X(t)$ whose state space is $0, 1, \ldots, N$ and for which

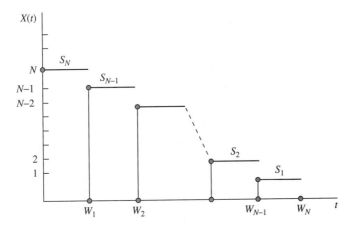

Figure 6.1 A typical sample path of a pure death process, showing the sojourn times $S_N, \ldots, S_1$ and the waiting times $W_1, W_2, \ldots, W_N$.

(i) $\Pr\{X(t+h) = k-1|X(t) = k\} = \mu_k h + o(h), k = 1, \ldots, N;$
(ii) $\Pr\{X(t+h) = k|X(t) = k\} = 1 - \mu_k h + o(h), k = 1, \ldots, N;$ (6.12)
(iii) $\Pr\{X(t+h) > k|X(t) = k\} = 0, k = 0, 1, \ldots, N.$

The parameter μ_k is the "death rate" operating or in effect while the process sojourns in state k. It is a common and useful convention to assign $\mu_0 = 0$.

When the death parameters $\mu_1, \mu_2, \ldots, \mu_N$ are distinct, i.e., $\mu_j \neq \mu_k$ if $j \neq k$, then we have the explicit transition probabilities

$$P_N(t) = e^{-\mu_N t};$$

and for $n < N$,

$$P_n(t) = \Pr\{X(t) = n|X(0) = N\}$$
$$= \mu_{n+1}\mu_{n+2}\cdots\mu_N \left[A_{n,n}e^{-\mu_n t} + \cdots + A_{N,n}e^{-\mu_N t}\right], \quad (6.13)$$

where

$$A_{k,n} = \frac{1}{(\mu_N - \mu_k)\cdots(\mu_{k+1} - \mu_k)(\mu_{k-1} - \mu_k)\cdots(\mu_n - \mu_k)}.$$

6.2.1 The Linear Death Process

As an example, consider a pure death process in which the death rates are proportional to population size. This process, which we will call the *linear death process*, complements the Yule, or linear birth, process. The parameters are $\mu_k = k\alpha$, where α is the individual death rate in the population. Then,

$$A_{n,n} = \frac{1}{(\mu_N - \mu_n)(\mu_{N-1} - \mu_n) \cdots (\mu_{n+1} - \mu_n)}$$

$$= \frac{1}{\alpha^{N-n-1}(N-n)(N-n-1)\cdots(2)(1)},$$

$$A_{n+1,n} = \frac{1}{(\mu_N - \mu_{n+1}) \cdots (\mu_{n+2} - \mu_{n+1})(\mu_n - \mu_{n+1})}$$

$$= \frac{1}{\alpha^{N-n-1}(N-n-1)\cdots(1)(-1)},$$

$$A_{k,n} = \frac{1}{(\mu_N - \mu_k) \cdots (\mu_{k+1} - \mu_k)(\mu_{k-1} - \mu_k) \cdots (\mu_n - \mu_k)}$$

$$= \frac{1}{\alpha^{N-n-1}(N-k)\cdots(1)(-1)(-2)\cdots(n-k)}$$

$$= \frac{1}{\alpha^{N-n-1}(-1)^{k-n}(N-k)!\,(k-n)!}.$$

Then,

$$P_n(t) = \mu_{n+1}\mu_{n+2}\cdots\mu_N \sum_{k=n}^{N} A_{k,n} e^{-\mu_k t}$$

$$= \alpha^{N-n-1} \frac{N!}{n!} \sum_{k=n}^{N} \frac{e^{-k\alpha t}}{\alpha^{N-n-1}(-1)^{k-n}(N-k)!\,(k-n)!}$$

$$= \frac{N!}{n!} e^{-n\alpha t} \sum_{j=0}^{N-n} \frac{(-1)^j e^{-j\alpha t}}{(N-n-j)!\,j!} \tag{6.14}$$

$$= \frac{N!}{n!\,(N-n)!} e^{-n\alpha t} \left(1 - e^{-\alpha t}\right)^{N-n}, \quad n = 0, \dots, N.$$

Let T be the time of population extinction. Formally, $T = \min\{t \geq 0; X(t) = 0\}$. Then, $T \leq t$ if and only if $X(t) = 0$, which leads to the cumulative distribution function of T via

$$F_T(t) = \Pr\{T \leq t\} = \Pr\{X(t) = 0\}$$

$$= P_0(t) = \left(1 - e^{-\alpha t}\right)^N, \quad t \geq 0. \tag{6.15}$$

The linear death process can be viewed in yet another way, a way that again confirms the intimate connection between the exponential distribution and a continuous time parameter Markov chain. Consider a population consisting of N individuals, each of whose lifetimes is an independent exponentially distributed random variable with

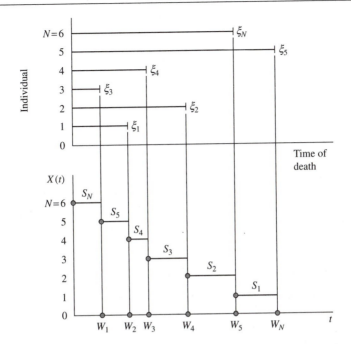

Figure 6.2 The linear death process. As depicted here, the third individual is the first to die, the first individual is the second to die, etc.

parameter α. Let $X(t)$ be the number of survivors in this population at time t. Then, $X(t)$ is the linear pure death process whose parameters are $\mu_k = k\alpha$ for $k = 0, 1, \ldots, N$. To help understand this connection, let $\xi_1, \xi_2, \ldots, \xi_N$ denote the times of death of the individuals labeled $1, 2, \ldots, N$, respectively. Figure 6.2 shows the relation between the individual lifetimes $\xi_1, \xi_2, \ldots, \xi_N$ and the death process $X(t)$.

The sojourn time in state N, denoted by S_N, equals the time of the earliest death, or $S_N = \min\{\xi_1, \ldots, \xi_N\}$. Since the lifetimes are independent and have the same exponential distribution,

$$
\begin{aligned}
\Pr\{S_N > t\} &= \Pr\{\min\{\xi_1, \ldots, \xi_N\} > t\} \\
&= \Pr\{\xi_1 > t, \ldots, \xi_N > t\} \\
&= [\Pr\{\xi_1 > t\}]^N \\
&= e^{-N\alpha t}.
\end{aligned}
$$

That is, S_N has an exponential distribution with parameter $N\alpha$. Similar reasoning applies when there are k members alive in the population. The memoryless property of the exponential distribution implies that the remaining lifetime of each of these k individuals is exponentially distributed with parameter α. Then, the sojourn time S_k is the minimum of these k remaining lifetimes and hence is exponentially distributed with parameter $k\alpha$. To give one more approach in terms of transition rates, each individual in the population has a constant death rate of α in the sense that

$$
\begin{aligned}
\Pr\{t < \xi_1 < t + h | t < \xi_1\} &= \frac{\Pr\{t < \xi_1 < t + h\}}{\Pr\{t < \xi_1\}} \\
&= \frac{e^{-\alpha t} - e^{-\alpha(t+h)}}{e^{-\alpha t}} \\
&= 1 - e^{-\alpha h} \\
&= \alpha h + o(h) \quad \text{as } h \downarrow 0.
\end{aligned}
$$

If each of k individuals alive in the population at time t has a constant death rate of α, then the total population death rate should be $k\alpha$, directly proportional to the population size. This shortcut approach to specifying appropriate death parameters is a powerful and often-used tool of stochastic modeling. The next example furnishes another illustration of its use.

6.2.2   Cable Failure Under Static Fatigue

A cable composed of parallel fibers under tension is being designed to support a high-altitude weather balloon. With a design load of 1000 kg and a design lifetime of 100 years, how many fibers should be used in the cable?

The low-weight, high-strength fibers to be used are subject to *static fatigue*, or eventual failure when subjected to a constant load. The higher the constant load, the shorter the life, and experiments have established a linear plot on log–log axes between average failure time and load that is shown in Figure 6.3.

The relation between mean life μ_T and load l that is illustrated in Figure 6.3 takes the analytic form

$$
\log_{10} \mu_T = 2 - 40 \log_{10} l.
$$

Were the cable to be designed on the basis of average life, to achieve the 100 year design target each fiber should carry 1 kg. Since the total load is 1000 kg, $N = \frac{1000}{1} = 1000$ fibers should be used in the cable.

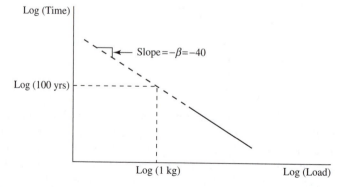

Figure 6.3 A linear relation between log mean failure time and log load.

One might suppose that this large number ($N = 1000$) of fibers would justify designing the cable based on average fiber properties. We shall see that such reasoning is dangerously wrong.

Let us suppose, however, as is the case with many modern high-performance structural materials, that there is a large amount of random scatter of individual fiber lifetimes about the mean. How does this randomness affect the design problem?

Some assumption must be made concerning the probability distribution governing individual fiber lifetimes. In practice, it is extremely difficult to gather sufficient data to determine this distribution with any degree of certainty. Most data do show, however, a significant degree of skewness, or asymmetry. Because it qualitatively matches observed data and because it leads to a pure death process model that is accessible to exhaustive analysis, we will assume that the probability distribution for the failure time T of a single fiber subjected to the time-varying tensile load $l(t)$ is given by

$$\Pr\{T \le t\} = 1 - \exp\left\{-\int_0^t K[l(s)]\mathrm{d}s\right\}, \quad t \ge 0.$$

This distribution corresponds to a *failure rate*, or *hazard rate*, of $r(t) = K[l(t)]$ wherein a single fiber, having not failed prior to time t and carrying the load $l(t)$, will fail during the interval $(t, t + \Delta t]$ with probability

$$\Pr\{t < T \le t + \Delta t | T > t\} = K[l(t)]\Delta t + o(\Delta t).$$

The function $K[l]$, called the *breakdown rule*, expresses how changes in load affect the failure probability. We are concerned with the *power law breakdown rule* in which $K[l] = l^\beta / A$ for some positive constants A and β. Assuming power law breakdown, under a constant load $l(t) = l$, the single fiber failure time is exponentially distributed with mean $\mu_T = E[T|l] = 1/K[l] = Al^{-\beta}$. A plot of mean failure time versus load is linear on log–log axes, matching the observed properties of our fiber type. For the design problem, we have $\beta = 40$ and $A = 100$.

Now, place N of these fibers in parallel and subject the resulting bundle or cable to a total load, constant in time, of NL, where L is the nominal load per fiber. What is the probability distribution of the time at which the cable fails? Since the fibers are in parallel, this system failure time equals the failure time of the last fiber.

Under the stated assumptions governing single-fiber behavior, $X(t)$, the number of unfailed fibers in the cable at time t, evolves as a pure death process with parameters $\mu_k = kK[NL/k]$ for $k = 1, 2, \ldots, N$. Given $X(t) = k$ surviving fibers at time t and assuming that the total bundle load NL is shared equally among them, then each carries load NL/k and has a corresponding failure rate of $K[NL/k]$. As there are k such survivors in the bundle, the bundle, or system, failure rate is $\mu_k = kK[NL/k]$ as claimed.

It was mentioned earlier that the system failure time was W_N, the waiting time to the Nth fiber failure. Then, $\Pr\{W_N \le t\} = \Pr\{X(t) = 0\} = P_0(t)$, where $P_n(t)$ is given explicitly by (2.13) in terms of $\mu_1, \ldots, \mu_N$. Alternatively, we may bring to bear the sojourn time description of the pure death process and, following Figure 6.1, write

$$W_N = S_N + S_{N-1} + \cdots + S_1,$$

where $S_N, S_{N-1}, \ldots, S_1$ are independent exponentially distributed random variables and S_k has parameter $\mu_k = kK[NL/k] = k(NL/k)^\beta / A$. The mean system failure time is readily computed to be

$$E[W_N] = E[S_N] + \cdots + E[S_1]$$

$$= AL^{-\beta} \sum_{k=1}^{N} \frac{1}{k} \left(\frac{k}{N} \right)^\beta \tag{6.16}$$

$$= AL^{-\beta} \sum_{k=1}^{N} \left(\frac{k}{N} \right)^{\beta-1} \left(\frac{1}{N} \right).$$

The sum in the expression for $E[W_N]$ seems formidable at first glance, but a very close approximation is readily available when N is large. Figure 6.4 compares the sum to an integral.

From Figure 6.4, we see that

$$\sum_{k=1}^{N} \left(\frac{k}{N} \right)^{\beta-1} \left(\frac{1}{N} \right) \approx \int_0^1 x^{\beta-1} dx = \frac{1}{\beta}.$$

Indeed, we readily obtain

$$\frac{1}{\beta} = \int_0^1 x^{\beta-1} dx \leq \sum_{k=1}^{N} \left(\frac{k}{N} \right)^{\beta-1} \left(\frac{1}{N} \right) \leq \int_{1/N}^{1+1/N} x^{\beta-1} dx$$

$$= \left(\frac{1}{\beta} \right) \left[\left(1 + \frac{1}{N} \right)^\beta - \left(\frac{1}{N} \right)^\beta \right].$$

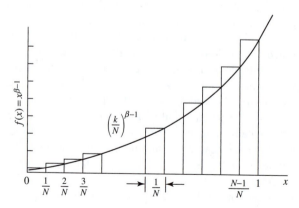

Figure 6.4 The sum $\Sigma_{k=1}^{N} (k/N)^{\beta-1} (1/N)$ is a Riemann approximation to $\int_0^1 x^{\beta-1} dx = 1/\beta$.

When $N = 1000$ and $\beta = 40$, the numerical bounds are

$$\left(\frac{1}{40}\right) \leq \sum_{k=1}^{N}\left(\frac{k}{N}\right)^{\beta-1}\left(\frac{1}{N}\right) \leq \left(\frac{1}{40}\right)(1.0408),$$

which shows that the integral determines the sum to within about 4%.

Substituting $1/\beta$ for the sum in (6.16) gives the average cable life

$$E[W_N] \approx \frac{A}{\beta L^\beta}$$

to be compared with the average fiber life of

$$\mu_T = \frac{A}{L^\beta}.$$

That is, a cable lasts only about $1/\beta$ as long as an average fiber under an equivalent load. With $A = 100$, $\beta = 40$, and $N = 1000$, the designed cable would last, on the average, $100/\left[40(1)^{40}\right] = 2.5$ years, far short of the desired life of 100 years. The cure is to increase the number of fibers in the cable, thereby decreasing the per fiber load. Increasing the number of fibers from N to N' decreases the nominal load per fiber from L to $L' = NL/N'$. To achieve parity in fiber-cable lifetimes, we equate

$$\frac{A}{L^\beta} = \frac{A}{\beta(NL/N')^\beta},$$

or

$$N' = N\beta^{1/\beta}.$$

For the given data, this calls for $N' = 1000(40)^{1/40} = 1097$ fibers. That is, the design lifetime can be restored by increasing the number of fibers in the cable by about 10%.

Exercises

6.2.1 A pure death process starting from $X(0) = 3$ has death parameters $\mu_0 = 0$, $\mu_1 = 3$, $\mu_2 = 2$, and $\mu_3 = 5$. Determine $P_n(t)$ for $n = 0, 1, 2, 3$.

6.2.2 A pure death process starting from $X(0) = 3$ has death parameters $\mu_0 = 0$, $\mu_1 = 3$, $\mu_2 = 2$, and $\mu_3 = 5$. Let W_3 be the random time that it takes the process to reach state 0.

  (a) Write W_3 as a sum of sojourn times and thereby deduce that the mean time is $E[W_3] = \frac{31}{30}$.

  (b) Determine the mean of $W_1 + W_2 + W_3$.

  (c) What is the variance of W_3?

6.2.3 Give the transition probabilities for the pure death process described by $X(0) = 3, \mu_3 = 1, \mu_2 = 2$, and $\mu_1 = 3$.

6.2.4 Consider the linear death process (Section 6.2.1) in which $X(0) = N = 5$ and $\alpha = 2$. Determine $\Pr\{X(t) = 2\}$.

Hint: Use equation (6.14).

Problems

6.2.1 Let $X(t)$ be a pure death process starting from $X(0) = N$. Assume that the death parameters are $\mu_1, \mu_2, \ldots, \mu_N$. Let T be an independent exponentially distributed random variable with parameter θ. Show that

$$\Pr\{X(T) = 0\} = \prod_{i=1}^{N} \frac{\mu_i}{\mu_i + \theta}.$$

6.2.2 Let $X(t)$ be a pure death process with constant death rates $\mu_k = \theta$ for $k = 1, 2, \ldots, N$. If $X(0) = N$, determine $P_n(t) = \Pr\{X(t) = n\}$ for $n = 0, 1, \ldots, N$.

6.2.3 A pure death process $X(t)$ with parameters $\mu_1, \mu_2, \ldots$ starts at $X(0) = N$ and evolves until it reaches the absorbing state 0. Determine the mean area under the $X(t)$ trajectory.

Hint: This is $E[W_1 + W_2 + \cdots + W_N]$.

6.2.4 A chemical solution contains N molecules of type A and M molecules of type B. An irreversible reaction occurs between type A and B molecules in which they bond to form a new compound AB. Suppose that in any small time interval of length h, any particular unbonded A molecule will react with any particular unbonded B molecule with probability $\theta h + o(h)$, where θ is a reaction rate. Let $X(t)$ denote the number of unbonded A molecules at time t.
(a) Model $X(t)$ as a pure death process by specifying the parameters.
(b) Assume that $N < M$ so that eventually all of the A molecules become bonded. Determine the mean time until this happens.

6.2.5 Consider a cable composed of fibers following the breakdown rule $K[l] = \sinh(l) = \frac{1}{2}\left(e^l - e^{-l}\right)$ for $l \geq 0$. Show that the mean cable life is given by

$$E[W_N] = \sum_{k=1}^{N} \{k \sinh(NL/k)\}^{-1} = \sum_{k=1}^{N} \left\{\frac{k}{N} \sinh\left(\frac{L}{k/N}\right)\right\}^{-1} \left(\frac{1}{N}\right)$$

$$\approx \int_0^1 \{x \sinh(L/x)\}^{-1} dx.$$

6.2.6 Let T be the time to extinction in the linear death process with parameters $X(0) = N$ and α (see Section 6.2.1).

(a) Using the sojourn time viewpoint, show that

$$E[T] = \frac{1}{\alpha}\left[\frac{1}{N} + \frac{1}{N-1} + \cdots + \frac{1}{1}\right].$$

(b) Verify the result of (a) by using equation (6.15) in

$$E[T] = \int_0^\infty \Pr\{T > t\}dt = \int_0^\infty [1 - F_T(t)]dt.$$

Hint: Let $y = 1 - e^{-\alpha t}$.

6.3 Birth and Death Processes

An obvious generalization of the pure birth and pure death processes discussed in Sections 6.1 and 6.2 is to permit $X(t)$ both to increase and to decrease. Thus, if at time t the process is in state n, it may, after a random sojourn time, move to either of the neighboring states $n+1$ or $n-1$. The resulting *birth and death process* can then be regarded as the continuous-time analog of a random walk (Chapter 3, Section 3.5.3).

Birth and death processes form a powerful tool in the kit of the stochastic modeler. The richness of the birth and death parameters facilitates modeling a variety of phenomena. At the same time, standard methods of analysis are available for determining numerous important quantities such as stationary distributions and mean first passage times. This section and later sections contain several examples of birth and death processes and illustrate how they are used to draw conclusions about phenomena in a variety of disciplines.

6.3.1 Postulates

As in the case of the pure birth processes, we assume that $X(t)$ is a Markov process on the states $0, 1, 2, \ldots$ and that its transition probabilities $P_{ij}(t)$ are stationary; that is

$$P_{ij}(t) = \Pr\{X(t+s) = j \mid X(s) = i\} \quad \text{for all} \quad s \geq 0.$$

In addition, we assume that the $P_{ij}(t)$ satisfy

1. $P_{i,i+1}(h) = \lambda_i h + o(h)$ as $h \downarrow 0, i \geq 0$;
2. $P_{i,i-1}(h) = \mu_i h + o(h)$ as $h \downarrow 0, i \geq 1$;
3. $P_{i,i}(h) = 1 - (\lambda_i + \mu_i)h + o(h)$ as $h \downarrow 0, i \geq 0$;
4. $P_{ij}(0) = \delta_{ij}$;
5. $\mu_0 = 0, \lambda_0 > 0, \mu_i, \lambda_i > 0, i = 1, 2, \ldots$.

The $o(h)$ in each case may depend on i. The matrix

$$\mathbf{A} = \begin{Vmatrix} -\lambda_0 & \lambda_0 & 0 & 0 & \cdots \\ \mu_1 & -(\lambda_1 + \mu_1) & \lambda_1 & 0 & \cdots \\ 0 & \mu_2 & -(\lambda_2 + \mu_2) & \lambda_2 & \cdots \\ 0 & 0 & \mu_3 & -(\lambda_3 + \mu_3) & \cdots \\ \vdots & \vdots & \vdots & \vdots & \end{Vmatrix} \tag{6.17}$$

is called the *infinitesimal generator* of the process. The parameters λ_i and μ_i are called, respectively, the infinitesimal birth and death rates. In Postulates (1) and (2), we are assuming that if the process starts in state i, then in a small interval of time the probabilities of the population increasing or decreasing by 1 are essentially proportional to the length of the interval.

Since the $P_{ij}(t)$ are probabilities, we have $P_{ij}(t) \geq 0$ and

$$\sum_{j=0}^{\infty} P_{ij}(t) \leq 1. \tag{6.18}$$

Using the Markov property of the process, we may also derive the so-called *Chapman–Kolmogorov equation*

$$P_{ij}(t+s) = \sum_{k=0}^{\infty} P_{ik}(t) P_{kj}(s). \tag{6.19}$$

This equation states that in order to move from state i to state j in time $t+s$, $X(t)$ moves to some state k in time t and then from k to j in the remaining time s. This is the continuous-time analog of formula (3.11) in Chapter 3.

Thus far, we have mentioned only the transition probabilities $P_{ij}(t)$. In order to obtain the probability that $X(t) = n$, we must specify where the process starts or more generally the probability distribution for the initial state. We then have

$$\Pr\{X(t) = n\} = \sum_{i=0}^{x} q_i P_{in}(t),$$

where

$$q_i = \Pr\{X(0) = i\}.$$

6.3.2 Sojourn Times

With the aid of the preceding assumptions, we may calculate the distribution of the random variable S_i, which is the sojourn time of $X(t)$ in state i; that is, given that the process is in state i, what is the distribution of the time S_i until it first leaves state i? If we let

$$\Pr\{S_i \geq t\} = G_i(t),$$

it follows easily by the Markov property that as $h \downarrow 0$,

$$G_i(t+h) = G_i(t)G_i(h) = G_i(t)[P_{ii}(h) + o(h)]$$
$$= G_i(t)[1 - (\lambda_i + \mu_i)h] + o(h),$$

or

$$\frac{G_i(t+h) - G_i(t)}{h} = -(\lambda_i + \mu_i)G_i(t) + o(1)$$

so that

$$G_i'(t) = -(\lambda_i + \mu_i)G_i(t). \tag{6.20}$$

If we use the conditions $G_i(0) = 1$, the solution of this equation is

$$G_i(t) = \exp[-(\lambda_i + \mu_i)t];$$

that is, S_i follows an exponential distribution with mean $(\lambda_i + \mu_i)^{-1}$. The proof presented here is not quite complete, since we have used the intuitive relationship

$$G_i(h) = P_{ii}(h) + o(h)$$

without a formal proof.

According to Postulates (1) and (2), during a time duration of length h, a transition occurs from state i to $i+1$ with probability $\lambda_i h + o(h)$ and from state i to $i-1$ with probability $\mu_i h + o(h)$. It follows intuitively that, given that a transition occurs at time t, the probability that this transition is to state $i+1$ is $\lambda_i/(\lambda_i + \mu_i)$ and to state $i-1$ is $\mu_i/(\lambda_i + \mu_i)$. The rigorous demonstration of this result is beyond the scope of this book.

It leads to an important characterization of a birth and death process, however, wherein the description of the motion of $X(t)$ is as follows: The process sojourns in a given state i for a random length of time whose distribution function is an exponential distribution with parameter $(\lambda_i + \mu_i)$. When leaving state i the process enters either state $i+1$ or state $i-1$ with probabilities $\lambda_i/(\lambda_i + \mu_i)$ and $\mu_i/(\lambda_i + \mu_i)$, respectively. The motion is analogous to that of a random walk except that transitions occur at random times rather than at fixed time periods.

The traditional procedure for constructing birth and death processes is to prescribe the birth and death parameters $\{\lambda_i, \mu_i\}_{i=0}^{\infty}$ and build the path structure by utilizing the preceding description concerning the waiting times and the conditional transition probabilities of the various states. We determine realizations of the process as follows.

Suppose $X(0) = i$; the particle spends a random length of time, exponentially distributed with parameter $(\lambda_i + \mu_i)$, in state i and subsequently moves with probability $\lambda_i/(\lambda_i + \mu_i)$ to state $i + 1$ and with probability $\mu_i/(\lambda_i + \mu_i)$ to state $i - 1$. Next, the particle sojourns a random length of time in the new state and then moves to one of its neighboring states and so on. More specifically, we observe a value t_1 from the exponential distribution with parameter $(\lambda_i + \mu_i)$ that fixes the initial sojourn time in state i. Then, we toss a coin with probability of heads $p_i = \lambda_i/(\lambda_i + \mu_i)$. If heads (tails) appear, we move the particle to state $i + 1(i - 1)$. In state $i + 1$, we observe a value t_2 from the exponential distribution with parameter $(\lambda_{i+1} + \mu_{i+1})$ that fixes the sojourn time in the seconds state visited. If the particle at the first transition enters state $i - 1$, the subsequent sojourn time t_2' is an observation from the exponential distribution with parameter $(\lambda_{i-1} + \mu_{i-1})$. After the second wait is completed, a Bernoulli trial is performed that chooses the next state to be visited, and the process continues in the same way.

A typical outcome of these sampling procedures determines a realization of the process. Its form might be, e.g.,

$$X(t) = \begin{cases} i, & \text{for } 0 < t < t_1, \\ i+1, & \text{for } t_1 < t < t_1 + t_2, \\ i, & \text{for } t_1 + t_2 < t < t_1 + t_2 + t_3, \\ \vdots & \qquad \vdots \end{cases}$$

Thus, by sampling from exponential and Bernoulli distributions appropriately, we construct typical sample paths of the process. Now, it is possible to assign to this set of paths (realizations of the process) a probability measure in a consistent way so that $P_{ij}(t)$ is determined satisfying (6.18) and (6.19). This result is rather deep, and its rigorous discussion is beyond the level of this book. The process obtained in this manner is called the minimal process associated with the infinitesimal matrix $\mathbf{A}$ defined in (6.17).

The preceding construction of the minimal process is fundamental, since the infinitesimal parameters need not determine a unique stochastic process obeying (6.18), (6.19), and Postulates 1 through 5 of Section 6.3.1. In fact, there could be several Markov processes that possess the same infinitesimal generator. Fortunately, such complications do not arise in the modeling of common phenomena. In the special case of birth and death processes for which $\lambda_0 > 0$, a sufficient condition that there exists a unique Markov process with transition probability function $P_{ij}(t)$ for which the infinitesimal relations (6.18) and (6.19) hold is that

$$\sum_{n=0}^{\infty} \frac{1}{\lambda_n \theta_n} \sum_{k=0}^{n} \theta_k = \infty, \tag{6.21}$$

where

$$\theta_0 = 1, \quad \theta_n = \frac{\lambda_0 \lambda_1 \cdots \lambda_{n-1}}{\mu_1 \mu_2 \cdots \mu_n}, \quad n = 1, 2, \ldots.$$

In most practical examples of birth and death processes, the condition (6.21) is met, and the birth and death process associated with the prescribed parameters is uniquely determined.

6.3.3 Differential Equations of Birth and Death Processes

As in the case of the pure birth and pure death processes, the transition probabilities $P_{ij}(t)$ satisfy a system of differential equations known as the backward Kolmogorov differential equations. These are given by

$$P'_{0j}(t) = -\lambda_0 P_{0j}(t) + \lambda_0 P_{1j}(t), \tag{6.22}$$
$$P'_{ij}(t) = \mu_i P_{i-1,j}(t) - (\lambda_i + \mu_i)P_{ij}(t) + \lambda_i P_{i+1,j}(t), \quad i \geq 1,$$

and the boundary condition $P_{ij}(0) = \delta_{ij}$.

To derive these, we have, from equation (6.19),

$$\begin{aligned}
P_{ij}(t+h) &= \sum_{k=0}^{\infty} P_{ik}(h)P_{kj}(t) \\
&= P_{i,i-1}(h)P_{i-1,j}(t) + P_{i,i}(h)P_{ij}(t) + P_{i,i+1}(h)P_{i+1,j}(t) \\
&\quad + {\sum_{k}}' P_{ik}(h)P_{kj}(t),
\end{aligned} \tag{6.23}$$

where the last summation is over all $k \neq i-1, i, i+1$. Using Postulates (1), (2), and (3) of Section 6.3.1, we obtain

$$\begin{aligned}
{\sum_{k}}' P_{ik}(h)P_{kj}(t) &\leq {\sum_{k}}' P_{ik}(h) \\
&= 1 - [P_{i,i}(h) + P_{i,i-1}(h) + P_{i,i+1}(h)] \\
&= 1 - [1 - (\lambda_i + \mu_i)h + o(h) + \mu_i h + o(h) + \lambda_i h + o(h)] \\
&= o(h)
\end{aligned}$$

so that

$$P_{ij}(t+h) = \mu_i h P_{i-1,j}(t) + [1 - (\lambda_i + \mu_i)h]P_{ij}(t) + \lambda_i h P_{i+1,j}(t) + o(h).$$

Transposing the term $P_{ij}(t)$ to the left-hand side and dividing the equation by h, we obtain, after letting $h \downarrow 0$,

$$P'_{ij}(t) = \mu_i P_{i-1,j}(t) - (\lambda_i + \mu_i)P_{ij}(t) + \lambda_i P_{i+1,j}(t).$$

The backward equations are deduced by decomposing the time interval $(0, t+h)$, where h is positive and small, into the two periods

$$(0, h), \quad (h, t+h)$$

and examining the transition in each period separately. In this sense, the backward equations result from a "first step analysis," the first step being over the short time interval of duration h.

A different result arises from a "last step analysis," which proceeds by splitting the time interval $(0, t + h)$ into the two periods

$$(0, t), \quad (t, t + h)$$

and adapting the preceding reasoning. From this viewpoint, under more stringent conditions, we can derive a further system of differential equations

$$
\begin{aligned}
P'_{i0}(t) &= -\lambda_0 P_{i,0}(t) + \mu_1 P_{i,1}(t), \\
P'_{ij}(t) &= \lambda_{j-1} P_{i,j-1}(t) - (\lambda_j + \mu_j) P_{ij}(t) + \mu_{j+1} P_{i,j+1}(t), \quad j \geq 1,
\end{aligned}
\tag{6.24}
$$

with the same initial condition $P_{ij}(0) = \delta_{ij}$. These are known as the forward Kolmogorov differential equations. To derive these equations, we interchange t and h in equation (6.23), and under stronger assumptions in addition to Postulates (1), (2), and (3), it can be shown that the last term is again $o(h)$. The remainder of the argument is the same as before. The usefulness of the differential equations will become apparent in the examples that we study in this and the next section.

A sufficient condition that (6.24) hold is that $[P_{kj}(h)]/h = o(1)$ for $k \neq j, j-1$, $j+1$, where the $o(1)$ term apart from tending to zero uniformly bounded with respect to k for fixed j as $h \to 0$. In this case, it can be proved that $\sum'_k P_{ik}(t) P_{kj}(h) = o(h)$.

Example *Linear Growth with Immigration* A birth and death process is called a linear growth process if $\lambda_n = \lambda n + a$ and $\mu_n = \mu n$ with $\lambda > 0$, $\mu > 0$, and $a > 0$. Such processes occur naturally in the study of biological reproduction and population growth. If the state n describes the current population size, then the average instantaneous rate of growth is $\lambda n + a$. Similarly, the probability of the state of the process decreasing by one after the elapse of a small duration of time h is $\mu n h + o(h)$. The factor λn represents the natural growth of the population owing to its current size, while the second factor a may be interpreted as the infinitesimal rate of increase of the population due to an external source such as immigration. The component μn, which gives the mean infinitesimal death rate of the present population, possesses the obvious interpretation.

If we substitute the above values of λ_n and μ_n in (6.24), we obtain

$$
\begin{aligned}
P'_{i0}(t) &= -a P_{i0}(t) + \mu P_{i1}(t), \\
P'_{ij}(t) &= [\lambda(j-1) + a] P_{i,j-1}(t) - [(\lambda + \mu)j + a] P_{ij}(t) \\
&\quad + \mu(j+1) P_{i,j+1}(t), \quad j \geq 1.
\end{aligned}
$$

Now, if we multiply the jth equation by j and sum, it follows that the expected value

$$
E[X(t)] = M(t) = \sum_{j=1}^{\infty} j P_{ij}(t)
$$

satisfies the differential equation

$$M'(t) = a + (\lambda - \mu)M(t),$$

with initial condition $M(0) = i$, if $X(0) = i$. The solution of this equation is

$$M(t) = at + i \quad \text{if } \lambda = \mu$$

and

$$M(t) = \frac{a}{\lambda - \mu} \left\{ e^{(\lambda-\mu)t} - 1 \right\} + ie^{(\lambda-\mu)t} \quad \text{if } \lambda \neq \mu. \tag{6.25}$$

The second moment, or variance, may be calculated in a similar way. It is interesting to note that $M(t) \to \infty$ as $t \to \infty$ if $\lambda \geq \mu$, while if $\lambda < \mu$, the mean population size for large t is approximately

$$\frac{a}{\mu - \lambda}.$$

These results suggest that in the second case, wherein $\lambda < \mu$, the population stabilizes in the long run in some form of statistical equilibrium. Indeed, it can be shown that a limiting probability distribution $\{\pi_j\}$ exists for which $\lim_{t \to \infty} P_{ij}(t) = \pi_j, j = 0, 1, \ldots$. Such limiting distributions for general birth and death processes are the subject of Section 6.4.

Example *The Two-State Markov Chain* Consider a Markov chain $\{X(t)\}$ with state $\{0, 1\}$ whose infinitesimal matrix is

$$\mathbf{A} = \begin{array}{c} \\ 0 \\ 1 \end{array} \begin{array}{c} 0 \quad\quad 1 \\ \left\| \begin{array}{cc} -\alpha & \alpha \\ \beta & -\beta \end{array} \right\| \end{array}. \tag{6.26}$$

The process alternates between states 0 and 1. The sojourn times in state 0 are independent and exponentially distributed with parameter α. Those in state 1 are independent and exponentially distributed with parameter β. This is a finite-state birth and death process for which $\lambda_0 = \alpha, \lambda_1 = 0, \mu_0 = 0$, and $\mu_1 = \beta$. The first Kolmogorov forward equation in (6.24) becomes

$$P'_{00}(t) = -\alpha P_{00}(t) + \beta P_{01}(t). \tag{6.27}$$

Now, $P_{01}(t) = 1 - P_{00}(t)$, which placed in (6.27) gives

$$P'_{00}(t) = \beta - (\alpha + \beta)P_{00}(t).$$

Let $Q_{00}(t) = e^{(\alpha+\beta)t} P_{00}(t)$. Then,

$$
\begin{aligned}
\frac{dQ_{00}(t)}{dt} &= e^{(\alpha+\beta)t} P'_{00}(t) + (\alpha+\beta) e^{(\alpha+\beta)t} P_{00}(t) \\
&= e^{(\alpha+\beta)t} \left[P'_{00}(t) + (\alpha+\beta) P_{00}(t) \right] \\
&= \beta e^{(\alpha+\beta)t},
\end{aligned}
$$

which can be integrated immediately to yield

$$
\begin{aligned}
Q_{00}(t) &= \beta \int e^{(\alpha+\beta)t} dt + C \\
&= \left(\frac{\beta}{\alpha+\beta} \right) e^{(\alpha+\beta)t} + C.
\end{aligned}
$$

The initial condition $Q_{00}(0) = 1$ determines the constant of integration to be $C = \alpha/(\alpha+\beta)$. Thus,

$$
Q_{00}(t) = e^{(\alpha+\beta)t} P_{00}(t) = \left(\frac{\beta}{\alpha+\beta} \right) e^{(\alpha+\beta)t} + \left(\frac{\alpha}{\alpha+\beta} \right) \tag{6.28}
$$

and

$$
P_{00}(t) = \frac{\beta}{\alpha+\beta} + \frac{\alpha}{\alpha+\beta} e^{-(\alpha+\beta)t}. \tag{6.29a}
$$

Since $P_{01}(t) = 1 - P_{00}(t)$, we have

$$
P_{01}(t) = \frac{\alpha}{\alpha+\beta} - \frac{\alpha}{\alpha+\beta} e^{-(\alpha+\beta)t}, \tag{6.29b}
$$

and by symmetry,

$$
P_{11}(t) = \frac{\alpha}{\alpha+\beta} + \frac{\beta}{\alpha+\beta} e^{-(\alpha+\beta)t}, \tag{6.29c}
$$

$$
P_{10}(t) = \frac{\beta}{\alpha+\beta} - \frac{\beta}{\alpha+\beta} e^{-(\alpha+\beta)t}. \tag{6.29d}
$$

These transition probabilities assume a more succinct form if we reparametrize according to $\pi = \alpha/(\alpha+\beta)$ and $\tau = \alpha+\beta$. Then,

$$
P_{00}(t) = (1-\pi) + \pi e^{-\tau t}, \tag{6.30a}
$$

$$
P_{01}(t) = \pi - \pi e^{-\tau t}, \tag{6.30b}
$$

$$
P_{10}(t) = (1-\pi) - (1-\pi) e^{-\tau t}, \tag{6.30c}
$$

and

$$P_{11}(t) = \pi + (1 - \pi)e^{-\tau t}. \tag{6.30d}$$

Observe that

$$\lim_{t \to \infty} P_{01}(t) = \lim_{t \to \infty} P_{11}(t) = \pi$$

so that π is the long run probability of finding the process in state 1 independently of where the process began. The long run behavior of general birth and death processes is the subject of the next section.

Exercises

6.3.1 Particles are emitted by a radioactive substance according to a Poisson process of rate λ. Each particle exists for an exponentially distributed length of time, independent of the other particles, before disappearing. Let $X(t)$ denote the number of particles alive at time t. Argue that $X(t)$ is a birth and death process and determine the parameters.

6.3.2 Patients arrive at a hospital emergency room according to a Poisson process of rate λ. The patients are treated by a single doctor on a first come, first served basis. The doctor treats patients more quickly when the number of patients waiting is higher. An industrial engineering time study suggests that the mean patient treatment time when there are k patients in the system is of the form $m_k = \alpha - \beta k/(k+1)$, where α and β are constants with $\alpha > \beta > 0$. Let $N(t)$ be the number of patients in the system at time t (waiting and being treated). Argue that $N(t)$ might be modeled as a birth and death process with parameters $\lambda_k = \lambda$ for $k = 0, 1, \ldots$ and $\mu_k = k/m_k$ for $k = 0, 1, \ldots$. State explicitly any necessary assumptions.

6.3.3 Let $\{V(t)\}$ be the two-state Markov chain whose transition probabilities are given by (6.30a–d). Suppose that the initial distribution is $(1 - \pi, \pi)$. That is, assume that $\Pr\{V(0) = 0\} = 1 - \pi$ and $\Pr\{V(0) = 1\} = \pi$. In this case, show that $\Pr\{V(t) = 1\} = \pi$ for all times $t > 0$.

Problems

6.3.1 Let $\xi_n, n = 0, 1, \ldots$, be a two-state Markov chain with transition probability matrix

$$\mathbf{P} = \begin{array}{c} \\ 0 \\ 1 \end{array} \begin{array}{cc} 0 & 1 \\ \left\| \begin{array}{cc} 0 & 1 \\ 1 - \alpha & \alpha \end{array} \right\|. \end{array}$$

Let $\{N(t); t \geq 0\}$ be a Poisson process with parameter λ. Show that

$$X(t) = \xi_{N(t)}, \quad t \geq 0,$$

is a two-state birth and death process and determine the parameters λ_0 and μ_1 in terms of α and λ.

6.3.2 Collards were planted equally spaced in a single row in order to provide an experimental setup for observing the chaotic movements of the flea beetle (*Phyllotreta cruciferae*). A beetle at position k in the row remains on that plant for a random length of time having mean m_k (which varies with the "quality" of the plant) and then is equally likely to move right $(k + 1)$ or left $(k - 1)$. Model the position of the beetle at time t as a birth and death process having parameters $\lambda_k = \mu_k = 1/(2m_k)$ for $k = 1, 2, \ldots, N - 1$, where the plants are numbered $0, 1, \ldots, N$. What assumptions might be plausible at the ends 0 and N?

6.3.3 Let $\{V(t)\}$ be the two-state Markov chain whose transition probabilities are given by (6.30a–d). Suppose that the initial distribution is $(1 - \pi, \pi)$. That is, assume that $\Pr\{V(0) = 0\} = 1 - \pi$ and $\Pr\{V(0) = 1\} = \pi$. For $0 < s < t$, show that

$$E[V(s)V(t)] = \pi - \pi P_{10}(t - s),$$

whence

$$\text{Cov}[V(s), V(t)] = \pi(1 - \pi)e^{-(\alpha+\beta)|t-s|}.$$

6.3.4 *A Stop-and-Go Traveler* The velocity $V(t)$ of a stop-and-go traveler is described by the two-state Markov chain whose transition probabilities are given by (6.30a–d). The distance traveled in time t is the integral of the velocity:

$$S(t) = \int_0^t V(u)du.$$

Assuming that the velocity at time $t = 0$ is $V(0) = 0$, determine the mean of $S(t)$. Take for granted the interchange of integral and expectation in

$$E[S(t)] = \int_0^t E[V(u)]du.$$

6.4 The Limiting Behavior of Birth and Death Processes

For a general birth and death process that has no absorbing states, it can be proved that the limits

$$\lim_{t \to \infty} P_{ij}(t) = \pi_j \geq 0 \tag{6.31}$$

exist and are independent of the initial state i. It may happen that $\pi_j = 0$ for all states j. When the limits π_j are strictly positive, however, and satisfy

$$\sum_{j=0}^{\infty} \pi_j = 1, \tag{6.32}$$

they form a probability distribution that is called, naturally enough, the *limiting distribution* of the process. The limiting distribution is also a *stationary distribution* in that

$$\pi_j = \sum_{i=0}^{\infty} \pi_i P_{ij}(t), \tag{6.33}$$

which tells us that if the process starts in state i with probability π_i, then at any time t it will be in state i with the same probability π_i. The proof of (6.33) follows from (6.19) and (6.31) if we let $t \to \infty$ and use the fact that $\Sigma_{i=0}^{\infty}\pi_i = 1$.

The general importance of birth and death processes as models derives in large part from the availability of standard formulas for determining if a limiting distribution exists and what its values are when it does. These formulas follow from the Kolmogorov forward equations (6.24) that were derived in Section 6.3.3:

$$P'_{i,0}(t) = -\lambda_0 P_{i,0}(t) + \mu_1 P_{i,1}(t),$$
$$P'_{i,j}(t) = \lambda_{j-1} P_{i,j-1}(t) - (\lambda_j + \mu_j) P_{ij}(t) + \mu_{j+1} P_{i,j+1}(t), \quad j \geq 1, \tag{6.34}$$

with the initial condition $P_{iJ}(0) = \delta_{ij}$. Now pass to the limit as $t \to \infty$ in (6.34) and observe first that the limit of the right side of (6.34) exists according to (6.31). Therefore, the limit of the left side, the derivatives $P'_{ij}(t)$, exists as well. Since the probabilities are converging to a constant, the limit of these derivatives must be zero. In summary, passing to the limit in (6.34) produces

$$0 = -\lambda_0 \pi_0 + \mu_1 \pi_1,$$
$$0 = \lambda_{j-1} \pi_{j-1} - (\lambda_j + \mu_j) \pi_j + \mu_{j+1} \pi_{j+1}, \quad j \geq 1. \tag{6.35}$$

The solution to (6.35) is obtained by induction. Letting

$$\theta_0 = 1 \quad \text{and} \quad \theta_j = \frac{\lambda_0 \lambda_1 \cdots \lambda_{j-1}}{\mu_1 \mu_2 \cdots \mu_J} \quad \text{for } j \geq 1, \tag{6.36}$$

we have $\pi_1 = \lambda_0 \pi_0 / \mu_1 = \theta_1 \pi_0$. Then, assuming that $\pi_k = \theta_k \pi_0$ for $k = 1, \ldots, j$, we obtain

$$\mu_{j+1} \pi_{j+1} = (\lambda_j + \mu_j)\theta_j \pi_0 - \lambda_{j-1}\theta_{j-1}\pi_0$$
$$= \lambda_j \theta_j \pi_0 + (\mu_j \theta_j - \lambda_{j-1}\theta_{j-1})\pi_0$$
$$= \lambda_j \theta_j \pi_0,$$

and finally

$$\pi_{j+1} = \theta_{j+1}\pi_0.$$

In order that the sequence $\{\pi_j\}$ define a distribution, we must have $\Sigma_j \pi_j = 1$. If $\Sigma \theta_j < \infty$, then we may sum the following,

$$\pi_0 = \theta_0 \pi_0$$
$$\pi_1 = \theta_1 \pi_0$$
$$\pi_2 = \theta_2 \pi_0$$

$$\vdots \qquad \vdots$$

$$\overline{1 = (\Sigma \theta_k)\pi_0}$$

to see that $\pi_0 = 1/\Sigma_{k=0}^{\infty}\theta_k$, and then

$$\pi_j = \theta_j \pi_0 = \frac{\theta_j}{\Sigma_{k=0}^{\infty}\theta_k} \quad \text{for } j = 0, 1, \ldots. \tag{6.37}$$

If $\Sigma\theta_k = \infty$, then necessarily $\pi_0 = 0$, and then $\pi_j = \theta_j\pi_0 = 0$ for all j, and there is no limiting distribution ($\lim_{t\to\infty} P_{ij}(t) = 0$ for all j).

Example *Linear Growth with Immigration* As described in the example at the end of Section 6.3.3, this process has birth parameters $\lambda_n = a + \lambda n$ and death parameters $\mu_n = \mu n$ for $n = 0, 1, \ldots$, where $\lambda > 0$ is the individual birth rate, $a > 0$ is the rate of immigration into the population, and $\mu > 0$ is the individual death rate.

Suppose $\lambda < \mu$. It was shown in Section 6.3.3 that the population mean $M(t)$ converges to $a/(\mu - \lambda)$ as $t \to \infty$. Here, we will determine the limiting distribution of the process under the same condition $\lambda < \mu$.

Then, $\theta_0 = 1$, $\theta_1 = a/\mu$, $\theta_2 = a(a + \lambda)/[\mu(2\mu)]$, $\theta_3 = a(a + \lambda)(a + 2\lambda)/[\mu(2\mu)(3\mu)]$, and, in general,

$$\theta_k = \frac{a(a+\lambda)\cdots[a+(k-1)\lambda]}{\mu^k(k)!}$$

$$= \frac{(a/\lambda)[(a/\lambda)+1]\cdots[(a/\lambda)+k-1]}{k!}\left(\frac{\lambda}{\mu}\right)^k$$

$$= \binom{(a/\lambda)+k-1}{k}\left(\frac{\lambda}{\mu}\right)^k.$$

Now, use the infinite binomial formula (Chapter 1, equation (1.71)),

$$(1-x)^{-N} = \sum_{k=0}^{\infty}\binom{N+k-1}{k}x^k \quad \text{for} |x| < 1,$$

to determine that

$$\sum_{k=0}^{\infty} \theta_k = \sum_{k=0}^{\infty} \binom{(a/\lambda)+k-1}{k} \left(\frac{\lambda}{\mu}\right)^k = \left(1 - \frac{\lambda}{\mu}\right)^{-(a/\lambda)}$$

when $\lambda < \mu$. Thus, $\pi_0 = (1 - \lambda/\mu)^{a/\lambda}$, and

$$\pi_k = \left(\frac{\lambda}{\mu}\right)^k \frac{(a/\lambda)[(a/\lambda)+1]\cdots[(a/\lambda)+k-1]}{k!}(1-\lambda/\mu)^{a/\lambda} \quad \text{for } k > 1.$$

Example *Repairman Models* A system is composed of N machines, of which at most $M \le N$ can be operating at any one time. The rest are "spares." When a machine is operating, it operates a random length of time until failure. Suppose this failure time is exponentially distributed with parameter μ.

When a machine fails, it undergoes repair. At most R machines can be "in repair" at any one time. The repair time is exponentially distributed with parameter λ. Thus, a machine can be in any of four states: (1) operating; (2) "up," but not operating, i.e., a spare; (3) in repair; and (4) waiting for repair. There is a total of N machines in the system. At most M can be operating. At most R can be in repair.

The action is diagrammed in Figure 6.5.

Let $X(t)$ be the number of machines "up" at time t, either operating or spare. Then, (we assume) the number operating is $\min\{X(t), M\}$, and the number of spares is $\max\{0, X(t) - M\}$. Let $Y(t) = N - X(t)$ be the number of machines "down." Then, the number in repair is $\min\{Y(t), R\}$, and the number waiting for repair is $\max\{0, Y(t) - R\}$.

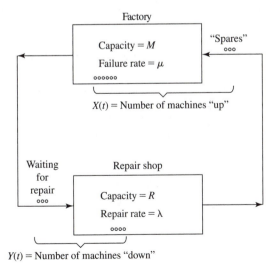

Figure 6.5 Repairman model.

The foregoing formulas permit us to determine the number of machines in any category, once $X(t)$ is known.

Then, $X(t)$ is a finite-state birth and death process[1] with parameters

$$\lambda_n = \lambda \times \min\{N - n, R\}$$

$$= \begin{cases} \lambda R & \text{for } n = 0, 1, \ldots, N - R, \\ \lambda(N - n) & \text{for } n = N - R + 1, \ldots, N, \end{cases}$$

and

$$\mu_n = \mu \times \min\{n, M\} = \begin{cases} \mu n & \text{for } n = 0, 1, \ldots, M, \\ \mu M & \text{for } n = M + 1, \ldots, N. \end{cases}$$

It is now a routine task to determine the limiting probability distribution for any values of λ, μ, N, M, and R. (See Problems 6.4.1 and 6.4.7.) In terms of the limiting probabilities $\pi_0, \pi_1, \ldots, \pi_N$, some quantities of interest are the following:

$$\textit{Average Machines Operating} = \pi_1 + 2\pi_2 + \cdots + M\pi_M$$
$$+ M(\pi_{M+1} + \cdots + \pi_N);$$

$$\textit{Long Run Utilization} = \frac{\text{Average Machines Operating}}{\text{Capacity}}$$
$$= \frac{\pi_1 + 2\pi_2 + \cdots + M\pi_M}{M}$$
$$+ (\pi_{M+1} + \cdots + \pi_N);$$

$$\textit{Average Idle Repair Capacity} = 1\pi_{N-R+1} + 2\pi_{N-R+2} + \cdots + R\pi_N.$$

These and other similar quantities can be used to evaluate the desirability of adding additional repair capability, additional spare machines, and other possible improvements.

The stationary distribution assumes quite simple forms in certain special cases. For example, consider the special case in which $M = N = R$. The situation arises, for instance, when each machine's operator becomes its repairman upon its failure. Then, $\lambda_n = \lambda(N - n)$ and $\mu_n = \mu n$ for $n = 0, 1, \ldots, N$, and following (6.36), we determine $\theta_0 = 1, \theta_1 = \lambda N/\mu, \theta_2 = (\lambda N)\lambda(N - 1)/\mu(2\mu)$, and, in general,

$$\theta_k = \frac{N(N-1)\cdots(N-k+1)}{(1)(2)\cdots(k)}\left(\frac{\lambda}{\mu}\right)^k = \binom{N}{k}\left(\frac{\lambda}{\mu}\right)^k.$$

---

[1] The definition of birth and death processes was given for an infinite number of states. The adjustments in the definitions and analyses for the case of a finite number of states are straightforward and even simpler than the original definitions and are left to the reader.

The binomial formula $(1 + x)^N = \Sigma_{k=0}^N \binom{N}{k} x^k$ applies to yield

$$\sum_{k=0}^N \theta_k = \sum_{k=0}^N \binom{N}{k} \left(\frac{\lambda}{\mu} \right)^k = \left(1 + \frac{\lambda}{\mu} \right)^N.$$

Thus, $\pi_0 = [1 + (\lambda/\mu)]^{-N} = [\mu/(\lambda + \mu)]^N$, and

$$\pi_k = \binom{N}{k} \left(\frac{\lambda}{\mu} \right)^k [\mu/(\lambda + \mu)]^N$$

$$= \binom{N}{k} \left(\frac{\lambda}{\lambda + \mu} \right)^k \left(\frac{\mu}{\lambda + \mu} \right)^{N-k}. \tag{6.38}$$

We recognize (6.38) as the familiar binomial distribution.

Example *Logistic Process* Suppose we consider a population whose size $X(t)$ ranges between two fixed integers N and $M(N < M)$ for all $t \geq 0$. We assume that the birth and death rates per individual at time t are given by

$$\lambda = \alpha(M - X(t)) \quad \text{and} \quad \mu = \beta(X(t) - N)$$

and that the individual members of the population act independently of each other. The resulting birth and death rates for the population then become

$$\lambda_n = \alpha n(M - n) \quad \text{and} \quad \mu_n = \beta n(n - N).$$

To see this, we observe that if the population size $X(t)$ is n, then each of the n individuals has an infinitesimal birth rate λ so that $\lambda_u = \alpha n(M - n)$. The same rationale applies in the interpretation of the μ_n.

Under such conditions, one would expect the process to fluctuate between the two constants N and M, since, e.g., if $X(t)$ is near M, the death rate is high and the birth rate is low, and then $X(t)$ will tend toward N. Ultimately, the process should display stationary fluctuations between the two limits N and M.

The stationary distribution in this case is

$$\pi_{N+m} = \frac{c}{N+m} \binom{M-N}{m} \left(\frac{\alpha}{\beta} \right)^m, \quad m = 0, 1, 2, \dots, M - N,$$

where c is an appropriate constant determined so that $\Sigma_m \pi_{N+m} = 1$. To see this, we observe that

$$\theta_{N+m} = \frac{\lambda_N \lambda_{N+1} \cdots \lambda_{N+m-1}}{\mu_{N+1} \mu_{N+2} \cdots \mu_{N+m}}$$

$$= \frac{\alpha^m N(N+1) \cdots (N+m-1)(M-N) \cdots (M-N-m+1)}{\beta^m (N+1) \cdots (N+m)m!}$$

$$= \frac{N}{N+m} \binom{M-N}{m} \left(\frac{\alpha}{\beta} \right)^m.$$

Example *Some Genetic Models* Consider a population consisting of N individuals who are either of gene type **a** or gene type **A**. The state of the process $X(t)$ represents the number of **a** individuals at time t. We assume that the probability that any individual dies and is replaced by another during the time interval $(t, t+h)$ is $\lambda h + o(h)$ independent of the values of $X(t)$ and that the probability of two or more changes occurring in a time interval h is $o(h)$.

The changes in the population structure are affected as follows. An individual is to be replaced by another chosen randomly from the population; that is, if $X(t) = j$, then an **a**-type is selected to be replaced with probability j/N and an **A**-type with probability $1 - j/N$. We refer to this stage as death. Next, birth takes place by the following rule. Another selection is made randomly from the population to determine the type of the new individual replacing the one who died. The model introduces mutation pressures that admit the possibility that the type of the new individual may be altered upon birth. Specifically, let γ_1 denote the probability that an **a**-type mutates to an **A**-type, and let γ_2 denote the probability of an **A**-type mutating to an **a**-type.

The probability that the new individual added to the population is of type **a** is

$$\frac{j}{N}(1 - \gamma_1) + \left(1 - \frac{j}{N}\right)\gamma_2. \tag{6.39}$$

We deduce this formula as follows: The probability that we select an **a**-type and that no mutation occurs is $(j/N)(1 - \gamma_1)$. Moreover, the final type may be an **a**-type if we select an **A**-type that subsequently mutates into an **a**-type. The probability of this contingency is $(1 - j/N)\gamma_2$. The combination of these two possibilities gives (6.39).

We assert that the conditional probability that $X(t+) - X(t) = 1$ when a change of state occurs is

$$\left(1 - \frac{j}{N}\right)\left[\frac{j}{N}(1 - \gamma_1) + \left(1 - \frac{j}{N}\right)\gamma_2\right], \quad \text{where } X(t) = j. \tag{6.40}$$

In fact, the **a**-type population size can increase only if an **A**-type dies (is replaced). This probability is $1 - (j/N)$. The second factor is the probability that the new individual is of type **a** as in (6.39).

In a similar way, we find that the conditional probability that $X(t+) - X(t) = -1$ when a change of state occurs is

$$\frac{j}{N}\left[\left(1 - \frac{j}{N}\right)(1 - \gamma_2) + \frac{j}{N}\gamma_1\right], \quad \text{where } X(t) = j.$$

The number of type **a** individuals in the population is thus a birth and death process with a finite number of states and infinitesimal birth and death rates

$$\lambda_j = \lambda\left(1 - \frac{j}{N}\right)\left[\frac{j}{N}(1 - \gamma_1) + \left(1 - \frac{j}{N}\right)\gamma_2\right]N$$

and

$$\mu_j = \lambda \frac{j}{N}\left[\frac{j}{N}\gamma_1 + \left(1 - \frac{j}{N}\right)(1 - \gamma_2)\right]N, \quad 0 \le j \le N.$$

Although these parameters seem rather complicated, it is interesting to see what happens to the stationary measure $\{\pi_k\}_{k=0}^{N}$ if we let the population size $N \to \infty$ and the probabilities of mutation per individual γ_1 and γ_2 tend to zero in such a way that $\gamma_1 N \to \kappa_1$ and $\gamma_2 N \to \kappa_2$, where $0 < \kappa_1, \kappa_2 < \infty$. At the same time, we shall transform the state of the process to the interval $[0, 1]$ by defining new states j/N, i.e., the fraction of **a**-types in the population. To examine the stationary density at a fixed fraction x, where $0 < x < 1$, we shall evaluate π_k as $k \to \infty$ in such a way that $k = [xN]$, where $[xN]$ is the greatest integer less than or equal to xN.

Keeping these relations in mind, we write

$$\lambda_j = \frac{\lambda(N-j)}{N}(1 - \gamma_1 - \gamma_2)j\left(1 + \frac{a}{j}\right), \quad \text{where } a = \frac{N\gamma_2}{1 - \gamma_1 - \gamma_2},$$

and

$$\mu_j = \frac{\lambda(N-j)}{N}(1 - \gamma_1 - \gamma_2)j\left(1 + \frac{b}{N-j}\right), \quad \text{where } b = \frac{N\gamma_1}{1 - \gamma_1 - \gamma_2}.$$

Then,

$$\begin{aligned}
\log \theta_k &= \sum_{j=0}^{k-1} \log \lambda_j - \sum_{j=1}^{k} \log \mu_j \\
&= \sum_{j=1}^{k-1} \log\left(1 + \frac{a}{j}\right) - \sum_{j=1}^{k-1} \log\left(1 + \frac{b}{N-j}\right) + \log Na \\
&\quad - \log(N-k)k\left(1 + \frac{b}{N-k}\right).
\end{aligned}$$

Now, using the expansion

$$\log(1 + x) = x - \frac{x^2}{2} + \frac{x^3}{3} - \cdots, \quad |x| < 1,$$

it is possible to write

$$\sum_{j=1}^{k-1} \log\left(1 + \frac{a}{j}\right) = a\sum_{j=1}^{k-1} \frac{1}{j} + c_k,$$

where c_k approaches a finite limit as $k \to \infty$. Therefore, using the relation

$$\sum_{j=1}^{k-1} \frac{1}{j} \sim \log k \quad \text{as } k \to \infty,$$

we have

$$\sum_{j=1}^{k-1} \log\left(1 + \frac{a}{j}\right) \sim \log k^a + c_k \quad \text{as } k \to \infty.$$

In a similar way, we obtain

$$\sum_{j=1}^{k-1} \log\left(1 + \frac{b}{N-j}\right) \sim \log \frac{N^b}{(N-k)^b} + d_k \quad \text{as } k \to \infty,$$

where d_k approaches a finite limit as $k \to \infty$. Using the above relations, we have

$$\log \theta_k \sim \log\left(C_k \frac{k^a (N-k)^b Na}{N^b (N-k) k}\right) \quad \text{as } k \to \infty, \tag{6.41}$$

where $\log C_k = c_k + d_k$, which approaches a limit, say C, as $k \to \infty$. Notice that $a \to \kappa_2$ and $b \to \kappa_1$ as $N \to \infty$. Since $k = [Nx]$, we have, for $N \to \infty$,

$$\theta_k \sim C\kappa_2 N^{\kappa_2 - 1} x^{\kappa_2 - 1} (1-x)^{\kappa_1 - 1}.$$

Now, from (6.41), we have

$$\theta_k \sim aC_k k^{a-1}\left(1 - \frac{k}{N}\right)^{b-1}.$$

Therefore,

$$\frac{1}{N^a} \sum_{k=0}^{N-1} \theta_k \sim \frac{a}{N} \sum_{k=0}^{N-1} C_k \left(\frac{k}{N}\right)^{a-1} \left(1 - \frac{k}{N}\right)^{b-1}.$$

Since $C_k \to C$ as k tends to ∞, we recognize the right side as a Riemann sum approximation of

$$\kappa_2 C \int_0^1 x^{\kappa_2 - 1} (1-x)^{\kappa_1 - 1} dx.$$

Thus,

$$\sum_{i=0}^{N} \theta_i \sim N^{\kappa_2} \kappa_2 C \int_0^1 x^{\kappa_2-1}(1-x)^{\kappa_1-1} dx$$

so that the resulting density on $[0, 1]$ is

$$\frac{\theta_k}{\Sigma \theta_i} \sim \frac{1}{N} \frac{x^{\kappa_2-1}(1-x)^{\kappa_1-1}}{\int_0^1 x^{\kappa_2-1}(1-x)^{\kappa_1-1} dx} = \frac{x^{\kappa_2-1}(1-x)^{\kappa_1-1} dx}{\int_0^1 x^{\kappa_2-1}(1-x)^{\kappa_1-1} dx},$$

since $dx \sim 1/N$. This is a beta distribution with parameters κ_1 and κ_2.

Exercises

6.4.1 In a birth and death process with birth parameters $\lambda_n = \lambda$ for $n = 0, 1, \ldots$ and death parameters $\mu_n = \mu n$ for $n = 0, 1, \ldots$, we have

$$P_{0j}(t) = \frac{(\lambda p)^j e^{-\lambda p}}{j!},$$

where

$$p = \frac{1}{\mu} \left[1 - e^{-\mu t}\right].$$

Verify that these transition probabilities satisfy the forward equations (6.34), with $i = 0$.

6.4.2 Let $X(t)$ be a birth and death process where the possible states are $0, 1, \ldots, N$, and the birth and death parameters are, respectively, $\lambda_n = \alpha(N - n)$ and $\mu_n = \beta n$. Determine the stationary distribution.

6.4.3 Determine the stationary distribution for a birth and death process having infinitesimal parameters $\lambda_n = \alpha(n + 1)$ and $\mu_n = \beta n^2$ for $n = 0, 1, \ldots$, where $0 < \alpha < \beta$.

6.4.4 Consider two machines, operating simultaneously and independently, where both machines have an exponentially distributed time to failure with mean $1/\mu$ (μ is the failure rate). There is a single repair facility, and the repair times are exponentially distributed with rate λ.

(a) In the long run, what is the probability that no machines are operating?

(b) How does your answer in (a) change if at most one machine can operate, and thus be subject to failure, at any time?

6.4.5 Consider the birth and death parameters $\lambda_n = \theta < 1$ and $\mu_n = n/(n + 1)$ for $n = 0, 1, \ldots$. Determine the stationary distribution.

6.4.6 A birth and death process has parameters $\lambda_n = \lambda$ and $\mu_n = n\mu$, for $n = 0, 1, \ldots$. Determine the stationary distribution.

Problems

6.4.1 For the repairman model of the second example of this section, suppose that $M = N = 5, R = 1, \lambda = 2$, and $\mu = 1$. Using the limiting distribution for the system, determine
   (a) The average number of machines operating.
   (b) The equipment utilization.
   (c) The average idle repair capacity.
   How do these system performance measures change if a second repairman is added?

6.4.2 Determine the stationary distribution, when it exists, for a birth and death process having constant parameters $\lambda_n = \lambda$ for $n = 0, 1, \ldots$ and $\mu_n = \mu$ for $n = 1, 2, \ldots$.

6.4.3 A factory has five machines and a single repairman. The operating time until failure of a machine is an exponentially distributed random variable with parameter (rate) 0.20 per hour. The repair time of a failed machine is an exponentially distributed random variable with parameter (rate) 0.50 per hour. Up to five machines may be operating at any given time, their failures being independent of one another, but at most one machine may be in repair at any time. In the long run, what fraction of time is the repairman idle?

6.4.4 This problem considers a continuous time Markov chain model for the changing pattern of relationships among members in a group. The group has four members: a, b, c, and d. Each pair of the group may or may not have a certain relationship with each other. If they have the relationship, we say that they are *linked*. For example, being linked may mean that the two members are communicating with each other. The following graph illustrates links between a and b, between a and c, and between b and d:

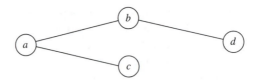

Figure 6.6

Suppose that any pair of unlinked individuals will become linked in a small time interval of length h with probability $\alpha h + o(h)$. Any pair of linked individuals will lose their link in a small time interval of length h with probability $\beta h + o(h)$. Let $X(t)$ denote the number of linked pairs of individuals in the group at time t. Then, $X(t)$ is a birth and death process.
   (a) Specify the birth and death parameters λ_k and μ_k for $k = 0, 1, \ldots$.
   (b) Determine the stationary distribution for the process.

6.4.5 A chemical solution contains N molecules of type A and an equal number of molecules of type B. A reversible reaction occurs between type A and B molecules in which they bond to form a new compound AB. Suppose that in any small time interval of length h, any particular unbonded A molecule will react with any particular unbonded B molecule with probability $\alpha h + o(h)$, where α is a reaction rate of formation. Suppose also that in any small time interval of length h, any particular AB molecule disassociates into its A and B constituents with probability $\beta h + o(h)$, where β is a reaction rate of dissolution. Let $X(t)$ denote the number of AB molecules at time t. Model $X(t)$ as a birth and death process by specifying the parameters.

6.4.6 A time-shared computer system has three terminals that are attached to a central processing unit (CPU) that can simultaneously handle at most two active users. If a person logs on and requests service when two other users are active, then the request is held in a buffer until it can receive service. Let $X(t)$ be the total number of requests that are either active or in the buffer at time t. Suppose that $X(t)$ is a birth and death process with parameters

$$\lambda_k = \begin{cases} \lambda & \text{for } k = 0, 1, 2, \\ 0 & \text{for } k \geq 3 \end{cases}$$

and

$$\mu_k = \begin{cases} k\mu & \text{for } k = 0, 1, 2, \\ 2\mu & \text{for } k = 3. \end{cases}$$

Determine the long run probability that the computer is fully loaded.

6.4.7 A system consists of three machines and two repairmen. At most two machines can operate at any time. The amount of time that an operating machine works before breaking down is exponentially distributed with mean 5. The amount of time that it takes a single repairman to fix a machine is exponentially distributed with mean 4. Only one repairman can work on a failed machine at any given time. Let $X(t)$ be the number of machines in operating condition at time t.

(a) Calculate the long run probability distribution for $X(t)$.

(b) If an operating machine produces 100 units of output per hour, what is the long run output per hour of the system?

6.4.8 A birth and death process has parameters

$$\lambda_k = \alpha(k+1) \quad \text{for } k = 0, 1, 2, \ldots,$$

and

$$\mu_k = \beta(k+1) \quad \text{for } k = 1, 2, \ldots.$$

Assuming that $\alpha < \beta$, determine the limiting distribution of the process. Simplify your answer as much as possible.

6.5   Birth and Death Processes with Absorbing States

Birth and death processes in which $\lambda_0 = 0$ arise frequently and are correspondingly important. For these processes, the zero state is an absorbing state. A central example is the linear-growth birth and death process without immigration (cf. Section 6.3.3). In this case, $\lambda_n = n\lambda$ and $\mu_n = n\mu$. Since growth of the population results exclusively from the existing population, it is clear that when the population size becomes zero it remains zero thereafter; that is, 0 is an absorbing state.

6.5.1   Probability of Absorption into State 0

It is of interest to compute the probability of absorption into state 0 starting from state $i(i \geq 1)$. This is not, a priori, a certain event, since conceivably the particle (i.e., state variable) may wander forever among the states $(1, 2, \ldots)$ or possibly drift to infinity.

Let $u_i(i = 1, 2, \ldots)$ denote the probability of absorption into state 0 from the initial state i. We can write a recursion formula for u_i by considering the possible states after the first transition. We know that the first transition entails the movements

$$i \to i+1 \qquad \text{with probability } \frac{\lambda_i}{\mu_i + \lambda_i},$$

$$i \to i-1 \qquad \text{with probability } \frac{\mu_i}{\mu_i + \lambda_i}.$$

Invoking the familiar first step analysis, we directly obtain

$$u_i = \frac{\lambda_i}{\mu_i + \lambda_i} u_{i+1} + \frac{\mu_i}{\mu_i + \lambda_i} u_{i-1}, \quad i \geq 1, \tag{6.42}$$

where $u_0 = 1$.

Another method for deriving (6.42) is to consider the "embedded random walk" associated with a given birth and death process. Specifically, we examine the birth and death process only at the transition times. The discrete time Markov chain generated in this manner is denoted by $\{Y_n\}_{n=0}^{\infty}$, where $Y_0 = X_0$ is the initial state and $Y_n(n \geq 1)$ is the state at the nth transition. Obviously, the transition probability matrix has the form

$$\mathbf{P} = \begin{Vmatrix} 1 & 0 & 0 & 0 & \cdots \\ q_1 & 0 & p_1 & 0 & \cdots \\ 0 & q_2 & 0 & p_2 & \cdots \\ \vdots & \vdots & & & \end{Vmatrix},$$

where

$$p_i = \frac{\lambda_i}{\lambda_i + \mu_i} = 1 - q_i \quad \text{for } i \geq 1.$$

The probability of absorption into state 0 for the embedded random walk is the same as for the birth and death processes, since both processes execute the same transitions. A closely related problem (gambler's ruin) for a random walk was examined in Chapter 3, Section 3.6.1.

We turn to the task of solving (6.42) subject to the conditions $u_0 = 1$ and $0 \leq u_i \leq 1 (i \geq 1)$. Rewriting (6.42), we have

$$(u_{i+1} - u_i) = \frac{\mu_i}{\lambda_i}(u_i - u_{i-1}), \quad i \geq 1.$$

Defining $v_i = u_{i+1} - u_i$, we obtain

$$v_i = \frac{\mu_i}{\lambda_i} v_{i-1}, \quad i \geq 1.$$

Iteration of the last relation yields the formula $v_i = \rho_i v_0$, where

$$\rho_0 = 1 \quad \text{and} \quad \rho_i = \frac{\mu_1 \mu_2 \cdots \mu_i}{\lambda_1 \lambda_2 \cdots \lambda_i} \quad \text{for } i \geq 1;$$

and with $u_{i+1} - u_i = v_i$,

$$u_{i+1} - u_i = v_i = \rho_i v_0 = \rho_i(u_1 - u_0) = \rho_i(u_1 - 1) \quad \text{for } i \geq 1.$$

Summing these last equations from $i = 1$ to $i = m - 1$, we have

$$u_m - u_1 = (u_1 - 1) \sum_{i=1}^{m-1} \rho_i, \quad m > 1. \tag{6.43}$$

Since u_m by its very meaning is bounded by 1, we see that if

$$\sum_{i=1}^{\infty} \rho_i = \infty, \tag{6.44}$$

then necessarily $u_1 = 1$ and $u_m = 1$ for all $m \geq 2$. In other words, if (6.44) holds, then ultimate absorption into state 0 is certain from any initial state.

Suppose $0 < u_1 < 1$; then, of course,

$$\sum_{i=1}^{\infty} \rho_i < \infty.$$

Obviously, u_m is decreasing in m, since passing from state m to state 0 requires entering the intermediate states in the intervening time. Furthermore, it can be shown that $u_m \to 0$ as $m \to \infty$. Now, letting $m \to \infty$ in (6.43) permits us to solve for u_1; thus,

$$u_1 = \frac{\sum_{i=1}^{\infty} \rho_i}{1 + \sum_{i=1}^{\infty} \rho_i};$$

and then from (6.43), we obtain

$$u_m = \frac{\Sigma_{i=m}^{\infty} \rho_i}{1 + \Sigma_{i=1}^{\infty} \rho_i}, \quad m \geq 1.$$

6.5.2 Mean Time Until Absorption

Consider the problem of determining the mean time until absorption, starting from state m.

We assume that condition (6.44) holds so that absorption is certain. Notice that we cannot reduce our problem to a consideration of the embedded random walk, since the actual time spent in each state is relevant for the calculation of the mean absorption time.

Let w_i be the mean absorption time starting from state i (this could be infinite). Considering the possible states following the first transition, instituting a first step analysis, and recalling the fact that the mean waiting time in state i is $(\lambda_i + \mu_i)^{-1}$ (it is actually exponentially distributed with parameter $\lambda_i + \mu_i$), we deduce the recursion relation

$$w_i = \frac{1}{\lambda_i + \mu_i} + \frac{\lambda_i}{\lambda_i + \mu_i} w_{i+1} + \frac{\mu_i}{\lambda_i + \mu_i} w_{i-1}, \quad i \geq 1, \tag{6.45}$$

where $w_0 = 0$. Letting $z_i = w_i - w_{i+1}$ and rearranging (6.45) leads to

$$z_i = \frac{1}{\lambda_i} + \frac{\mu_i}{\lambda_i} z_{i-1}, \quad i \geq 1. \tag{6.46}$$

Iterating this relation gives

$$z_1 = \frac{1}{\lambda_1} + \frac{\mu_1}{\lambda_1} z_0,$$

$$z_2 = \frac{1}{\lambda_2} + \frac{\mu_2}{\lambda_2} z_1 = \frac{1}{\lambda_2} + \frac{\mu_2}{\lambda_2 \lambda_1} + \frac{\mu_2 \mu_1}{\lambda_2 \lambda_1} z_0,$$

$$z_3 = \frac{1}{\lambda_3} + \frac{\mu_3}{\lambda_3 \lambda_2} + \frac{\mu_3 \mu_2}{\lambda_3 \lambda_2 \lambda_1} + \frac{\mu_3 \mu_2 \mu_1}{\lambda_3 \lambda_2 \lambda_1} z_0,$$

and finally

$$z_m = \sum_{i=1}^{m} \frac{1}{\lambda_i} \prod_{j=i+1}^{m} \frac{\mu_j}{\lambda_j} + \left(\prod_{j=1}^{m} \frac{\mu_j}{\lambda_j} \right) z_0.$$

(The product $\Pi_{m+1}^{m} \mu_j/\lambda_j$ is interpreted as 1.) Using the notation

$$\rho_0 = 1 \quad \text{and} \quad \rho_i = \frac{\mu_1 \mu_2 \cdots \mu_i}{\lambda_1 \lambda_2 \cdots \lambda_i}, \quad i \geq 1,$$

the expression for z_m becomes

$$z_m = \sum_{i=1}^{m} \frac{1}{\lambda_i} \frac{\rho_m}{\rho_i} + \rho_m z_0,$$

or, since $z_m = w_m - w_{m+1}$ and $z_0 = w_0 - w_1 = -w_1$, then

$$\frac{1}{\rho_m}(w_m - w_{m+1}) = \sum_{i=1}^{m} \frac{1}{\lambda_i \rho_i} - w_1. \qquad (6.47)$$

If $\sum_{i=1}^{\infty}(1/\lambda_i \rho_i) = \infty$, then inspection of (6.47) reveals that necessarily $w_1 = \infty$. Indeed, it is probabilistically evident that $w_m < w_{m+1}$ for all m, and this property would be violated for m large if we assume to the contrary that w_1 is finite.

Now, suppose $\sum_{i=1}^{\infty}(1/\lambda_i \rho_i) < \infty$; then, letting $m \to \infty$ in (6.47) gives

$$w_1 = \sum_{i=1}^{\infty} \frac{1}{\lambda_i \rho_i} - \lim_{m \to \infty} \frac{1}{\rho_m}(w_m - w_{m+1}).$$

It is more involved but still possible to prove that

$$\lim_{m \to \infty} \frac{1}{\rho_m}(w_m - w_{m+1}) = 0,$$

and then,

$$w_1 = \sum_{i=1}^{\infty} \frac{1}{\lambda_i \rho_i}.$$

We summarize the discussion of this section in the following theorem:

Theorem 6.1. *Consider a birth and death process with birth and death parameters λ_n and $\mu_n, n \geq 1$, where $\lambda_0 = 0$ so that 0 is an absorbing state. The probability of absorption into state 0 from the initial state m is*

$$u_m = \begin{cases} \dfrac{\sum_{i=m}^{\infty} \rho_i}{1 + \sum_{i=1}^{\infty} \rho_i} & \text{if } \sum_{i=1}^{\infty} \rho_i < \infty, \\[2ex] 1 & \text{if } \sum_{i=1}^{\infty} \rho_i = \infty. \end{cases} \qquad (6.48)$$

The mean time to absorption is

$$w_m = \begin{cases} \infty & \text{if } \sum_{i=1}^{\infty} \dfrac{1}{\lambda_i \rho_i} = \infty, \\[2ex] \sum_{i=1}^{\infty} \dfrac{1}{\lambda_i \rho_i} + \sum_{k=1}^{m-1} \rho_k \sum_{j=k+1}^{\infty} \dfrac{1}{\lambda_j \rho_j} & \text{if } \sum_{i=1}^{\infty} \dfrac{1}{\lambda_i \rho_i} < \infty, \end{cases} \qquad (6.49)$$

where $\rho_0 = 1$ and $\rho_i = (\mu_1 \mu_2 \cdots \mu_i)/(\lambda_1 \lambda_2 \cdots \lambda_i)$.

Example *Population Processes* Consider the linear growth birth and death process without immigration (cf. Section 6.3.3) for which $\mu_n = n\mu$ and $\lambda_n = n\lambda, n = 0, 1, \ldots$. During a short time interval of length h, a *single individual* in the population dies with probability $\mu h + o(h)$ and gives birth to a new individual with probability $\lambda h + o(h)$, and thus, $\mu > 0$ and $\lambda > 0$ represent the *individual* death and birth rates, respectively.

Substitution of $a = 0$ and $i = m$ in equation (6.25) determines the mean population size at time t for a population starting with $X(0) = m$ individuals. This mean population size is $M(t) = me^{(\lambda-\mu)t}$, exhibiting exponential growth or decay depending on whether $\lambda > \mu$ or $\lambda < \mu$.

Let us now examine the extinction phenomenon and determine the probability that the population eventually dies out. This phenomenon corresponds to absorption in state 0 for the birth and death process.

When $\lambda_n = n\lambda$ and $\mu_n = n\mu$, a direct calculation yields $\rho_i = (\mu/\lambda)^i$, and then,

$$\sum_{i=m}^{\infty} \rho_i = \sum_{i=m}^{\infty} (\mu/\lambda)^i = \begin{cases} \dfrac{(\mu/\lambda)^m}{1 - (\mu/\lambda)} & \text{when } \lambda > \mu, \\ \infty & \text{when } \lambda \leq \mu. \end{cases}$$

From Theorem 6.1, the probability of eventual extinction starting with m individuals is

$$\Pr\{\text{Extinction}|X(0) = m\} = \begin{cases} (\mu/\lambda)^m & \text{when } \lambda > \mu, \\ 1 & \text{when } \lambda \leq \mu. \end{cases} \tag{6.50}$$

When $\lambda = \mu$, the process is sure to vanish eventually. Yet, in this case, the mean population size remains constant at the initial population level. Similar situations where mean values do not adequately describe population behavior frequently arise when stochastic elements are present.

We turn attention to the mean time to extinction assuming that extinction is certain, i.e., when $\lambda \leq \mu$. For a population starting with a single individual, then, from (6.49) with $m = 1$, we determine this mean time to be

$$\begin{aligned}
\sum_{i=1}^{\infty} \frac{1}{\lambda_i \rho_i} &= \frac{1}{\lambda} \sum_{i=1}^{\infty} \frac{1}{i} \left(\frac{\lambda}{\mu}\right)^i \\
&= \frac{1}{\lambda} \sum_{i=1}^{\infty} \int_0^{(\lambda/\mu)} x^{i-1} dx \\
&= \frac{1}{\lambda} \int_0^{(\lambda/\mu)} \sum_{i=1}^{\infty} x^{i-1} dx \\
&= \frac{1}{\lambda} \int_0^{(\lambda/\mu)} \frac{dx}{(1-x)}
\end{aligned} \tag{6.51}$$

$$
= -\frac{1}{\lambda} \ln(1-x) \Big|_0^{(\lambda/\mu)}
$$

$$
= \begin{cases} \dfrac{1}{\lambda} \ln\left(\dfrac{\mu}{\mu-\lambda}\right) & \text{when } \mu > \lambda, \\[2mm] \infty & \text{when } \mu = \lambda. \end{cases}
$$

When the birth rate λ exceeds the death rate μ, a linear growth birth and death process can, with strictly positive probability, grow without limit. In contrast, many natural populations exhibit density-dependent behavior wherein the individual birth rates decrease or the individual death rates increase or both changes occur as the population grows. These changes are ascribed to factors including limited food supplies, increased predation, crowding, and limited nesting sites. Accordingly, we introduce a notion of environmental *carrying capacity* K, an upper bound that the population size cannot exceed.

Since all individuals have a chance of dying, with a finite carrying capacity, all populations will eventually become extinct. Our measure of population fitness will be the mean time to extinction, and it is of interest to population ecologists studying colonization phenomena to examine how the capacity K, the birth rate λ, and the death rate μ affect this mean population lifetime.

The model should have the properties of exponential growth (on the average) for small populations, as well as the ceiling K beyond which the population cannot grow. There are several ways of approaching the population size K and staying there at equilibrium. Since all such models give more or less the same qualitative results, we stipulate the simplest model, in which the birth parameters are

$$
\lambda_n = \begin{cases} n\lambda & \text{for } n = 0, 1, \ldots, K-1, \\ 0 & \text{for } n > K. \end{cases}
$$

Theorem 6.1 yields w_1, the mean time to population extinction starting with a single individual, as given by

$$
w_1 = \sum_{i=1}^{\infty} \frac{1}{\lambda_i \rho_i} = \sum_{i=1}^{\infty} \frac{\lambda_1 \lambda_2 \cdots \lambda_{i-1}}{\mu_1 \mu_2 \cdots \mu_i} = \frac{1}{\mu} \sum_{i=1}^{K} \frac{1}{i} \left(\frac{\lambda}{\mu}\right)^{i-1}. \tag{6.52}
$$

Equation (6.52) isolates the distinct factors influencing the mean time to population extinction. The first factor is $1/\mu$, the mean lifetime of an individual, since μ is the individual death rate. Thus, the sum in (6.52) represents the mean *generations*, or mean lifespans, to population extinction, a dimensionless quantity that we denote by

$$
M_g = \mu w_1 = \sum_{i=1}^{K} \frac{1}{i} \theta^{i-1}, \quad \text{where } \theta = \frac{\lambda}{\mu}. \tag{6.53}
$$

Next, we examine the influence of the birth–death, or reproduction, ratio $\theta = \lambda/\mu$, and the carrying capacity K on the mean time to extinction. Since λ represents the individual birth rate and $1/\mu$ is the mean lifetime of a single member in the population, we may interpret the reproduction ratio $\theta = \lambda(1/\mu)$ as the mean number of offspring of an arbitrary individual in the population. Accordingly, we might expect significantly different behavior when $\theta < 1$ as opposed to when $\theta > 1$, and this is indeed the case. A carrying capacity of $K = 100$ is small. When K is on the order of 100 or more, we have the following accurate approximations, their derivations being sketched in Exercises 6.5.1 and 6.5.2:

$$M_g \approx \begin{cases} \dfrac{1}{\theta} \ln\left(\dfrac{1}{1-\theta}\right) & \text{for } \theta < 1, \\[2mm] 0.5772157 + \ln K & \text{for } \theta = 1, \\[2mm] \dfrac{1}{K}\left(\dfrac{\theta^K}{\theta - 1}\right) & \text{for } \theta > 1. \end{cases} \qquad (6.54)$$

The contrast between $\theta < 1$ and $\theta > 1$ is vivid. When $\theta < 1$, the mean generations to extinction M_g is almost independent of carrying capacity K and approaches the asymptotic value $\theta^{-1} \ln(1-\theta)^{-1}$ quite rapidly. When $\theta > 1$, the mean generations to extinction M_g grows exponentially in K. Some calculations based on (6.54) are given in Table 6.1.

Example *Sterile Male Insect Control*   The screwworm fly, a cattle pest in warm climates, was eliminated from the southeastern United States by the release into the environment of sterilized adult male screwworm flies. When these males, artificially sterilized by radiation, mate with native females, there are no offspring, and in this manner, part of the reproductive capacity of the natural population is nullified by their presence. If the sterile males are sufficiently plentiful so as to cause even a small decline in the population level, then this decline accelerates in succeeding generations even if the number of sterile males is maintained at approximately the same level because the ratio of sterile to fertile males will increase as the natural population drops. Because of this compounding effect, if the sterile male control method works at all, it works to such an extent as to drive the native population to extinction in the area in which it is applied.

Recently, a multibillion-dollar effort involving the sterile male technique has been proposed for the control of the cotton boll weevil. In this instance, it was felt that

Table 6.1 Mean generations to extinction for a population starting with a single parent and where θ is the reproduction rate and K is the environmental capacity

K	$\theta = 0.8$	$\theta = 1$	$\theta = 1.2$
10	1.96	2.88	3.10
100	2.01	5.18	4.14×10^4
1000	2.01	7.48	7.59×10^{76}

pretreatment with a pesticide could reduce the natural population size to a level such that the sterile male technique would become effective. Let us examine this assumption, first with a deterministic model and then in a stochastic setting.

For both models, we suppose that sexes are present in equal numbers, that sterile and fertile males are equally competitive, and that a constant number S of sterile males is present in each generation. In the deterministic case, if N_0 fertile males are in the parent generation and the N_0 fertile females choose mates equally likely from the entire male population, then the fraction $N_0/(N_0 + S)$ of these matings will be with fertile males and will produce offspring. Letting θ denote the number of male offspring that results from a fertile mating, we calculate the size N of the next generation according to

$$N_1 = \theta N_0 \left(\frac{N_0}{N_0 + S} \right). \tag{6.55}$$

For a numerical example, suppose that there are $N_0 = 100$ fertile males (and an equal number of fertile females) in the parent generation of the native population, and that $S = 100$ sterile male insects are released. If $\theta = 4$, meaning that a fertile mating produces four males (and four females) for the succeeding generation, then the number of both sexes in the first generation is

$$N_1 = 4(100) \left(\frac{100}{100 + 100} \right) = 200;$$

the population has increased, and the sterile male control method has failed.

On the other hand, if a pesticide can be used to reduce the initial population size to $N_0 = 20$, or 20% of its former level, and $S = 100$ sterile males are released, then

$$N_1 = 4(20) \left(\frac{20}{20 + 100} \right) = 13.33,$$

and the population is declining. The succeeding population sizes are given in Table 6.2, above. With the pretreatment, the population becomes extinct by the fourth generation.

Often deterministic or average value models will adequately describe the evolution of large populations. But extinction is a small population phenomenon, and even in the presence of significant long-term trends, small populations are strongly influenced by

Table 6.2 The trend of an insect population subject to sterile male releases

Generation	Number of Insects Natural Population	Number of Sterile Insects	Ratio Sterile to Fertile	Number of Progeny
Parent	20	100	5:1	13.33
F_1	13.33	100	7.5:1	6.27
F_2	6.27	100	16:1	1.48
F_3	1.48	100	67.5:1	0.09
F_4	0.09	100	1156:1	—

the chance fluctuations that determine which of extinction or recolonization will occur. This fact motivates us to examine a stochastic model of the evolution of a population in the presence of sterile males. The factors in our model are

λ, the individual birth rate;
μ, the individual death rate;
$\theta = \lambda/\mu$, the mean offspring per individual;
K, the carrying capacity of the environment;
S, the constant number of sterile males in the population;
m, the initial population size.

We assume that both sexes are present in equal numbers in the natural population and that $X(t)$, the number of either sex present at time t, evolves as a birth and death process with parameters

$$\lambda_n = \begin{cases} \lambda n\left(\dfrac{n}{n+S}\right) & \text{if } 0 \leq n < K, \\ 0 & \text{for } n \geq K, \end{cases}$$

and   (6.56)

$$\mu_n = \mu n \quad \text{for } n = 0, 1, \ldots.$$

This is the colonization model of the *Population Processes* example, modified in analogy with (6.55) by including in the birth rate the factor $n/(n+S)$ to represent the probability that a given mating will be fertile.

To calculate the mean time to extinction w_m as given in (6.49), we first use (6.56) to determine

$$\rho_k = \frac{\mu_1 \mu_2 \cdots \mu_k}{\lambda_1 \lambda_2 \cdots \lambda_k} = \left(\frac{\mu}{\lambda}\right)^k \frac{(k+S)!}{k! \, S!} \quad \text{for } k = 1, \ldots, K-1,$$

$$\rho_0 = 1, \quad \text{and} \quad \rho_K = \infty, \text{ or } 1/\rho_K = 0,$$

and then substitute these expressions for ρ_k into (6.49) to obtain

$$\begin{aligned} w_m &= \sum_{j=1}^{K} \frac{1}{\lambda_j \rho_j} + \sum_{k=1}^{m-1} \rho_k \sum_{j=k+1}^{K} \frac{1}{\lambda_j \rho_j} \\ &= \sum_{k=0}^{m-1} \rho_k \sum_{j=k+1}^{K} \frac{1}{\lambda_j \rho_j} = \sum_{k=0}^{m-1} \rho_k \sum_{j=k+1}^{K} \frac{1}{\mu_j \rho_{j-1}} \\ &= \frac{1}{\mu} \left\{ \sum_{k=0}^{m-1} \sum_{j=k}^{K-1} \frac{1}{j+1} \theta^{j-k} \frac{j! \, (S+k)!}{k! \, (S+j)!} \right\}. \end{aligned}$$ (6.57)

Because of the factorials, equation (6.57) presents numerical difficulties when direct computations are attempted. A simple iterative scheme works to provide accurate and effective computation, however. We let

$$\alpha_k = \sum_{j=k}^{K-1} \frac{1}{j+1} \theta^{j-k} \frac{j!\,(S+k)!}{k!\,(S+j)!}$$

so that $w_m = (\alpha_0 + \cdots + \alpha_{m-1})/\mu$. But, it is easily verified that

$$\alpha_{k-1} = \frac{1}{k} + \theta\left(\frac{k}{S+k}\right)\alpha_k.$$

Beginning with $\alpha_K = 0$, one successively computes $\alpha_{K-1}, \alpha_{K-2}, \ldots, \alpha_0$, and then $w_m = (\alpha_0 + \cdots + \alpha_{m-1})/\mu$.

Using this method, we have computed the mean generations to extinction in the stochastic model for comparison with the deterministic model as given in Table 6.3. Table 6.3 lists the mean generations to extinction for various initial population sizes m when $K = S = 100, \lambda = 4$, and $\mu = 1$ so that $\theta = 4$. Instead of the four generations to extinction as predicted by the deterministic model when $m = 20$, we now estimate that the population will persist for over 8 billion generations!

What is the explanation for the dramatic difference between the predictions of the deterministic model and the predictions of the stochastic model? The stochastic model allows the small but positive probability that the population will not die out but will recolonize and return to a higher level near the environmental capacity K and then persist for an enormous length of time.

While both models are qualitative, the practical implications cannot be dismissed. In any large-scale control effort, a wide range of habitats and microenvironments is bound to be encountered. The stochastic model suggests the likely possibility that some subpopulation in some pocket might persist and later recolonize the entire area.

Table 6.3 The mean lifespans to extinction in a birth and death model of a population containing a constant number $S = 100$ of sterile males

Initial Population Size	Mean Lifespans to Extinction
20	8,101,227,748
10	4,306,531
5	3,822
4	566
3	65
2	6.3
1	1.2

A sterile male program that depends on a pretreatment with an insecticide for its success is chancy at best.

Exercises

6.5.1 Assuming $\theta < 1$, verify the following steps in the approximation to M_g, the mean generation to extinction as given in (6.53):

$$M_g = \sum_{i=1}^{K} \frac{1}{i}\theta^{i-1} = \theta^{-1} \sum_{i=1}^{K} \int_0^\theta x^{i-1}\,dx$$

$$= \theta^{-1} \int_0^\theta \frac{1-x^K}{1-x}\,dx = \theta^{-1} \int_0^\theta \frac{dx}{1-x} - \theta^{-1} \int_0^\theta \frac{x^K}{1-x}\,dx$$

$$= \frac{1}{\theta}\ln\frac{1}{1-\theta} - \theta^{-1} \int_0^\theta x^K\left(1+x+x^2+\cdots\right)dx$$

$$= \frac{1}{\theta}\ln\frac{1}{1-\theta} - \frac{1}{\theta}\left(\frac{\theta^{K+1}}{K+1} + \frac{\theta^{K+2}}{K+2} + \cdots\right)$$

$$= \frac{1}{\theta}\ln\frac{1}{1-\theta} - \frac{\theta^K}{K+1}\left(1 + \frac{K+1}{K+2}\theta + \frac{K+1}{K+3}\theta^2 + \cdots\right)$$

$$\approx \frac{1}{\theta}\ln\frac{1}{1-\theta} - \frac{\theta^K}{(K+1)(1-\theta)}.$$

6.5.2 Assume that $\theta > 1$ and verify the following steps in the approximation to M_g, the mean generation to extinction as given in (6.53):

$$M_g = \sum_{i=1}^{K} \frac{1}{i}\theta^{i-1} = \theta^K \sum_{i=1}^{K} \frac{1}{i}\theta^{K-i+1}$$

$$= \theta^K \sum_{j=1}^{K} \frac{1}{K-j+1}\left(\frac{1}{\theta}\right)^j$$

$$= \frac{\theta^{K-1}}{K}\left[1 + \frac{K}{K-1}\left(\frac{1}{\theta}\right) + \frac{K}{K-2}\left(\frac{1}{\theta}\right)^2 + \cdots + \frac{K}{1}\left(\frac{1}{\theta}\right)^{K-1}\right]$$

$$\approx \frac{\theta^{K-1}}{K}\left[\frac{1}{1-(1/\theta)}\right] = \frac{\theta^K}{K(\theta-1)}.$$

Problems

6.5.1 Consider the sterile male control model as described in the example entitled *"Sterile Male Insect Control"* and let u_m be the probability that the population becomes extinct before growing to size K starting with $X(0) = m$ individuals. Show that

$$u_m = \frac{\sum_{i=m}^{K-1} \rho_i}{\sum_{i=0}^{K-1} \rho_i} \quad \text{for } m = 1, \ldots, K,$$

where

$$\rho_i = \theta^{-i} \frac{(S+i)!}{i!}.$$

6.5.2 Consider a birth and death process on the states $0, 1, \ldots, 5$ with parameters

$$
\begin{aligned}
&\lambda_0 = \mu_0 = \lambda_5 = \mu_5 = 0, \\
&\lambda_1 = 1, \quad \lambda_2 = 2, \quad \lambda_3 = 3, \quad \lambda_4 = 4, \\
&\mu_1 = 4, \quad \mu_2 = 3, \quad \mu_3 = 2, \quad \mu = 1.
\end{aligned}
$$

Note that 0 and 5 are absorbing states. Suppose the process begins in state $X(0) = 2$.
(a) What is the probability of eventual absorption in state 0?
(b) What is the mean time to absorption?

6.6 Finite-State Continuous Time Markov Chains

A continuous time Markov chain $X(t)(t > 0)$ is a Markov process on the states $0, 1, 2, \ldots$. We assume as usual that the transition probabilities are stationary; that is,

$$P_{ij}(t) = \Pr\{X(t+s) = j | X(s) = i\}. \tag{6.58}$$

In this section, we consider only the case where the state spaces S is finite, labeled as $\{0, 1, 2, \ldots, N\}$.

The Markov property asserts that $P_{ij}(t)$ satisfies

(a) $P_{ij}(t) \geq 0$,

(b) $\sum_{j=0}^{N} P_{ij}(t) = 1, \quad i, j = 0, 1, \ldots, N$, and

(c) $P_{ik}(s+t) = \sum_{j=0}^{N} P_{ij}(s) P_{jk}(t) \quad \text{for } t, s \geq 0$ (Chapman–Kolmogorov relation),

and we postulate in addition that

(d) $\lim_{t \to 0+} P_{ij}(t) = \begin{cases} 1, & i = j, \\ 0, & i \neq j. \end{cases}$

If $\mathbf{P}(t)$ denotes the matrix $\|P_{ij}(t)\|_{i,j=0}^{N}$, then property (c) can be written compactly in matrix notation as

$$\mathbf{P}(t+s) = \mathbf{P}(t)\mathbf{P}(s), \quad t, s \geq 0. \tag{6.59}$$

Property (d) asserts that $\mathbf{P}(t)$ is continuous at $t = 0$, since the fact $\mathbf{P}(0) = \mathbf{I}$ (= identity matrix) is implied by (6.59). It follows simply from (6.59) that $\mathbf{P}(t)$ is continuous for all $t > 0$. In fact, if $s = h > 0$ in (6.59), then because of (d), we have

$$\lim_{h \to 0+} \mathbf{P}(t+h) = \mathbf{P}(t) \lim_{h \to 0+} \mathbf{P}(h) = \mathbf{P}(t)\mathbf{I} = \mathbf{P}(t). \tag{6.60}$$

On the other hand, for $t > 0$ and $0 < h < t$, we write (6.59) in the form

$$\mathbf{P}(t) = \mathbf{P}(t-h)\mathbf{P}(h). \tag{6.61}$$

But $\mathbf{P}(h)$ is near the identity when h is sufficiently small, and so $\mathbf{P}(h)^{-1}$ [the inverse of $\mathbf{P}(h)$] exists and also approaches the identity $\mathbf{I}$. Therefore,

$$\mathbf{P}(t) = \mathbf{P}(t) \lim_{h \to 0+} (\mathbf{P}(h))^{-1} = \lim_{h \to 0+} \mathbf{P}(t-h). \tag{6.62}$$

The limit relations (6.60) and (6.62) together show that $\mathbf{P}(t)$ is continuous.

Actually, $\mathbf{P}(t)$ is not only continuous but also differentiable in that the limits

$$\lim_{h \to 0+} \frac{1 - P_{ii}(h)}{h} = q_i,$$

$$\lim_{h \to 0+} \frac{P_{ij}(h)}{h} = q_{ij}, \quad i \neq j, \tag{6.63}$$

exist, where $0 \leq q_{ij} < \infty (i \neq j)$ end $0 \leq q_i < \infty$. Starting with the relation

$$1 - P_{ii}(h) = \sum_{j=0, j \neq i}^{N} P_{ij}(h),$$

dividing by h, and letting h decrease to zero yields directly the relation

$$q_i = \sum_{j=0, j \neq i}^{N} q_{ij}.$$

The rates q_i and q_{ij} furnish an infinitesimal description of the process with

$$\Pr\{X(t+h) = j | X(t) = i\} = q_{ij}h + o(h) \quad \text{for } i \neq j,$$
$$\Pr\{X(t+h) = i | X(t) = i\} = 1 - q_i h + o(h).$$

In contrast to the infinitesimal description, the sojourn description of the process proceeds as follows: Starting in state i, the process sojourns there for a duration that is exponentially distributed with parameter q_i. The process then jumps to state $j \neq i$ with probability $p_{ij} = q_{ij}/q_i$; the sojourn time in state j is exponentially distributed with parameter q_j, and so on. The sequence of states visited by the process, denoted by $\xi_0, \xi_1, \ldots$, is a Markov chain with discrete parameter, called the *embedded Markov chain*. Conditioned on the state sequence $\xi_0, \xi_1, \ldots$, the successive sojourn times $S_0, S_1, \ldots$ are independent exponentially distributed random variables with parameters $q_{\xi_0}, q_{\xi_1}, \ldots$, respectively.

Assuming that (6.63) has been verified, we now derive an explicit expression for $P_{ij}(t)$ in terms of the infinitesimal matrix

$$
\mathbf{A} = \begin{Vmatrix}
-q_0 & q_{01} & \cdots & q_{0N} \\
q_{10} & -q_1 & & q_{1N} \\
\vdots & & & \\
q_{N0} & q_{N1} & \cdots & -q_N
\end{Vmatrix}.
$$

The limit relations (6.63) can be expressed concisely in matrix form:

$$
\lim_{h \to 0+} \frac{\mathbf{P}(h) - \mathbf{I}}{h} = \mathbf{A}, \tag{6.64}
$$

which shows that $\mathbf{A}$ is the matrix derivative of $\mathbf{P}(t)$ at $t = 0$. Formally, $\mathbf{A} = \mathbf{P}'(0)$.

With the aid of (6.64) and referring to (6.59), we have

$$
\frac{\mathbf{P}(t+h) - \mathbf{P}(t)}{h} = \frac{\mathbf{P}(t)[\mathbf{P}(h) - \mathbf{I}]}{h} = \frac{\mathbf{P}(h) - \mathbf{I}}{h}\mathbf{P}(t). \tag{6.65}
$$

The limit on the right exists, and this leads to the matrix differential equation

$$
\mathbf{P}'(t) = \mathbf{P}(t)\mathbf{A} = \mathbf{A}\mathbf{P}(t), \tag{6.66}
$$

where $\mathbf{P}'(t)$ denotes the matrix whose elements are $P'_{ij}(t) = dP_{ij}(t)/dt$. The existence of $P'_{ij}(t)$ is an obvious consequence of (6.64) and (6.65). The differential equations (6.66) under the initial condition $\mathbf{P}(0) = \mathbf{I}$ can be solved by standard methods to yield the formula

$$
\mathbf{P}(t) = e^{\mathbf{A}t} = \mathbf{I} + \sum_{n=1}^{\infty} \frac{\mathbf{A}^n t^n}{n!}. \tag{6.67}
$$

Example *The Two-State Markov Chain* Consider a Markov chain $\{X(t)\}$ with states $\{0, 1\}$ whose infinitesimal matrix is

$$
\mathbf{A} = \begin{matrix} & \begin{matrix} 0 & \ 1 \end{matrix} \\ \begin{matrix} 0 \\ 1 \end{matrix} & \begin{Vmatrix} -\alpha & \alpha \\ \beta & -\beta \end{Vmatrix} \end{matrix}.
$$

The process alternates between states 0 and 1. The sojourn times in state 0 are independent and exponentially distributed with parameter α. Those in state 1 are independent and exponentially distributed with parameter β. We carry out the matrix multiplication

$$\begin{Vmatrix} -\alpha & \alpha \\ \beta & -\beta \end{Vmatrix} \times \begin{Vmatrix} -\alpha & \alpha \\ \beta & -\beta \end{Vmatrix} = \begin{Vmatrix} \alpha^2 + \alpha\beta & -\alpha^2 - \alpha\beta \\ -\beta^2 - \alpha\beta & \beta^2 + \alpha\beta \end{Vmatrix}$$

$$= -(\alpha + \beta) \begin{Vmatrix} -\alpha & \alpha \\ \beta & -\beta \end{Vmatrix}$$

to see that $\mathbf{A}^2 = -(\alpha + \beta)\mathbf{A}$. Repeated multiplication by $\mathbf{A}$ then yields

$$\mathbf{A}^n = [-(\alpha + \beta)]^{n-1}\mathbf{A},$$

which when inserted into (6.67) simplifies the sum according to

$$\mathbf{P}(t) = \mathbf{I} - \frac{1}{\alpha + \beta} \sum_{n=1}^{\infty} \frac{[-(\alpha + \beta)t]^n}{n!} A$$

$$= \mathbf{I} - \frac{1}{\alpha + \beta} \left[e^{-(\alpha+\beta)t} - 1 \right] \mathbf{A}$$

$$= \mathbf{I} + \frac{1}{\alpha + \beta} \mathbf{A} - \frac{1}{\alpha + \beta} \mathbf{A} e^{-(\alpha+\beta)t}.$$

And with $\pi = \alpha/(\alpha + \beta)$ and $\tau = \alpha + \beta$,

$$\mathbf{P}(t) = \begin{Vmatrix} 1 - \pi & \pi \\ 1 - \pi & \pi \end{Vmatrix} + \begin{Vmatrix} \pi & -\pi \\ -(1 - \pi) & (1 - \pi) \end{Vmatrix} e^{-\tau t},$$

which is the matrix expression for equations (6.30a–d).

Returning to the general Markov chain on states $\{0, 1, \ldots, N\}$, when the chain is irreducible (all states communicate), then $P_{ij}(t) > 0$ for $i, j = 0, 1, \ldots, N$ and $\lim_{t \to \infty} P_{ij}(t) = \pi_j > 0$ exists independently of the initial state i. The limiting distribution may be found by passing to the limit in (6.66), noting that $\lim_{t \to \infty} \mathbf{P}'(t) = 0$. The resulting equations for $\boldsymbol{\pi} = (\pi_0, \pi_1, \ldots, \pi_N)$ are

$$0 = \boldsymbol{\pi}\mathbf{A} = (\pi_0, \pi_1, \ldots, \pi_N) \begin{Vmatrix} -q_0 & q_{01} & \cdots & q_{0N} \\ q_{10} & -q_1 & \cdots & q_{1N} \\ \vdots & \vdots & \cdots & \vdots \\ q_{N0} & q_{N1} & & -q_N \end{Vmatrix},$$

which is the same as

$$\pi_j q_j = \sum_{i \neq j} \pi_i q_{ij}, \quad j = 0, 1, \ldots, N. \tag{6.68}$$

Equation (6.68) together with

$$\pi_0 + \pi_1 + \cdots + \pi_N = 1 \tag{6.69}$$

determines the limiting distribution.

Equation (6.68) has a mass balance interpretation that aids us in understanding it. The left side $\pi_j q_j$ represents the long run rate at which particles executing the Markov process leave state j. This rate must equal the long run rate at which particles arrive at state j if equilibrium is to be maintained. Such arriving particles must come from some state $i \neq j$, and a particle moves from state $i \neq j$ to state j at rate q_{ij}. Therefore, the right side $\Sigma_{i \neq j} \pi_i q_{ij}$ represents the total rate of arriving particles.

Example *Industrial Mobility and the Peter Principle* Let us suppose that a draftsman position at a large engineering firm can be occupied by a worker at any of three levels: T = Trainee, J = Junior draftsman, and S = Senior draftsman. Let $X(t)$ denote the level of the person in the position at time t, and suppose that $X(t)$ evolves as a Markov chain whose infinitesimal matrix is

$$\mathbf{A} = \begin{array}{c} \\ T \\ J \\ S \end{array} \begin{array}{c} \begin{array}{ccc} T & J & S \end{array} \\ \left\| \begin{array}{ccc} -a_T & a_T & 0 \\ a_{JT} & -a_J & a_{JS} \\ a_S & 0 & -a_S \end{array} \right\| \end{array}.$$

Thus, a Trainee stays at that rank for an exponentially distributed time having parameter a_T and then becomes a Junior draftsman. A Junior draftsman stays at that level for an exponentially distributed length of time having parameter $a_J = a_{JT} + a_{JS}$. Then, the Junior draftsman leaves the position and is replaced by a Trainee with probability a_{JT}/a_J or is promoted to a Senior draftsman with probability a_{JS}/a_J and so on.

Alternatively, we may describe the model by specifying the movements during short time intervals according to

$$\Pr\{X(t+h) = J | X(t) = T\} = a_T h + o(h),$$
$$\Pr\{X(t+h) = T | X(t) = J\} = a_{JT} h + o(h),$$
$$\Pr\{X(t+h) = S | X(t) = J\} = a_{JS} h + o(h),$$
$$\Pr\{X(t+h) = T | X(t) = S\} = a_S h + o(h),$$

and

$$\Pr\{X(t+h) = i | X(t) = i\} = 1 - a_i h + o(h) \quad \text{for } i = T, J, S.$$

The equations for the equilibrium distribution (π_T, π_J, π_S) are, according to (6.68),

$$a_T \pi_T = \qquad a_{JT} \pi_J + a_S \pi_S,$$
$$a_J \pi_J = a_T \pi_T,$$
$$a_S \pi_S = \qquad a_{JS} \pi_J,$$
$$1 = \quad \pi_T + \qquad \pi_J + \quad \pi_S,$$

and the solution is

$$\pi_T = \frac{a_S a_J}{a_S a_J + a_S a_T + a_T a_{JS}},$$

$$\pi_J = \frac{a_S a_T}{a_S a_J + a_S a_T + a_T a_{JS}},$$

$$\pi_S = \frac{a_T a_{JS}}{a_S a_J + a_S a_T + a_T a_{JS}}.$$

Let us consider a numerical example for comparison with an alternative model to be developed later. We suppose that the mean times in the three states are

State	Mean Time
T	0.1
J	0.2
S	1.0

and that a Junior draftsman leaves and is replaced by a Trainee with probability $\frac{2}{5}$ and is promoted to a Senior draftsman with probability $\frac{3}{5}$. These suppositions lead to the prescription $a_T = 10, a_{JT} = 2, a_{JS} = 3$, and $a_S = 1$. The equilibrium probabilities are

$$\pi_T = \frac{1(5)}{1(5) + 1(10) + 10(3)} = \frac{5}{45} = 0.11,$$

$$\pi_J = \frac{10}{45} = 0.22,$$

$$\pi_S = \frac{30}{45} = 0.67.$$

But the duration that people spend in any given position is not exponentially distributed in general. A bimodal distribution is often observed in which many people leave rather quickly, while others persist for a substantial time. A possible explanation for this phenomenon is found in the "Peter Principle,"[2] which asserts that a worker is promoted until finally reaching a position in which he or she is incompetent. When this happens, the worker stays in that job until retirement. Let us modify the industrial mobility model to accommodate the Peter Principle by considering two types of Junior draftsmen, *Competent* and *Incompetent*. We suppose that a fraction p of Trainees are Competent and $q = 1 - p$ are Incompetent. We assume that a competent Junior draftsman stays at that level for an exponentially distributed duration with parameter a_C and then is promoted to Senior draftsman. Finally, an incompetent Junior draftsman stays in that position until retirement, an exponentially distributed sojourn with parameter

[2] Laurence, J.P., & Hull, R. (1969). The Peter Principle. Cutchogue, NY: Buccanear Books.

a_I, and then he or she is replaced by a Trainee. The relevant infinitesimal matrix is given by

$$
\mathbf{A} = \begin{array}{c} \\ T \\ I \\ C \\ S \end{array}
\begin{array}{c} \begin{array}{cccc} T & I & C & S \end{array} \\
\left\| \begin{array}{cccc}
-a_T & qa_T & pa_T & \\
a_I & -a_I & & \\
& & -a_C & a_C \\
a_S & & & -a_S
\end{array} \right\|.
\end{array}
$$

The duration in the Junior draftsman position now follows a probability law that is a mixture of exponential densities. To compare this model with the previous model, suppose that $p = \frac{3}{5}, q = \frac{2}{5}, a_I = 2.86$, and $a_C = 10$. These numbers were chosen so as to make the mean duration as a Junior draftsman,

$$
p\left(\frac{1}{a_C}\right) + q\left(\frac{1}{a_I}\right) = \left(\frac{3}{5}\right)(0.10) + \left(\frac{2}{5}\right)(0.35) = 0.20,
$$

the same as in the previous calculations. The probability density of this duration is

$$
f(t) = \frac{3}{5}(10)e^{-10t} + \frac{2}{5}(2.86)e^{-2.86t} \quad \text{for } t \geq 0.
$$

This density is plotted in Figure 6.7, for comparison with the exponential density $g(t) = 5e^{-5t}$, which has the same mean. The bimodal tendency is indicated in that $f(t) > g(t)$ when t is near zero and when t is very large.

With the numbers as given and $a_T = 10$ and $a_S = 1$ as before, the stationary distribution $(\pi_T, \pi_I, \pi_C, \pi_S)$ is found by solving

$$
\begin{aligned}
10\pi_T &= \quad 2.86\pi_I, \quad\quad 1\pi_S, \\
2.86\pi_I &= 4\pi_T, \\
10\pi_C &= 6\pi_T, \\
1\pi_S &= \quad\quad\quad 10\pi_C, \\
1 &= \pi_T + \quad \pi_I + \quad \pi_C + \pi_S.
\end{aligned}
$$

The solution is

$$
\begin{aligned}
\pi_T &= 0.111, \quad \pi_I = 0.155, \\
\pi_S &= 0.667, \quad \pi_C = 0.067.
\end{aligned}
$$

Let us make two observations before leaving this example. First, the limiting probabilities π_T, π_S, and $\pi_J = \pi_I + \pi_C$ agree between the two models. This is a common occurrence in stochastic modeling, wherein the limiting behavior of a process is rather insensitive to certain details of the model and depends only on the first moments, or means. When this happens, the model assumptions can be chosen for their mathematical convenience with no loss.

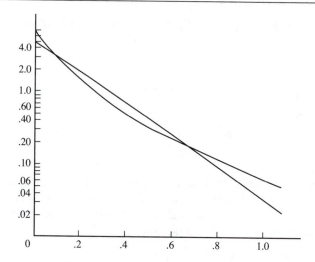

Figure 6.7 The exponential density (straight line) versus the mixed exponential density (curved line). Both distributions have the same mean. A logarithmic scale was used to accentuate the differences.

The second observation is specific to the Peter Principle. We have assumed that $p = \frac{3}{5}$ of Trainees are competent Junior draftsmen and only $q = \frac{2}{5}$ are Incompetent. Yet in the long run, a Junior draftsman is found to be Incompetent with probability $\pi_I/(\pi_I + \pi_C) = 0.155/(0.155 + 0.067) = 0.70!$

Example *Redundancy and the Burn-In Phenomenon* An airlines reservation system has two computers, one online and one backup. The operating computer fails after an exponentially distributed duration having parameter μ and is replaced by the standby. There is one repair facility, and repair times are exponentially distributed with parameter λ. Let $X(t)$ be the number of computers in operating condition at time t. Then, $X(t)$ is a Markov chain whose infinitesimal matrix is

$$\mathbf{A} = \begin{array}{c} \\ 0 \\ 1 \\ 2 \end{array} \begin{array}{c} \begin{array}{ccc} 0 & 1 & 2 \end{array} \\ \left\| \begin{array}{ccc} -\lambda & \lambda & 0 \\ \mu & -(\lambda+\mu) & \lambda \\ 0 & \mu & -\mu \end{array} \right\| \end{array}.$$

The stationary distribution (π_0, π_1, π_2) satisfies

$$\begin{aligned} \lambda\pi_0 &= \mu\pi_1, \\ (\lambda+\mu)\pi_1 &= \lambda\pi_0, \quad +\mu\pi_2, \\ \mu\pi_2 &= \lambda\pi_1, \\ 1 &= \pi_0 + \pi_1 + \pi_2, \end{aligned}$$

and the solution is

$$\pi_0 = \frac{1}{1 + (\lambda/\mu) + (\lambda/\mu)^2},$$

$$\pi_1 = \frac{\lambda/\mu}{1 + (\lambda/\mu) + (\lambda/\mu)^2},$$

$$\pi_2 = \frac{(\lambda/\mu)^2}{1 + (\lambda/\mu) + (\lambda/\mu)^2}.$$

The availability, or probability that at least one computer is operating, is $1 - \pi_0 = \pi_1 + \pi_2$.

Often, in practice, the assumption of exponentially distributed operating times is not realistic because of the so-called *burn-in phenomenon*. This idea is best explained in terms of the *hazard* rate $r(t)$ associated with a probability density function $f(t)$ of a nonnegative failure time T. Recall that $r(t)\Delta t$ measures the conditional probability that the item fails in the next time interval $(t, t + \Delta t)$ given that it has survived up to time t, and therefore, we have

$$r(t) = \frac{f(t)}{1 - F(t)} \quad \text{for } t \geq 0,$$

where $F(t)$ is the cumulative distribution function associated with the probability density function $f(t)$.

A constant hazard rate $r(t) = \lambda$ for all t corresponds to the exponential density function $f(t) = \lambda e^{-\lambda t}$ for $t \geq 0$. The burn-in phenomenon is described by a hazard rate that is initially high and then decays to a constant level, where it persists, possibly later to rise again (aging). It corresponds to a situation in which a newly manufactured or newly repaired item has a significant probability of failing early in its use. If the item survives this test period, however, it then operates in an exponential or memoryless manner. The early failures might correspond to incorrect manufacture or faulty repair, or might be a property of the materials used.

Anyone familiar with automobile repairs has experienced the burn-in phenomenon.

One of many possible ways to model the burn-in phenomenon is to use a mixture of exponential densities

$$f(t) = p\alpha e^{-\alpha t} + q\beta e^{-\beta t}, \quad t \geq 0, \tag{6.70}$$

where $0 < p = 1 - q < 1$ and α, β are positive. The density function for which $p = 0.1, \alpha = 10, q = 0.9$, and $\beta = 0.909 \cdots = 1/1.1$ has mean one. Its hazard rate is plotted in Figure 6.8, where the higher initial burn-in level is evident.

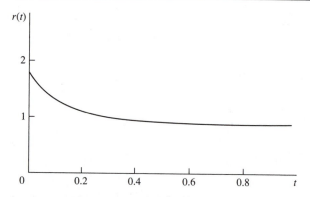

Figure 6.8 The hazard rate corresponding to the density given in (6.70). The higher hazard rate at the initial t values represents the burn-in phenomenon.

We may incorporate the burn-in phenomenon corresponding to the mixed exponential density (6.70) by expanding the state space according to the following table:

Notation	State
0	Both computers down
1_A	One operating computer, current up time has parameter α
1_B	One operating computer, current up time has parameter β
2_A	Two operating computers, current up time has parameter α
2_B	Two operating computers, current up time has parameter β

Equation (6.70) corresponds to a probability p that a computer beginning operation will have an exponentially distributed up time with parameter α and a probability q that the parameter is β. Accordingly, we have the infinitesimal matrix

$$
\mathbf{A} = \begin{array}{c} \\ 0 \\ 1_A \\ 1_B \\ 2_A \\ 2_B \end{array}
\begin{array}{c} \begin{array}{ccccc} 0 & 1_A & 1_B & 2_A & 2_B \end{array} \\
\left\| \begin{array}{ccccc}
-\lambda & p\lambda & q\lambda & & \\
\alpha & -(\lambda+\alpha) & & \lambda & \\
\beta & & -(\lambda+\beta) & & \lambda \\
& p\alpha & q\alpha & -\alpha & \\
& p\beta & q\beta & & -\beta
\end{array} \right\|
\end{array}.
$$

The stationary distribution can be determined in the usual way by applying (6.68).

Exercises

6.6.1 A certain type component has two states: $0 = $ OFF and $1 = $ OPERATING. In state 0, the process remains there a random length of time, which is exponentially distributed with parameter α, and then moves to state 1. The time in state 1 is exponentially distributed with parameter β, after which the process returns to state 0.

The *system* has two of these components, A and B, with distinct parameters:

Component	Operating Failure Rate	Repair Rate
A	β_A	α_A
B	β_B	α_B

In order for the *system* to operate, at least one of components A and B must be operating (a parallel system). Assume that the component stochastic processes are independent of one another. Determine the long run probability that the system is operating by

(a) Considering each component separately as a two-state Markov chain and using their statistical independence;

(b) Considering the system as a four-state Markov chain and solving equations (6.68).

6.6.2 Let $X_1(t)$ and $X_2(t)$ be independent two-state Markov chains having the same infinitesimal matrix

$$
\mathbf{A} = \begin{array}{c} \\ 0 \\ 1 \end{array} \begin{array}{cc} 0 & 1 \\ \left\| \begin{array}{cc} -\lambda & \lambda \\ \mu & -\mu \end{array} \right\| \end{array}.
$$

Argue that $Z(t) = X_1(t) + X_2(t)$ is a Markov chain on the state space $S = \{0, 1, 2\}$ and determine the transition probability matrix $\mathbf{P}(t)$ for $Z(t)$.

Problems

6.6.1 Let $Y_n, n = 0, 1, \ldots$, be a discrete time Markov chain with transition probabilities $\mathbf{P} = \|P_{ij}\|$, and let $\{N(t); t \geq 0\}$ be an independent Poisson process of rate λ. Argue that the compound process

$$
X(t) = Y_{N(t)}, \quad t \geq 0,
$$

is a Markov chain in continuous time and determine its infinitesimal parameters.

6.6.2 A certain type component has two states: $0 = $ OFF and $1 = $ OPERATING. In state 0, the process remains there a random length of time, which is exponentially distributed with parameter α, and then moves to state 1. The time in state 1 is exponentially distributed with parameter β, after which the process returns to state 0.

The system has three of these components, A, B, and C, with distinct parameters:

Component	Operating Failure Rate	Repair Rate
A	β_A	α_A
B	β_B	α_B
C	β_C	α_C

In order for the *system* to operate, component A must be operating, and at least one of components B and C must be operating. In the long run, what fraction of time does the system operate? Assume that the component stochastic processes are independent of one another.

6.6.3 Let $X_1(t), X_2(t), \dots, X_N(t)$ be independent two-state Markov chains having the same infinitesimal matrix

$$\mathbf{A} = \begin{array}{c} \\ 0 \\ 1 \end{array} \begin{array}{cc} 0 & 1 \\ \left\| \begin{array}{cc} -\lambda & \lambda \\ \mu & -\mu \end{array} \right\|. \end{array}$$

Determine the infinitesimal matrix for the Markov chain $Z(t) = X_1(t) + \cdots + X_N(t)$.

6.6.4 A system consists of two units, both of which may operate simultaneously, and a single repair facility. The probability that an operating system will fail in a short time interval of length Δt is $\mu(\Delta t) + o(\Delta t)$. Repair times are exponentially distributed, but the parameter depends on whether the failure was *regular* or *severe*. The fraction of regular failures is p, and the corresponding exponential parameter is α. The fraction of severe failure is $q = 1 - p$, and the exponential parameter is $\beta < \alpha$.

Model the system as a continuous time Markov chain by taking as states the pairs (x, y), where $x = 0, 1, 2$ is the number of units operating and $y = 0, 1, 2$ is the number of units undergoing repair for a severe failure. The possible states are $(2, 0), (1, 0), (1, 1), (0, 0), (0, 1)$, and $(0, 2)$. Specify the infinitesimal matrix $\mathbf{A}$. Assume that the units enter the repair shop on a first come, first served basis.

6.7 A Poisson Process with a Markov Intensity[3]

Consider "points" scattered in some manner along the semi-infinite interval $[0, \infty)$, and for an interval of the form $I = (a, b]$, with $0 \le a < b < \infty$, let $N(I)$ count the number of "points" in the interval I. Then $N(I)$, as I ranges over the half-open intervals $I = (a, b]$, is a point process. (See Chapter 5, Section 5.5 for generalizations to higher dimensions.) Suppose that, conditional on a given intensity, $N(I)$ is a nonhomogeneous Poisson process, but where the intensity function $\{\lambda(t), t \ge 0\}$ is itself a stochastic process. Such point processes were introduced in Chapter 5, Section 5.1.4, where they were called *Cox processes* in honor of their discoverer. While Cox processes are sufficiently general to describe a plethora of phenomena, they remain simple enough to permit explicit calculation, at least in some instances. As an illustration, we will derive the probability of no points in an interval for a Cox process in which the intensity function is a two-state Markov chain in continuous time. The Cox process alternates between being "ON" and "OFF." When the underlying intensity is "ON," points occur according to a Poisson process of constant intensity λ. When the

---

[3] Starred sections contain material of a more specialized or advanced nature.

underlying process is "OFF," no points occur. We will call this basic model a $(0, \lambda)$ *Cox process* to distinguish it from a later generalization. The $(0, \lambda)$ Cox process might describe bursts of rainfall in a locale that alternates between dry spells and wet ones, or arrivals to a queue from a supplier that randomly shuts down. Rather straightforward extensions of the techniques that we will now use in this simple case can be adapted to cover more complex models and computations, as we will subsequently show.

We assume that the intensity process $\{\lambda(t); t \geq 0\}$ is a two-state Markov chain in continuous time for which

$$\Pr\{\lambda(t+h) = \lambda | \lambda(t) = 0\} = \alpha h + o(h), \tag{6.71}$$

and

$$\Pr\{\lambda(t+h) = 0 | \lambda(t) = \lambda\} = \beta h + o(h). \tag{6.72}$$

This intensity is merely the constant λ times the two-state birth and death process of Section 6.3. As may be seen by allowing $t \to \infty$ in (6.29), such a process has the limiting distribution $\Pr\{\lambda(\infty) = 0\} = \beta/(\alpha + \beta)$ and $\Pr\{\lambda(\infty) = \lambda\} = \alpha/(\alpha + \beta)$. We will assume that the intensity process begins with this limiting distribution, or explicitly, that $\Pr\{\lambda(0) = 0\} = \beta/(\alpha + \beta)$ and $\Pr\{\lambda(0) = \lambda\} = \alpha/(\alpha + \beta)$. With this assumption, the intensity process is stationary in the sense that $\Pr\{\lambda(t) = 0\} = \beta/(\alpha + \beta)$ and $\Pr\{\lambda(t) = \lambda\} = \alpha/(\alpha + \beta)$ for all $t \geq 0$. This stationarity carries over to the Cox process $N(I)$ to imply that $\Pr\{N((0, t]) = k\} = \Pr\{N((s, s + t]) = k\}$ for all $s, t \geq 0$ and $k = 0, 1, \ldots$. We are interested in determining

$$f(t; \lambda) = \Pr\{N((0, t]) = 0\}.$$

Let

$$\Lambda(t) = \int_0^t \lambda(s) \, ds \tag{6.73}$$

and note the conditional Poisson probability

$$\Pr\{N((0, t]) = 0 | \lambda(s) \quad \text{for } s \leq t\} = e^{-\Lambda(t)}$$

so that upon removing the conditioning via the law of total probability, we obtain

$$f(t; \lambda) = E\left[e^{-\Lambda(t)}\right] = f_0(t) + f_1(t), \tag{6.74}$$

where

$$f_0(t) = \Pr\{N((0, t]) = 0 \quad \text{and} \quad \lambda(t) = 0\}, \tag{6.75}$$

and

$$f_1(t) = \Pr\{N((0, t]) = 0 \quad \text{and} \quad \lambda(t) = \lambda\}. \tag{6.76}$$

Using an infinitesimal "last step analysis" similar to that used to derive the Kolmogorov forward equations, we will derive a pair of first-order linear differential equations for $f_0(t)$ and $f_1(t)$. To this end, by analyzing the possibilities at time t and using the law of total probability, we begin with

$$\begin{aligned}
f_0(t + h) &= f_0(t) \Pr\{N((t, t + h]) = 0 | \lambda(t) = 0\} \Pr\{\lambda(t + h) = 0 | \lambda(t) = 0\} \\
&\quad + f_1(t) \Pr\{N((t, t + h]) = 0 | \lambda(t) = \lambda\} \Pr\{\lambda(t + h) = 0 | \lambda(t) = \lambda\} \\
&= f_0(t)[1 - \alpha h + o(h)] + f_1(t) e^{-\lambda h} \beta h + o(h),
\end{aligned}$$

and

$$\begin{aligned}
f_1(t + h) &= f_1(t) \Pr\{N((t, t + h]) = 0 | \lambda(t) = \lambda\} \Pr\{\lambda(t + h) = \lambda | \lambda(t) = \lambda\} \\
&\quad + f_0(t) \Pr\{N((t, t + h]) = 0 | \lambda(t) = 0\} \Pr\{\lambda(t + h) = \lambda | \lambda(t) = 0\} \\
&= f_1(t) e^{-\lambda h}[1 - \beta h + o(h)] + f_0(t) \alpha h + o(h).
\end{aligned}$$

We rearrange the terms and use $e^{-\lambda h} = 1 - \lambda h + o(h)$ to get

$$f_0(t + h) - f_0(t) = -\alpha f_0(t) h + \beta f_1(t) h + o(h)$$

and

$$f_1(t + h) - f_1(t) = -(\beta + \lambda) f_1(t) h + \alpha f_0(t) h + o(h),$$

which, after dividing by h and letting h tend to zero, become the differential equations

$$\frac{df_0(t)}{dt} = -\alpha f_0(t) + \beta f_1(t) \tag{6.77}$$

and

$$\frac{df_1(t)}{dt} = -(\beta + \lambda) f_1(t) + \alpha f_0(t). \tag{6.78}$$

The initial conditions are

$$f_0(0) = \Pr\{\lambda(0) = 0\} = \beta/(\alpha + \beta) \tag{6.79}$$

and

$$f_1(0) = \Pr\{\lambda(0) = \lambda\} = \alpha/(\alpha + \beta). \tag{6.80}$$

Such coupled first-order linear differential equations are readily solved. In our case, after carrying out the solution and simplifying the result, the answer is $\Pr\{N((0,t]) = 0\} = f_0(t) + f_1(t) = f(t; \lambda)$, where

$$f(t; \lambda) = c_+ \exp\{-\mu_+ t\} + c_- \exp\{-\mu_- t\} \tag{6.81}$$

with

$$\mu_\pm = \frac{1}{2} \left\{ (\lambda + \alpha + \beta) \pm \sqrt{(\lambda + \alpha + \beta)^2 - 4\alpha\lambda} \right\}, \tag{6.82}$$

$$c_+ = \frac{[\alpha\lambda/(\alpha + \beta)] - \mu_-}{\mu_+ - \mu_-} \tag{6.83}$$

and

$$c_- = \frac{\mu_+ - [\alpha\lambda/(\alpha + \beta)]}{\mu_+ - \mu_-}. \tag{6.84}$$

A Generalization Let N be a Cox process driven by a 0–1 Markov chain $\lambda(t)$, but now suppose that when the intensity process is in state 0, the Cox process is, conditionally, a Poisson process of rate λ_0, and when the intensity process is in state 1, then the Cox process is a Poisson process of rate λ_1. The earlier Cox process had $\lambda_0 = 0$ and $\lambda_1 = \lambda$. Without loss of generality, we assume $0 < \lambda_0 < \lambda_1$.

In order to evaluate $\Pr\{N((0,t]) = 0\}$, we write N as the sum $N = N_1 + N_2$ of two independent processes, where N_1 is a Poisson process of constant rate λ_0 and N_2 is a $(0, \lambda)$ Cox process with $\lambda = \lambda_1 - \lambda_0$. Then, N is zero if and only if both N_1 and N_2 are zero, whence

$$\Pr\{N((0,t]) = 0\} = \Pr\{N_1((0,t]) = 0\} \cdot \Pr\{N_2((0,t]) = 0\}$$
$$= e^{-\lambda_0 t} f(t, \lambda_1 - \lambda_0). \tag{6.85}$$

Example The tensile strength $S(t)$ of a single fiber of length t is often assumed to follow a Weibull distribution of the form

$$\Pr\{S(t) > x\} = \exp\left\{-t\sigma x^\delta\right\}, \quad \text{for } x > 0, \tag{6.86}$$

where δ and σ are positive material constants. The explicit appearance of the length t in the exponent is an expression of a weakest-link size effect, in which the fiber strength is viewed as the minimum strength of independent sections. This theory suggests that the survivorship probability of strength for a fiber of length t should satisfy the relation

$$\Pr\{S(t) > x\} = [\Pr\{S(1) > x\}]^t, \quad t > 0. \tag{6.87}$$

The Weibull distribution is the only type of distribution that is concentrated on $0 \leq x < \infty$ and satisfies (6.87).

However, a fiber under stress may fail from a surface flaw such as a notch or scratch, or from an internal flaw such as a void or inclusion. Where the diameter d of the fiber varies along its length, the relative magnitude of these two types of flaws will also vary, since the surface of the fiber is proportional to d, while the volume is proportional to d^2. As a simple generalization, suppose that the two types of flaws alternate and that the changes from one flaw type to the other follow a two-state Markov chain along the continuous length of the fiber. Further, suppose that a fiber of constant type i flaw, for $i = 0, 1$, will support the load x with probability

$$\Pr\{S(t) > x\} = \exp\{-t\sigma_i x^{\delta_i}\}, \quad x > 0,$$

where σ_i and δ_i are positive constants.

We can evaluate the survivorship probability for the fiber having Markov varying flaw types by bringing in an appropriate (λ_0, λ_1) Cox process. For a fixed $x > 0$, suppose that flaws that are weaker than x will occur along a fiber of constant flaw type i according to a Poisson process of rate $\lambda_i(x) = \sigma_i x^{\delta_i}$, for $i = 0, 1$. A fiber of length t and having Markov varying flaw types will carry a load of x if and only if there are no flaws weaker than x along the fiber. Accordingly, for the random flaw type fiber, using (6.85), we have

$$\Pr\{S(t) > x\} = e^{-\lambda_0(x)t} f(t; \lambda_1(x) - \lambda_0(x)). \tag{6.88}$$

Equation (6.88) may be evaluated numerically under a variety of assumptions for comparison with observed fiber tensile strengths. Where fibers having two flaw types are tested at several lengths, (6.88) may be used to extrapolate and predict strengths at lengths not measured.

It is sometimes more meaningful to reparametrize according to $\pi = \alpha/(\alpha + \beta)$ and $\tau = \alpha + \beta$. Here, π is the long run fraction of fiber length for which the applicable flaw distribution is of type 1, and $1 - \pi$ is the similar fraction of type 0 flaw behavior. The second parameter τ is a measure of the rapidity with which the flaw types alternate. In particular, when $\tau = 0$, the diameter or flaw type remains in whichever state it began, and the survivor probability reduces to the mixture

$$\Pr\{S(t) > x\} = \pi e^{-\lambda_1(x)t} + (1 - \pi)e^{-\lambda_0(x)t}. \tag{6.89}$$

On the other hand, at $\tau = \infty$, the flaw type process alternates instantly, and the survivor probability simplifies to

$$\Pr\{S(t) > x\} = \exp\{-t[\pi\lambda_1(x) + (1 - \pi)\lambda_0(x)]\}. \tag{6.90}$$

The probability distribution for $N((0, t])$ Let $\Lambda(t)$ be the cumulative intensity for a Cox process and suppose that we have evaluated

$$g(t; \theta) = E\left[e^{-(1-\theta)\Lambda(t)}\right], \quad 0 < \theta < 1.$$

For a $(0, \lambda)$ Cox process, for instance, $g(t, \theta) = f(t; (1 - \theta)\lambda)$, where f is defined in (6.81). Upon expanding as a power series in θ, according to

$$
\begin{aligned}
g(t; \theta) &= E\left[e^{-\Lambda(t)} \sum_{k=0}^{\infty} \frac{\Lambda(t)^k}{k!} \theta^k \right] \\
&= \sum_{k=0}^{\infty} E\left[e^{-\Lambda(t)} \frac{\Lambda(t)^k}{k!} \right] \theta^k \\
&= \sum_{k=0}^{\infty} \Pr\{N((0, t]) = k\} \theta^k,
\end{aligned}
$$

we see that the coefficient of θ^k in the power series is $\Pr\{N((0, t]) = k\}$. In principle then, the probability distribution for the points in an interval in a Cox process can be determined in any particular instance.

Exercises

6.7.1 Suppose that a $(0, \lambda)$ Cox process has $\alpha = \beta = 1$ and $\lambda = 2$. Show that $\mu_{\pm} = 2 \pm \sqrt{2}$ and $c_- = \frac{1}{4}(2 + \sqrt{2}) = 1 - c_+$, whence

$$
\Pr\{N((0, t]) = 0\} = e^{-2t}\left[\cosh(\sqrt{2}t) + \frac{\sqrt{2}}{2} \sinh(\sqrt{2}t) \right].
$$

6.7.2 Suppose that a $(0, \lambda)$ Cox process has $\alpha = \beta = 1$ and $\lambda = 2$. Show that

$$
f_0(t) = \frac{1 + \sqrt{2}}{4} e^{-(2 - \sqrt{2})t} + \frac{1 - \sqrt{2}}{4} e^{-(2 + \sqrt{2})t}
$$

and

$$
f_1(t) = \frac{1}{4} e^{-(2 - \sqrt{2})t} + \frac{1}{4} e^{-(2 + \sqrt{2})t}
$$

satisfy the differential equations (6.77) and (6.78) with the initial conditions (6.79) and (6.80).

Problems

6.7.1 Consider a stationary Cox process driven by a two-state Markov chain. Let $\pi = \alpha/(\alpha + \beta)$ be the probability that the process begins in state λ.
  (a) By using the transition probabilities given in (6.30a–d), show that $\Pr\{\lambda(t) = \lambda\} = \pi$ for all $t > 0$.
  (b) Show that $E[N((0, t])] = \pi \lambda t$ for all $t > 0$.

6.7.2 The *excess life* $\gamma(t)$ in a point process is the random length of the duration from time t until the next event. Show that the cumulative distribution function for the excess life in a Cox process is given by $\Pr\{\gamma(t) \leq x\} = 1 - \Pr\{N((t, t + x]) = 0\}$.

6.7.3 Let T be the time to the first event in a stationary $(0, \lambda)$ Cox process. Find the probability density function $\phi(t)$ for T. Show that when $\alpha = \beta = 1$ and $\lambda = 2$, this density function simplifies to $\phi(t) = \exp\{-2t\} \cosh(\sqrt{2}t)$.

6.7.4 Let T be the time to the first event in a stationary $(0, \lambda)$ Cox process. Find the expected value $E[T]$. Show that $E[T] = \frac{3}{2}$ when $\alpha = \beta = 1$ and $\lambda = 2$. What is the average duration between events in this process?

6.7.5 Determine the conditional probability of no points in the interval $(t, t + s]$, given that there are no points in the interval $(0, t]$ for a stationary Cox process driven by a two-state Markov chain. Establish the limit

$$\lim_{t \to \infty} \Pr\{N((t, t + s]) = 0 | N((0, t]) = 0\} = e^{-\mu - s}, \quad s > 0.$$

6.7.6 Show that the Laplace transform

$$\phi(s; \lambda) = \int_0^\infty e^{-st} f(t; \lambda) dt$$

is given by

$$\phi(s; \lambda) = \frac{s + (1 - \pi)\lambda + \tau}{s^2 + (\tau + \lambda)s + \pi\tau\lambda},$$

where $\tau = \alpha + \beta$ and $\pi = \alpha/(\alpha + \beta)$. Evaluate the limit (a) as $\tau \to \infty$, and (b) as $\tau \to 0$.

6.7.7 Consider a $(0, \lambda)$ stationary Cox process with $\alpha = \beta = 1$ and $\lambda = 2$. Show that $g(t; \theta) = f(t; (1 - \theta)\lambda)$ is given by

$$g(t; \theta) = e^{-(2 - \theta)t} \left\{ \cosh(Rt) + \frac{1}{R} \sinh(Rt) \right\},$$

where

$$R = \sqrt{\theta^2 - 2\theta + 2}.$$

Use this to evaluate $\Pr\{N((0, 1]) = 1\}$.

6.7.8 Consider a stationary $(0, \lambda)$ Cox process. A long duration during which no events were observed would suggest that the intensity process is in state 0. Show that

$$\Pr\{\lambda(t) = 0 | N((0, t]) = 0\} = \frac{f_0(t)}{f(t)},$$

where $f_0(t)$ is defined in (6.75).

6.7.9 Show that

$$f_0(t) = a_+ e^{-\mu_+ t} + a_- e^{-\mu_- t}$$

and

$$f_1(t) = b_+ e^{-\mu_+ t} + b_- e^{-\mu_- t}$$

with

$$a_\pm = \frac{1}{2}(1 - \pi)\left[1 \mp \frac{(\alpha + \beta + \lambda)}{R}\right], R = \sqrt{(\alpha + \beta + \lambda)^2 - 4\alpha\lambda}$$

$$b_\pm = \frac{1}{2}\pi\left[1 \pm \frac{(\lambda - \alpha - \beta)}{R}\right]$$

satisfy the differential equations (6.77) and (6.78) subject to the initial conditions (6.79) and (6.80).

6.7.10 Consider a stationary $(0, \lambda)$ Cox process.

(a) Show that

$$\Pr\{N((0, h]) > 0, N((h, h + t]) = 0\} = f(t; \lambda) - f(t + h; \lambda),$$

whence

$$\Pr\{N((h, h + t]) = 0 | N((0, h]) > 0\} = \frac{f(t; \lambda) - f(t + h; \lambda)}{1 - f(h; \lambda)}.$$

(b) Establish the limit

$$\lim_{h \to 0} \Pr\{N((h, h + t]) = 0 | N((0, h]) > 0\} = \frac{f'(t; \lambda)}{f'(0; \lambda)},$$

where

$$f'(t; \lambda) = \frac{df(t; \lambda)}{dt}.$$

(c) We interpret the limit in (b) as the conditional probability

$$\Pr\{N((0, t]) = 0 | \text{Event occurs at time } 0\}.$$

Show that

$$\Pr\{N((0, t]) = 0 | \text{Event at time } 0\} = p_+ e^{-\mu_+ t} + p_- e^{-\mu_- t},$$

where

$$p_+ = \frac{c_+ \mu_+}{c_+ \mu_+ + c_- \mu_-}, \quad p_- = \frac{c_- \mu_-}{c_+ \mu_+ + c_- \mu_-}.$$

(d) Let τ be the time to the first event in $(0, \infty)$ in a stationary $(0, \lambda)$ Cox process with $\alpha = \beta = 1$ and $\lambda = 2$. Show that

$$E[\tau | \text{Event at time } 0] = 1.$$

Why does this differ from the result in Problem 6.7.4?

6.7.11 *A Stop-and-Go Traveler* The velocity $V(t)$ of a stop-and-go traveler is described by a two-state Markov chain. The successive durations in which the traveler is stopped are independent and exponentially distributed with parameter α, and they alternate with independent exponentially distributed sojourns, parameter β, during which the traveler moves at unit speed. Take the stationary case in which $\Pr\{V(0) = 1\} = \pi = \alpha/(\alpha + \beta)$. The distance traveled in time t is the integral of the velocity:

$$S(t) = \int_0^t V(u)du.$$

Show that

$$E\left[e^{-\theta S(t)}\right] = f(t; \theta), \quad \theta \text{ real}.$$

(This is the Laplace transform of the probability density function of $S(t)$.)

6.7.12 Let τ be the time of the first event in a $(0, \lambda)$ Cox process. Let the 0 and λ states represent "OFF" and "ON," respectively.

(a) Show that the total duration in the $(0, \tau]$ interval that the system is ON is exponentially distributed with parameter λ and does not depend on α, β, or the starting state.

(b) Assume that the process begins in the OFF state. Show that the total duration in the $(0, \tau]$ interval that the system is OFF has the same distribution as

$$\sum_{k=0}^{N(\vartheta)} \eta_k,$$

where ζ is exponentially distributed with parameter λ, $N(t)$ is a Poisson process with parameter β, and $\eta_0, \eta_1, \dots$ are independent and exponentially distributed with parameter α.

7 Renewal Phenomena

7.1 Definition of a Renewal Process and Related Concepts

Renewal theory began with the study of stochastic systems whose evolution through time was interspersed with renewals or regeneration times when, in a statistical sense, the process began anew. Today, the subject is viewed as the study of general functions of independent, identically distributed, nonnegative random variables representing the successive intervals between renewals. The results are applicable in a wide variety of both theoretical and practical probability models.

A *renewal (counting) process* $\{N(t), t \geq 0\}$ is a nonnegative integer-valued stochastic process that registers the successive occurrences of an event during the time interval $(0, t]$, where the times between consecutive events are *positive, independent, identically distributed* random variables. Let the successive durations between events be $\{X_k\}_{k=1}^{\infty}$ (often representing the lifetimes of some units successively placed into service) such that X_i is the elapsed time from the $(i-1)$st event until the occurrence of the ith event. We write

$$F(x) = \Pr\{X_k \leq x\}, \quad k = 1, 2, 3, \ldots,$$

for the common probability distribution of $X_1, X_2, \ldots$. A basic stipulation for renewal processes is $F(0) = 0$, signifying that $X_1, X_2, \ldots$ are positive random variables. We refer to

$$W_n = X_1 + X_2 + \cdots + X_n, \quad n \geq 1 \tag{7.1}$$
$$(W_0 = 0 \text{ by convention}),$$

as the *waiting time* until the occurrence of the nth event.

The relation between the interoccurrence times $\{X_k\}$ and the renewal counting process $\{N(t), t \geq 0\}$ is depicted in Figure 7.1. Note formally that

$$N(t) = \text{number of indices } n \text{ for which } 0 < W_n \leq t. \tag{7.2}$$

In common practice, the counting process $\{N(t), t \geq 0\}$ and the partial sum process $\{W_n, n \geq 0\}$ are interchangeably called the "renewal process." The prototypical renewal model involves successive replacements of lightbulbs. A bulb is installed for service at time $W_0 = 0$, fails at time $W_1 = X_1$, and is then exchanged for a fresh bulb. The second bulb fails at time $W_2 = X_1 + X_2$ and is replaced by a third bulb. In general,

An Introduction to Stochastic Modeling
© 2011 Elsevier Inc. All rights reserved.

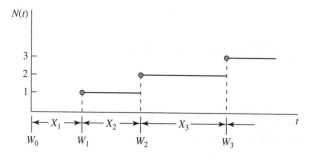

Figure 7.1 The relation between the interoccurrence times X_k and the renewal counting process $N(t)$.

the nth bulb burns out at time $W_n = X_1 + \cdots + X_n$ and is immediately replaced, and the process continues. It is natural to assume that the successive lifetimes are statistically independent, with probabilistically identical characteristics in that

$$\Pr\{X_k \leq x\} = F(x) \quad \text{for } k = 1, 2, \ldots.$$

In this process, $N(t)$ records the number of lightbulb replacements up to time t.

The principal objective of renewal theory is to derive properties of certain random variables associated with $\{N(t)\}$ and $\{W_n\}$ from knowledge of the interoccurrence distribution F. For example, it is of significance and relevance to compute the expected number of renewals for the time duration $(0, t]$:

$$E[N(t)] = M(t)$$

is called the *renewal function*. To this end, several pertinent relationships and formulas are worth recording. In principle, the probability law of $W_n = X_1 + \cdots + X_n$ can be calculated in accordance with the convolution formula

$$\Pr\{W_n \leq x\} = F_n(x),$$

where $F_1(x) = F(x)$ is assumed known or prescribed, and then

$$F_n(x) = \int_0^\infty F_{n-1}(x - y) \, dF(y) = \int_0^x F_{n-1}(x - y) \, dF(y).$$

Such convolution formulas were reviewed in Chapter 1, Section 1.2.5.

The fundamental connecting link between the waiting time process $\{W_n\}$ and the renewal counting process $\{N(t)\}$ is the observation that

$$N(t) \geq k \quad \text{if and only if} \quad W_k \leq t. \tag{7.3}$$

In words, equation (7.3) asserts that the number of renewals up to time t is at least k if and only if the kth renewal occurred on or before time t. Since this equivalence is the basis for much that follows, the reader should verify instances of it by referring to Figure 7.1.

It follows from (7.3) that

$$\Pr\{N(t) \geq k\} = \Pr\{W_k \leq t\} \tag{7.4}$$
$$= F_k(t), \quad t \geq 0, k = 1, 2, \ldots,$$

and consequently,

$$\Pr\{N(t) = k\} = \Pr\{N(t) \geq k\} - \Pr\{N(t) \geq k+1\}$$
$$= F_k(t) - F_{k+1}(t), \quad t \geq 0, k = 1, 2, \ldots. \tag{7.5}$$

For the renewal function $M(t) = E[N(t)]$, we sum the tail probabilities in the manner $E[N(t)] = \Sigma_{k=1}^{\infty} \Pr\{N(t) \geq k\}$, as derived in Chapter 1, equation (1.49), and then use (7.4) to obtain

$$M(t) = E[N(t)] = \sum_{k=1}^{\infty} \Pr\{N(t) \geq k\}$$
$$= \sum_{k=1}^{\infty} \Pr\{W_k \leq t\} = \sum_{k=1}^{\infty} F_k(t). \tag{7.6}$$

There are a number of other random variables of interest in renewal theory. Three of these are the *excess life* (also called the excess random variable), the *current life* (also called the age random variable), and the *total life*, defined, respectively, by

$$\gamma_t = W_{N(t)+1} - t \qquad \text{(excess or residual lifetime),}$$
$$\delta_t = t - W_{N(t)} \qquad \text{(current life or age random variable),}$$
$$\beta_t = \gamma_t + \delta_t \qquad \text{(total life).}$$

A pictorial description of these random variables is given in Figure 7.2.

An important identity enables us to evaluate the mean of $W_{N(t)+1}$ in terms of the mean lifetime $\mu = E[X_1]$ of each unit and the renewal function $M(t)$. Namely, it is true for every renewal process that

$$E[W_{N(t)+1}] = E[X_1 + \cdots + X_{N(t)+1}]$$
$$= E[X_1]\{E[N(t) + 1]\},$$

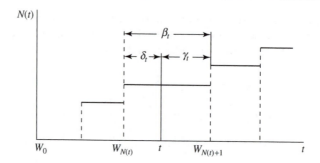

Figure 7.2 The excess life γ_t, the current life δ_t, and the total life β_t.

or

$$E[W_{N(t)+1}] = \mu\{M(t) + 1\}. \tag{7.7}$$

At first glance, this identity resembles the formula given in Chapter 2, equation (2.30) for the mean of a random sum, which asserts that $E[X_1 + \cdots + X_N] = E[X_1]E[N]$ when N is an integer-valued random variable *that is independent of* $X_1, X_2, \ldots$. The random sum approach does not apply in the current context, however, the crucial difference being that the random number of summands $N(t) + 1$ is *not* independent of the summands themselves. Indeed, in Section 7.3, on the Poisson process viewed as a renewal process, we will show that the last summand $X_{N(t)+1}$ has a mean that approaches twice the unconditional mean $\mu = E[X_1]$ for t large. For this reason, it is *not* correct, in particular, that $E[W_{N(t)}]$ can be evaluated as the product of $E[X_1]$ and $E[N(t)]$. In view of these comments, the identity expressed in equation (7.7) becomes more intriguing and remarkable.

To derive (7.7), we will use the fundamental equivalence (7.3) in the form

$$N(t) \geq j - 1 \quad \text{if and only if} \quad X_1 + \cdots + X_{j-1} \leq t,$$

which expressed in terms of indicator random variables becomes

$$\mathbf{1}\{N(t) \geq j - 1\} = \mathbf{1}\{X_1 + \cdots + X_{j-1} \leq t\}.$$

Since this indicator random variable is a function only of the random variables $X_1, \ldots, X_{j-1}$, it is independent of X_j, and thus we may evaluate

$$\begin{aligned}
E[X_j \mathbf{1}\{X_1 + \cdots + X_{j-1} \leq t\}] &= E[X_j]E[\mathbf{1}\{X_1 + \cdots + X_{j-1} \leq t\}] \\
&= E[X_j]\Pr\{X_1 + \cdots + X_{j-1} \leq t\} \tag{7.8} \\
&= \mu F_{j-1}(t).
\end{aligned}$$

With (7.8) in hand, the evaluation of the equivalence expressed in (7.7) becomes straightforward. We have

$$E[W_{N(t)+1}] = E[X_1 + \cdots + X_{N(t)+1}]$$

$$= E[X_1] + E\left[\sum_{j=2}^{N(t)+1} X_j\right]$$

$$= \mu + E\left[\sum_{j=2}^{\infty} X_j \mathbf{1}\{N(t) + 1 \geq j\}\right]$$

$$= \mu + \sum_{j=2}^{\infty} E[X_j \mathbf{1}\{X_1 + \cdots + X_{j-1} \leq t\}]$$

$$= \mu + \mu \sum_{j=2}^{\infty} F_{j-1}(t) \qquad \text{(using (7.8))}$$

$$= \mu[1 + M(t)] \qquad \text{(using (7.6))}.$$

Some examples of the use of the identity (7.7) will appear in the exercises, and an alternative proof in the case of a discrete renewal process can be found in Section 7.6.

Exercises

7.1.1 Verify the following equivalences for the age and the excess life in a renewal process $N(t)$:

$$\gamma_t > x \quad \text{if and only if} \quad N(t+x) - N(t) = 0;$$

and for $0 < x < t$,

$$\delta_t > x \quad \text{if and only if} \quad N(t) - N(t-x) = 0.$$

Why is the condition $x < t$ important in the second case but not the first?

7.1.2 Consider a renewal process in which the interoccurrence times have an exponential distribution with parameter λ:

$$f(x) = \lambda e^{-\lambda x}, \quad \text{and} \quad F(x) = 1 - e^{-\lambda x} \quad \text{for } x > 0.$$

Calculate $F_2(t)$ by carrying out the appropriate convolution [see the equation just prior to (7.3)], and then determine $\Pr\{N(t) = 1\}$ from equation (7.5).

7.1.3 Which of the following are true statements?
(a) $N(t) < k$ if and only if $W_k > t$.
(b) $N(t) \leq k$ if and only if $W_k \geq t$.
(c) $N(t) > k$ if and only if $W_k < t$.

7.1.4 Consider a renewal process for which the lifetimes $X_1, X_2, \ldots$ are discrete random variables having the Poisson distribution with mean λ. That is,

$$\Pr\{X_k = n\} = \frac{e^{-\lambda}\lambda^n}{n!} \quad \text{for } n = 0, 1, \ldots.$$

(a) What is the distribution of the waiting time W_k?
(b) Determine $\Pr\{N(t) = k\}$.

Problems

7.1.1 Verify the following equivalences for the age and the excess life in a renewal process $N(t)$: (Assume $t > x$.)

$$\Pr\{\delta_t \geq x, \gamma_t > y\} = \Pr\{N(t-x) = N(t+y)\}$$
$$= \sum_{k=0}^{\infty} \Pr\{W_k < t-x, W_{k+1} > t+y\}$$
$$= [1 - F(t+y)]$$
$$+ \sum_{k=1}^{\infty} \int_0^{t-x} [1 - F(t+y-z)]dF_k(z).$$

Carry out the evaluation when the interoccurrence times are exponentially distributed with parameter λ, so that dF_k is the gamma density

$$dF_k(z) = \frac{\lambda^k z^{k-1}}{(k-1)!}e^{-\lambda z}dz \quad \text{for } z > 0.$$

7.1.2 From equation (7.5), and for $k \geq 1$, verify that

$$\Pr\{N(t) = k\} = \Pr\{W_k \leq t < W_{k+1}\}$$
$$= \int_0^t [1 - F(t-x)]dF_k(x),$$

and carry out the evaluation when the interoccurrence times are exponentially distributed with parameter λ, so that dF_k is the gamma density

$$dF_k(z) = \frac{\lambda^k z^{k-1}}{(k-1)!}e^{-\lambda z}dz \quad \text{for } z > 0.$$

7.1.3 A fundamental identity involving the renewal function, valid for all renewal processes, is

$$E[W_{N(t)+1}] = E[X_1](M(t) + 1).$$

See equation (7.7). Using this identity, show that the mean excess life can be evaluated in terms of the renewal function via the relation

$$E[\gamma_t] = E[X_1](1 + M(t)) - t.$$

7.1.4 Let γ_t be the excess life and δ_t the age in a renewal process having interoccurrence distribution function $F(x)$. Determine the conditional probability $\Pr\{\gamma_t > y | \delta_t = x\}$ and the conditional mean $E[\gamma_t | \delta_t = x]$.

7.2 Some Examples of Renewal Processes

Stochastic models often contain random times at which they, or some part of them, begin afresh in a statistical sense. These renewal instants form natural embedded renewal processes, and they are found in many diverse fields of applied probability including branching processes, insurance risk models, phenomena of population growth, evolutionary genetic mechanisms, engineering systems, and econometric structures. When a renewal process is discovered embedded within a model, the powerful results of renewal theory become available for deducing implications.

7.2.1 Brief Sketches of Renewal Situations

The synopses that follow suggest the wide scope and diverse contexts in which renewal processes arise. Several of the examples will be studied in more detail in later sections.

(a) *Poisson Processes* A Poisson process $\{N(t), t \geq 0\}$ with parameter λ is a renewal counting process having the exponential interoccurrence distribution

$$F(x) = 1 - e^{-\lambda x}, \quad x \geq 0,$$

as established in Chapter 5, Theorem 5.5. This particular renewal process possesses a host of special features, highlighted later in Section 7.3.

(b) *Counter Processes* The times between successive electrical impulses or signals impinging on a recording device (counter) are often assumed to form a renewal process. Most physically realizable counters lock for some duration immediately upon registering an impulse and will not record impulses arriving during this dead period. Impulses are recorded only when the counter is free (i.e., unlocked). Under quite reasonable assumptions, the sequence of events of the times of recorded impulses forms a renewal process, but it should be emphasized that the renewal process of recorded impulses is a secondary renewal process derived from the original renewal process comprising the totality of all arriving impulses.

(c) *Traffic Flow* The distances between successive cars on an indefinitely long single-lane highway are often assumed to form a renewal process. So also are the time durations between consecutive cars passing a fixed location.

(d) *Renewal Processes Associated with Queues* In a single-server queueing process, there are embedded many natural renewal processes. We cite two examples:

(i) If customer arrival times form a renewal process, then the times of the starts of successive busy periods generate a second renewal process.

(ii) For the situation in which the input process (the arrival pattern of customers) is Poisson, the successive moments in which the server passes from a busy to a free state determine a renewal process.

(e) *Inventory Systems* In the analysis of most inventory processes, it is customary to assume that the pattern of demands forms a renewal process. Most of the standard inventory policies induce renewal sequences, e.g., the times of replenishment of stock.

(f) *Renewal Processes in Markov Chains* Let $Z_0, Z_1, \ldots$ be a recurrent Markov chain. Suppose $Z_0 = i$, and consider the durations (elapsed number of steps) between successive visits to state i. Specifically, let $W_0 = 0$,

$$W_1 = \min\{n > 0; Z_n = i\},$$

and

$$W_{k+1} = \min\{n > W_k; Z_n = i\}, \quad k = 1, 2, \ldots.$$

Since each of these times is computed from the same starting state i, the Markov property guarantees that $X_k = W_k - W_{k-1}$ are independent and identically distributed, and thus $\{X_k\}$ generates a renewal process.

7.2.2   Block Replacement

Consider a lightbulb whose life, measured in discrete units, is a random variable X, where $\Pr\{X = k\} = p_k$ for $k = 1, 2, \ldots$. Assuming that one starts with a fresh bulb and that each bulb is replaced by a new one when it burns out, let $M(n) = E[N(n)]$ be the expected number of replacements up to time n.

Because of economies of scale, in a large building such as a factory or office it is often cheaper, on a per bulb basis, to replace all the bulbs, failed or not, than it is to replace a single bulb. A *block replacement policy* attempts to take advantage of this reduced cost by fixing a block period K and then replacing bulbs as they fail during periods $1, 2, \ldots, K - 1$, and replacing all bulbs, failed or not, in period K. This strategy is also known as "group relamping." If c_1 is the per bulb block replacement cost and c_2 is the per bulb failure replacement cost ($c_1 < c_2$), then the mean total cost during the block replacement cycle is $c_1 + c_2 M(K - 1)$, where $M(K - 1) = E[N(K - 1)]$ is the mean number of failure replacements. Since the block replacement cycle consists of K periods, the mean total cost per bulb per unit time is

$$\theta(K) = \frac{c_1 + c_2 M(K - 1)}{K}.$$

If we can determine the renewal function $M(n)$ from the life distribution $\{p_k\}$, then we can choose the block period $K = K^*$ so as to minimize the cost rate $\theta(K)$. Of course, this cost must be compared to the cost of replacing only upon failure.

The renewal function $M(n)$, or expected number of replacements up to time n, solves the equation

$$M(n) = F_X(n) + \sum_{k=1}^{n-1} p_k M(n-k) \quad \text{for } n = 1, 2, \ldots.$$

To derive this equation, condition on the life X_1 of the first bulb. If it fails after time n, there are no replacements during periods $[1, 2, \ldots, n]$. On the other hand, if it fails at time $k < n$, then we have its failure plus, on the average, $M(n-k)$ additional replacements during the interval $[k+1, k+2, \ldots, n]$. Using the law of total probability to sum these contributions, we obtain

$$M(n) = \sum_{k=n+1}^{\infty} p_k(0) + \sum_{k=1}^{n} p_k[1 + M(n-k)]$$

$$= F_X(n) + \sum_{k=1}^{n-1} p_k M(n-k) \quad [\text{because } M(0) = 0],$$

as asserted.

Thus, we determine

$$M(1) = F_X(1),$$

$$M(2) = F_X(2) + p_1 M(1),$$

$$M(3) = F_X(3) + p_1 M(2) + p_2 M(1),$$

and so on.

To consider a numerical example, suppose that

$$p_1 = 0.1, \quad p_2 = 0.4, \quad p_3 = 0.3, \quad \text{and} \quad p_4 = 0.2,$$

and

$$c_1 = 2 \quad \text{and} \quad c_2 = 3.$$

Then,

$$M(1) = p_1 = 0.1,$$

$$M(2) = (p_1 + p_2) + p_1 M(1) = (0.1 + 0.4) + 0.1(0.1) = 0.51,$$

$$M(3) = (p_1 + p_2 + p_3) + p_1 M(2) + p_2 M(1)$$

$$= (0.1 + 0.4 + 0.3) + 0.1(0.51) + 0.4(0.1) = 0.891,$$

$$M(4) = (p_1 + p_2 + p_3 + p_4) + p_1 M(3) + p_2 M(2) + p_3 M(1)$$

$$= 1 + 0.1(0.891) + 0.4(0.51) + 0.3(0.1) = 1.3231.$$

The average costs are shown in the following table:

Block Period K	Cost $= \dfrac{c_1 + c_2 M(K-1)}{K} = \theta(K)$
1	2.00000
2	1.15000
3	1.17667
4	1.16825
5	1.19386

The minimum cost block period is $K^* = 2$.

We wish to elicit one more insight from this example. Forgetting about block replacement, we continue to calculate

$$M(5) = 1.6617,$$
$$M(6) = 2.0647,$$
$$M(7) = 2.4463,$$
$$M(8) = 2.8336,$$
$$M(9) = 3.2136,$$
$$M(10) = 3.6016.$$

Let u_n be the probability that a replacement occurs in period n. Then, $M(n) = M(n-1) + u_n$ asserts that the mean replacements up to time n is the mean replacements up to time $n - 1$ plus the probability that a replacement occurs in period n. The calculations are shown in the following table:

n	$u_n = M(n) - M(n-1)$
1	0.1000
2	0.4100
3	0.3810
4	0.4321
5	0.3386
6	0.4030
7	0.3816
8	0.3873
9	0.3800
10	0.3880

The probability of a replacement in period n seems to be converging. This is indeed the case, and the limit is the reciprocal of the mean bulb lifetime:

$$\frac{1}{E[X_1]} = \frac{1}{0.1(1) + 0.4(2) + 0.3(3) + 0.2(4)}$$

$$= 0.3846 \cdots .$$

This calculation makes sense. If a lightbulb lasts, on the average, $E[X_1]$ time units, then the probability that it will need to be replaced in any period should approximate $1/E[X_1]$. Actually, the relationship is not as simple as just stated. Further discussion takes place in Sections 7.4 and 7.6.

Exercises

7.2.1 Let $\{X_n; n = 0, 1, \ldots\}$ be a two-state Markov chain with the transition probability matrix

$$\mathbf{P} = \begin{array}{c} \\ 0 \\ 1 \end{array} \begin{array}{cc} 0 & 1 \\ \left\| \begin{array}{cc} 1-a & a \\ b & 1-b \end{array} \right\|. \end{array}$$

State 0 represents an *operating* state of some system, while state 1 represents a *repair* state. We assume that the process begins in state $X_0 = 0$, and then the successive returns to state 0 from the repair state form a renewal process. Determine the mean duration of one of these renewal intervals.

7.2.2 A certain type component has two states: $0 = \text{OFF}$ and $1 = \text{OPERATING}$. In state 0, the process remains there a random length of time, which is exponentially distributed with parameter α, and then moves to state 1. The time in state 1 is exponentially distributed with parameter β, after which the process returns to state 0.

The *system* has two of these components, A and B, with distinct parameters:

Component	Operating Failure Rate	Repair Rate
A	β_A	α_A
B	β_B	α_B

In order for the *system* to operate, at least one of components A and B must be operating (a parallel system). Assume that the component stochastic processes are independent of one another. Consider the successive instants that the *system* enters the failed state from an operating state. Use the memoryless property of the exponential distribution to argue that these instants form a renewal process.

7.2.3 Calculate the mean number of renewals $M(n) = E[N(n)]$ for the renewal process having interoccurrence distribution

$$p_1 = 0.4, \quad p_2 = 0.1, \quad p_3 = 0.3, \quad p_4 = 0.2$$

for $n = 1, 2, \ldots, 10$. Also calculate $u_n = M(n) - M(n-1)$.

Problems

7.2.1 For the block replacement example of this section for which $p_1 = 0.1, p_2 = 0.4, p_3 = 0.3$, and $p_4 = 0.2$, suppose the costs are $c_1 = 4$ and $c_2 = 5$. Determine the minimal cost block period K^* and the cost of replacing upon failure alone.

7.2.2 Let $X_1, X_2, \ldots$ be the interoccurrence times in a renewal process. Suppose $\Pr\{X_k = 1\} = p$ and $\Pr\{X_k = 2\} = q = 1 - p$. Verify that

$$M(n) = E[N(n)] = \frac{n}{1+q} - \frac{q^2}{(1+q)^2}\left[1 - (-q)^n\right]$$

for $n = 2, 4, 6, \ldots$.

7.2.3 Determine $M(n)$ when the interoccurrence times have the geometric distribution

$$\Pr\{X_1 = k\} = p_k = \beta(1 - \beta)^{k-1} \quad \text{for } k = 1, 2, \ldots,$$

where $0 < \beta < 1$.

7.3   The Poisson Process Viewed as a Renewal Process

As mentioned earlier, the Poisson process with parameter λ is a renewal process whose interoccurrence times have the exponential distribution $F(x) = 1 - e^{-\lambda x}, x \geq 0$. The memoryless property of the exponential distribution (see Sections 1.4.2, 1.5.2 of Chapter 1, and Chapter 5) serves decisively in yielding the explicit computation of a number of functionals of the Poisson renewal process.

The Renewal Function
Since $N(t)$ has a Poisson distribution, then

$$\Pr\{N(t) = k\} = \frac{(\lambda t)^k e^{-\lambda t}}{k!}, \quad k = 0, 1, \ldots,$$

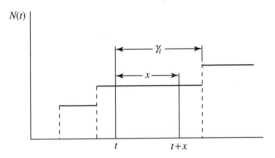

Figure 7.3 The excess life γ_t exceeds x if and only if there are no renewals in the interval $(t, t+x]$.

and

$$M(t) = E[N(t)] = \lambda t.$$

Excess Life

Observe that the excess life at time t exceeds x if and only if there are no renewals in the interval $(t, t+x]$ (Figure 7.3). This event has the same probability as that of no renewals in the interval $(0, x]$, since a Poisson process has stationary independent increments. In formal terms, we have

$$\Pr\{\gamma_t > x\} = \Pr\{N(t+x) - N(t) = 0\}$$
$$= \Pr\{N(x) = 0\} = e^{-\lambda x}. \tag{7.9}$$

Thus, in a Poisson process, the excess life possesses the same exponential distribution

$$\Pr\{\gamma_t \le x\} = 1 - e^{-\lambda x}, \quad x \ge 0, \tag{7.10}$$

as every life, another manifestation of the memoryless property of the exponential distribution.

Current Life

The current life δ_t, of course, cannot exceed t, while for $x < t$ the current life exceeds x if and only if there are no renewals in $(t-x, t]$, which again has probability $e^{-\lambda x}$. Thus, the current life follows the truncated exponential distribution

$$\Pr\{\delta_t \le x\} = \begin{cases} 1 - e^{-\lambda x} & \text{for } 0 \le x < t, \\ 1 & \text{for } t \le x. \end{cases} \tag{7.11}$$

Mean Total Life

Using the evaluation of equation (1.50) in Chapter 1 for the mean of a nonnegative random variable, we have

$$E[\beta_t] = E[\gamma_t] + E[\delta_t]$$

$$= \frac{1}{\lambda} + \int_0^t \Pr\{\delta_t > x\}dx$$

$$= \frac{1}{\lambda} + \int_0^t e^{-\lambda x}dx$$

$$= \frac{1}{\lambda} + \frac{1}{\lambda}\left(1 - e^{-\lambda t}\right).$$

Observe that the mean total life is significantly larger than the mean life $1/\lambda = E[X_k]$ of any particular renewal interval. A more striking expression of this phenomenon is revealed when t is large, where the process has been in operation for a long duration. Then, the mean total life $E[\beta_t]$ is approximately twice the mean life. These facts appear at first paradoxical.

Let us reexamine the definition of the total life β_t with a view to explaining on an intuitive basis the seeming discrepancy. First, an arbitrary time point t is fixed. Then, β_t measures the length of the renewal interval containing the point t. Such a procedure will tend with higher likelihood to favor a lengthy renewal interval rather than one of short duration. The phenomenon is known as length-biased sampling and occurs, well disguised, in a number of sampling situations.

Joint Distribution of γ_t and δ_t

The joint distribution of γ_t and δ_t is determined in the same manner as the marginals. In fact, for any $x > 0$ and $0 < y < t$, the event $\{\gamma_t > x, \delta_t > y\}$ occurs if and only if there are no renewals in the interval $(t - y, t + x]$, which has probability $e^{-\lambda(x+y)}$. Thus,

$$\Pr\{\gamma_t > x, \delta_t > y\} = \begin{cases} e^{-\lambda(x+y)} & \text{if } x > 0, 0 < y < t, \\ 0 & \text{if } y \geq t. \end{cases} \tag{7.12}$$

For the Poisson process, observe that γ_t and δ_t are independent, since their joint distribution factors as the product of their marginal distributions.

Exercises

7.3.1 Let $W_1, W_2, \ldots$ be the event times in a Poisson process $\{X(t); t \geq 0\}$ of rate λ. Evaluate

$$\Pr\{W_{N(t)+1} > t + s\} \quad \text{and} \quad E[W_{N(t)+1}].$$

7.3.2 Particles arrive at a counter according to a Poisson process of rate λ. An arriving particle is recorded with probability p and lost with probability $1 - p$ independently of the other particles. Show that the sequence of recorded particles is a Poisson process of rate λp.

7.3.3 Let $W_1, W_2, \ldots$ be the event times in a Poisson process $\{N(t); t \geq 0\}$ of rate λ. Show that

$$N(t) \quad \text{and} \quad W_{N(t)+1}$$

are independent random variables by evaluating

$$\Pr\{N(t) = n \quad \text{and} \quad W_{N(t)+1} > t + s\}.$$

Problems

7.3.1 In another form of *sum quota sampling* (see Chapter 5, Section 5.4.2), a sequence of nonnegative independent and identically distributed random variables $X_1, X_2, \ldots$ is observed, the sampling continuing until the first time that the sum of the observations *exceeds* the quota t. In renewal process terminology, the sample size is $N(t) + 1$. The sample mean is

$$\frac{W_{N(t)+1}}{N(t)+1} = \frac{X_1 + \cdots + X_{N(t)+1}}{N(t)+1}.$$

An important question in statistical theory is whether or not this sample mean is unbiased. That is, how does the expected value of this sample mean relate to the expected value of, say, X_1? Assume that the individual X summands are exponentially distributed with parameter λ, so that $N(t)$ is a Poisson process, and evaluate the expected value of the foregoing sample mean and show that

$$E\left[\frac{W_{N(t)+1}}{N(t)+1}\right] = \frac{1}{\lambda}\left[1 - e^{-\lambda t}\right]\left(1 + \frac{1}{\lambda t}\right).$$

Hint: Use the result of the previous exercise, that

$$W_{N(t)+1} \quad \text{and} \quad N(t)$$

are independent, and then evaluate separately

$$E[W_{N(t)+1}] \quad \text{and} \quad E\left[\frac{1}{N(t)+1}\right].$$

7.3.2 A fundamental identity involving the renewal function, valid for all renewal processes, is

$$E[W_{N(t)+1}] = E[X_1](M(t) + 1).$$

See equation (7.7). Evaluate the left side and verify the identity when the renewal counting process is a Poisson process.

7.3.3 Pulses arrive at a counter according to a Poisson process of rate λ. All physically realizable counters are imperfect, incapable of detecting all signals that enter their detection chambers. After a particle or signal arrives, a counter must recuperate, or renew itself, in preparation for the next arrival. Signals arriving during the readjustment period, called dead time or locked time, are lost. We must distinguish between the arriving particles and the recorded particles. The experimenter observes only the particles recorded; from this observation he desires to infer the properties of the arrival process.

Suppose that each arriving pulse locks the counter for a fixed time τ. Determine the probability $p(t)$ that the counter is free at time t.

7.3.4 This problem is designed to aid in the understanding of length-biased sampling. Let X be a uniformly distributed random variable on $[0, 1]$. Then, X divides $[0, 1]$ into the subintervals $[0, X]$ and $(X, 1]$. By symmetry, each subinterval has mean length $\frac{1}{2}$. Now pick one of these subintervals at random in the following way: Let Y be independent of X and uniformly distributed on $[0, 1]$, and pick the subinterval $[0, X]$ or $(X, 1]$ that Y falls in. Let L be the length of the subinterval so chosen. Formally,

$$L = \begin{cases} X & \text{if } Y \leq X, \\ 1 - X & \text{if } Y > X. \end{cases}$$

Determine the mean of L.

7.3.5 Birds are perched along a wire as shown according to a Poisson process of rate λ per unit distance:

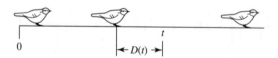

At a fixed point t along the wire, let $D(t)$ be the random distance to the nearest bird. What is the mean value of $D(t)$? What is the probability density function $f_t(x)$ for $D(t)$?

7.4 The Asymptotic Behavior of Renewal Processes

A large number of the functionals that have explicit expressions for Poisson renewal processes are far more difficult to compute for other renewal processes. There are, however, many simple formulas that describe the asymptotic behavior, for large values of t, of a general renewal process. We summarize some of these asymptotic results in this section.

7.4.1 The Elementary Renewal Theorem

The Poisson process is the only renewal process (in continuous time) whose renewal function $M(t) = E[N(t)]$ is exactly linear. All renewal functions are asymptotically linear, however, in the sense that

$$\lim_{t \to \infty} \frac{M(t)}{t} = \lim_{t \to \infty} \frac{E[N(t)]}{t} = \frac{1}{\mu}, \tag{7.13}$$

where $\mu = E[X_k]$ is the mean interoccurrence time. This fundamental result, known as the *elementary renewal theorem*, is invoked repeatedly to compute functionals describing the long run behavior of stochastic models having renewal processes associated with them.

The elementary renewal theorem (7.4.1) holds even when the interoccurrence times have infinite mean, and then $\lim_{t \to \infty} M(t)/t = 1/\infty = 0$.

The elementary renewal theorem is so intuitively plausible that it has often been viewed as obvious. The left side, $\lim_{t \to \infty} M(t)/t$, describes the long run mean number of renewals or replacements per unit time. The right side, $1/\mu$, is the reciprocal of the mean life of a component. Isn't it obvious that if a component lasts, on the average, μ time units, then in the long run these components will be replaced at the rate of $1/\mu$ per unit time? However plausible and convincing this argument may be, it is not obvious, and to establish the elementary renewal theorem requires several steps of mathematical analysis, beginning with the law of large numbers. As our main concern is stochastic modeling, we omit this derivation, as well as the derivations of the other asymptotic results summarized in this section, in order to give more space to their application.

Example *Age Replacement Policies* Let $X_1, X_2, \ldots$ represent the lifetimes of items (lightbulbs, transistor cards, machines, etc.) that are successively placed in service, the next item commencing service immediately following the failure of the previous one. We stipulate that $\{X_k\}$ are independent and identically distributed positive random variables with finite mean $\mu = E[X_k]$. The elementary renewal theorem tells us to expect to replace items over the long run at a mean rate of $1/\mu$ per unit time.

In the long run, any replacement strategy that substitutes items prior to their failure will use more than $1/\mu$ items per unit time. Nonetheless, where there is some benefit in avoiding failure in service, and where units deteriorate, in some sense, with age, there may be an economic or reliability advantage in considering alternative replacement strategies. Telephone or utility poles serve as good illustrations of this concept. Clearly, it is disadvantageous to allow these poles to fail in service because of the damage to the wires they carry, the damage to adjoining property, overtime wages paid for emergency replacements, and revenue lost while service is down. Therefore, an attempt is usually made to replace older utility poles before they fail. Other instances of planned replacement occur in preventative maintenance strategies for aircraft, where "time" is now measured by operating hours.

An age replacement policy calls for replacing an item upon its failure or upon its reaching age T, whichever occurs first. Arguing intuitively, we would expect that the long run fraction of failure replacements, items that fail before age T, will be $F(T)$, and the corresponding fraction of (conceivably less expensive) planned replacements will be $1 - F(T)$. A renewal interval for this modified age replacement policy obviously follows a distribution law

$$F_T(x) = \begin{cases} F(x) & \text{for } x < T, \\ 1 & \text{for } x \geq T, \end{cases}$$

and the mean renewal duration is

$$\mu_T = \int_0^\infty \{1 - F_T(x)\}\,dx = \int_O^T \{1 - F(x)\}dx < \mu.$$

The elementary renewal theorem indicates that the long run mean replacement rate under age replacement is increased to $1/\mu_T$.

Now, let $Y_1, Y_2, \ldots$ denote the times between actual successive failures. The random variable Y_1 is composed of a random number of time periods of length T (corresponding to replacements not associated with failures), plus a last time period in which the distribution is that of a failure conditioned on failure before age T; i.e., Y_1 has the distribution of $NT + Z$, where

$$\Pr\{N \geq k\} = \{1 - F(T)\}^k, \quad k = 0, 1, \ldots,$$

and

$$\Pr\{Z \leq z\} = \frac{F(z)}{F(T)}, \quad 0 \leq z \leq T.$$

Hence,

$$E[Y_1] = \frac{1}{F(T)} \left\{ T[1 - F(T)] + \int_0^T (F(T) - F(x))dx \right\}$$

$$= \frac{1}{F(T)} \int_0^T \{1 - F(x)\}dx = \frac{\mu_T}{F(T)}.$$

The sequence of random variables for interoccurrence times of the bona fide failure $\{Y_i\}$ generates a renewal process whose mean rate of failures per unit time in the long run is $1/E[Y_1]$. This inference again relies on the elementary renewal theorem.

Depending on F, the modified failure rate $1/E[Y_1]$ may possibly yield a lower failure rate than $1/\mu$, the rate when replacements are made only upon failure.

Let us suppose that each replacement, whether planned or not, costs K dollars, and that each failure incurs an additional penalty of c dollars. Multiplying these costs by the appropriate rates gives the long run mean cost per unit time as a function of the replacement age T:

$$
\begin{aligned}
C(T) &= \frac{K}{\mu_T} + \frac{c}{E[Y_1]} \\
&= \frac{K + cF(T)}{\int_0^T [1 - F(x)]dx}.
\end{aligned}
$$

In any particular situation, a routine calculus exercise or recourse to numerical computation produces the value of T that minimizes the long run cost rate. For example, if $K = 1, c = 4$, and lifetimes are uniformly distributed on $[0, 1]$, then $F(x) = x$ for $0 \leq x \leq 1$, and

$$
\int_0^T [1 - F(x)]dx = T\left(1 - \frac{1}{2}T\right)
$$

and

$$
C(T) = \frac{1 + 4T}{T(1 - T/2)}.
$$

To obtain the cost minimizing T, we differentiate $C(T)$ with respect to T and equate to zero, thereby obtaining

$$
\frac{dC(T)}{dT} = 0 = \frac{4T(1 - T/2) - (1 + 4T)(1 - T)}{[T(1 - T/2)]^2},
$$

$$
0 = 4T - 2T^2 - 1 + T - 4T + 4T^2,
$$

$$
0 = 2T^2 + T - 1,
$$

$$
T = \frac{-1 \pm \sqrt{1 + 8}}{4} = \left(\frac{1}{2}, -1\right),
$$

and the optimal choice is $T^* = \frac{1}{2}$. Routine calculus will verify that this choice leads to a minimum cost, and not a maximum or inflection point.

7.4.2 The Renewal Theorem for Continuous Lifetimes

The elementary renewal theorem asserts that

$$
\lim_{t \to \infty} \frac{M(t)}{t} = \frac{1}{\mu}.
$$

It is tempting to conclude from this that $M(t)$ behaves like t/μ as t grows large, but the precise meaning of the phrase "behaves like" is rather subtle. For example, suppose that all of the lifetimes are deterministic, say $X_k = 1$ for $k = 1, 2, \ldots$. Then, it is straightforward to calculate

$$
\begin{aligned}
M(t) = N(t) &= 0 && \text{for } 0 \le t < 1, \\
&= 1 && \text{for } 1 \le t < 2, \\
&= k && \text{for } k \le t < k+1.
\end{aligned}
$$

That is, $M(t) = [t]$, where $[t]$ denotes the greatest integer not exceeding t. Since $\mu = 1$ in this example, then $M(t) - t/\mu = [t] - t$, a function that oscillates indefinitely between 0 and -1. While it remains true in this illustration that $M(t)/t = [t]/t \to 1 = 1/\mu$, it is not clear in what sense $M(t)$ "behaves like" t/μ. If we rule out the periodic behavior that is exemplified in the extreme by this deterministic example, then $M(t)$ behaves like t/μ in the sense described by the *renewal theorem*, which we now explain. Let $M(t, t+h] = M(t+h) - M(t)$ denote the mean number of renewals in the interval $(t, t+h]$. The renewal theorem asserts that when periodic behavior is precluded, then

$$
\lim_{t \to \infty} M(t, t+h] = h/\mu \quad \text{for any fixed } h > 0. \tag{7.14}
$$

In words, asymptotically, the mean number of renewals in an interval is proportional to the interval's length, with proportionality constant $1/\mu$.

A simple and prevalent situation in which the renewal theorem (7.4.2) is valid occurs when the lifetimes $X_1, X_2, \ldots$ are continuous random variables having the probability density function $f(x)$. In this circumstance, the renewal function is differentiable, and

$$
m(t) = \frac{dM(t)}{dt} = \sum_{n=1}^{\infty} f_n(t), \tag{7.15}
$$

where $f_n(t)$ is the probability density function for $W_n = X_1 + \cdots + X_n$. Now (7.14) may be written in the form

$$
\frac{M(t+h) - M(t)}{h} \to \frac{1}{\mu} \quad \text{as } t \to \infty,
$$

which, when h is small, suggests that

$$
\lim_{t \to \infty} m(t) = \lim_{t \to \infty} \frac{dM(t)}{dt} = \frac{1}{\mu}, \tag{7.16}
$$

and indeed, this is the case in all but the most pathological of circumstances when $X_1, X_2, \ldots$ are continuous random variables.

If in addition to being continuous, the lifetimes $X_1, X_2, \ldots$ have a finite mean μ and finite variance σ^2, then the renewal theorem can be refined to include a second term.

Under the stated conditions, we have

$$\lim_{t \to \infty} \left[M(t) - \frac{t}{\mu} \right] = \frac{\sigma^2 - \mu^2}{2\mu^2}. \tag{7.17}$$

Example When the lifetimes $X_1, X_2, \ldots$ have the gamma density function

$$f(x) = xe^{-x} \quad \text{for } x > 0, \tag{7.18}$$

then the waiting times $W_n = X_1 + \cdots + X_n$ have the gamma density

$$f_n(x) = \frac{x^{2n-1}}{(2n-1)!} e^{-x} \quad \text{for } x > 0,$$

as may be verified by performing the appropriate convolutions. (See Chapter 1, Section 1.2.5.) Substitution into (7.15) yields

$$
\begin{aligned}
m(x) &= \sum_{n=1}^{\infty} f_n(x) = e^{-x} \sum_{n=1}^{\infty} \frac{x^{2n-1}}{(2n-1)!} \\
&= e^{-x} \frac{e^x - e^{-x}}{2} = \frac{1}{2} \left(1 - e^{-2x} \right),
\end{aligned}
$$

and

$$M(t) = \int_0^t m(x)\mathrm{d}x = \frac{1}{2}t - \frac{1}{4}\left[1 - e^{-2t} \right].$$

Since the gamma density in (7.18) has moments $\mu = 2$ and $\sigma^2 = 2$, we verify that $m(t) \to 1/\mu$ as $t \to \infty$ and $M(t) - t/\mu \to -\frac{1}{4} = \left(\sigma^2 - \mu^2 \right)/2\mu^2$, in agreement with (7.16) and (7.17).

7.4.3 The Asymptotic Distribution of $N(t)$

The elementary renewal theorem

$$\lim_{t \to x} \frac{E[N(t)]}{t} = \frac{1}{\mu} \tag{7.19}$$

implies that the asymptotic mean of $N(t)$ is approximately t/μ. When $\mu = E[X_k]$ and $\sigma^2 = \text{Var}[X_k] = E\left[(X_k - \mu)^2 \right]$ are finite, then the asymptotic variance of $N(t)$ behaves according to

$$\lim_{t \to \infty} \frac{\text{Var}[N(t)]}{t} = \frac{\sigma^2}{\mu^3}. \tag{7.20}$$

That is, the asymptotic variance of $N(t)$ is approximately $t\sigma^2/\mu^3$. If we standardize $N(t)$ by subtracting its asymptotic mean and dividing by its asymptotic standard deviation, we get the following convergence to the normal distribution:

$$\lim_{t\to\infty} \Pr\left\{ \frac{N(t) - t/\mu}{\sqrt{t\sigma^2/\mu^3}} \le x \right\} = \frac{1}{\sqrt{2\pi}} \int_{-\infty}^{x} e^{-y^2/2} dy.$$

In words, for large values of t, the number of renewals $N(t)$ is approximately normally distributed with mean and variance given by (7.19) and (7.20), respectively.

7.4.4 The Limiting Distribution of Age and Excess Life

Again we assume that the lifetimes $X_1, X_2, \ldots$ are continuous random variables with finite mean μ. Let $\gamma_t = W_{N(t)+1} - t$ be the excess life at time t. The excess life has the limiting distribution

$$\lim_{t\to\infty} \Pr\{\gamma_t \le x\} = \frac{1}{\mu} \int_0^x [1 - F(y)] dy. \tag{7.21}$$

The reader should verify that the right side of (7.21) defines a valid distribution function, which we denote by $H(x)$. The corresponding probability density function is $h(y) = \mu^{-1}[1 - F(y)]$. The mean of this limiting distribution is determined according to

$$\int_0^\infty y h(y) dy = \frac{1}{\mu} \int_0^\infty y[1 - F(y)] dy$$

$$= \frac{1}{\mu} \int_0^\infty y \left\{ \int_y^\infty f(t) dt \right\} dy$$

$$= \frac{1}{\mu} \int_0^\infty f(t) \left\{ \int_0^t y\,dy \right\} dt$$

$$= \frac{1}{2\mu} \int_0^\infty t^2 f(t) dt$$

$$= \frac{\sigma^2 + \mu^2}{2\mu},$$

where σ^2 is the common variance of the lifetimes $X_1, X_2, \ldots$.

$$\{\gamma_t \ge x \text{ and } \delta_t \ge y\} \quad \text{if and only if} \quad \{\gamma_{t-y} \ge x+y\}. \tag{7.22}$$

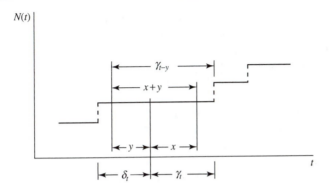

Figure 7.4 $\{\delta_t \geq y$ and $\gamma_t \geq x\}$ if and only if $\{\gamma_{t-y} \geq x + y\}$.

It follows that

$$
\lim_{t \to \infty} \Pr\{\gamma_t \geq x, \delta_t \geq y\} = \lim_{t \to \infty} \Pr\{\gamma_{t-y} \geq x + y\}
$$

$$
= \mu^{-1} \int_{x+y}^{\infty} [1 - F(z)] \, dz,
$$

exhibiting the joint limiting distribution of (γ_t, δ_t). In particular,

$$
\lim_{t \to \infty} \Pr\{\delta_t \geq y\} = \lim_{t \to \infty} \Pr\{\gamma_t \geq 0, \delta_t \geq y\}
$$

$$
= \mu^{-1} \int_{y}^{\infty} [1 - F(z)] \, dz
$$

$$
= 1 - H(y).
$$

The limiting distribution for the current life, or age, $\delta_t = t - W_{N(t)}$ can be deduced from the corresponding result (7.21) for the excess life. With the aid of Figure 7.4, corroborate the equivalence.

Exercises

7.4.1 Consider the triangular lifetime density $f(x) = 2x$ for $0 < x < 1$. Determine an asymptotic expression for the expected number of renewals up to time t.

    Hint: Use equation (7.17).

7.4.2 Consider the triangular lifetime density $f(x) = 2x$ for $0 < x < 1$. Determine an asymptotic expression for the probability distribution of excess life. Using this distribution, determine the limiting mean excess life and compare with the general result following equation (7.21).

7.4.3 Consider the triangular lifetime density function $f(x) = 2x$, for $0 < x < 1$. Determine the optimal replacement age in an age replacement model with replacement cost $K = 1$ and failure penalty $c = 4$ (cf. the example in Section 7.4.1).

7.4.4 Show that the optimal age replacement policy is to replace upon failure alone when lifetimes are exponentially distributed with parameter λ. Can you provide an intuitive explanation?

7.4.5 What is the limiting distribution of excess life when renewal lifetimes have the uniform density $f(x) = 1$, for $0 < x < 1$?

7.4.6 A machine can be in either of two states: "up" or "down." It is up at time zero and thereafter alternates between being up and down. The lengths $X_1, X_2, \ldots$ of successive up times are independent and identically distributed random variables with mean α, and the lengths $Y_1, Y_2, \ldots$ of successive down times are independent and identically distributed with mean β.

(a) In the long run, what fraction of time is the machine up?

(b) If the machine earns income at a rate of $13 per unit time while up, what is the long run total rate of income earned by the machine?

(c) If each down time costs $7, regardless of how long the machine is down, what is the long run total down time cost per unit time?

Problems

7.4.1 Suppose that a renewal function has the form $M(t) = t + [1 - \exp(-at)]$. Determine the mean and variance of the interoccurrence distribution.

7.4.2 A system is subject to failures. Each failure requires a repair time that is exponentially distributed with rate parameter α. The operating time of the system until the next failure is exponentially distributed with rate parameter β. The repair times and the operating times are all statistically independent. Suppose that the system is operating at time 0. Using equation (7.17), determine an approximate expression for the mean number of failures up to time t, the approximation holding for $t \gg 0$.

7.4.3 Suppose that the life of a lightbulb is a random variable X with hazard rate $h(x) = \theta x$ for $x > 0$. Each failed lightbulb is immediately replaced with a new one. Determine an asymptotic expression for the mean age of the lightbulb in service at time t, valid for $t \gg 0$.

7.4.4 A developing country is attempting to control its population growth by placing restrictions on the number of children each family can have. This society places a high premium on female children, and it is felt that any policy that ignores the desire to have female children will fail. The proposed policy is to allow any married couple to have children up to the first female baby, at which point they must cease having children. Assume that male and female children are equally likely. The number of children in any family is a random variable N. In the population as a whole, what fraction of children are female? Use the elementary renewal theorem to justify your answer.

7.4.5 A Markov chain $X_0, X_1, X_2, \ldots$ has the transition probability matrix

$$\mathbf{P} = \begin{array}{c} \\ 0 \\ 1 \\ 2 \end{array} \begin{array}{ccc} 0 & 1 & 2 \\ \left\| \begin{array}{ccc} 0.3 & 0.7 & 0 \\ 0.6 & 0 & 0.4 \\ 0 & 0.5 & 0.5 \end{array} \right\| \end{array}.$$

A *sojourn* in a state is an uninterrupted sequence of consecutive visits to that state.

(a) Determine the mean duration of a typical sojourn in state 0.

(b) Using renewal theory, determine the long run fraction of time that the process is in state 1.

7.5 Generalizations and Variations on Renewal Processes

7.5.1 Delayed Renewal Processes

We continue to assume that $\{X_k\}$ are all independent positive random variables, but only $X_2, X_3, \ldots$ (from the second on) are identically distributed with distribution function F, while X_1 has possibly a different distribution function G. Such a process is called a *delayed renewal process*. We have all the ingredients for an ordinary renewal process except that the initial time to the first renewal has a distribution different from that of the other interoccurrence times.

A delayed renewal process will arise when the component in operation at time $t = 0$ is not new, but all subsequent replacements are new. For example, suppose that the time origin is taken y time units after the start of an ordinary renewal process. Then, the time to the first renewal after the origin in the delayed process will have the distribution of the excess life at time y of an ordinary renewal process.

As before, let $W_0 = 0$ and $W_n = X_1 + \cdots + X_n$, and let $N(t)$ count the number of renewals up to time t. But now it is essential to distinguish between the mean number of renewals in the delayed process

$$M_D(t) = E[N(t)], \tag{7.23}$$

and the renewal function associated with the distribution F,

$$M(t) = \sum_{k=1}^{\infty} F_k(t). \tag{7.24}$$

For the delayed process, the elementary renewal theorem is

$$\lim_{t \to \infty} \frac{M_D(t)}{t} = \frac{1}{\mu}, \quad \text{where } \mu = E[X_2], \tag{7.25}$$

and the renewal theorem states that

$$\lim_{t \to \infty} [M_D(t) - M_D(t - h)] = \frac{h}{\mu},$$

provided $X_2, X_3, \ldots$ are continuous random variables.

7.5.2   Stationary Renewal Processes

A delayed renewal process for which the first life has the distribution function

$$G(x) = \mu^{-1} \int_0^x \{1 - F(y)\} dy$$

is called a stationary renewal process. We are attempting to model a renewal process that began indefinitely far in the past, so that the remaining life of the item in service at the origin has the limiting distribution of the excess life in an ordinary renewal process. We recognize G as this limiting distribution.

It is anticipated that such a process exhibits a number of stationary, or time-invariant, properties. For a stationary renewal process,

$$M_D(t) = E[N(t)] = \frac{t}{\mu} \tag{7.26}$$

and

$$\Pr\{\gamma_t^D \leq x\} = G(x),$$

for all t. Thus, what is in general only an asymptotic renewal relation becomes an identity, holding for all t, in a stationary renewal process.

7.5.3   Cumulative and Related Processes

Suppose associated with the ith unit, or lifetime interval, is a second random variable Y_i ($\{Y_i\}$ identically distributed) in addition to the lifetime X_i. We allow X_i and Y_i to be dependent but assume that the pairs (X_1, Y_1), $(X_2, Y_2), \ldots$ are independent. We use the notation $F(x) = \Pr\{X_i \leq x\}$, $G(y) = \Pr\{Y_i \leq y\}$, $\mu = E[X_i]$, and $v = E[Y_i]$.

A number of problems of practical and theoretical interest have a natural formulation in those terms.

Renewal Processes Involving Two Components to Each Renewal interval

Suppose that Y_i represents a portion of the duration X_i. Figure 7.5 illustrates the model. There we have depicted the Y portion occurring at the beginning of the interval, but this assumption is not essential for the results that follow.

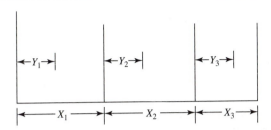

Figure 7.5 A renewal process in which an associated random variable Y_i, represents a portion of the ith renewal interval.

Let $p(t)$ be the probability that t falls in a Y portion of some renewal interval. When $X_1, X_2, \ldots$ are continuous random variables, the renewal theorem implies the following important asymptotic evaluation:

$$\lim_{t \to \infty} p(t) = \frac{E[Y_1]}{E[X_1]}. \tag{7.27}$$

Here are some concrete examples.

A Replacement Model

Consider a replacement model in which replacement is not instantaneous. Let Y_i be the operating time and Z_i the lag period preceding installment of the $(i+1)$st operating unit. (The delay in replacement can be conceived as a period of repair of the service unit.) We assume that the sequence of times between successive replacements $X_k = Y_k + Z_k, k = 1, 2, \ldots$, constitutes a renewal process. Then $p(t)$, the probability that the system is in operation at time t, converges to $E[Y_1]/E[X_1]$.

A Queueing Model

A queueing process is a process in which customers arrive at some designated place where a service of some kind is being rendered, e.g., at the teller's window in a bank or beside the cashier at a supermarket. It is assumed that the time between arrivals, or interarrival time, and the time that is spent in providing service for a given customer are governed by probabilistic laws.

If arrivals to a queue follow a Poisson process of intensity λ, then the successive times X_k from the commencement of the kth busy period to the start of the next busy period form a renewal process. (A busy period is an uninterrupted duration when the queue is not empty.) Each X_k is composed of a busy portion Z_k and an idle portion Y_k. Then $p(t)$, the probability that the queue is empty at time t, converges to $E[Y_1]/E[X_1]$. This example is treated more fully in Chapter 9, which is devoted to queueing systems.

The Peter Principle

The "Peter Principle" asserts that a worker will be promoted until finally reaching a position in which he or she is incompetent. When this happens, the person stays in that job until retirement. Consider the following single job model of the Peter Principle: A person is selected at random from the population and placed in the job. If the person is competent, he or she remains in the job for a random time having cumulative distribution function F and mean μ and is promoted. If incompetent, the person remains for a random time having cumulative distribution function G and mean $v > \mu$ and retires. Once the job is vacated, another person is selected at random and the process repeats. Assume that the infinite population contains the fraction p of competent people and $q = 1 - p$ incompetent ones.

In the long run, what fraction of time is the position held by an incompetent person?

A renewal occurs every time the position is filled, and therefore the mean duration of a renewal cycle is

$$E[X_k] = p\mu + (1 - p)v.$$

To answer the question, we let $Y_k = X_k$ if the kth person is incompetent, and $Y_k = 0$ if the kth person is competent. Then, the long run fraction of time that the position is held by an incompetent person is

$$\frac{E[Y_1]}{E[X_1]} = \frac{(1 - p)v}{p\mu + (1 - p)v}.$$

Suppose that $p = \frac{1}{2}$ of the people are competent, and that $v = 10$, while $\mu = 1$. Then,

$$\frac{E[Y_1]}{E[X_1]} = \frac{(1/2)(10)}{(1/2)(10) + (1/2)(1)} = \frac{10}{11} = 0.91.$$

Thus, while half of the people in the population are competent, the job is filled by a competent person only 9% of the time!

Cumulative Processes

Interpret Y_i as a cost or value associated with the ith renewal cycle. A class of problems with a natural setting in this general context of pairs (X_i, Y_i), where X_i generates a renewal process, will now be considered. Interest here focuses on the so-called cumulative process

$$W(t) = \sum_{k=1}^{N(t)+1} Y_k,$$

the accumulated costs or value up to time t (assuming that transactions are made at the beginning of a renewal cycle).

The elementary renewal theorem asserts in this case that

$$\lim_{t \to \infty} \frac{1}{t} E[W(t)] = \frac{E[Y_1]}{\mu}. \tag{7.28}$$

This equation justifies the interpretation of $E[Y_1]/\mu$ as a long run mean cost or value per unit time, an interpretation that was used repeatedly in the examples of Section 7.2.

Here are some examples of cumulative processes.

Replacement Models

Suppose Y_i is the cost of the ith replacement. Let us suppose that under an age replacement strategy (see Section 7.3 and the example entitled "Age Replacement Policies" in Section 7.4) a planned replacement at age T costs c_1 dollars, while a failure replaced at time $x < T$ costs c_2 dollars. If Y_k is the cost incurred at the kth replacement cycle, then

$$Y_k = \begin{cases} c_1 & \text{with probability } 1 - F(T), \\ c_2 & \text{with probability } F(T), \end{cases}$$

and $E[Y_k] = c_1[1 - F(T)] + c_2 F(T)$. Since the expected length of a replacement cycle is

$$E[\min\{X_k, T\}] = \int_0^T [1 - F(x)]dx,$$

we have that the long run cost per unit time is

$$\frac{c_1[1 - F(T)] + c_2 F(T)}{\int_0^T [1 - F(x)]dx},$$

and in any particular situation a routine calculus exercise or recourse to numerical computation produces the value of T that minimizes the long run cost per unit time.

Under a block replacement policy, there is one planned replacement every T units of time and, on the average, $M(T)$ failure replacements, so the expected cost is $E[Y_k] = c_1 + c_2 M(T)$, and the long run mean cost per unit time is $\{c_1 + c_2 M(T)\}/T$.

Risk Theory

Suppose claims arrive at an insurance company according to a renewal process with interoccurrence times $X_1, X_2, \ldots$. Let Y_k be the magnitude of the kth claim. Then, $W(t) = \sum_{k=1}^{N(t)+1} Y_k$ represents the cumulative amount claimed up to time t, and the long run mean claim rate is

$$\lim_{t \to \infty} \frac{1}{t} E[W(t)] = \frac{E[Y_1]}{E[X_1]}.$$

Maintaining Current Control of a Process

A production process produces items one by one. At any instance, the process is in one of two possible states, which we label *in-control* and *out-of-control*. These states are not directly observable. Production begins with the process in-control, and it remains in-control for a random and unobservable length of time before a breakdown occurs, after which the process is out-of-control. A control chart is to be used to help detect when the out-of-control state occurs, so that corrective action may be taken.

To be more specific, we assume that the quality of an individual item is a normally distributed random variable having an unknown mean and a known variance σ^2. If the process is in-control, the mean equals a standard target, or design value, μ_0. Process breakdown takes the form of shift in mean away from standard to $\mu_1 = \mu_0 \pm \delta\sigma$, where δ is the amount of the shift in standard deviation units.

The Shewhart control chart method for maintaining process control calls for measuring the qualities of the items as they are produced, and then plotting these qualities versus time on a chart that has lines drawn at the target value μ_0 and above and below this target value at $\mu_0 \pm k\sigma$, where k is a parameter of the control scheme being used. As long as the plotted qualities fall inside these so-called *action lines* at $\mu_0 \pm k\sigma$, the process is assumed to be operating in-control, but if ever a point falls outside these lines, the process is assumed to have left the in-control state, and investigation and repair are instituted. There are obviously two possible types of errors that can be made while, thus, controlling the process: (1) needless investigation and repair when the process is in-control yet an observed quality purely by chance falls outside the action lines and (2) continued operation with the process out-of-control because the observed qualities are falling inside the action lines, again by chance.

Our concern is the rational choice of the parameter k, i.e., the rational spacing of the action lines, so as to balance, in some sense, these two possible errors.

The probability that a single quality will fall outside the action lines when the process is in-control is given by an appropriate area under the normal density curve. Denoting this probability by α, we have

$$\alpha = \Phi(-k) + 1 - \Phi(k) = 2\Phi(-k),$$

where $\Phi(x) = (2\pi)^{-1/2} \int_{-\infty}^{x} \exp(-y^2/2)\,dy$ is the standard cumulative normal distribution function. Representative values are given in the following table:

k	α
1.645	0.10
1.96	0.05

Similarly, the probability, denoted by p, that a single point will fall outside the action lines when the process is out-of-control is given by

$$p = \Phi(-\delta - k) + 1 - \Phi(-\delta + k).$$

Let S denote the number of items inspected before an out-of-control signal arises assuming that the process is out-of-control. Then, $\Pr\{S=1\}=p$, $\Pr\{S=2\}=(1-p)p$, and in general, $\Pr\{S = n\} = (1-p)^{n-1}p$. Thus, S has a geometric distribution, and

$$E[S] = \frac{1}{p}.$$

Let T be the number of items produced while the process is in-control. We suppose that the mean operating time in-control $E[T]$ is known from past records.

The sequence of durations between detected and repaired out-of-control conditions forms a renewal process because each such duration begins with a newly repaired process and is a probabilistic replica of all other such intervals. It follows from the general elementary renewal theorem that the long run fraction of time spent out-of-control (O.C.) is

$$\text{O.C.} = \frac{E[S]}{E[S] + E[T]} = \frac{1}{1 + pE[T]}.$$

The long run number of repairs per unit time is

$$R = \frac{1}{E[S] + E[T]} = \frac{p}{1 + pE[T]}.$$

Let N be the random number of "false alarms" while the process is in-control, i.e., during the time up to T, the first out-of-control. Then, conditioned on T, the random variable N has a binomial distribution with probability parameter α, and thus $E[N|T\} = \alpha T$ and $E[N] = \alpha E[T]$. Again, it follows from the general elementary renewal theorem that the long run false alarms per unit time (F.A.) is

$$\text{F.A.} = \frac{E[N]}{E[S] + E[T]} = \frac{\alpha p E[T]}{1 + pE[T]}.$$

If each false alarm costs c dollars, each repair cost K dollars, and the cost rate while operating out-of-control is C dollars, then we have the long run average cost per unit time of

$$\begin{aligned}
\text{A.C.} &= C(\text{O.C.}) + K(R) + c(\text{F.A.}) \\
&= \frac{C + Kp + c\alpha p E[T]}{1 + pE[T]}.
\end{aligned}$$

By trial and error one may now choose k, which determines α and p, so as to minimize this average cost expression.

Exercises

7.5.1 Jobs arrive at a certain service system according to a Poisson process of rate λ. The server will accept an arriving customer only if it is idle at the time of arrival. Potential customers arriving when the system is busy are lost. Suppose that the service times are independent random variables with mean service time μ. Show that the long run fraction of time that the server is idle is $1/(1 + \lambda\mu)$. What is the long run fraction of potential customers that are lost?

7.5.2 The weather in a certain locale consists of alternating wet and dry spells. Suppose that the number of days in each rainy spell is Poisson distributed with parameter 2, and that a dry spell follows a geometric distribution with a mean of 7 days. Assume that the successive durations of rainy and dry spells are statistically independent random variables. In the long run, what is the probability on a given day that it will be raining?

7.5.3 Consider a lightbulb whose life is a continuous random variable X with probability density function $f(x)$, for $x > 0$. Assuming that one starts with a fresh bulb and that each failed bulb is immediately replaced by a new one, let $M(t) = E[N(t)]$ be the expected number of renewals up to time t. Consider a *block replacement policy* (see Section 7.2.1) that replaces each failed bulb immediately at a cost of c per bulb and replaces all bulbs at the fixed times $T, 2T, 3T, \ldots$. Let the block replacement cost per bulb be $b < c$. Show that the long run total mean cost per bulb per unit time is

$$\Theta(T) = \frac{b + cM(T)}{T}.$$

Investigate the choice of a cost minimizing value T^* when $M(t) = t + 1 - \exp(-at)$.

Problems

7.5.1 A certain type component has two states: $0 = $ OFF and $1 = $ OPERATING. In state 0, the process remains there a random length of time, which is exponentially distributed with parameter α, and then moves to state 1. The time in state 1 is exponentially distributed with parameter β, after which the process returns to state 0.

The *system* has two of these components, A and B, with distinct parameters:

Component	Operating Failure Rate	Repair Rate
A	β_A	α_A
B	β_B	α_B

In order for the *system* to operate, at least one of components A and B must be operating (a parallel system). Assume that the component stochastic processes are independent of one another.

(a) In the long run, what fraction of time is the system inoperational (not operating)?

(b) Once the system enters the failed state, what is the mean duration there prior to returning to operation?

(c) Define a cycle as the time between the instant that the system first enters the failed state and the next such instant. Using renewal theory, find the mean duration of a cycle.

(d) What is the mean system operating duration between successive system failures?

7.5.2 The random lifetime X of an item has a distribution function $F(x)$. What is the mean total life $E[X|X > x]$ of an item of age x?

7.5.3 At the beginning of each period, customers arrive at a taxi stand at times of a renewal process with distribution law $F(x)$. Assume an unlimited supply of cabs, such as might occur at an airport. Suppose that each customer pays a random fee at the stand following the distribution law $G(x)$, for $x > 0$. Write an expression for the sum $W(t)$ of money collected at the stand by time t, and then determine the limit expectation

$$\lim_{t \to \infty} \frac{E[W(t)]}{t}.$$

7.5.4 A lazy professor has a ceiling fixture in his office that contains two light-bulbs. To replace a bulb, the professor must fetch a ladder, and being lazy, when a single bulb fails, he waits until the second bulb fails before replacing them both. Assume that the length of life of the bulbs are independent random variables.

   (a) If the lifetimes of the bulbs are exponentially distributed, with the same parameter, what fraction of time, in the long run, is our professor's office half lit?

   (b) What fraction of time, in the long run, is our professor's office half lit if the bulbs that he buys have the same uniform $(0, 1)$ lifetime distribution?

7.6 Discrete Renewal Theory*

In this section, we outline the renewal theory that pertains to nonnegative integer-valued lifetimes. We emphasize renewal equations, the renewal argument, and the renewal theorem (Theorem 7.1).

   Consider a lightbulb whose life, measured in discrete units, is a random variable X where $\Pr\{X = k\} = p_k$ for $k = 0, 1, \ldots$. If one starts with a fresh bulb and if each bulb when it burns out is replaced by a new one, then $M(n)$, the expected number of renewals (not including the initial bulb) up to time n, solves the equation

$$M(n) = F_X(n) + \sum_{k=0}^{n} p_k M(n - k), \tag{7.29}$$

where $F_X(n) = p_0 + \cdots + p_n$ is the cumulative distribution function of the random variable X. A vector or functional equation of the form (7.29) in the unknowns $M(0), M(1), \ldots$ is termed a *renewal equation*. The equation is established by a *renewal argument*, a first step analysis that proceeds by conditioning on the life of the first bulb and then invoking the law of total probability. In the case of (7.29), e.g., if the first bulb fails at time $k \leq n$, then we have its failure plus, on the average, $M(n - k)$ additional failures in the interval $[k, k+1, \ldots, n]$. We weight this conditional mean by

---

* The discrete renewal model is a special case in the general renewal theory presented in Sections 7.1–7.5 and does not arise in later chapters.

the probability $p_k = \Pr\{X_1 = k\}$ and sum according to the law of total probability to obtain

$$M(n) = \sum_{k=0}^{n} [1 + M(n-k)] p_k$$

$$= F_X(n) = \sum_{k=0}^{n} p_k M(n-k).$$

Equation (7.29) is only a particular instance of what is called a renewal equation. In general, a renewal equation is prescribed by a given bounded sequence $\{b_k\}$ and takes the form

$$v_n = b_n + \sum_{k=0}^{n} p_k v_{n-k} \qquad \text{for } n = 0, 1, \ldots. \tag{7.30}$$

The unknown variables are $v_0, v_1, \ldots$, and $p_0, p_1, \ldots$ is a probability distribution for which, to avoid trivialities, we always assume $p_0 < 1$.

Let us first note that there is one and only one sequence $v_0, v_1, \ldots$ satisfying a renewal equation, because we may solve (7.30) successively to get

$$v_0 = \frac{b_0}{1 - p_0},$$

$$v_1 = \frac{b_1 + p_1 v_0}{1 - p_0}, \tag{7.31}$$

and so on.

Let u_n be the mean number of renewals that take place exactly in period n. When $p_0 = 0$, so that the lifetimes are strictly positive and at most one renewal can occur in any period, then u_n is the probability that a single renewal occurs in period n. The sequence $u_0, u_1, \ldots$ satisfies a renewal equation that is of fundamental importance in the general theory. Let

$$\delta_n = \begin{cases} 1 & \text{for } n = 0, \\ 0 & \text{for } n > 0. \end{cases} \tag{7.32}$$

Then, $\{u_n\}$ satisfies the renewal equation

$$u_n = \delta_n + \sum_{k=0}^{n} p_k u_{n-k} \qquad \text{for } n = 0, 1, \ldots. \tag{7.33}$$

Again, equation (7.33) is established via a renewal argument. First, observe that δ_n counts the initial bulb, the renewal at time 0. Next, condition on the lifetime of this first bulb. If it fails in period $k \leq n$, which occurs with probability p_k, then the process begins afresh and the conditional probability of a renewal in period n becomes u_{n-k}. Weighting the contingency represented by u_{n-k} by its respective probability p_k and summing according to the law of total probability then yields (7.33).

The next lemma shows how the solution $\{v_n\}$ to the general renewal equation (7.30) can be expressed in terms of the solution $\{u_n\}$ to the particular equation (7.33).

Lemma 7.1. *If* $\{v_n\}$ *satisfies (7.30) and* $\{u_n\}$ *satisfies (7.33), then*

$$v_n = \sum_{k=0}^{n} b_{n-k} u_k \qquad for \ n = 0, 1, \ldots.$$

Proof. In view of our remarks on the existence and uniqueness of solutions to equation (7.30), we need only verify that $v_n = \sum_{k=0}^{n} b_{n-k} u_k$ satisfies (7.30). We have

$$v_n = \sum_{k=0}^{n} b_{n-k} u_k$$

$$= \sum_{k=0}^{n} b_{n-k} \left\{ \delta_k + \sum_{l=0}^{k} p_{k-l} u_l \right\}$$

$$= b_n + \sum_{k=0}^{n} \sum_{l=0}^{k} b_{n-k} p_{k-l} u_l$$

$$= b_n + \sum_{l=0}^{n} \sum_{k=l}^{n} b_{n-k} p_{k-l} u_l$$

$$= b_n + \sum_{l=0}^{n} \sum_{j=0}^{n-l} p_j b_{n-l-j} u_l$$

$$= b_n + \sum_{j=0}^{n} \sum_{l=0}^{n-j} p_j b_{n-j-l} u_l$$

$$= b_n + \sum_{j=0}^{n} p_j v_{n-j}.$$

∎

Example Let $X_1, X_2, \ldots$ be the successive lifetimes of the bulbs and let $W_0 = 0$ and $W_n = X_1 + \cdots + X_n$ be the replacement times. We assume that $p_0 = \Pr\{X_1 = 0\} = 0$. The number of replacements (not including the initial bulb) up to time n is given by

$$N(n) = k \quad \text{for } W_k \leq n < W_{k+1}.$$

The $M(n) = E[N(n)]$ satisfies the renewal equation (7.29)

$$M(n) = p_0 + \cdots + p_n + \sum_{k=0}^{n} p_k M(n-k),$$

and elementary algebra shows that $m_n = E[N(n) + 1] = M(n) + 1$ satisfies

$$m_n = 1 + \sum_{k=0}^{n} p_k m_{n-k} \quad \text{for } n = 0, 1, \ldots. \tag{7.34}$$

Then, (7.34) is a renewal equation for which $b_n \equiv 1$ for all n. In view of Lemma 7.1, we conclude that

$$m_n = \sum_{k=0}^{n} 1 u_k = u_0 + \cdots + u_n.$$

Conversely, $u_n = m_n - m_{n-1} = M(n) - M(n-1)$.

To continue with the example, let $g_n = E[W_{N(n)+1}]$. The definition is illustrated in Figure 7.6. We will argue that g_n satisfies a certain renewal equation. As shown in Figure 7.6, $W_{N(n)+1}$ always includes the first renewal duration X_1. In addition, if $X_1 = k \leq n$, which occurs with probability p_k, then the conditional mean of the added lives constituting $W_{N(n)+1}$ is g_{n-k}. Weighting these conditional means by their respective probabilities and summing according to the law of total probability then gives

$$g_n = E[X_1] + \sum_{k=0}^{n} g_{n-k} p_k.$$

Hence, by Lemma 7.1,

$$g_n = \sum_{k=0}^{n} E[X_1] u_k = E[X_1] m_n.$$

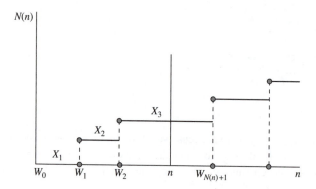

Figure 7.6 $W_{N(n)+1}$ always contains X_1 and contains additional durations when $X_1 = k \leq n$.

We get the interesting formula [see (7.7)]

$$E[X_1 + \cdots + X_{N(n)+1}] = E[X_1] \times E[N(n) + 1]. \tag{7.35}$$

Note that $N(n)$ is not independent of $\{X_k\}$, and yet (7.35) still prevails.

7.6.1 The Discrete Renewal Theorem

The renewal theorem provides conditions under which the solution $\{v_n\}$ to a renewal equation will converge as n grows large. Certain periodic behavior, such as failures occurring only at even ages, must be precluded, and the simplest assumption assuring this preclusion is that $p_1 > 0$.

Theorem 7.1. *Suppose that $0 < p_1 < 1$ and that $\{u_n\}$ and $\{v_n\}$ are the solutions to the renewal equations (7.33) and (7.30), respectively. Then (a) $\lim_{n \to \infty} u_n = 1/\Sigma_{k=0}^{\infty} k p_k$; and (b) if $\Sigma_{k=0}^{\infty} |b_k| < \infty$, then $\lim_{n \to \infty} v_n = \left\{ \Sigma_{k=0}^{\infty} b_k \right\} / \left\{ \Sigma_{k=0}^{\infty} k p_k \right\}$.*

We recognize that $\Sigma_{k=0}^{\infty} k p_k = E[X_1]$ is the mean lifetime of a unit. Thus, (a) in Theorem 7.1 asserts that in the long run, the probability of a renewal occurring in a given interval is one divided by the mean life of a unit.

Remark Theorem 7.1 holds in certain circumstances when $p_1 = 0$. It suffices to assume that the greatest common divisor of the integers k for which $p_k > 0$ is one.

Example Let $\gamma_n = W_{N(n)+1} - n$ be the excess life at time n. For a fixed integer m, let $f_n(m) = \Pr\{\gamma_n = m\}$. We will establish a renewal equation for $f_n(m)$ by conditioning on the first life X_1. For $m \geq 1$,

$$\Pr\{\gamma_n = m | X_1 = k\} = \begin{cases} f_{n-k}(m) & \text{if } 0 \leq k \leq n, \\ 1 & \text{if } k = n + m, \\ 0 & \text{otherwise.} \end{cases}$$

(The student is urged to diagram the alternatives arising in $\Pr\{\gamma_n = m | X_1 = k\}$.) Then, by the law of total probability,

$$f_n(m) = \Pr\{\gamma_n = m\} = \sum_{k=0}^{\infty} \Pr\{\gamma_n = m | X_1 = k\} p_k \tag{7.36}$$

$$= p_{m+n} + \sum_{k=0}^{n} f_{n-k}(m) p_k.$$

We apply Theorem 7.1 with $b_n = p_{m+n}$ to conclude that

$$\lim_{n \to \infty} \Pr\{\gamma_n = m\} = \frac{\Sigma_{k=0}^{\infty} p_{m+k}}{\Sigma_{k=0}^{\infty} k p_k}$$

$$= \frac{\Pr\{X_1 \geq m\}}{E[X_1]}, \quad m = 1, 2, \ldots.$$

The limit is a bona fide probability mass function, since its terms sum to one:

$$\frac{\sum_{m=1}^{\infty} \Pr\{X_1 \geq m\}}{E[X_1]} = \frac{E[X_1]}{E[X_1]} = 1.$$

7.6.2 Deterministic Population Growth with Age Distribution

In this section, we will discuss a simple deterministic model of population growth that takes into account the age structure of the population. Surprisingly, the discrete renewal theorem (Theorem 7.1) will play a role in the analysis. As the language will suggest, the deterministic model that we treat may be viewed as describing the mean population size in a more elaborate stochastic model that is beyond our scope to develop fully.

A Simple Growth Model

Let us set the stage by reviewing a simple model that has no age structure. We consider a single species evolving in discrete time $t = 0, 1, 2, \ldots$, and we let N_t be the population size at time t. We assume that each individual present in the population at time t gives rise to a constant number λ of offspring that form the population at time $t+1$. (If death does not occur in the model, then we include the parent as one of the offspring, and then necessarily $\lambda \geq 1$.) If N_0 is the initial population size, and each individual gives rise to λ offspring, then

$$N_1 = \lambda N_0,$$
$$N_2 = \lambda N_1 = \lambda^2 N_0,$$

and in general,

$$N_t = \lambda^t N_0. \tag{7.37}$$

If $\lambda > 1$, then the population grows indefinitely in time; if $\lambda < 1$, then the population dies out; while if $\lambda = 1$, then the population size remains constant at $N_t = N_0$ for all $t = 0, 1, \ldots$.

The Model with Age Structure

We shall now introduce an age structure in the population. We need the following notation:

$n_{u,t}$ = the number of individuals of age u in the population at time t;

$N_t = \sum_{u=0}^{\infty} n_{u,t}$ = the total number of individuals in the population at time t;

b_t = the number of new individuals created in the population at time t, the number of births;

β_u = the expected number of progeny of a single individual of age u in one time period;

l_u = the probability that an individual will survive, from birth, at least to age u.

The conditional probability that an individual survives at least to age u, given that he has survived to age $u - 1$, is simply the ration l_u/l_{u-1}. The *net maternity function* is the product

$$m_u = l_u \beta_u$$

and is the birth rate adjusted for the death of some fraction of the population. That is, m_u is the expected number of offspring at age u of an individual now of age 0.

Let us derive the total progeny of a single individual during its lifespan. An individual survives at least to age u with probability l_u, and then during the next unit of time gives rise to β_u offspring. Summing $l_u \beta_u = m_u$ over all ages u then gives the total progeny of a single individual:

$$M = \sum_{u=0}^{\infty} l_u \beta_u = \sum_{u=0}^{\infty} m_u. \tag{7.38}$$

If $M > 1$, then we would expect the population to increase over time; if $M < 1$, then we would expect the population to decrease; while if $M = 1$, then the population size should neither increase nor decrease in the long run. This is indeed the case, but the exact description of the population evolution is more complex, as we will now determine.

In considering the effect of age structure on a growing population, our interest will center on b_t, the number of new individuals created in the population at time t. We regard β_u, l_u, and $n_{u,0}$ as known, and the problem is to determine b_t for $t \geq 0$. Once b_t is known, then $n_{u,t}$ and N_t may be determined according to, e.g.,

$$n_{0,1} = b_1, \tag{7.39}$$

$$n_{u,1} = n_{u-1,0} \left[\frac{l_u}{l_{u-1}} \right] \quad \text{for } u \geq 1, \tag{7.40}$$

and

$$N_1 = \sum_{u=0}^{\infty} n_{u,1}. \tag{7.41}$$

In the first of these simple relations, $n_{0,1}$, is the number in the population at time 1 of age 0, which obviously is the same as b_1, those born in the population at time 1. For the second equation, $n_{u,1}$ is the number in the population at time 1 of age u. These individuals must have survived from the $n_{u-1,0}$ individuals in the population at time 0 of age $u - 1$; the conditional probability of survivorship is $[l_u/l_{u-1}]$, which explains the second equation. The last relation simply asserts that the total population size results by summing the numbers of individuals of all ages. The generalizations of (7.39) through (7.41) are

$$n_{0,t} = b_t, \tag{7.42}$$

$$n_{u,t} = n_{u-1,t-1}\left[\frac{l_u}{l_{u-1}}\right] \quad \text{for } u \geq 1, \tag{7.43}$$

and

$$N_t = \sum_{u=0}^{\infty} n_{u,t} \quad \text{for } t \geq 1. \tag{7.44}$$

Having explained how $n_{u,t}$ and N_t are found once b_t is known, we turn to determining b_t. The number of individuals created at time t has two components. One component, a_t, say, counts the offspring of those individuals in the population at time t who already existed at time 0. In the simplest case, the population begins at time $t = 0$ with a single ancestor of age $u = 0$, and then the number of offspring of this individual at time t is $a_t = m_t$, the net maternity function. More generally, assume that there were $n_{u,0}$ individuals of age u at time 0. The probability that an individual of age u at time 0 will survive to time t (at which time it will be of age $t + u$) is l_{t+u}/l_u. Hence the number of individuals of age u at time 0 that survive to time t is $n_{u,0}(l_{t+u}/l_u)$, and each of these individuals, now of age $t + u$, will produce β_{t+u} new offspring. Adding over all ages we obtain

$$
\begin{aligned}
a_t &= \sum_{u=0}^{\infty} \beta_{t+u} n_{u,0} \frac{l_{t+u}}{l_u} \\
&= \sum_{u=0}^{\infty} \frac{m_{t+u} n_{u,0}}{l_u}.
\end{aligned} \tag{7.45}
$$

The second component of b_t counts those individuals created at time t whose parents were not initially in the population but were born after time 0. Now, the number of individuals created at time τ is b_τ. The probability that one of these individuals survives to time t, at which time he will be of age $t - \tau$, is $l_{t-\tau}$. The rate of births for individuals of age $t - \tau$ is $\beta_{t-\tau}$. The second component results from summing over τ and gives

$$
\begin{aligned}
b_t &= a_t + \sum_{\tau=0}^{t} \beta_{t-\tau} l_{t-\tau} b_\tau \\
&= a_t + \sum_{\tau=0}^{t} m_{t-\tau} b_\tau.
\end{aligned} \tag{7.46}
$$

Example Consider an organism that produces two offspring at age 1, and two more at age 2, and then dies. The population begins with a single organism of age 0 at time 0.

We have the data

$$n_{0,0} = 1, \; n_{u,0} = 0 \quad \text{for } u \geq 1,$$
$$b_1 = b_2 = 2,$$
$$l_0 = l_1 = l_2 = 1 \quad \text{and} \quad l_u = 0 \quad \text{for} \quad u > 2.$$

We calculate from (7.45) that

$$a_0 = 0, \quad a_1 = 2, \quad a_2 = 2, \quad \text{and} \quad a_t = 0, \quad \text{for } t > 2.$$

Finally, (7.46) is solved recursively as

$$b_0 = 0,$$
$$b_1 = a_1 + m_0 b_1 + m_1 b_0$$
$$\quad = 2 + 0 + 0 = 2,$$
$$b_2 = a_2 + m_0 b_2 + m_1 b_1 + m_2 b_0$$
$$\quad = 2 + 0 + (2)(2) + 0 = 6,$$
$$b_3 = a_3 + m_0 b_3 + m_1 b_2 + m_2 b_1 + m_3 b_0$$
$$\quad = 0 + 0 + (2)(6) + (2)(2) + 0 = 16.$$

Thus, e.g., an individual of age 0 at time 0 gives rise to 16 new individuals entering the population at time 3.

The Long Run Behavior

Somewhat surprisingly, since no "renewals" are readily apparent, the discrete renewal theorem (Theorem 7.1) will be invoked to deduce the long run behavior of this age-structured population model. Observe that (7.46)

$$b_t = a_t + \sum_{\tau=0}^{t} m_{t-\tau} b_\tau$$

$$= a_t + \sum_{v=0}^{t} m_v b_{t-v}$$

$$(7.47)$$

has the form of a renewal equation except that $\{m_v\}$ is not necessarily a bona fide probability distribution in that, typically, $\{m_v\}$ will not sum to one. Fortunately, there is a trick that overcomes this difficulty. We introduce a variable s, whose value will be chosen later, and let

$$m_v^{\#} = m_v s^v, \quad b_v^{\#} = b_v s^v, \quad \text{and} \quad a_v^{\#} = a_v s^v.$$

Now multiply (7.47) by s^t and observe that $s^t m_v b_{t-v} = (m_v s^v)(b_{t-v} s^{t-v}) = m_v^\# b_{t-v}^\#$ to get

$$b_t^\# = a_t^\# + \sum_{v=0}^{t} m_v^\# b_{t-v}^\#. \tag{7.48}$$

This renewal equation holds no matter what value we choose for s. We, therefore, choose s such that $\{m_v^\#\}$ is a bona fide probability distribution. That is, we fix the value of s such that

$$\sum_{v=0}^{\infty} m_v^\# = \sum_{v=0}^{\infty} m_v s^v = 1.$$

There is always a unique such s whenever $1 < \sum_{v=0}^{\infty} m_v < \infty$. We may now apply the renewal theorem to (7.48), provided that its hypothesis concerning nonperiodic behavior is satisfied. For this it suffices, e.g., that $m_1 > 0$. Then, we conclude that

$$\lim_{t \to \infty} b_t^\# = \lim_{t \to \infty} b_t s^t = \frac{\sum_{v=0}^{\infty} a_v^\#}{\sum_{v=0}^{\infty} v m_v^\#}. \tag{7.49}$$

We set $\lambda = 1/s$ and $K = \sum_{v=0}^{\infty} a_v^\# / \sum_{v=0}^{\infty} v m_v^\#$ to write (7.49) in the form

$$b_t \sim K\lambda^t \quad \text{for } t \text{ large.}$$

In words, asymptotically, the population grows at rate λ where $\lambda = 1/s$ is the solution to

$$\sum_{v=0}^{\infty} m_v \lambda^{-v} = 1.$$

When t is large ($t > u$), then (7.43) may be iterated in the manner

$$n_{u,t} = n_{u-1,t-1} \left[\frac{l_u}{l_{u-1}} \right]$$

$$= n_{u-2,t-2} \left[\frac{l_{u-1}}{l_{u-2}} \right] \left[\frac{l_u}{l_{u-1}} \right]$$

$$= n_{u-2,t-2} \left[\frac{l_u}{l_{u-2}} \right]$$

$$\vdots$$

$$= n_{0,t-u} \left[\frac{l_u}{l_0} \right] = b_{t-u} l_u.$$

This simply expresses that those of age u at time t were born $t - u$ time units ago and survived. Since for large t we have $b_{t-u} \sim K\lambda^{t-u}$, then

$$n_{u,t} \sim Kl_u\lambda^{t-u} = K(l_u\lambda^{-u})\lambda^t,$$

$$N_t = \sum_{u=0}^{\infty} n_{u,t} \sim K \sum_{u=0}^{\infty} (l_u\lambda^{-u})\lambda^t,$$

and

$$\lim_{t\to\infty} \frac{n_{u,t}}{N_t} = \frac{l_u\lambda^{-u}}{\sum_{v=0}^{\infty} l_v\lambda^{-v}}.$$

This last expression furnishes the asymptotic, or *stable, age distribution* in the population.

Example Continuing the example in which $m_1 = m_2 = 2$ and $m_k = 0$ otherwise, then we have

$$\sum_{v=0}^{\infty} m_v s^v = 2s + 2s^2 = 1,$$

which we solve to obtain

$$s = \frac{-2 \pm \sqrt{4+8}}{4} = \frac{-1 \pm \sqrt{3}}{2}$$
$$= (0.366, -1.366).$$

The relevant root is $s = 0.366$, whence $\lambda = 1/s = 2.732$. Thus asymptotically, the population grows geometrically at rate $\lambda = 2.732\cdots$, and the stable age distribution is as shown in the following table:

Age	Fraction of Population
0	$1/(1+s+s^2) = 0.6667$
1	$s/(1+s+s^2) = 0.2440$
2	$s^2/(1+s+s^2) = 0.0893$

Exercises

7.6.1 Solve for v_n for $n = 0, 1, \ldots, 10$ in the renewal equation

$$v_n = b_n + \sum_{k=0}^{n} p_k v_{n-k} \quad \text{for } n = 0, 1, \ldots,$$

where $b_0 = b_1 = \frac{1}{2}, b_2 = b_3 = \cdots = 0$, and $p_0 = \frac{1}{4}, p_1 = \frac{1}{2}$, and $p_2 = \frac{1}{4}$.

7.6.2 (Continuation of Exercise 7.6.1)

    (a) Solve for u_n for $n = 0, 1, \ldots, 10$ in the renewal equation

$$u_n = \delta_n + \sum_{k=0}^{n} p_k u_{n-k} \quad \text{for } n = 0, 1, \ldots,$$

    where $\delta_0 = 1, \delta_1 = \delta_2 = \cdots = 0$, and $\{p_k\}$ is as defined in Exercise 7.6.1.

    (b) Verify that the solution v_n in Exercise 7.6.1 and u_n are related according to $v_n = \sum_{k=0}^{n} b_k u_{n-k}$.

7.6.3 Using the data of Exercises 7.6.1 and 7.6.2, determine

    (a) $\lim_{n \to \infty} u_n$.

    (b) $\lim_{n \to \infty} v_n$.

Problems

7.6.1 Suppose the lifetimes $X_1, X_2, \ldots$ have the geometric distribution

$$\Pr\{X_1 = k\} = \alpha(1 - \alpha)^{k-1} \quad \text{for } k = 1, 2, \ldots,$$

where $0 < \alpha < 1$.

    (a) Determine u_n for $n = 1, 2, \ldots$.

    (b) Determine the distribution of excess life γ_n by using Lemma 7.1 and (7.36).

7.6.2 Marlene has a fair die with the usual six sides. She throws the die and records the number. She throws the die again and adds the second number to the first. She repeats this until the cumulative sum of all the tosses first exceeds a prescribed number n. (a) When $n = 10$, what is the probability that she stops at a cumulative sum of 13? (b) When n is large, what is the approximate probability that she stops at a sum of $n + 3$?

7.6.3 Determine the long run population growth rate for a population whose individual net maternity function is $m_2 = m_3 = 2$, and $m_k = 0$ otherwise. Why does delaying the age at which offspring are first produced cause a reduction in the population growth rate? (The population growth rate when $m_1 = m_2 = 2$, and $m_k = 0$ otherwise was determined in the last example of this section.)

7.6.4 Determine the long run population growth rate for a population whose individual net maternity function is $m_0 = m_1 = 0$ and $m_2 = m_3 = \cdots = a > 0$. Compare this with the population growth rate when $m_2 = a$, and $m_k = 0$ for $k \neq 2$.

8 Brownian Motion and Related Processes

8.1 Brownian Motion and Gaussian Processes

The Brownian motion stochastic process arose early in this century as an attempt to explain the ceaseless irregular motions of tiny particles suspended in a fluid, such as dust motes floating in air. Today, the Brownian motion process and its many generalizations and extensions occur in numerous and diverse areas of pure and applied science such as economics, communication theory, biology, management science, and mathematical statistics.

8.1.1 A Little History

The story begins in the summer of 1827, when the English botanist Robert Brown observed that microscopic pollen grains suspended in a drop of water moved constantly in haphazard zigzag trajectories. Following the reporting of his findings, other scientists verified the strange phenomenon. Similar *Brownian motion* was apparent whenever very small particles were suspended in a fluid medium, e.g., smoke particles in air. Over time, it was established that finer particles move more rapidly, that the motion is stimulated by heat, and that the movement becomes more active with a decrease in fluid viscosity.

A satisfactory explanation had to wait until the next century, when in 1905, Einstein would assert that the Brownian motion originates in the continual bombardment of the pollen grains by the molecules of the surrounding water, with successive molecular impacts coming from different directions and contributing different impulses to the particles. Einstein argued that as a result of the continual collisions, the particles themselves had the same average kinetic energy as the molecules. Belief in molecules and atoms was far from universal in 1905, and the success of Einstein's explanation of the well-documented existence of Brownian motion did much to convince a number of distinguished scientists that such things as atoms actually exist. Incidentally, 1905 is the same year in which Einstein set forth his theory of relativity and his quantum explanation for the photoelectric effect. Any single one of his 1905 contributions would have brought him recognition by his fellow physicists. Today, a search in a university library under the subject heading "Brownian motion" is likely to turn up dozens of books on the stochastic process called Brownian motion and few, if any, on the irregular movements observed by Robert Brown. The literature on the model has far surpassed and overwhelmed the literature on the phenomenon itself!

Brownian motion is complicated because the molecular bombardment of the pollen grain is itself a complicated process, so it is not surprising that it took more than another decade to get a clear picture of the Brownian motion stochastic process. It was not until 1923 that Norbert Wiener set forth the modern mathematical foundation. The reader may also encounter "Wiener process" or "Wiener–Einstein process" as names for the stochastic process that we will henceforth simply call "Brownian motion."

Predating Einstein by several years, in 1900 in Paris, Louis Bachelier proposed what we would now call a "Brownian motion model" for the movement of prices in the French bond market. While Bachelier's paper was largely ignored by academics for many decades, his work now stands as the innovative first step in a mathematical theory of stock markets that has greatly altered the financial world of today. Later in this chapter, we will have much to say about Brownian motion and related models in finance.

8.1.2   The Brownian Motion Stochastic Process

In terms of our general framework of stochastic processes (cf. Chapter 1, Section 1.1.1), the Brownian motion process is an example of a continuous-time, continuous-state-space Markov process. Let $B(t)$ be the y component (as a function of time) of a particle in Brownian motion. Let x be the position of the particle at time t_0; i.e., $B(t_0) = x$. Let $p(y, t|x)$ be the probability density function, in y, of $B(t_0 + t)$, given that $B(t_0) = x$. We postulate that the probability law governing the transitions is stationary in time, and therefore $p(y, t|x)$ does not depend on the initial time t_0.

Since $p(y, t|x)$ is a probability density function in y, we have the properties

$$p(y, t|x) \geq 0 \quad \text{and} \quad \int_{-\infty}^{\infty} p(y, t|x) dy = 1. \tag{8.1}$$

Further, we stipulate that $B(t_0 + t)$ is likely to be near $B(t_0) = x$ for small values of t. This is done formally by requiring that

$$\lim_{t \to 0} p(y, t|x) = 0 \quad \text{for} \quad y \neq x. \tag{8.2}$$

From physical principles, Einstein showed that $p(y, t|x)$ must satisfy the partial differential equation

$$\frac{\partial p}{\partial t} = \frac{1}{2} \sigma^2 \frac{\partial^2 p}{\partial x^2}. \tag{8.3}$$

This is called the *diffusion equation*, and σ^2 is the *diffusion coefficient*, which Einstein showed to be given by $\sigma^2 = RT/Nf$, where R is the gas constant, T is the temperature, N is Avogadro's number, and f is a coefficient of friction. By choosing the proper scale, we may take $\sigma^2 = 1$. With this choice, we can verify directly (see Exercise 8.1.3) that

$$p(y, t|x) = \frac{1}{\sqrt{2\pi t}} \exp\left(-\frac{1}{2t}(y - x)^2\right) \tag{8.4}$$

is a solution of (8.3). In fact, it is the only solution under the conditions (8.1) and (8.2). We recognize (8.4) as a normal probability density function whose mean is x and whose variance is t. That is, the position of the particle t time units after observations begin is normally distributed. The mean position is the initial location x, and the variance is the time of observation t.

Because the normal distribution will appear over and over in this chapter, we are amply justified in standardizing some notation to deal with it. Let

$$\phi(z) = \frac{1}{\sqrt{2\pi}} e^{-\frac{1}{2}z^2}, \quad -\infty < z < \infty, \tag{8.5}$$

be the standard normal probability density function, and let

$$\Phi(z) = \int_{-\infty}^{z} \phi(x)\,dx \tag{8.6}$$

be the corresponding cumulative distribution function. A small table (Table 8.1) of the cumulative normal distribution appears at the end of this section. Let

$$\phi_t(z) = \frac{1}{\sqrt{t}}\phi(z/\sqrt{t}), \tag{8.7}$$

Table 8.1 The Cumulative Normal Distribution

$$\Phi(x) = \int_{-\infty}^{x} \frac{1}{\sqrt{2\pi}} e^{-\frac{1}{2}u^2}\,du$$

x	$\Phi(x)$
-3	0.00135
-2	0.02275
-1	0.1587
0	0.5000
1	0.8413
2	0.97725
3	0.99865
-2.326	0.01
-1.96	0.025
-1.645	0.05
-1.282	0.10
1.282	0.90
1.645	0.95
1.96	0.975
2.236	0.99

and

$$\Phi_t(z) = \int\limits_{-\infty}^{z} \phi_t(x)\mathrm{d}x = \Phi(z/\sqrt{t}) \tag{8.8}$$

be the probability density function and cumulative distribution function, respectively, for the normal distribution with mean zero and variance t. In this notation, the transition density in (8.4) is given by

$$p(y, t|x) = \phi_t(y - x), \tag{8.9}$$

and

$$\Pr\{B(t) \le y | B(0) = x\} = \Phi\left(\frac{y - x}{\sqrt{t}}\right).$$

The transition probability density function in (8.4) or (8.9) gives only the probability distribution of $B(t) - B(0)$. The complete description of the Brownian motion process with diffusion coefficient σ^2 is given by the following definition.

Definition Brownian motion with diffusion coefficient σ^2 is a stochastic process $\{B(t); t \ge 0\}$ with the properties:

(a) Every increment $B(s + t) - B(s)$ is normally distributed with mean zero and variance $\sigma^2 t$; $\sigma^2 > 0$ is a fixed parameter.
(b) For every pair of disjoint time intervals $(t_1, t_2]$, $(t_3, t_4]$, with $0 \le t_1 < t_2 \le t_3 < t_4$, the increments $B(t_4) - B(t_3)$ and $B(t_2) - B(t_1)$ are independent random variables, and similarly for n disjoint time intervals, where n is an arbitrary positive integer.
(c) $B(0) = 0$, and $B(t)$ is continuous as a function of t.

The definition says that a displacement $B(s + t) - B(s)$ is independent of the past, or alternatively, if we know $B(s) = x$, then no further knowledge of the values of $B(\tau)$ for past times $\tau < s$ has any effect on our knowledge of the probability law governing the future movement $B(s + t) - B(s)$. This is a statement of the Markov character of the process. We emphasize, however, that the independent increment assumption (b) is actually more restrictive than the Markov property. A typical Brownian motion path is illustrated in Figure 8.1.

The choice $B(0) = 0$ is arbitrary, and we often consider *Brownian motion starting at* x, for which $B(0) = x$ for some fixed point x. For Brownian motion starting at x, the variance of $B(t)$ is $\sigma^2 t$, and σ^2 is termed the *variance parameter* in the stochastic process literature. The process $\tilde{B}(t) = B(t)/\sigma$ is a Brownian motion process whose variance parameter is one, the so-called *standard Brownian motion*. By this device, we may always reduce an arbitrary Brownian motion to a standard Brownian motion; for the most part, we derive results only for the latter. By part (a) of the definition, for

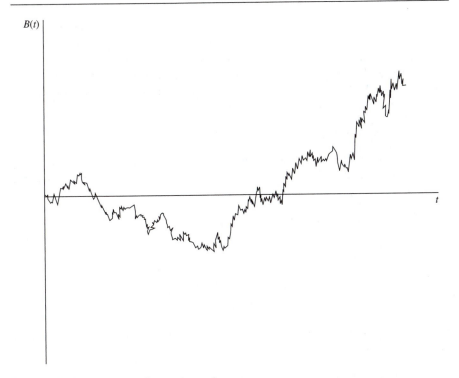

Figure 8.1 A typical Brownian motion path.

a standard Brownian motion ($\sigma^2 = 1$), we have

$$\Pr\{B(s+t) \le y | B(s) = x\} = \Pr\{B(s+t) - B(s) \le y - x\}$$

$$= \Phi\left(\frac{y-x}{\sqrt{t}}\right).$$

Remark Let us look, for the moment, at a Brownian displacement $B(\Delta t)$ after a small elapsed time Δt. The mean displacement is zero, and the variance of the displacement is Δt itself. It is much more common in practical work to use a standard deviation, the square root of the variance, to measure variability. For the normal distribution, for instance, the probability of an observation more than 2 standard deviations away from the mean is about 5%, and the standard deviation is in the same units as the original measurement, and not (units)^2. The standard deviation of the Brownian displacement is $\sqrt{\Delta t}$, which is much larger than Δt itself when Δt is small. Indeed, StdDev$[B(\Delta t)]/\Delta t = \sqrt{\Delta t}/\Delta t = 1/\sqrt{\Delta t} \to \infty$ as $\Delta t \to 0$. This is simply another manifestation of the erratic movements of the Brownian particle, yet it is a point that Bachelier and others had difficulty in handling. *But the variance, being linear in time, and not the standard deviation, is the only possibility if displacements over disjoint time intervals are to be stationary and independent.* Write a total displacement $B(s+t) - B(0)$ as the sum of two incremental steps in the form

$B(s+t) - B(0) = \{B(s) - B(0)\} + \{B(s+t) - B(s)\}$. The incremental steps being statistically independent, their variances must add. The stationary assumption is that the statistics of the second step $B(t+s) - B(t)$ do not depend on the time t when the step began, but only on the duration s of the movement. We must have, then, $\text{Var}[B(t+s)] = \text{Var}[B(t)] + \text{Var}[B(s)]$, and the only nonnegative solution to such an equation is to have the variance of the displacement a linear function of time.

The Covariance Function

Using the independent increments assumption (b), we will determine the covariance of the Brownian motion. Recall that $E[B(t)] = 0$ and that $E[B(t)^2] = \sigma^2 t$. Then, for $0 \le s < t$,

$$
\begin{aligned}
\text{Cov}[B(s), B(t)] &= E[B(s)B(t)] \\
&= E[B(s)\{B(t) - B(s) + B(s)\}] \\
&= E[B(s)^2] + E[B(s)\{B(t) - B(s)\}] \\
&= \sigma^2 s + E[B(s)]E[B(t) - B(s)] \quad \text{(by (b))} \\
&= \sigma^2 s \quad \text{(since } E[B(s)] = 0).
\end{aligned}
$$

Similarly, if $0 \le t < s$, we obtain $\text{Cov}[B(s), B(t)] = \sigma^2 t$. Both cases may be treated in a single expression via

$$
\text{Cov}[B(s), B(t)] = \sigma^2 \min\{s, t\}, \quad \text{for } s, t \ge 0. \tag{8.10}
$$

8.1.3 The Central Limit Theorem and the Invariance Principle

Let $S_n = \xi_1 + \cdots + \xi_n$ be the sum of n independent and identically distributed random variables $\xi_1, \ldots, \xi_n$ having zero means and unit variances. In this case, the central limit theorem asserts that

$$
\lim_{n \to \infty} \Pr\left\{\frac{S_n}{\sqrt{n}} \le x\right\} = \Phi(x) \quad \text{for all } x.
$$

The central limit theorem is stated as a limit. In stochastic modeling, it is used to justify the normal distribution as appropriate for a random quantity whose value results from numerous small random effects, all acting independently and additively. It also justifies the approximate calculation of probabilities for the sum of independent and identically distributed summands in the form $\Pr\{S_n \le x\} \approx \Phi(x/\sqrt{n})$, the approximation known to be excellent even for moderate values of n in most cases in which the distribution of the summands is not too skewed.

In a similar manner, functionals computed for a Brownian motion can often serve as excellent approximations for analogous functionals of a partial sum *process*, as we now explain. As a function of the continuous variable t, define

$$
B_n(t) = \frac{S_{[nt]}}{\sqrt{n}}, \quad t \ge 0, \tag{8.11}
$$

where $[x]$ is the greatest integer less than or equal to x. Observe that

$$B_n(t) = \frac{S_k}{\sqrt{n}} = \frac{\sqrt{[nt]}}{\sqrt{n}} \frac{S_k}{\sqrt{k}}, \quad \text{for } \frac{k}{n} \le t < \frac{k}{n} + \frac{1}{n}.$$

Because $S_k/\sqrt{k}$ has unit variance, the variance of $B_n(t)$ is $[nt]/n$, which converges to t as $n \to \infty$. When n is large, then $k = [nt]$ is large, and $S_k/\sqrt{k}$ is approximately normally distributed by the central limit theorem, and, finally, $B_n(t)$ inherits the independent increments property (b) from the postulated independence of the summands. It is reasonable, then, to believe that $B_n(t)$ should behave much like a standard Brownian motion process, at least when n is large. This is indeed true, and while we cannot explain the precise way in which it holds in an introductory text such as this, the reader should leave with some intuitive feeling for the usefulness of the result and, we hope, a motivation to learn more about stochastic processes. The convergence of the *sequence of stochastic processes* defined in (8.11) to a standard Brownian motion is termed the *invariance principle*. It asserts that some functionals of a partial sum process of independent and identically distributed zero mean and unit variance random variables should not depend too heavily on (should be invariant of) the actual distribution of the summands, but be approximately given by the analogous functional of a standard Brownian motion, provided only that the summands are not too badly behaved.

Example Suppose that the summands have the distribution in which $\xi = \pm 1$, each with probability $\frac{1}{2}$. Then, the partial sum process S_n is a simple random walk for which we calculated in Chapter 3, Section 3.5.3 (using a different notation)

$$\Pr\{S_n \text{ reaches } -a < 0 \text{ before } b > 0 | S_0 = 0\}$$
$$= \frac{b}{a+b}. \tag{8.12}$$

Upon changing the scale in accordance with (8.11), we have

$$\Pr\{B_n(t) \text{ reaches } -a < 0 \text{ before } b > 0 | S_0 = 0\}$$
$$= \frac{b\sqrt{n}}{a\sqrt{n} + b\sqrt{n}} = \frac{b}{a+b},$$

and invoking the invariance principle, it should be, and is, the case that for a standard Brownian motion we have

$$\Pr\{B(t) \text{ reaches } -a < 0 \text{ before } b > 0 | B(0) = 0\}$$
$$= \frac{b}{a+b}. \tag{8.13}$$

Finally, invoking the invariance principle for a second time, the evaluation in (8.12) should hold approximately for an arbitrary partial sum process, provided only that the independent and identically distributed summands have zero means and unit variances.

8.1.4 Gaussian Processes

A random vector $X_1, \ldots, X_n$ is said to have a *multivariate normal distribution,* or a *joint normal distribution,* if every linear combination $\alpha_1 X_1 + \cdots + \alpha_n X_n$, α_i real, has a univariate normal distribution. Obviously, if $X_1, \ldots, X_n$ has a joint normal distribution, then so does the random vector $Y_1, \ldots, Y_m$, defined by the linear transformation in which

$$Y_j = \alpha_{j1} X_1 + \cdots + \alpha_{jn} X_n, \quad \text{for } j = 1, \ldots, m,$$

for arbitrary constants α_{ji}.

The multivariate normal distribution is specified by two parameters, the mean values $\mu_i = E[X_i]$ and the covariance matrix whose entries are $\Gamma_{ij} = \text{Cov}[X_i, X_j]$. In the joint normal distribution, $\Gamma_{ij} = 0$ is sufficient to imply that X_i and X_j are independent random variables.

Let T be an abstract set and $\{X(t); t \text{ in } T\}$ a stochastic process. We call $\{X(t); t \text{ in } T\}$ a *Gaussian process* if for every $n = 1, 2, \ldots$ and every finite subset $\{t_1, \ldots, t_n\}$ of T, the random vector $(X(t_1), \ldots, X(t_n))$ has a multivariate normal distribution. Equivalently, the process is Gaussian if every linear combination

$$\alpha_1 X(t_1) + \cdots + \alpha_n X(t_n), \quad \alpha_i \text{ real,}$$

has a univariate normal distribution. Every Gaussian process is described uniquely by its two parameters, the mean and covariance functions, given respectively by

$$\mu(t) = E[X(t)], \quad t \text{ in } T,$$

and

$$\Gamma(s, t) = E[\{X(s) - \mu(s)\}\{X(t) - \mu(t)\}], \quad s, t \text{ in } T.$$

The covariance function is positive definite in the sense that for every $n = 1, 2, \ldots,$ real numbers $\alpha_1, \ldots, \alpha_n$, and elements $t_1, \ldots, t_1$ in T,

$$\sum_{i=1}^{n} \sum_{j=1}^{n} \alpha_i \alpha_j \Gamma(t_i, t_j) \geq 0.$$

One need only evaluate the expected value of $\left(\sum_{i=1}^{n} \alpha_i \{X(t_i) - \mu(t_i)\} \right)^2 \geq 0$ in terms of the covariance function in order to verify this.

Conversely, given an arbitrary mean value function $\mu(t)$ and a positive definite covariance function $\Gamma(s, t)$, there, then, exists a corresponding Gaussian process. Brownian motion is the unique Gaussian process having continuous trajectories, zero mean, and covariance function (8.10). We shall use this feature, that the mean value and covariance functions define a Gaussian process, several times in what follows.

We have seen how the invariance principle leads to the Gaussian process called Brownian motion. Gaussian processes also arise as the limits of normalized sums of independent and identically distributed random *functions*. To sketch out this idea, let $\xi_1(t), \xi_2(t), \ldots$ be independent and identically distributed random functions, or stochastic processes. Let $\mu(t) = E[\xi(t)]$ and $\Gamma(s, t) = \text{Cov}[\xi(s), \xi(t)]$ be the mean value and covariance functions, respectively. Motivated by the central limit theorem, we define

$$X_N(t) = \frac{\sum_{i=1}^{N}\{\xi_i(t) - \mu(t)\}}{\sqrt{N}}.$$

The central limit theorem tells us that the distribution of $X_N(t)$ converges to the normal distribution for each fixed time point t. A multivariate extension of the central limit theorem asserts that for any finite set of time points $(t_1, \ldots, t_n)$, the random vector

$$(X_N(t_1), \ldots, X_N(t_n))$$

has, in the limit for large N, a multivariate normal distribution. It is not difficult to believe, then, that under ordinary circumstances, the stochastic processes $\{X_N(t); t \geq 0\}$ would converge, in an appropriate sense, to a Gaussian process $\{X(t); t \geq 0\}$ whose mean is zero and whose covariance function is $\Gamma(s, t)$. We call this the *central limit principle for random functions*. Several instances of its application appear in this chapter, the first of which is next.

Example *Cable Strength Under Equal Load Sharing*   Consider a cable constructed from N wires in parallel. Suspension bridge cables are usually built this way. A section of the cable is clamped at each end and elongated by increasing the distance between the clamps. The problem is to determine the maximum tensile load that the cable will sustain in terms of the probabilistic and mechanical characteristics of the individual wires.

Let L_0 be the reference length of an unstretched unloaded strand of wire, and let L be the length after elongation. The *nominal strain* is defined to be $t = (L - L_0)/L_0$. Steadily increasing t causes the strand to stretch and exert a force $\xi(t)$ on the clamps, up to some random failure strain ζ, at which point the wire breaks. Hooke's law of elasticity asserts that the wire force is proportional to wire strain, with Young's modulus K as the proportionality constant. Taken all together, we write the force on the wire as a function of the nominal strain as

$$\xi(t) = \begin{cases} Kt, & \text{for } 0 \leq t < \zeta, \\ 0 & \text{for } \zeta \leq t. \end{cases} \tag{8.14}$$

A typical load function is depicted in Figure 8.2.

We will let $F(x) = \Pr\{\zeta \leq x\}$ be the cumulative distribution function of the failure strain. We easily determine the mean load on the wire to be

$$\mu(t) = E[\xi(t)] = E[Kt\mathbf{1}\{t < \zeta\}] = Kt[1 - F(t)]. \tag{8.15}$$

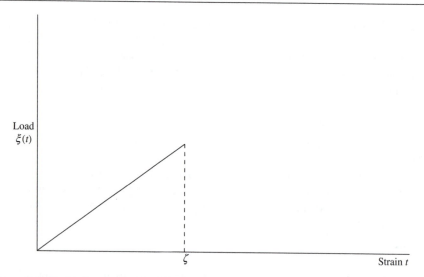

Figure 8.2 The load on an elastic wire as a function of nominal strain. At a strain of ζ the wire fails, and the load carried drops to zero.

The higher moments are, for $0 < s < t$,

$$E[\xi(s)\xi(t)] = K^2 st E[\mathbf{1}\{s < \zeta\}\mathbf{1}\{t < \zeta\}]$$
$$= K^2 st[1 - F(t)]$$

and

$$\Gamma(s, t) = E[\xi(s)\xi(t)] - E[\xi(s)]E[\xi(t)]$$
$$= K^2 st F(s)[1 - F(t)], \quad \text{for } 0 < s < t. \tag{8.16}$$

Turning to the cable, if it is clamped at the ends and elongated, then each wire within it is elongated the same amount. The total force $S_N(t)$ on the cable is the sum of the forces exerted by the individual wires. If we assume that the wires are independent and *a priori* identical, then these wire forces $\xi_1(t), \xi_2(t), \ldots$ are independent and identically distributed random functions, and

$$S_N(t) = \sum_{i=1}^{N} \xi_i(t)$$

is the random load experienced by the cable as a function of the cable strain. An illustration when $N = 5$ is given in Figure 8.3.

We are interested in the maximum load that the cable could carry without failing. This is

$$Q_N = \max\{S_N(t); t \ge 0\}.$$

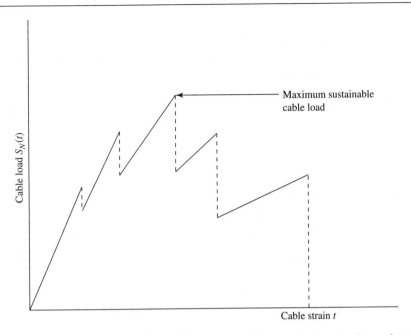

Figure 8.3 The load experienced by a cable composed of five elastic wire strands as a function of the cable strain t.

To obtain an approximation to the distribution of Q_N, we apply the central limit principle for random functions. This leads us to believe that

$$X_N(t) = \frac{S_N(t) - N\mu(t)}{\sqrt{N}}$$

should, for large N, be approximately a Gaussian process $X(t)$ with mean zero and covariance function given by (8.16). We write this approximation in the form

$$S_N(t) \approx N\mu(t) + \sqrt{N}X(t). \tag{8.17}$$

When N is large, the dominant term on the right of (8.17) is $N\mu(t)$. Let t^* be the value of t that maximizes $\mu(t)$. We assume that t^* is unique and that the second derivative of $\mu(t)$ is strictly negative at $t = t^*$. We would then expect that

$$Q_N = \max S_N(t) \approx N\mu\left(t^*\right) + \sqrt{N}X\left(t^*\right). \tag{8.18}$$

That is, we would expect that the cable strength would be approximately normally distributed with mean $N\mu(t^*)$ and variance $N\Gamma(t^*, t^*)$. To carry out a numerical example, suppose that $F(x) = 1 - \exp\{-x^5\}$, a Weibull distribution with shape parameter 5. It is easily checked that $t^* = 1/5^{0.2} = 0.7248$, that $\mu(t^*) = 0.5934K$, and $\Gamma(t^*, t^*) = (0.2792)^2 K^2$. A cable composed of $N = 30$ wires would have a strength that is approximately normally distributed with mean $30(0.5934)K = 17.8K$ and standard deviation $0.2792\sqrt{30}K = 1.5292K$.

The above heuristics can be justified, and, indeed, significant refinements in the approximation have been made. We have referred to the approach as the central limit principle for random functions because we have not supplied sufficient details to label it a theorem. Nevertheless, we will see several more applications of the principle in subsequent sections of this chapter.

Exercises

8.1.1 Let $\{B(t); t \geq 0\}$ be a standard Brownian motion.
  (a) Evaluate $\Pr\{B(4) \leq 3 | B(0) = 1\}$.
  (b) Find the number c for which $\Pr\{B(9) > c | B(0) = 1\} = 0.10$.
8.1.2 Let $\{B(t); t \geq 0\}$ be a standard Brownian motion and $c > 0$ a constant. Show that the process defined by $W(t) = cB\left(t/c^2\right)$ is a standard Brownian motion.
8.1.3 (a) Show that

$$\frac{d\phi(x)}{dx} = \phi'(x) = -x\phi(x),$$

where $\phi(x)$ is given in (8.5).
  (b) Use the result in (a) together with the chain rule of differentiation to show that

$$p(y, t|x) = \phi_t(y - x) = \frac{1}{\sqrt{t}}\phi\left(\frac{y - x}{\sqrt{t}}\right)$$

satisfies the diffusion equation (8.3).
8.1.4 Consider a standard Brownian motion $\{B(t); t \geq 0\}$ at times $0 < u < u + v < u + v + w$, where $u, v, w > 0$.
  (a) Evaluate the product moment $E[B(u)B(u + v)B(u + v + w)]$.
  (b) Evaluate the product moment

$$E[B(u)B(u + v)B(u + v + w)B(u + v + w + x)]$$

where $x > 0$.
8.1.5 Determine the covariance functions for the stochastic processes
  (a) $U(t) = e^{-t}B\left(e^{2t}\right)$, for $t \geq 0$.
  (b) $V(t) = (1 - t)B(t/(1 - t))$, for $0 < t < 1$.
  (c) $W(t) = tB(1/t)$, with $W(0) = 0$.
  $B(t)$ is standard Brownian motion.
8.1.6 Consider a standard Brownian motion $\{B(t); t \geq 0\}$ at times $0 < u < u + v < u + v + w$, where $u, v, w > 0$.
  (a) What is the probability distribution of $B(u) + B(u + v)$?
  (b) What is the probability distribution of $B(u) + B(u + v) + B(u + v + w)$?

8.1.7 Suppose that in the absence of intervention, the cash on hand for a certain corporation fluctuates according to a standard Brownian motion $\{B(t); t \geq 0\}$. The company manages its cash using an (s, S) policy: If the cash level ever drops to zero, it is instantaneously replenished up to level s; If the cash level ever rises up to S, sufficient cash is invested in long-term securities to bring the cash-on-hand down to level s. In the long run, what fraction of cash interventions are investments of excess cash?

Hint: Use equation (8.13).

Problems

8.1.1 Consider the simple random walk

$$S_n = \xi_1 + \cdots + \xi_n, \quad S_0 = 0,$$

in which the summands are independent with $\Pr\{\xi = \pm 1\} = \frac{1}{2}$. In Chapter 3, Section 3.5.3, we showed that the mean time for the random walk to first reach $-a < 0$ or $b > 0$ is ab. Use this together with the invariance principle to show that $E[T] = ab$, where

$$T = T_{a,b} = \min\{t \geq 0; B(t) = -a \text{ or } B(t) = b\},$$

and $B(t)$ is standard Brownian motion.

Hint: The approximate Brownian motion (8.11) rescales the random walk in both time and space.

8.1.2 Evaluate $E\left[e^{\lambda B(t)}\right]$ for an arbitrary constant λ and standard Brownian motion $B(t)$.

8.1.3 For a positive constant ϵ, show that

$$\Pr\left\{\frac{|B(t)|}{t} > \epsilon\right\} = 2\{1 - \Phi(\epsilon\sqrt{t})\}.$$

How does this behave when t is large $(t \to \infty)$? How does it behave when t is small $(t \approx 0)$?

8.1.4 Let $\alpha_1, \ldots, \alpha_n$ be real constants. Argue that

$$\sum_{i=1}^{n} \alpha_i B(t_i)$$

is normally distributed with mean zero and variance

$$\sum_{i=1}^{n} \sum_{j=1}^{n} \alpha_i \alpha_j \min\{t_i, t_j\}.$$

8.1.5 Consider the simple random walk

$$S_n = \xi_1 + \cdots + \xi_n, \quad S_0 = 0,$$

in which the summands are independent with $\Pr\{\xi = \pm 1\} = \frac{1}{2}$. We are going to stop this random walk when it first drops a units below its maximum to date. Accordingly, let

$$M_n = \max_{0 \leq k \leq n} S_k, \quad Y_n = M_n - S_n, \quad \text{and}$$

$$\tau = \tau_a = \min\{n \geq 0; Y_n = a\}.$$

(a) Use a first step analysis to show that

$$\Pr\{M_\tau = 0\} = \frac{1}{1+a}.$$

(b) Why is $\Pr\{M_\tau \geq 2\} = \Pr\{M_\tau \geq 1\}^2$, and

$$\Pr\{M_\tau \geq k\} = \left(\frac{a}{1+a}\right)^k ?$$

Identify the distribution of M_τ.

(c) Let $B(t)$ be standard Brownian motion, $M(t) = \max\{B(u); 0 \leq u \leq t\}$, $Y(t) = M(t) - B(t)$, and $\tau = \min\{t \geq 0; Y(t) = a\}$. Use the invariance principle to argue that $M(\tau)$ has an exponential distribution with mean a.

Note: τ is a popular strategy for timing the sale of a stock. It calls for keeping the stock as long as it is going up, but to sell it the first time that it drops a units from its best price to date. We have shown that $E[M(\tau)] = a$, whence $E[B(\tau)] = E[M(\tau)] - a = 0$, so that the strategy does not gain a profit, on average, in the Brownian motion model for stock prices.

8.1.6 Manufacturers of crunchy munchies such as cheese crisps use compression testing machines to gauge product quality. The crisp, or whatever, is placed between opposing plates, which then move together. As the crisp is crunched, the force is measured as a function of the distance that the plates have moved. The output of the compression testing machine is a graph of force versus distance that is much like Figure 8.3. What aspects of the graph might be measures of product quality? Model the test as a row of tiny balloons between parallel plates. Each single balloon might follow a force–distance behavior of the form $\sigma = Ke(1 - q(e))$, where σ is the force, K is Young's modulus, e is strain or distance, and $q(e)$ is a function that measures departures from Hooke's law, to allow for soggy crisps. Each balloon obeys this relationship up until the random strain ζ at which it bursts. Determine the mean force as a function of strain. Use $F(x)$ for the cumulative distribution function of failure strain.

8.1.7 For $n = 0, 1, \ldots$, show that (a) $B(n)$ and (b) $B(n)^2 - n$ are martingales (see Chapter 2, Section 2.5).

8.1.8 *Computer Challenge* A problem of considerable contemporary importance is how to simulate a Brownian motion stochastic process. The invariance principle provides one possible approach. An infinite series expression that N. Wiener introduced may provide another approach. Let $Z_0, Z_1, \ldots$ be a series of independent standard normal random variables. The infinite series

$$B(t) = \frac{t}{\sqrt{\pi}} Z_0 + \sqrt{\frac{2}{\pi}} \sum_{m=1}^{\infty} \frac{\sin mt}{m} Z_m, \quad 0 \le t \le 1,$$

is a standard Brownian motion for $0 \le t \le 1$. Try to simulate a Brownian motion stochastic process, at least approximately, by using finite sums of the form

$$B_N(t) = \frac{t}{\sqrt{\pi}} Z_0 + \sqrt{\frac{2}{\pi}} \sum_{m=1}^{N} \frac{\sin mt}{m} Z_m, \quad 0 \le t \le 1.$$

If $B(t), 0 \le t \le 1$, is a standard Brownian motion on the interval $[0, 1]$, then $B'(t) = (1+t)B(1/(1+t)), 0 \le t < \infty$, is a standard Brownian motion on the interval $[0, \infty)$. This suggests

$$B'_N(t) = (1+t)B_N\left(\frac{1}{1+t}\right), \quad 0 \le t < \infty,$$

as an approximate standard Brownian motion. In what ways do these finite approximations behave like Brownian motion? Clearly, they are zero mean Gaussian processes. What is the covariance function, and how does it compare to that of Brownian motion? Do the gambler's ruin probabilities of (8.13) accurately describe their behavior? It is known that the squared variation of a Brownian motion stochastic process is not random, but constant:

$$\lim_{n \to \infty} \sum_{k=1}^{n} \left| B\left(\frac{k}{n}\right) - B\left(\frac{k-1}{n}\right) \right|^2 = 1.$$

This is a further consequence of the variance relation $E\left[(\Delta B)^2\right] = \Delta t$ (see the remark in Section 8.1.2). To what degree do the finite approximations meet this criterion?

8.2 The Maximum Variable and the Reflection Principle

Using the continuity of the trajectories of Brownian motion and the symmetry of the normal distribution, we will determine a variety of interesting probability expressions for the Brownian motion process. The starting point is the *reflection principle*.

8.2.1   The Reflection Principle

Let $B(t)$ be a standard Brownian motion. Fix a value $x > 0$ and a time $t > 0$. Bearing in mind the continuity of the Brownian motion, property (c) of the definition, consider the collection of sample paths $B(u)$ for $u \geq 0$ with $B(0) = 0$ and for which $B(t) > x$. Since $B(u)$ is continuous and $B(0) = 0$, there exists a time τ, itself a random variable depending on the particular sample trajectory, at which the Brownian motion $B(u)$ first attains the value x.

We next describe a new path $B^*(u)$ obtained from $B(u)$ by reflection. For $u > \tau$, we reflect $B(u)$ about the horizontal line at height $x > 0$ to obtain

$$B^*(u) = \begin{cases} B(u), & \text{for } u \leq \tau, \\ x - [B(u) - x], & \text{for } u > \tau. \end{cases}$$

Figure 8.4 illustrates the construction. Note that $B^*(t) < x$ because $B(t) > x$.

Because the conditional probability law of the path for $u > \tau$, given that $B(\tau) = x$, is symmetric with respect to the values $y > x$ and $y < x$, and independent of the history prior to τ,* the reflection argument displays for every sample path with $B(t) > x$ two

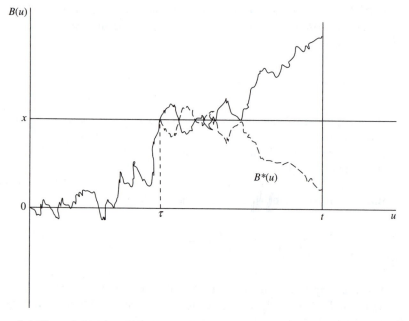

Figure 8.4 The path $B(u)$ is reflected about the horizontal line at x, showing that for every path ending at $B(t) > x$, there are two paths, $B(u)$ and $B^*(u)$, that attain the value x somewhere in the interval $0 \leq u \leq t$.

---

* The argument is not quite complete because the definition asserts that an increment in a Brownian motion after a fixed time t is independent of the past, whereas here we are restarting from the random time τ. While the argument is incomplete, the assertion is true: A Brownian path begins afresh from *hitting times* such as τ.

equally likely sample paths, $B(u)$ and $B^*(u)$, for which both

$$\max_{0 \leq u \leq t} B(u) > x \quad \text{and} \quad \max_{0 \leq u \leq t} B^*(u) > x.$$

Conversely, by the nature of this correspondence, every sample path $B(u)$ for which $\max_{0 \leq u \leq t} B(u) > x$ results from either of two equally likely sample paths, exactly one of which is such that $B(t) > x$. The two-to-one correspondence fails only if $B(t) = x$, but because $B(t)$ is a continuous random variable (normal distribution), we have $\Pr\{B(t) = x\} = 0$, and this case can be safely ignored. Thus, we conclude that

$$\Pr\left\{ \max_{0 \leq u \leq t} B(u) > x \right\} = 2\Pr\{B(t) > x\}.$$

In terms of the maximum process defined by

$$M(t) = \max_{0 \leq u \leq t} B(u), \tag{8.19}$$

and using the notation set forth in (8.8), we have

$$\Pr\{M(t) > x\} = 2[1 - \Phi_t(x)]. \tag{8.20}$$

8.2.2 The Time to First Reach a Level

With the help of (8.20), we may determine the probability distribution of the random time τ_x at which the Brownian motion first attains a prescribed value $x > 0$ starting from $B(0) = 0$. Formally, define the *hitting time*

$$\tau_x = \min\{u \geq 0; B(u) = x\}. \tag{8.21}$$

Clearly, $\tau_x \leq t$ if and only if $M(t) \geq x$. In words, the Brownian motion attains the level $x > 0$ before time t if and only if at time t the maximum of the process is at least x. If the two events are equivalent, then their probabilities must be the same. That is,

$$\Pr\{\tau_x \leq t\} = \Pr\{M(t) \geq x\} = 2[1 - \Phi_t(x)] \tag{8.22}$$

$$= \frac{2}{\sqrt{2\pi t}} \int_x^\infty e^{-\xi^2/(2t)} d\xi.$$

The change of variable $\xi = \eta\sqrt{t}$ leads to

$$\Pr\{\tau_x \leq t\} = \sqrt{\frac{2}{\pi}} \int_{x/\sqrt{t}}^\infty e^{-\eta^2/2} d\eta. \tag{8.23}$$

The probability density function of the random time τ is obtained by differentiating (8.23) with respect to t, giving

$$f_{\tau_x}(t) = \frac{x t^{-3/2}}{\sqrt{2\pi}} e^{-x^2/(2t)} \quad \text{for} \quad 0 < t < \infty. \tag{8.24}$$

8.2.3 The Zeros of Brownian Motion

As a final illustration of the far-reaching consequences of the reflection principle and equation (8.24), we will determine the probability that a standard Brownian motion $B(t)$, with $B(0) = 0$, will cross the t axis at least once in the time interval $(t, t+s]$ for $t, s > 0$. Let us denote this quantity by $\vartheta(t, t+s)$. The result is

$$
\begin{aligned}
\vartheta(t, t+s) &= \Pr\{B(u) = 0 \text{ for some } u \text{ in } (t, t+s]\} \\
&= \frac{2}{\pi} \arctan \sqrt{s/t} \\
&= \frac{2}{\pi} \arccos \sqrt{t/(t+s)}.
\end{aligned}
\tag{8.25}
$$

First, let us define some notation concerning the hitting time τ_x defined in (8.21). Let

$$H_t(z, x) = \Pr\{\tau_x \le t | B(0) = z\}$$

be the probability that a standard Brownian motion starting from $B(0) = z$ will reach the level x before time t. In equation (8.22), we gave an integral that evaluated

$$H_t(0, x) = \Pr\{\tau_x \le t | B(0) = 0\}, \quad \text{for } x > 0.$$

The symmetry and spatial homogeneity of the Brownian motion make it clear that $H_t(0, x) = H_t(x, 0)$. That is, the probability of reaching $x > 0$ starting from $B(0) = 0$ before time t is the same as the probability of reaching 0 starting from $B(0) = x$. Consequently, from (8.24) we have

$$
\begin{aligned}
H_t(0, x) = H_t(x, 0) &= \Pr\{\tau_0 \le t | B(0) = x\} \\
&= \int_0^t \frac{x}{\sqrt{2\pi}} \xi^{-3/2} e^{-x^2/(2\xi)} d\xi.
\end{aligned}
\tag{8.26}
$$

We will condition on the value of the Brownian motion at time t and use the law of total probability to derive (8.25). Accordingly, we have

$$\vartheta(t, t+s) = \int_{-\infty}^{\infty} \Pr\{B(u) = 0 \text{ for some } u \text{ in } (t, t+s] | B(t) = x\} \phi_t(x) dx,$$

where $\phi_t(x)$ is the probability density function for $B(t)$ as given in (8.7). Then, using (8.26),

$$\vartheta(t, t+s) = 2 \int_0^\infty H_s(x, 0)\phi_t(x)\,dx$$

$$= 2 \int_0^\infty \left\{ \int_0^s \frac{x}{\sqrt{2\pi}} \xi^{-3/2} e^{-x^2/(2\xi)}\,d\xi \right\} \frac{1}{\sqrt{2\pi t}} e^{-x^2/2t}\,dx$$

$$= \frac{1}{\pi\sqrt{t}} \int_0^s \left\{ \int_0^\infty x e^{-x^2/(2\xi) - x^2/(2t)}\,dx \right\} \xi^{-3/2}\,d\xi.$$

To evaluate the inner integral, we let

$$v = \frac{x^2}{2}\left(\frac{1}{\xi} + \frac{1}{t}\right),$$

whence

$$dv = x\left(\frac{1}{\xi} + \frac{1}{t}\right)dx,$$

and so

$$\vartheta(t, t+s) = \frac{1}{\pi\sqrt{t}} \int_0^s \left(\frac{1}{\xi} + \frac{1}{t}\right)^{-1} \left\{ \int_0^\infty v e^{-v}\,dv \right\} \xi^{-3/2}\,d\xi$$

$$= \frac{\sqrt{t}}{\pi} \int_0^s \frac{d\xi}{(t+\xi)\sqrt{\xi}}.$$

The change of variable $\eta = \sqrt{\xi/t}$ gives

$$\vartheta(t, t+s) = \frac{2}{\pi} \int_0^{\sqrt{s/t}} \frac{d\eta}{1+\eta^2} = \frac{2}{\pi} \arccos\sqrt{t/(t+s)}.$$

Finally, Exercise 8.2.2 asks the student to use standard trigonometric identities to show the equivalence $\arctan\sqrt{s/t} = \arccos\sqrt{t/(t+s)}$.

Exercises

8.2.1 Let $\{B(t); t \geq 0\}$ be a standard Brownian motion, with $B(0) = 0$, and let $M(t) = \max\{B(u); 0 \leq u \leq t\}$.

(a) Evaluate $\Pr\{M(4) \leq 2\}$.

(b) Find the number c for which $\Pr\{M(9) > c\} = 0.10$.

8.2.2 Show that

$$\arctan \sqrt{s/t} = \arccos \sqrt{t/(s+t)}.$$

8.2.3 Suppose that net inflows to a reservoir are described by a standard Brownian motion. If at time 0, the reservoir has $x = 3.29$ units of water on hand, what is the probability that the reservoir never becomes empty in the first $t = 4$ units of time?

8.2.4 Consider the simple random walk

$$S_n = \xi_1 + \cdots + \xi_n, \quad S_0 = 0,$$

in which the summands are independent with $\Pr\{\xi = \pm 1\} = \frac{1}{2}$. Let $M_n = \max_{0 \leq k \leq n} S_k$. Use a reflection argument to show that

$$\Pr\{M_n \geq a\} = 2\Pr\{S_n > a\} + \Pr\{S_n = a\}, \quad a > 0.$$

8.2.5 Let τ_0 be the largest zero of a standard Brownian motion not exceeding $a > 0$. That is, $\tau_0 = \max\{u \geq 0; B(u) = 0 \text{ and } u \leq a\}$. Show that

$$\Pr\{\tau_0 < t\} = \frac{2}{\pi} \arcsin \sqrt{t/a}.$$

8.2.6 Let τ_1 be the smallest zero of a standard Brownian motion that exceeds $b > 0$. Show that

$$\Pr\{\tau_1 < t\} = \frac{2}{\pi} \arccos \sqrt{b/t}.$$

Problems

8.2.1 Find the conditional probability that a standard Brownian motion is not zero in the interval $(t, t+b]$ given that it is not zero in the interval $(t, t+a]$, where $0 < a < b$ and $t > 0$.

8.2.2 Find the conditional probability that a standard Brownian motion is not zero in the interval $(0, b]$ given that it is not zero in the interval $(0, a]$, where $0 < a < b$.

Hint: Let $t \to 0$ in the result of Problem 8.2.1.

8.2.3 For a fixed $t > 0$, show that $M(t)$ and $|B(t)|$ have the same marginal probability distribution, whence

$$f_{M(t)}(z) = \frac{2}{\sqrt{t}} \phi\left(\frac{z}{\sqrt{t}}\right) \quad \text{for} \quad z > 0.$$

(Here $M(t) = \max_{0 \leq u \leq t} B(u)$.) Show that

$$E[M(t)] = \sqrt{2t/\pi}.$$

For $0 < s < t$, do $(M(s), M(t))$ have the same joint distribution as $(|B(s)|,$ $|B(t)|)$?

8.2.4 Use the reflection principle to obtain

$$\Pr\{M(t) \geq z, B(t) \leq x\} = \Pr\{B(t) \geq 2z - x\}$$
$$= 1 - \Phi\left(\frac{2z - x}{\sqrt{t}}\right) \quad \text{for} \quad 0 < x < m.$$

($M(t)$ is the maximum defined in (8.19).) Differentiate with respect to x, and then with respect to z, to obtain the joint density function for $M(t)$ and $B(t)$:

$$f_{M(t),B(t)}(z, x) = \frac{2z - x}{t} \frac{2}{\sqrt{t}} \phi\left(\frac{2z - x}{\sqrt{t}}\right).$$

8.2.5 Show that the joint density function for $M(t)$ and $Y(t) = M(t) - B(t)$ is given by

$$f_{M(t),Y(t)}(z, y) = \frac{z + y}{t} \frac{2}{\sqrt{t}} \phi\left(\frac{z + y}{\sqrt{t}}\right).$$

8.2.6 Use the result of Problem 8.2.5 to show that $Y(t) = M(t) - B(t)$ has the same distribution as $|B(t)|$.

8.3 Variations and Extensions

A variety of processes derived from Brownian motion find relevance and application in stochastic modeling. We briefly describe a few of these.

8.3.1 Reflected Brownian Motion

Let $\{B(t); t \geq 0\}$ be a standard Brownian motion process. The stochastic process

$$R(t) = |B(t)| = \begin{cases} B(t), & \text{if } B(t) \geq 0, \\ -B(t), & \text{if } B(t) < 0, \end{cases}$$

is called *Brownian motion reflected at the origin*, or, more briefly, *reflected Brownian motion*. Reflected Brownian motion reverberates back to positive values whenever it reaches the zero level and, thus, might be used to model the movement of a pollen grain in the vicinity of a container boundary that the grain cannot cross.

Since the moments of $R(t)$ are the same as those of $|B(t)|$, the mean and variance of reflected Brownian motion are easily determined. Under the condition that $R(0) = 0$, e.g., we have

$$E[R(t)] = \int\limits_{-\infty}^{\infty} |x| \phi_t(x) dx$$

$$= 2 \int\limits_{0}^{\infty} \frac{x}{\sqrt{2\pi t}} \exp\left(-x^2/2t\right) dx \qquad (8.27)$$

$$= \sqrt{2t/\pi}.$$

The integral was evaluated through the change of variable $y = x/\sqrt{t}$. Also,

$$\text{Var}[R(t)] = E\left[R(t)^2\right] - \{E[R(t)]\}^2$$

$$= E\left[B(t)^2\right] - 2t/\pi \qquad (8.28)$$

$$= \left(1 - \frac{2}{\pi}\right)t.$$

Reflected Brownian motion is a second example of a continuous-time, continuous-state-space Markov process. Its transition density $p(y, t|x)$ is derived from that of Brownian motion by differentiating

$$\Pr\{R(t) \le y | R(0) = x\} = \Pr\{-y \le B(t) \le y | B(0) = x\}$$

$$= \int\limits_{-y}^{y} \phi_t(z - x) dz$$

with respect to y to get

$$p(y, t|x) = \phi_t(y - x) + \phi_t(-y - x)$$
$$= \phi_t(y - x) + \phi_t(y + x).$$

8.3.2 Absorbed Brownian Motion

Suppose that the initial value $B(0) = x$ of a standard Brownian motion process is positive, and let τ be the first time that the process reaches zero. The stochastic process

$$A(t) = \begin{cases} B(t) & \text{for } t \le \tau, \\ 0 & \text{for } t > \tau \end{cases}$$

is called *Brownian motion absorbed at the origin*, which we will shorten to *absorbed Brownian motion*. Absorbed Brownian motion might be used to model the price of a share of stock in a company that becomes bankrupt at some future instant. We can evaluate the transition probabilities for absorbed Brownian motion by another use of the reflection principle introduced in Section 8.2. For $x > 0$ and $y > 0$, let

$$G_t(x, y) = \Pr\{A(t) > y | A(0) = x\} \tag{8.29}$$
$$= \Pr\left\{B(t) > y, \min_{0 \leq u \leq t} B(u) > 0 | B(0) = x\right\}.$$

To determine (8.29), we start with the obvious relation

$$\Pr\{B(t) > y | B(0) = x\} = G_t(x, y) + \Pr\left\{B(t) > y, \min_{0 \leq u \leq t} B(u) \leq 0 | B(0) = x\right\}.$$

The reflection principle is applied to the last term; Figure 8.5 is the appropriate picture to guide the analysis. We will argue that

$$\Pr\left\{B(t) > y, \min_{0 \leq u \leq t} B(u) \leq 0 | B(0) = x\right\}$$
$$= \Pr\left\{B(t) < -y, \min_{0 \leq u \leq t} B(u) \leq 0 | B(0) = x\right\} \tag{8.30}$$
$$= \Pr\{B(t) < -y | B(0) = x\} = \Phi_t(-y - x).$$

The reasoning behind (8.30) goes as follows: Consider a path starting at $x > 0$, satisfying $B(t) > y$, and which reaches zero at some intermediate time τ. By reflecting such

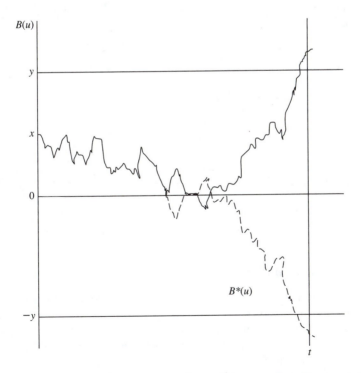

Figure 8.5 For every path $B(u)$ starting at x, ending at $B(t) > y$, and reaching zero in the interval, there is another path $B^*(u)$ starting at x and ending at $B^*(t) < -y$.

a path about zero after time τ, we obtain an equally likely path starting from x and assuming a value below $-y$ at time t. This implies the equality of the first two terms of (8.30). The equality of the last terms is clear from their meaning, since the condition that the minimum be below zero is superfluous in view of the requirement that the path end below $-y (y > 0)$. Inserting (8.30) into (8.29) yields

$$
\begin{aligned}
G_t(x, y) &= \Pr\{B(t) > y | B(0) = x\} - \Pr\{B(t) < -y | B(0) = x\} \\
&= 1 - \Phi_t(y - x) - \Phi_t(-(y + x)) \\
&= \Phi_t(y + x) - \Phi_t(y - x) \\
&= \int_{y-x}^{y+x} \phi_t(z) dz = \Phi\left(\frac{y+x}{\sqrt{t}}\right) - \Phi\left(\frac{y-x}{\sqrt{t}}\right).
\end{aligned}
\tag{8.31}
$$

From (8.29) and (8.31), we obtain the transition distribution for absorbed Brownian motion:

$$
\Pr\{A(t) > y | A(0) = x\} = \Phi\left(\frac{y+x}{\sqrt{t}}\right) - \Phi\left(\frac{y-x}{\sqrt{t}}\right).
\tag{8.32}
$$

Under the condition that $A(0) = x > 0$, $A(t)$ is a random variable that has both discrete and continuous parts. The discrete part is

$$
\begin{aligned}
\Pr\{A(t) = 0 | A(0) = x\} &= 1 - G_t(x, 0) \\
&= 1 - \int_{-x}^{x} \phi_t(z) dz \\
&= 2[1 - \Phi_t(x)].
\end{aligned}
$$

In the region $y > 0$, $A(t)$ is a continuous random variable whose transition density $p(y, t | x)$ is obtained by differentiating with respect to y in (8.32) and suitably changing the sign:

$$
p(y, t | x) = \phi_t(y - x) - \phi_t(y + x).
$$

8.3.3   The Brownian Bridge

The *Brownian bridge* $\{B^0(t); t \geq 0\}$ is constructed from a standard Brownian motion $\{B(t); t \geq 0\}$ by conditioning on the event $\{B(0) = B(1) = 0\}$. The Brownian bridge is used to describe certain random functionals arising in nonparametric statistics, and as a model for the publicly traded prices of bonds having a specified redemption value on a fixed expiration date.

   We will determine the probability distribution for $B^0(t)$ by using the conditional density formula for jointly normally distributed random variables derived in Chapter 2,

Problem 2.4.8. First, for $0 < t < 1$, the random variables $B(t)$ and $B(1) - B(t)$ are independent and normally distributed according to the definition of Brownian motion. It follows that $X = B(t)$ and $Y = B(1) = B(t) + \{B(1) - B(t)\}$ have a joint normal distribution (see Chapter 1, Section 1.4.6) for which we have determined $\mu_X = \mu_Y = 0$, $\sigma_X = \sqrt{t}$, $\sigma_Y = 1$, and $\rho = \text{Cov}[X, Y]/\sigma_X\sigma_Y = \sqrt{t}$. Using the results of Chapter 2, Problem 2.4.8, it then follows that given $Y = B(1) = y$, the conditional distribution of $X = B(t)$ is normal, with

$$\mu_{X|Y} = \mu_X + \frac{\rho\sigma_X}{\sigma_Y}(y - \mu_Y) = y\sqrt{t} = 0 \quad \text{when} \quad y = 0,$$

and

$$\sigma_{X|Y} = \sigma_X\sqrt{1 - \rho^2} = \sqrt{t(1 - t)}.$$

For the Brownian bridge, $B^0(t)$ is normally distributed with $E\left[B^0(t)\right] = 0$ and $\text{Var}\left[B^0(t)\right] = t(1 - t)$. Notice how the condition $B(0) = B(1) = 0$ causes the variance of $B^0(t)$ to vanish at $t = 0$ and $t = 1$.

The foregoing calculation of the variance can be extended to determine the covariance function. Consider times s, t with $0 < s < t < 1$. By first obtaining the joint distribution of $(B(s), B(t), B(1))$, and then the conditional joint distribution of $(B(s), B(t))$, given that $B(1) = 0$, one can verify that the Brownian bridge is a normally distributed stochastic process with mean zero and covariance function $\Gamma(s, t) = \text{Cov}\left[B^0(s), B^0(t)\right] = s(1 - t)$, for $0 < s < t < 1$. (See Problem 8.3.3 for an alternative approach.)

Example *The Empirical Distribution Function* Let $X_1, X_2, \ldots$ be independent and identically distributed random variables. The *empirical cumulative distribution function* corresponding to a sample of size N is defined by

$$F_N(t) = \frac{1}{N}\#\{X_i \leq t \quad \text{for } i = 1, \ldots, N\}$$

$$= \frac{1}{N}\sum_{i=1}^{N}\xi_i(t), \tag{8.33}$$

where

$$\xi_i(t) = \begin{cases} 1, & \text{if } X_i \leq t, \\ 0, & \text{if } X_i > t. \end{cases}$$

The empirical distribution function is an estimate, based on the observed sample, of the true distribution function $F(t) = \Pr\{X \leq t\}$. We will use the central limit principle for random functions (Section 8.1.4) to approximate the empirical distribution function by a Brownian bridge, assuming that the observations are uniformly distributed over the interval $(0, 1)$. (Problem 8.3.9 calls for the student to explore the case of a general distribution.) In the uniform case, $F(t) = t$ for $0 < t < 1$, and $\mu(t) = E[\xi(t)] = F(t) = t$. For the higher moments, when $0 < s < t < 1$, $E[\xi(s)\xi(t)] = F(s) = s$, and $\Gamma(s, t) = \text{Cov}[\xi(s), \xi(t)] = E[\xi(s)\xi(t)] - E[\xi(s)]E[\xi(t)] = s - st = s(1 - t)$.

In view of equation (8.33), which expresses the empirical distribution function in terms of a sum of independent and identically distributed random functions, we might expect the central limit principle for random functions to yield an approximation in terms of a Gaussian limit. Following the guidelines in Section 8.1.4, we would expect that

$$X_N(t) = \frac{\sum_{i=1}^{N}\{\xi_i(t) - \mu(t)\}}{\sqrt{N}}$$

$$= \frac{NF_N(t) - Nt}{\sqrt{N}}$$

$$= \sqrt{N}\{F_N(t) - t\}$$

would converge, in an appropriate sense, to a Gaussian process with zero mean and covariance $\Gamma(s, t) = s(1 - t)$, for $0 < s < t < 1$. As we have just seen, this process is a Brownian bridge. Therefore, we would expect the approximation

$$F_N(t) \approx t + \frac{1}{\sqrt{N}}B^0(t), \quad 0 < t < 1.$$

Such approximations are heavily used in the theory of nonparametric statistics.

8.3.4 Brownian Meander

Brownian meander $\{B^+(t); t \geq 0\}$ is Brownian motion conditioned to be positive. Recall (8.29) and (8.31):

$$G_t(x, y) = \Pr\{B(t) > y, \min_{0 \leq u \leq t}B(u) > 0|B(0) = x\}$$

$$= \Phi\left(\frac{y+x}{\sqrt{t}}\right) - \Phi\left(\frac{y-x}{\sqrt{t}}\right),$$

so that

$$G_t(x, 0) = \Pr\{\min_{0 \leq u \leq t}B(u) > 0|B(0) = x\}$$

$$= \Phi\left(\frac{x}{\sqrt{t}}\right) - \Phi\left(\frac{-x}{\sqrt{t}}\right).$$

The transition law for Brownian meander is

$$\Pr\{B^+(t) > y|B^+(0) = x\} = \Pr\{B(t) > y|\min_{0 \leq u \leq t}B(u) > 0, B(0) = x\},$$

whence

$$\Pr\left\{B^+(t) > y | B^+(0) = x\right\} = \frac{G_t(x, y)}{G_t(x, 0)}$$

$$= \frac{\Phi\left(\frac{y+x}{\sqrt{t}}\right) - \Phi\left(\frac{y-x}{\sqrt{t}}\right)}{\Phi\left(\frac{x}{\sqrt{t}}\right) - \Phi\left(\frac{-x}{\sqrt{t}}\right)}.$$

Of most interest is the limiting case as $x \to 0$,

$$\Pr\left\{B^+(t) > y | B^+(0) = 0\right\} = \lim_{x \to 0} \frac{\Phi\left(\frac{y+x}{\sqrt{t}}\right) - \Phi\left(\frac{y-x}{\sqrt{t}}\right)}{\Phi\left(\frac{x}{\sqrt{t}}\right) - \Phi\left(\frac{-x}{\sqrt{t}}\right)}$$

$$= \frac{\phi\left(\frac{y}{\sqrt{t}}\right)}{\phi(0)}$$

$$= e^{-\frac{1}{2}y^2/t}.$$

A simple integration yields the mean

$$E\left[B^+(t) | B^+(0) = 0\right] = \int_0^\infty \Pr\left\{B^+(t) > y | B^+(0) = 0\right\} dy$$

$$= \int_0^\infty e^{-\frac{1}{2}y^2/t} dy$$

$$= \frac{1}{2}\sqrt{2\pi t} \int_{-\infty}^\infty \frac{1}{\sqrt{(2\pi t)}} e^{-\frac{1}{2}y^2/t} dy$$

$$= \sqrt{\pi t/2}.$$

Exercises

8.3.1 Show that the cumulative distribution function for reflected Brownian motion is

$$\Pr\{R(t) < y | R(0) = x\} = \Phi\left(\frac{y-x}{\sqrt{t}}\right) - \Phi\left(\frac{-y-x}{\sqrt{t}}\right)$$

$$= \Phi\left(\frac{y-x}{\sqrt{t}}\right) + \Phi\left(\frac{y+x}{\sqrt{t}}\right) - 1$$

$$= \Phi\left(\frac{x+y}{\sqrt{t}}\right) - \Phi\left(\frac{x-y}{\sqrt{t}}\right).$$

Evaluate this probability when $x = 1, y = 3$, and $t = 4$.

8.3.2 The price fluctuations of a share of stock of a certain company are well described by a Brownian motion process. Suppose that the company is bankrupt if ever the share price drops to zero. If the starting share price is $A(0) = 5$, what is the probability that the company is bankrupt at time $t = 25$? What is the probability that the share price is above 10 at time $t = 25$?

8.3.3 The net inflow to a reservoir is well described by a Brownian motion. Because a reservoir cannot contain a negative amount of water, we suppose that the water level $R(t)$ at time t is a reflected Brownian motion. What is the probability that the reservoir contains more than 10 units of water at time $t = 25$? Assume that the reservoir has unlimited capacity and that $R(0) = 5$.

8.3.4 Suppose that the net inflows to a reservoir follow a Brownian motion. Suppose that the reservoir was known to be empty 25 time units ago but has never been empty since. Use a Brownian meander process to evaluate the probability that there is more than 10 units of water in the reservoir today.

8.3.5 Is reflected Brownian motion a Gaussian process? Is absorbed Brownian motion (cf. Section 8.1.4)?

Problems

8.3.1 Let $B_1(t)$ and $B_2(t)$ be independent standard Brownian motion processes. Define

$$R(t) = \sqrt{B_1(t)^2 + B_2(t)^2}, \quad t \geq 0.$$

$R(t)$ is the radial distance to the origin in a *two-dimensional Brownian motion*. Determine the mean of $R(t)$.

8.3.2 Let $B(t)$ be a standard Brownian motion process. Determine the conditional mean and variance of $B(t), 0 < t < 1$, given that $B(1) = b$.

8.3.3 Let $B(t)$ be a standard Brownian motion. Show that $B(u) - uB(1), 0 < u < 1$, is independent of $B(1)$.

(a) Use this to show that $B^0(t) = B(t) - tB(1), 0 \leq t \leq 1$, is a Brownian bridge.

(b) Use the representation in (a) to evaluate the covariance function for a Brownian bridge.

8.3.4 Let $B(t)$ be a standard Brownian motion. Determine the covariance function for

$$W^0(s) = (1 - s)B\left(\frac{s}{1 - s}\right), \quad 0 < s < 1,$$

and compare it to that for a Brownian bridge.

8.3.5 Determine the expected value for absorbed Brownian motion $A(t)$ at time $t = 1$ by integrating the transition density (8.32) according to

$$E[A(1)|A(0) = x] = \int_0^\infty yp(y, 1|x)dy$$

$$= \int_0^\infty y[\phi(y - x) - \phi(y + x)]dy.$$

The answer is $E[A(1)|A(0) = x] = x$. Show that $E[A(t)|A(0) = x] = x$ for all $t > 0$.

8.3.6 Let $M = \max\{A(t); t \geq 0\}$ be the largest value assumed by an absorbed Brownian motion $A(t)$. Show that $\Pr\{M > z|A(0) = x\} = x/z$ for $0 < x < z$.

8.3.7 Let $t_0 = 0 < t_1 < t_2 < \cdots$ be time points, and define $X_n = A(t_n)$, where $A(t)$ is absorbed Brownian motion starting from $A(0) = x$. Show that $\{X_n\}$ is a nonnegative martingale. Compare the maximal inequality (2.53) in Chapter 2 with the result in Problem 8.3.6.

8.3.8 Show that the transition densities for both reflected Brownian motion and absorbed Brownian motion satisfy the diffusion equation (8.3) in the region $0 < x < \infty$.

8.3.9 Let $F(t)$ be a cumulative distribution function and $B^0(u)$ a Brownian bridge.
   (a) Determine the covariance function for $B^0(F(t))$.
   (b) Use the central limit principle for random functions to argue that the empirical distribution functions for random variables obeying $F(t)$ might be approximated by the process in (a).

8.4 Brownian Motion with Drift

Let $\{B(t); t \geq 0\}$ be a standard Brownian motion process, and let μ and $\sigma > 0$ be fixed. The *Brownian motion with drift parameter* μ *and variance parameter* σ^2 is the process

$$X(t) = \mu t + \sigma B(t) \quad \text{for} \quad t \geq 0. \tag{8.34}$$

Alternatively, Brownian motion with drift parameter μ and variance parameter σ^2 is the process whose increments over disjoint time intervals are independent (property (b) of the definition of standard Brownian motion) and whose increments $X(t + s) - X(t), t, s > 0$, are normally distributed with mean μs and variance $\sigma^2 s$. When $X(0) = x$, we have

$$\Pr\{X(t) \leq y|X(0) = x\} = \Pr\{\mu t + \sigma B(t) \leq y|\sigma B(0) = x\}$$

$$= \Pr\left\{B(t) \leq \frac{y - \mu t}{\sigma} \,\middle|\, B(0) = \frac{x}{\sigma}\right\}$$

$$= \Phi_t\left(\frac{y - x - \mu t}{\sigma}\right) = \Phi\left(\frac{y - x - \mu t}{\sigma\sqrt{t}}\right).$$

Brownian motion with drift is not symmetric when $\mu \neq 0$, and the reflection principle cannot be used to compute the distribution of the maximum of the process. We will use an infinitesimal first step analysis to determine some properties of Brownian motion with drift. To set this up, let us introduce some notation to describe changes in the Brownian motion with drift over small time intervals of length Δt. We let $\Delta X = X(t + \Delta t) - X(t)$ and $\Delta B = B(t + \Delta t) - B(t)$. Then, $\Delta X = \mu \Delta t + \sigma \Delta B$, and

$$X(t + \Delta t) = X(t) + \Delta X = X(t) + \mu \Delta t + \sigma \Delta B. \tag{8.35}$$

We observe that the conditional moments of Δt, given $X(t) = x$, are

$$E[\Delta X | X(t) = x] = \mu \Delta t + \sigma E[\Delta B] = \mu \Delta t, \tag{8.36}$$

$$\text{Var}[\Delta X | X(t) = x] = \sigma^2 E\left[(\Delta B)^2\right] = \sigma^2 \Delta t, \tag{8.37}$$

and

$$E\left[(\Delta X)^2 | X(t) = x\right] = \sigma^2 \Delta t + (\mu \Delta t)^2 = \sigma^2 \Delta t + o(\Delta t), \tag{8.38}$$

while

$$E\left[(\Delta X)^c\right] = o(\Delta t) \quad \text{for} \quad c > 2. \tag{8.39}$$

8.4.1 The Gambler's Ruin Problem

Let us suppose that $X(0) = x$, and that $a < x$ and $b > x$ are fixed quantities. We will be interested in some properties of the random time T at which the process first assumes one of the values a or b. This so-called *hitting time* is formally defined by

$$T = T_{ab} = \min\{t \geq 0; X(t) = a \text{ or } X(t) = b\}.$$

Analogous to the gambler's ruin problem in a random walk (Chapter 3, Section 3.5.3), we will determine the probability that when the Brownian motion exits the interval (a, b), it does so at the point b. The solution for a standard Brownian motion was obtained in Section 8.1 by using the invariance principle. Here we solve the problem for Brownian motion with drift by instituting an infinitesimal first step analysis.

Theorem 8.1. *For a Brownian motion with drift parameter μ and variance parameter σ^2, and $a < x < b$,*

$$u(x) = \Pr\{X(T_{ab}) = b | X(0) = x\} = \frac{e^{-2\mu x/\sigma^2} - e^{-2\mu a/\sigma^2}}{e^{-2\mu b/\sigma^2} - e^{-2\mu a/\sigma^2}}. \tag{8.40}$$

Proof. Our proof is not entirely complete in that we will assume (1) that $u(x)$ is twice continuously differentiable, and (2) that we can choose a time increment Δt so small that exiting the interval (a, b) prior to time Δt can be neglected. With these provisos, at time Δt the Brownian motion will be at the position $X(0) + \Delta X = x + \Delta X$, and the conditional probability of exiting at the upper point b is now $u(x + \Delta X)$. Invoking the law of total probability, it must be that $u(x) = \Pr\{X(T) = b | X(0) = x\} = E[\Pr\{X(T) = b | X(0) = x, X(\Delta t) = x + \Delta X\} | X(0) = x] = E[u(x + \Delta X)]$, where $a < x < b$. ∎

The next step is to expand $u(x + \Delta X)$ in a Taylor series, whereby $u(x + \Delta X) = u(x) + u'(x)\Delta X + \frac{1}{2}u''(x)(\Delta X)^2 + o(\Delta X)^2$. Then,

$$u(x) = E[u(x + \Delta X)]$$
$$= u(x) + u'(x)E[\Delta X] + \frac{1}{2}u''(x)E\left[(\Delta X)^2\right] + E[o(\Delta t)].$$

We use (8.36), (8.38), and (8.39) to evaluate the moments of ΔX, obtaining

$$u(x) = u(x) + u'(x)\mu\Delta t + \frac{1}{2}u''(x)\sigma^2\Delta t + o(\Delta t),$$

which, after subtracting $u(x)$, dividing by Δt, and letting $\Delta t \to 0$, becomes the differential equation

$$0 = \mu u'(x) + \frac{1}{2}\sigma^2 u''(x) \quad \text{for} \quad a < x < b. \tag{8.41}$$

The solution to (8.41) is

$$u(x) = Ae^{-2\mu x/\sigma^2} + B,$$

where A and B are constants of integration. These constants are determined by the conditions $u(a) = 0$ and $u(b) = 1$. In words, the probability of exiting at b if the process starts at a is zero, while the probability of exiting at b if the process starts at b is one. When these conditions are used to determine A and B, then (8.40) results.

Example Suppose that the fluctuations in the price of a share of stock in a certain company are well described by a Brownian motion with drift $\mu = 1/10$ and variance $\sigma^2 = 4$. A speculator buys a share of this stock at a price of \$100 and will sell if ever the price rises to \$110 (a profit) or drops to \$95 (a loss). What is the probability that the speculator sells at a profit? We apply (8.40) with $a = 95$, $x = 100$, $b = 110$, and $2\mu/\sigma^2 = 2(0.1)/4 = 1/20$. Then,

$$\Pr\{\text{Sell at profit}\} = \frac{e^{-100/20} - e^{-95/20}}{e^{-110/20} - e^{-95/20}} = 0.419.$$

The Mean Time to Exit an Interval

Using another infinitesimal first step analysis, the mean time to exit an interval may be determined for Brownian motion with drift.

Theorem 8.2. *For a Brownian motion with drift parameter μ and variance parameter σ^2, and $a < x < b$,*

$$E[T_{ab}|X(0) = x] = \frac{1}{\mu}[u(x)(b-a) - (x-a)], \tag{8.42}$$

where $u(x)$ is given in (8.40).

Proof. Let $v(x) = E[T_{ab}|X(0) = x]$. As in the proof of Theorem 8.1, we will assume (1) that $v(x)$ is twice continuously differentiable, and (2) that we can choose a time increment Δt so small that exiting the interval (a, b) prior to time Δt can be neglected. With these provisos, after time Δt the Brownian motion will be at the position $X(0) + \Delta X = x + \Delta X$, and the conditional mean time to exit the interval is now $\Delta t + v(x + \Delta X)$. Invoking the law of total probability, it must be that $v(x) = E[T|X(0) = x] = E[\Delta t + E\{T - \Delta t|X(0) = x, X(\Delta t) = x + \Delta X\}|X(0) = x] = \Delta t + E[v(x + \Delta X)]$, where $a < x < b$. ∎

The next step is to expand $v(x + \Delta X)$ in a Taylor series, whereby $v(x + \Delta X) = v(x) + v'(x)\Delta X + \frac{1}{2}v''(x)(\Delta X)^2 + o(\Delta X)^2$. Then,

$$v(x) = \Delta t + E[v(x + \Delta X)]$$

$$= \Delta t + v(x) + v'(x)E[\Delta X] + \frac{1}{2}v''(x)E\left[(\Delta X)^2\right] + E\left[o(\Delta X)^2\right].$$

We use (8.36), (8.38), and (8.39) to evaluate the moments of ΔX, obtaining

$$v(x) = \Delta t + v(x) + v'(x)\mu\Delta t + \frac{1}{2}v''(x)\sigma^2\Delta t + o(\Delta t),$$

which, after subtracting $v(x)$, dividing by Δt, and letting $\Delta t \to 0$, becomes the differential equation

$$-1 = \mu v'(x) + \frac{1}{2}\sigma^2 v''(x) \quad \text{for} \quad a < x < b. \tag{8.43}$$

Since it takes no time to reach the boundary if the process starts at the boundary, the conditions are $v(a) = v(b) = 0$. Subject to these conditions, the solution to (8.43) is uniquely given by (8.42), as is easily verified (Problem 8.4.1).

Example *A Sequential Decision Procedure* A Brownian motion $X(t)$ either (1) has drift $\mu = +\frac{1}{2}\delta > 0$, or (2) has drift $\mu = -\frac{1}{2}\delta < 0$, and it is desired to determine which is the case by observing the process. The process will be monitored until it first reaches the level $b > 0$, in which case we will decide that the drift is $\mu = +\frac{1}{2}\delta$, or until it first drops to the level $a < 0$, which occurrence will cause us to decide in favor of $\mu = -\frac{1}{2}\delta$. This decision procedure is, of course, open to error, but we can evaluate these error probabilities and choose a and b so as to keep the error probabilities acceptably small. We have

$$\alpha = \Pr\left\{\text{Decide } \mu = +\frac{1}{2}\delta|\mu = -\frac{1}{2}\delta\right\}$$

$$= \Pr\left\{X(T) = b|E[X(t)] = -\frac{1}{2}\delta t\right\} \tag{8.44}$$

$$= \frac{1 - e^{+\delta a/\sigma^2}}{e^{+\delta b/\sigma^2} - e^{+\delta a/\sigma^2}}, \quad \text{(using (8.40))}$$

and

$$1 - \beta = \Pr\left\{\text{Decide } \mu = -\frac{1}{2}\delta \mid \mu = +\frac{1}{2}\delta\right\}$$

$$= \Pr\left\{X(T) = b \mid E[X(t)] = +\frac{1}{2}\delta t\right\} \tag{8.45}$$

$$= \frac{1 - e^{-\delta a/\sigma^2}}{e^{-\delta b/\sigma^2} - e^{-\delta a/\sigma^2}}.$$

If acceptable levels of the error probabilities α and β are prescribed, then we can solve in (8.44) and (8.45) to determine the boundaries to be used in the decision procedure. The reader should verify that these boundaries are

$$a = -\frac{\sigma^2}{\delta} \log\left(\frac{1-\alpha}{\beta}\right), \quad \text{and} \quad b = \frac{\sigma^2}{\delta} \log\left(\frac{1-\beta}{\alpha}\right). \tag{8.46}$$

For a numerical example, if $\sigma^2 = 4$ and we are attempting to decide between $\mu = -\frac{1}{2}$ and $\mu = +\frac{1}{2}$, and the acceptable error probabilities are chosen to be $\alpha = 0.05$ and $\beta = 0.10$, then the decision boundaries that should be used are $a = -4\log(0.95/0.10) = -9.01$, and $b = 4\log(0.90/0.05) = 11.56$.

In the above procedure for deciding the drift of a Brownian motion, the observation duration until a decision is reached will be a random variable whose mean will depend upon the true value of the drift. Using (8.42) with $x = 0$ and μ replaced by $\pm\frac{1}{2}\delta$ gives us the mean observation interval, as a function of the true mean μ:

$$E\left[T \mid \mu = -\frac{1}{2}\delta\right] = 2\left(\frac{\sigma}{\delta}\right)^2\left[(1-\alpha)\log\left(\frac{1-\alpha}{\beta}\right) - \alpha\log\left(\frac{1-\beta}{\alpha}\right)\right]$$

and

$$E\left[T \mid \mu = +\frac{1}{2}\delta\right] = 2\left(\frac{\sigma}{\delta}\right)^2\left[(1-\beta)\log\left(\frac{1-\beta}{\alpha}\right) - \beta\log\left(\frac{1-\alpha}{\beta}\right)\right].$$

We have developed a sequential decision procedure for evaluating the drift of a Brownian motion. However, invoking the invariance principle leads us to believe that similar results should maintain, at least approximately, in analogous situations in which the Brownian motion is replaced by a partial sum process of independent and identically distributed summands. The result is known as *Wald's approximation* for his celebrated sequential probability ratio test of a statistical hypothesis.

The Maximum of a Brownian Motion with Negative Drift

Consider a Brownian motion with drift $\{X(t)\}$, where the drift parameter μ is negative. Over time, such a process will tend toward ever lower values, and its maximum $M = \max\{X(t) - X(0); t \geq 0\}$ will be a well-defined and finite random variable.

Theorem 8.1 will enable us to show that M has an exponential distribution with parameter $2|\mu|/\sigma^2$. To see this, let us suppose that $X(0) = 0$ and that $a < 0 < b$ are constants. Then, Theorem 8.1 states that

$$\Pr\{X(T_{ab}) = b | X(0) = x\} = \frac{1 - e^{-2\mu a/\sigma^2}}{e^{-2\mu b/\sigma^2} - e^{-2\mu a/\sigma^2}}, \qquad (8.47)$$

where T_{ab} is the random time at which the process first reaches $a < 0$ or $b > 0$. That is, the probability that the Brownian motion reaches $b > 0$ before it ever drops to $a < 0$ is given by the right side of (8.47). Because both $\mu < 0$ and $a < 0$, then $a\mu > 0$ and

$$e^{-2\mu a/\sigma^2} = e^{-2|\mu a|/\sigma^2} \to 0 \quad \text{as} \quad a \to -\infty, \quad \text{and then}$$

$$\lim_{a \to -\infty} \Pr\{X(T_{ab}) = b\} = \frac{1 - 0}{e^{-2\mu b/\sigma^2} - 0} = e^{-2|\mu|b/\sigma^2}.$$

But as $a \to -\infty$ the left side of (8.47) becomes the probability that the process ever reaches the point b, i.e., the probability that the maximum M of the process ever exceeds b. We have deduced, then, the desired exponential distribution

$$\Pr\{M > b\} = e^{-2|\mu|b/\sigma^2}, \quad b > 0. \qquad (8.48)$$

8.4.2 Geometric Brownian Motion

A stochastic process $\{Z(t); t \geq 0\}$ is called a *geometric Brownian motion* with drift parameter α if $X(t) = \log Z(t)$ is a Brownian motion with drift $\mu = \alpha - \frac{1}{2}\sigma^2$ and variance parameter σ^2. Equivalently, $Z(t)$ is geometric Brownian motion starting from $Z(0) = z$ if

$$Z(t) = ze^{X(t)} = ze^{\left(\alpha - \frac{1}{2}\sigma^2\right)t + \sigma B(t)}, \qquad (8.49)$$

where $B(t)$ is a standard Brownian motion starting from $B(0) = 0$.

Modern mathematical economists usually prefer geometric Brownian motion over Brownian motion as a model for prices of assets, say shares of stock, that are traded in a perfect market. Such prices are nonnegative and exhibit random fluctuations about a long-term exponential decay or growth curve. Both of these properties are possessed by geometric Brownian motion, but not by Brownian motion itself. More importantly, if $t_0 < t_1 < \cdots < t_n$ are time points, then the successive ratios

$$\frac{Z(t_1)}{Z(t_0)}, \frac{Z(t_2)}{Z(t_1)}, \ldots, \frac{Z(t_n)}{Z(t_{n-1})}$$

are independent random variables, so that crudely speaking, the percentage changes over nonoverlapping time intervals are independent.

We turn to determining the mean and variance of geometric Brownian motion. Let ξ be a normally distributed random variable with mean zero and variance one. We begin by establishing the formula

$$E\left[e^{\lambda\xi}\right] = e^{\frac{1}{2}\lambda^2}, \quad -\infty < \lambda < \infty,$$

which results immediately from

$$1 = \int_{-\infty}^{\infty} \frac{1}{\sqrt{2\pi}} e^{-\frac{1}{2}(u-\lambda)^2} du \quad \text{(area under normal density)}$$

$$= \int_{-\infty}^{\infty} \frac{1}{\sqrt{2\pi}} e^{-\frac{1}{2}(u^2 - 2\lambda u + \lambda^2)} du$$

$$= e^{-\frac{1}{2}\lambda^2} \int_{-\infty}^{\infty} e^{\lambda u} \frac{1}{\sqrt{2\pi}} e^{-\frac{1}{2}u^2} du$$

$$= e^{-\frac{1}{2}\lambda^2} E\left[e^{\lambda\xi}\right].$$

To obtain the mean of geometric Brownian motion $Z(t) = ze^{X(t)} = ze^{\left(\alpha - \frac{1}{2}\sigma^2\right)t + \sigma B(t)}$, we use the fact that $\xi = B(t)/\sqrt{t}$ is normally distributed with mean zero and variance one, whence

$$E[Z(t)|Z(0) = z] = zE\left[e^{\left(\alpha - \frac{1}{2}\sigma^2\right)t + \sigma B(t)}\right]$$

$$= ze^{\left(\alpha - \frac{1}{2}\sigma^2\right)t} E\left[e^{\sigma\sqrt{t}\xi}\right] \quad (\xi = B(t)/\sqrt{t}) \tag{8.50}$$

$$= ze^{\left(\alpha - \frac{1}{2}\sigma^2\right)t} e^{\frac{1}{2}\sigma^2 t} = ze^{\alpha t}.$$

Equation (8.50) has interesting economic implications in the case where α is positive but small relative to the variance parameter σ^2. On the one hand, if α is positive, then the mean $E[Z(t)] = z\exp(\alpha t) \to \infty$ as $t \to \infty$. On the other hand, if α is positive but $\alpha < \frac{1}{2}\sigma^2$, then $\alpha - \frac{1}{2}\sigma^2 < 0$, and $X(t) = \left(\alpha - \frac{1}{2}\sigma^2\right)t + \sigma B(t)$ is drifting in the negative direction. As a consequence of the law of large numbers, it can be shown that $X(t) \to -\infty$ as $t \to \infty$ under these circumstances, so that $Z(t) = z\exp[X(t)] \to \exp(-\infty) = 0$. The geometric Brownian motion process is drifting ever closer to zero, while simultaneously, its mean or expected value is continually increasing! Here is yet another stochastic model in which the mean value function is entirely misleading as a sole description of the process.

The variance of the geometric Brownian motion is derived in much the same manner as the mean. First

$$E\left[Z(t)^2 | Z(0) = z\right] = z^2 E\left[e^{2X(t)}\right] = z^2 E\left[e^{2\left(\alpha - \frac{1}{2}\sigma^2\right)t + 2\sigma B(t)}\right]$$

$$= z^2 e^{2\left(\alpha + \frac{1}{2}\sigma^2\right)t} \quad \text{(as in (8.50))},$$

and then

$$\mathrm{Var}[Z(t)] = E\left[Z(t)^2\right] - \{E[Z(t)]\}^2$$

$$= z^2 e^{2\left(\alpha + \frac{1}{2}\sigma^2\right)t} - z^2 e^{2\alpha t} \tag{8.51}$$

$$= z^2 e^{2\alpha t}\left(e^{\sigma^2 t} - 1\right).$$

Because of their close relation as expressed in the definition (8.49), many results for Brownian motion can be directly translated into analogous results for geometric Brownian motion. For example, let us translate the gambler's ruin probability in Theorem 8.1. For $A < 1$ and $B > 1$, define

$$T = T_{A,B} = \min\left\{t \geq 0; \frac{Z(t)}{Z(0)} = A \text{ or } \frac{Z(t)}{Z(0)} = B\right\}.$$

Theorem 8.3. *For a geometric Brownian motion with drift parameter α and variance parameter σ^2, and $A < 1 < B$,*

$$\mathrm{Pr}\left\{\frac{Z(T)}{Z(0)} = B\right\} = \frac{1 - A^{1 - 2\alpha/\sigma^2}}{B^{1 - 2\alpha/\sigma^2} - A^{1 - 2\alpha/\sigma^2}}. \tag{8.52}$$

Example Suppose that the fluctuations in the price of a share of stock in a certain company are well described by a geometric Brownian motion with drift $\alpha = 1/10$ and variance $\sigma^2 = 4$. A speculator buys a share of this stock at a price of $100 and will sell if ever the price rises to $110 (a profit) or drops to $95 (a loss). What is the probability that the speculator sells at a profit? We apply (8.52) with $A = 0.95$, $B = 1.10$, and $1 - 2\alpha/\sigma^2 = 1 - 2(0.1)/4 = 0.95$. Then,

$$\mathrm{Pr}\{\text{Sell at profit}\} = \frac{1 - 0.95^{0.95}}{1.10^{0.95} - 0.95^{0.95}} = 0.3342.$$

Example *The Black-Scholes Option Pricing Formula* A *call*, or *warrant*, is an option entitling the holder to buy a block of shares in a given company at a specified price at any time during a stated interval. Thus, the call listed in the financial section of the newspaper as

Hewlett   Aug $60 $6

means that for a price of $6 per share, one may purchase the privilege (option) of buying the stock of Hewlett-Packard at a price of $60 per share at any time between now and August (by convention, always the third Friday of the month). The $60 figure is called the *striking price*. Since the most recent closing price of Hewlett was $59, the option of choosing when to buy, or not to buy at all, carries a *premium* of $7 = $60 + $6 − $59 over a direct purchase of the stock today.

Should the price of Hewlett rise to, say, $70 between now and the third Friday of August, the owner of such an option could exercise it, buying at the striking price of $60 and immediately selling at the then current market price of $70 for a $10 profit, less, of course, the $6 cost of the option itself. On the other hand, should the price of Hewlett fall, the option owner's loss is limited to his $6 cost of the option. Note that the seller (technically called the "writer") of the option has a profit limited to the $6 that he receives for the option but could experience a huge loss should the price of Hewlett soar, say to $100. The writer would then either have to give up his own Hewlett shares or buy them at $100 on the open market in order to fulfill his obligation to sell them to the option holder at $60.

What should such an option be worth? Is $6 for this privilege a fair price? While early researchers had studied these questions using a geometric Brownian motion model for the price fluctuations of the stock, they all assumed that the option should yield a higher mean return than the mean return from the stock itself because of the unlimited potential risk to the option writer. This assumption of a higher return was shown to be false in 1973 when Fisher Black, a financial consultant with a Ph.D. in applied mathematics, and Myron Scholes, an assistant professor in finance at MIT, published an entirely new and innovative analysis. In an idealized setting that included no transaction costs and an ability to borrow or lend limitless amounts of capital at the same fixed interest rate, they showed that an owner, or a writer, of a call option could simultaneously buy or sell the underlying stock ("program trading") in such a way as to exactly match the returns of the option. Having available two investment opportunities with exactly the same return effectively eliminates all risk, or randomness, by allowing an investor to buy one while selling the other. The implications of their result are many. First, since writing an option potentially carries no risk, its return must be the same as that for other riskless investments in the economy. Otherwise, limitless profit opportunities bearing no risk would arise. Second, since owning an option carries no risk, one should not exercise it early, but hold it until its expiration date, when, if the market price exceeds the striking price, it should be exercised, and otherwise not. These two implications then lead to a third, a formula that established the worth, or value, of the option.

The Black-Scholes paper spawned hundreds, if not thousands, of further academic studies. At the same time, their valuation formula quickly invaded the financial world, where soon virtually all option trades were taking place at or near their Black-Scholes value. It is remarkable that the valuation formula was adopted so quickly in the real world in spite of the esoteric nature of its derivation and the ideal world of its assumptions.

In order to present the Black-Scholes formula, we need some notation. Let $S(t)$ be the price at time t of a share of the stock under study. We assume that $S(t)$ is described by a geometric Brownian motion with drift parameter α and variance parameter σ^2. Let $F(z, \tau)$ be the value of an option, where z is the current price of the stock and τ is the time remaining until expiration. Let a be the striking price. When $\tau = 0$, and there is no time remaining, one exercises the option for a profit of $z - a$ if $z > a$ (market price greater than striking price) and does not exercise the option, but lets it lapse, or expire, if $z \leq a$. This leads to the condition

$$F(z, 0) = (z - a)^+ = \max\{z - a, 0\}.$$

The Black-Scholes analysis resulted in the valuation

$$F(z, \tau) = e^{-r\tau} E\left[(Z(\tau) - a)^+ | Z(0) = z\right], \tag{8.53}$$

where r is the return rate for secure, or riskless, investments in the economy, and where $Z(t)$ is a second geometric Brownian motion having drift parameter r and variance parameter σ^2. Looking at (8.53), the careful reader will wonder whether we have made a mistake. No, the worth of the option does *not* depend on the drift parameter α of the underlying stock.

In order to put the valuation formula into a useful form, we write

$$Z(\tau) = z e^{\left(r - \frac{1}{2}\sigma^2\right)\tau + \sigma\sqrt{\tau}\xi}, \quad \xi = B(\tau)/\sqrt{\tau}, \tag{8.54}$$

and observe that

$$z e^{\left(r - \frac{1}{2}\sigma^2\right)\tau + \sigma\sqrt{\tau}\xi} > a$$

is the same as

$$\xi > v_0 = \frac{\log(a/z) - \left(r - \frac{1}{2}\sigma^2\right)\tau}{\sigma\sqrt{\tau}}. \tag{8.55}$$

Then,

$$e^{r\tau} F(z, \tau) = E\left[(Z(\tau) - a)^+ | Z(0) = z\right]$$

$$= E\left[\left(z e^{\left(r - \frac{1}{2}\sigma^2\right)\tau + \sigma\sqrt{\tau}\xi} - a\right)^+\right]$$

$$= \int_{v_0}^{\infty} \left[z e^{\left(r - \frac{1}{2}\sigma^2\right)\tau + \sigma\sqrt{\tau}v} - a\right]\phi(v)dv.$$

$$= z e^{\left(r - \frac{1}{2}\sigma^2\right)\tau} \int_{v_0}^{\infty} e^{\sigma\sqrt{\tau}v}\phi(v)dv - a\int_{v_0}^{\infty}\phi(v)dv.$$

Completing the square in the form

$$-\frac{1}{2}v^2 + \sigma\sqrt{\tau}v = -\frac{1}{2}\left[(v - \sigma\sqrt{\tau})^2 - \sigma^2\tau\right]$$

shows that

$$e^{\sigma\sqrt{\tau}v}\phi(v) = e^{\frac{1}{2}\sigma^2\tau}\phi(v - \sigma\sqrt{\tau}),$$

whence

$$e^{r\tau}F(z,\tau) = ze^{\left(r-\frac{1}{2}\sigma^2\right)\tau}e^{+\frac{1}{2}\sigma^2\tau}\int_{v_0}^{\infty}\phi(v - \sigma\sqrt{\tau})dv - a[1 - \Phi(v_0)]$$

$$= ze^{r\tau}[1 - \Phi(v_0 - \sigma\sqrt{\tau})] - a[1 - \Phi(v_0)].$$

Finally, note that

$$v_0 - \sigma\sqrt{\tau} = \frac{\log(a/z) - \left(r + \frac{1}{2}\sigma^2\right)\tau}{\sigma\sqrt{\tau}}$$

and that

$$1 - \Phi(x) = \Phi(-x) \quad \text{and} \quad \log(a/z) = -\log(z/a)$$

to get, after multiplying by $e^{-r\tau}$, the end result

$$F(z,\tau) = z\Phi\left(\frac{\log(z/a) + \left(r + \frac{1}{2}\sigma^2\right)\tau}{\sigma\sqrt{\tau}}\right)$$

$$- ae^{-r\tau}\Phi\left(\frac{\log(z/a) + \left(r - \frac{1}{2}\sigma^2\right)\tau}{\sigma\sqrt{\tau}}\right).$$

(8.56)

Equation (8.56) is the Black-Scholes valuation formula. Four of the five factors that go into it are easily and objectively evaluated: The current market prize z, the striking price a, the time τ until the option expires, and the rate r of return from secure investments such as short-term government securities. It is the fifth factor, σ, sometimes called the *volatility*, that presents problems. It is, of course, possible to estimate this parameter based on past records of price movements. However, it should be emphasized that it is the volatility in the future that will affect the profitability of the option, and when economic conditions are changing, past history may not accurately indicate the future. One way around this difficulty is to work backwards and use the Black-Scholes formula to impute a volatility from an existing market price of the option. For example, the Hewlett-Packard call option that expires in August, six months or, $\tau = \frac{1}{2}$ year, in the future, with a current price of Hewlett-Packard stock of $59, a striking price of $60, and secure investments returning about $r = 0.05$, a volatility of $\sigma = 0.35$ is consistent with the listed option price of $6. (When $\sigma = 0.35$ is used in the Black-Scholes formula, the resulting valuation is $6.03.) A volatility derived in this manner is called an *imputed* or *implied* volatility. Someone who believes that the future will be more variable might regard the option at $6 as a good buy. Someone who believes the future to be less variable than the imputed volatility of $\sigma = 0.35$ might be inclined to offer a Hewlett-Packard option at $6.

Striking Price a	Time to Expiration (Years) τ	Offered Price	Black-Scholes Valuation $F(z, \tau)$
130	1/12	$17.00	$17.45
130	2/12	19.25	18.87
135	1/12	13.50	13.09
135	2/12	15.13	14.92
140	1/12	8.50	9.26
140	2/12	12.00	11.46
145	1/12	5.50	6.14
145	2/12	9.13	8.52
145	5/12	13.63	13.51
150	1/12	3.13	3.80
150	2/12	6.38	6.14
155	1/12	1.63	2.18
155	2/12	4.00	4.28
155	5/12	9.75	9.05

The above table compares actual offering prices on February 26, 1997, for options in IBM stock with their Black-Scholes valuation using (8.56). The current market price of IBM stock is $146.50, and, in all cases, the same volatility $\sigma = 0.30$ was used.

The agreement between the actual option prices and their Black-Scholes valuations seems quite good.

Exercises

8.4.1 A Brownian motion $\{X(t)\}$ has parameters $\mu = -0.1$ and $\sigma = 2$. What is the probability that the process is above $y = 9$ at time $t = 4$, given that it starts at $x = 2.82$?

8.4.2 A Brownian motion $\{X(t)\}$ has parameters $\mu = 0.1$ and $\sigma = 2$. Evaluate the probability of exiting the interval $(a, b]$ at the point b starting from $X(0) = 0$ for $b = 1, 10$, and 100 and $a = -b$. Why do the probabilities change when a/b is the same in all cases?

8.4.3 A Brownian motion $\{X(t)\}$ has parameters $\mu = 0.1$ and $\sigma = 2$. Evaluate the mean time to exit the interval $(a, b]$ from $X(0) = 0$ for $b = 1, 10$, and 100 and $a = -b$. Can you guess how this mean time varies with b for b large?

8.4.4 A Brownian motion $X(t)$ either (1) has drift $\mu = +\frac{1}{2}\delta > 0$, or (2) has drift $\mu = -\frac{1}{2}\delta < 0$, and it is desired to determine which is the case by observing the process for a *fixed* duration τ. If $X(\tau) > 0$, then the decision will be that $\mu = +\frac{1}{2}\delta$; If $X(\tau) \leq 0$, then $\mu = -\frac{1}{2}\delta$ will be stated. What should be the length τ of the observation period if the design error probabilities are set at $\alpha = \beta = 0.05$? Use $\delta = 1$ and $\sigma = 2$. Compare this fixed duration with the average duration of the sequential decision plan in the example of Section 8.4.1.

8.4.5 Suppose that the fluctuations in the price of a share of stock in a certain company are well described by a geometric Brownian motion with drift $\alpha = -0.1$ and variance $\sigma^2 = 4$. A speculator buys a share of this stock at a price of $100 and will sell if ever the price rises to $110 (a profit) or drops to $95 (a loss). What is the probability that the speculator sells at a profit?

8.4.6 Let ξ be a standard normal random variable.

(a) For an arbitrary constant a, show that

$$E\left[(\xi - a)^+\right] = \phi(a) - a[1 - \Phi(a)].$$

(b) Let X be normally distributed with mean μ and variance σ^2. Show that

$$E\left[(X - b)^+\right] = \sigma\left\{\phi\left(\frac{b-\mu}{\sigma}\right) - \left(\frac{b-\mu}{\sigma}\right)\left[1 - \Phi\left(\frac{b-\mu}{\sigma}\right)\right]\right\}.$$

Problems

8.4.1 What is the probability that a standard Brownian motion $\{B(t)\}$ ever crosses the line $a + bt (a > 0, b > 0)$?

8.4.2 Show that

$$\Pr\left\{\max_{t \geq 0} \frac{b + B(t)}{1 + t} > a\right\} = e^{-2a(a-b)}, \quad a > 0, b < a.$$

8.4.3 If $B^0(s), 0 < s < 1$, is a Brownian bridge process, then

$$B(t) = (1 + t)B^0\left(\frac{t}{1+t}\right)$$

is a standard Brownian motion. Use this representation and the result of Problem 8.4.2 to show that for a Brownian bridge $B^0(t)$,

$$\Pr\left\{\max_{0 \leq u \leq 1} B^0(u) > a\right\} = e^{-2a^2}.$$

8.4.4 A Brownian motion $X(t)$ either (1) has drift $\mu = \mu_0$, or (2) has drift $\mu = \mu_1$, where $\mu_0 < \mu_1$ are known constants. It is desired to determine which is the case by observing the process. Derive a sequential decision procedure that meets prespecified error probabilities α and β.

Hint: Base your decision on the process $X'(t) = X(t) - \frac{1}{2}(\mu_0 + \mu_1)$.

8.4.5 Change a Brownian motion with drift $X(t)$ into an absorbed Brownian motion with drift $X^A(t)$ by defining

$$X^A(t) = \begin{cases} X(t), & \text{for } t < \tau, \\ 0, & \text{for } t \geq \tau, \end{cases}$$

where

$$\tau = \min\{t \geq 0; X(t) = 0\}.$$

(We suppose that $X(0) = x > 0$ and that $\mu < 0$, so that absorption is sure to occur eventually.) What is the probability that the absorbed Brownian motion ever reaches the height $b > x$?

8.4.6 What is the probability that a geometric Brownian motion with drift parameter $\alpha = 0$ ever rises to more than twice its initial value? (You buy stock whose fluctuations are described by a geometric Brownian motion with $\alpha = 0$. What are your chances to double your money?)

8.4.7 A call option is said to be "in the money" if the market price of the stock is higher than the striking price. Suppose that the stock follows a geometric Brownian motion with drift α, variance σ^2, and has a current market price of z What is the probability that the option is in the money at the expiration time τ? The striking price is a.

8.4.8 Verify the Hewlett-Packard option valuation of $6.03 stated in the text when $\tau = \frac{1}{2}, z = \$59, a = 60, r = 0.05$, and $\sigma = 0.35$. What is the Black-Scholes valuation if $\sigma = 0.30$?

8.4.9 Let τ be the first time that a standard Brownian motion $B(t)$ starting from $B(0) = x > 0$ reaches zero. Let λ be a positive constant. Show that

$$w(x) = E\left[e^{-\lambda\tau}|B(0) = x\right] = e^{-\sqrt{2\lambda}x}.$$

Hint: Develop an appropriate differential equation by instituting an infinitesimal first step analysis according to

$$w(x) = E\left[E\left\{e^{-\lambda\tau}|B(\Delta t)\right\}|B(0) = x\right] = E\left[e^{-\lambda\Delta t}w(x + \Delta B)\right].$$

8.4.10 Let $t_0 = 0 < t_1 < t_2 < \cdots$ be time points, and define $X_n = Z(t_n)\exp(-rt_n)$, where $Z(t)$ is geometric Brownian motion with drift parameters r and variance parameter σ^2 (see the geometric Brownian motion in the Black-Scholes formula (8.53)). Show that $\{X_n\}$ is a martingale.

8.5 The Ornstein–Uhlenbeck Process*

The Ornstein–Uhlenbeck process $\{V(t); t \geq 0\}$ has two parameters, a drift coefficient $\beta > 0$ and a diffusion parameter σ^2. The process, starting from $V(0) = v$, is defined in

---

* This section contains material of a more specialized nature.

terms of a standard Brownian motion $\{B(t)\}$ by scale changes in both space and time:

$$V(t) = ve^{-\beta t} + \frac{\sigma e^{-\beta t}}{\sqrt{2\beta}} B\left(e^{2\beta t} - 1\right), \quad \text{for } t \geq 0. \tag{8.57}$$

The first term on the right of (8.57) describes an exponentially decreasing trend towards the origin. The second term represents the fluctuations about this trend in terms of a rescaled Brownian motion. The Ornstein–Uhlenbeck process is another example of a continuous-state-space, continuous-time Markov process having continuous paths, inheriting these properties from the Brownian motion in the representation (8.57). It is a Gaussian process (see the discussion in Section 8.1.4), and (8.57) easily shows its mean and variance to be

$$E[V(t)|V(0) = v] = ve^{-\beta t}, \tag{8.58}$$

and

$$\text{Var}[V(t)|V(0) = x] = e^{-2\beta t} \frac{\sigma^2}{2\beta} \text{Var}\left[B\left(e^{2\beta t} - 1\right)\right] \tag{8.59}$$

$$= \sigma^2 \left(\frac{1 - e^{-2\beta t}}{2\beta}\right).$$

Knowledge of the mean and variance of a normally distributed random variable allows its cumulative distribution function to be written in terms of the standard normal distribution (8.6), and by this means we can immediately express the transition distribution for the Ornstein–Uhlenbeck process as

$$\Pr\{V(t) \leq y|V(0) = x\} = \Phi\left(\frac{\sqrt{2\beta}\left(y - xe^{-\beta t}\right)}{\sigma\sqrt{1 - e^{-2\beta t}}}\right). \tag{8.60}$$

The Covariance Function

Suppose that $0 < u < s$, and that $V(0) = x$. Upon subtracting the mean as given by (8.58), we obtain

$$\text{Cov}[V(u), V(s)] = E\left[\{V(u) - xe^{-\beta u}\}\{V(s) - xe^{-\beta s}\}\right]$$

$$= \frac{\sigma^2}{2\beta} e^{-\beta(u+s)} E\left[\{B\left(e^{2\beta u} - 1\right)\}\{B\left(e^{2\beta s} - 1\right)\}\right] \tag{8.61}$$

$$= \frac{\sigma^2}{2\beta} e^{-\beta(u+s)} \left(e^{2\beta u} - 1\right)$$

$$= \frac{\sigma^2}{2\beta} \left(e^{-\beta(s-u)} - e^{-\beta(s+u)}\right).$$

8.5.1   A Second Approach to Physical Brownian Motion

The path that we have taken to introduce the Ornstein–Uhlenbeck process is not faithful to the way in which the process came about. To begin an explanation, let us recognize that all models of physical phenomena have deficiencies, and the Brownian motion stochastic process as a model for the Brownian motion of a particle is no exception. If $B(t)$ is the position of a pollen grain at time t and if this position is changing over time, then the pollen grain must have a velocity. Velocity is the infinitesimal change in position over infinitesimal time, and where $B(t)$ is the position of the pollen grain at time t, the velocity of the grain would be the derivative $dB(t)/dt$. *But while the paths of the Brownian motion stochastic process are continuous, they are not differentiable.* This remarkable statement is difficult to comprehend. Indeed, many elementary calculus explanations implicitly tend to assume that all continuous functions are differentiable, and if we were to be asked to find an example of one that was not, we might consider it quite a challenge. Yet each path of a continuous Brownian motion stochastic process is (with probability one) differentiable at no point. We have encountered yet another intriguing facet of stochastic processes that we cannot treat in full detail but must leave for future study. We will attempt some motivation, however. Recall that the variance of the Brownian increment ΔB is Δt. But variations in the normal distribution are not scaled in terms of the variance, but in terms of its square root, the standard deviation, so that the Brownian increment ΔB is roughly on the order of $\sqrt{\Delta t}$, and the approximate derivative

$$\frac{\Delta B}{\Delta t} = \frac{\Delta B}{\sqrt{\Delta t}} \cdot \frac{1}{\sqrt{\Delta t}}$$

is roughly on the order of $1/\sqrt{\Delta t}$. This, of course, becomes infinite as $\Delta t \to 0$, which suggests that a derivative of Brownian motion, were it to exist, could only take the values $\pm\infty$. As a consequence, the Brownian path cannot have a derivative. The reader can see from our attempt at explanation that the topic is well beyond the scope of an introductory text.

Although its movements may be erratic, a pollen grain, being a physical object of positive mass, must have a velocity, and the Ornstein–Uhlenbeck process arose as an attempt to model this velocity directly. Two factors are postulated to affect the particle's velocity over a small time interval. First, the frictional resistance or viscosity of the surrounding medium is assumed to reduce the magnitude of the velocity by a deterministic proportional amount, the constant of proportionality being $\beta > 0$. Second, there are random changes in velocity caused by collisions with neighboring molecules, the magnitude of these random changes being measured by a variance coefficient σ^2. That is, if $V(t)$ is the velocity at time t, and ΔV is the change in velocity over $(t, t + \Delta t]$, we might express the viscosity factor as

$$E[\Delta V | V(t) = v] = -\beta v \Delta t + o(\Delta t) \tag{8.62}$$

and the random factor by

$$\text{Var}[\Delta V | V(t) = v] = \sigma^2 \Delta t + o(\Delta t). \tag{8.63}$$

The Ornstein–Uhlenbeck process was developed by taking (8.62) and (8.63) together with the Markov property as the postulates, and from them deriving the transition probabilities (8.60). While we have chosen not to follow this path, we will verify that the mean and variance given in (8.58) and (8.59) do satisfy (8.62) and (8.63) over small time increments. Beginning with (8.58) and the Markov property, the first step is

$$E[V(t + \Delta t)|V(t) = v] = ve^{-\beta \Delta t} = v[1 - \beta \Delta t + o(\Delta t)],$$

and then,

$$E[\Delta V|V(t) = v] = E[V(t + \Delta t)|V(t) = v] - v$$
$$= -\beta v \Delta t + o(\Delta t),$$

and over small time intervals the mean change in velocity is the proportional decrease desired in (8.62). For the variance, we have

$$Var[\Delta V|V(t) = v] = Var[V(t + \Delta t)|V(t) = v]$$
$$= \sigma^2 \left(\frac{1 - e^{-2\beta \Delta t}}{2\beta} \right)$$
$$= \sigma^2 \Delta t + o(\Delta t),$$

and the variance of the velocity increment behaves as desired in (8.63). In fact, (8.62) and (8.63) together with the Markov property can be taken as the definition of the Ornstein–Uhlenbeck process in much the same way, but involving far deeper analysis, that the infinitesimal postulates of Chapter 5, Section 5.2.1, serve to define the Poisson process.

Example *Tracking Error* Let $V(t)$ be the measurement error of a radar system that is attempting to track a randomly moving target. We assume $V(t)$ to be an Ornstein–Uhlenbeck process. The mean increment $E[\Delta V|V(t) = v] = -\beta v \Delta t + o(\Delta t)$ represents the controller's effort to reduce the current error, while the variance term reflects the unpredictable motion of the target. If $\beta = 0.1, \sigma = 2$, and the system starts on target ($v = 0$), the probability that the error is less than one at time $t = 1$ is, using (8.60),

$$Pr\{|V(t)| < 1\} = \Phi\left(\frac{\sqrt{2\beta}}{\sigma\sqrt{1 - e^{-2\beta t}}} \right) - \Phi\left(\frac{\sqrt{-2\beta}}{\sigma\sqrt{1 - e^{-2\beta t}}} \right)$$

$$= \Phi\left(\frac{1}{\sqrt{20(1 - e^{-0.2})}} \right) - \Phi\left(\frac{-1}{\sqrt{20(1 - e^{-0.2})}} \right)$$

$$= \Phi(0.53) - \Phi(-0.53) = 0.4038.$$

As time passes, this near-target probability drops to $\Phi(1/\sqrt{20}) - \Phi(-1/\sqrt{20}) = \Phi(0.22) - \Phi(-0.22) = 0.1742.$

Example *Genetic Fluctuations Under Mutation* In Chapter 6, Section 6.4, we introduced a model describing fluctuations in gene frequency in a population of N individuals, each either of gene type **a** or gene type **A**. With $X(t)$ being the number of type **a** individuals at time t, we reasoned that $X(t)$ would be a birth and death process with parameters

$$\lambda_j = \lambda N \left(1 - \frac{j}{N}\right) \left[\frac{j}{N}(1 - \gamma_1) + \left(1 - \frac{j}{N}\right)\gamma_2\right]$$

and

$$\mu_j = \lambda N \frac{j}{N} \left[\frac{j}{N}\gamma_1 + \left(1 - \frac{j}{N}\right)(1 - \gamma_2)\right].$$

The parameters γ_1 and γ_2 measured the rate of mutation from **a**-type to **A**-type, and **A**-type to **a**-type, respectively. Here we attempt a simplified description of the model when the population size N is large. The steady state fluctuations in the relative gene frequency $X(t)/N$ are centered on the mean

$$\pi = \frac{\gamma_2}{\gamma_1 + \gamma_2}.$$

Accordingly, we define the rescaled and centered process

$$V_N(t) = \sqrt{N} \left(\frac{X(t)}{N} - \pi\right).$$

With

$$\Delta V = V_N(t + \Delta t) - V_N(t), \quad \text{and} \quad \Delta X = X(t + \Delta t) - X(t),$$

we have

$$E[\Delta X | X(t) = j] = \left(\lambda_j - \mu_j\right)\Delta t + o(\Delta t),$$

which becomes, after substitution and simplification,

$$E[\Delta X | X(t) = j] = N\lambda \left[\left(1 - \frac{j}{N}\right)\gamma_2 - \frac{j}{N}\gamma_1\right]\Delta t + o(\Delta t).$$

More tedious calculations show that

$$E\left[\Delta X^2 | X(t) = j\right] = N\lambda \left[\frac{2\gamma_1\gamma_2}{(\gamma_1 + \gamma_2)^2} + o\left(\frac{1}{N}\right)\right]\Delta t.$$

Our next step is to rescale these in terms of v, using

$$\frac{j}{N} = \pi + \frac{v}{\sqrt{N}}.$$

In the rescaled variables,

$$E[\Delta V | V_N(t) = v] = \frac{1}{\sqrt{N}} E\left[\Delta X \left| \frac{X(t)}{N} = \frac{j}{N} = \pi + \frac{v}{\sqrt{N}} \right.\right]$$

$$= \lambda \sqrt{N}\left[\left(1 - \pi - \frac{v}{\sqrt{N}}\right)\gamma_2 - \left(\pi + \frac{v}{\sqrt{N}}\right)\gamma_1\right]\Delta t + o(\Delta t)$$

$$= -\lambda(\gamma_1 + \gamma_2)v\Delta t + o(\Delta t).$$

A similar substitution shows that

$$E\left[\Delta V^2 | V_N(t) = v\right] = \frac{2\lambda\gamma_1\gamma_2}{(\gamma_1 + \gamma_2)^2}\Delta t + o(\Delta t).$$

Similar computations show that the higher moments of ΔV are negligible. Since the relations (8.62) and (8.63) serve to characterize the Ornstein–Uhlenbeck process, the evidence is compelling that the rescaled gene processes $\{V_N(t)\}$ will converge in some appropriate sense to an Ornstein–Uhlenbeck process $V(t)$ with

$$\beta = \lambda(\gamma_1 + \gamma_2) \quad \text{and} \quad \sigma^2 = \frac{2\lambda\gamma_1\gamma_2}{(\gamma_1 + \gamma_2)^2},$$

and

$$X(t) \approx N\pi + \sqrt{N}V(t) \qquad \text{for large } N.$$

This is indeed the case, but it represents another topic that we must leave for future study.

8.5.2 The Position Process

If $V(t)$ is the velocity of the pollen grain, then its position at time t would be

$$S(t) = S(0) + \int_0^t V(u)du. \tag{8.64}$$

Because the Ornstein–Uhlenbeck process is continuous, the integral in (8.64) is well defined. The Ornstein–Uhlenbeck process is normally distributed, and so is each approximating sum to the integral in (8.64). It must be, then, that the position process $S(t)$ is normally distributed, and, to describe it, we need only evaluate its mean and covariance functions. To simplify the mathematics without losing any essentials, let us assume that $S(0) = V(0) = 0$. Then,

$$E[S(t)] = E\left[\int_0^t V(s)ds\right] = \int_0^t E[V(s)]ds = 0.$$

(The interchange of integral and expectation needs justification. Since the expected value of a sum is always the sum of the expected values, the interchange of expectation with Riemann approximating sums is clearly valid. What is needed is justification in the limit as the approximating sums converge to the integrals.)

$$\text{Var}[S(t)] = E[S(t)^2] = E\left[\left\{\int_0^t V(s)ds\right\}^2\right]$$

$$= E\left[\left\{\int_0^t V(u)du\right\}\left\{\int_0^t V(s)ds\right\}\right]$$

$$= \int_0^t \int_0^t E[V(s)V(u)]du\,ds$$

$$= 2\int_0^t \int_0^s E[V(s)V(u)]du\,ds \tag{8.65}$$

$$= \frac{\sigma^2}{\beta}\int_0^t \int_0^s \left(e^{-\beta(s-u)} - e^{-\beta(s+u)}\right)du\,ds \qquad \text{(Using (8.61))}$$

$$= \frac{\sigma^2}{\beta^2}\int_0^t e^{-\beta s}\left(e^{\beta s} - 1 - 1 + e^{-\beta s}\right)ds$$

$$= \frac{\sigma^2}{\beta^2}\left[t - \frac{2}{\beta}\left(1 - e^{-\beta t}\right) + \frac{1}{2\beta}\left(1 - e^{-2\beta t}\right)\right].$$

This variance behaves like that of a Brownian motion when t is large in the sense that

$$\frac{\text{Var}[S(t)]}{t} \to \frac{\sigma^2}{\beta^2} \qquad \text{as} \quad t \to \infty.$$

That is, observed over a long time span, the particle's position as modeled by an Ornstein–Uhlenbeck velocity behaves much like a Brownian motion with variance parameter σ^2/β^2. In this sense, the Ornstein–Uhlenbeck model agrees with the Brownian motion model over long time spans and improves upon it for short durations. Section 8.5.4 offers another approach.

Example *Stock Prices* It is sometimes assumed that the market price of a share of stock follows the position process under an Ornstein–Uhlenbeck velocity. The model is consistent with the Brownian motion model over long time spans. In the short term, the price changes are not independent but have an exponentially decreasing correlation meant to capture some notion of a market momentum. Suppose a call option is to be exercised, if profitable at a striking price of a, at some fixed time t in the future. If $V(0) = 0$ and $S(0) = z$ is the current stock price, then the expected value of the option is

$$E\left[(S(t) - a)^+\right] = \tau \left\{\phi\left(\frac{a - \mu}{\tau}\right) - \left(\frac{a - \mu}{\tau}\right)\left[1 - \Phi\left(\frac{a - \mu}{\tau}\right)\right]\right\}, \tag{8.66}$$

where

$$\mu = z,$$

and

$$\tau^2 = \frac{\sigma^2}{\beta^2}\left[t - \frac{2}{\beta}\left(1 - e^{-\beta t}\right) + \frac{1}{2\beta}\left(1 - e^{-2\beta t}\right)\right].$$

Note that μ and τ^2 are the mean and variance of $S(t)$. The derivation is left for Problem 8.5.4.

8.5.3 The Long Run Behavior

It is easily seen from (8.58) and (8.59) that for large values of t, the mean of the Ornstein–Uhlenbeck process converges to zero and the variance to $\sigma^2/2\beta$. This leads to a limiting distribution for the process in which

$$\lim_{t\to\infty} \Pr\{V(t) < y | V(0) = x\} = \Phi\left(\frac{\sqrt{2\beta}y}{\sigma}\right). \tag{8.67}$$

That is, the limiting distribution of the process is normal with mean zero and variance $\sigma^2/(2\beta)$. We now set forth a representation of a *stationary Ornstein–Uhlenbeck process*, a process for which the limiting distribution in (8.67) holds for all finite times as well as in the limit. The stationary Ornstein–Uhlenbeck process $\{V^s(t); -\infty < t < \infty\}$ is represented in terms of a Brownian motion by

$$V^s(t) = \frac{\sigma}{\sqrt{2\beta}}e^{-\beta t}B\left(e^{2\beta t}\right), \quad -\infty < t < \infty. \tag{8.68}$$

The stationary Ornstein–Uhlenbeck process is Gaussian (see Section 8.1.4) and has mean zero. The covariance calculation is

$$\begin{aligned}
\Gamma(s, t) &= \text{Cov}\left[V^s(s), V^s(t)\right] \\
&= \frac{\sigma^2}{2\beta}e^{-\beta(s+t)}\text{Cov}\left[B\left(e^{2\beta s}\right), B\left(e^{2\beta t}\right)\right] \\
&= \frac{\sigma^2}{2\beta}e^{-\beta(s+t)}\min\left\{e^{2\beta s}, e^{2\beta t}\right\} \\
&= \frac{\sigma^2}{2\beta}e^{-\beta|t-s|}.
\end{aligned} \tag{8.69}$$

The stationary Ornstein–Uhlenbeck process is the unique Gaussian process with mean zero and covariance (8.69).

The independence of the Brownian increments implies that the stationary Ornstein–Uhlenbeck process is a Markov process, and it is straightforward to verify that the transition probabilities are given by (8.60).

Example *An Ehrenfest Urn Model in Continuous Time* A single particle switches repeatedly between urn A and urn B. Suppose that the duration it spends in an urn before moving is an exponentially distributed random variable with parameter β, and that all durations are independent. Let $\xi(t) = 1$ if the particle is in urn A at time t, and $\xi(t) = -1$ if in urn B. Then, $\{\xi(t); t \geq 0\}$ is a two-state Markov process in continuous time for which (see Chapter 6, (6.30))

$$\Pr\{\xi(t+s) = 1|\xi(t) = 1\} = \frac{1}{2} + \frac{1}{2}e^{-2\beta s}. \tag{8.70}$$

Let us further stipulate that the particle is equally likely to be in either urn at time zero. It follows, then, that it is equally likely to be in either urn at all times, and that therefore, $E[\xi(t)] = 0$ for all t. Using (8.70) and the symmetry of the process, we may derive the covariance. We have

$$
\begin{aligned}
E[\xi(t)\xi(t+s)] = {} & \frac{1}{2}\Pr\{\xi(t+s) = 1|\xi(t) = 1\} \\
& + \frac{1}{2}\Pr\{\xi(t+s) = -1|\xi(t) = -1\} \\
& - \frac{1}{2}\Pr\{\xi(t+s) = -1|\xi(t) = 1\} \\
& - \frac{1}{2}\Pr\{\xi(t+s) = 1|\xi(t) = -1\} \\
= {} & e^{-2\beta s}.
\end{aligned}
\tag{8.71}
$$

Now consider N of these particles, each alternating between the urns independently of the others, and let $\xi_i(t)$ track the position of the ith particle at time t. The disparity between the numbers of particles in the two urns is measured by

$$S_N(t) = \sum_{i=1}^{N} \xi_i(t).$$

If $S_N(t) = 0$, then the urns contain equal numbers of particles. If $S_N(t) = k$, then there are $(N+k)/2$ particles in urn A. The central limit principle for random functions suggests that

$$V_N(t) = \frac{1}{\sqrt{N}} S_N(t)$$

should, for large N, behave similarly to a Gaussian process with mean zero and covariance $\Gamma(s, t) = \exp\{-2\beta|t - s|\}$. This limiting process is a stationary Ornstein–Uhlenbeck process with $\sigma^2 = 2\beta$. Thus, we have derived the approximation

$$S_N(t) \approx \sqrt{N}V^s(t), \qquad t > 0,$$

for the behavior of this continuous-time urn model when the number of particles is large.

8.5.4 Brownian Measure and Integration*

We state, in the form of a theorem without proof, an exceedingly useful formula for computing certain functionals of Gaussian processes. This theorem provides a tiny glimpse into a vast and rich area of stochastic process theory, and we included it in this elementary text in a blatant attempt to entice the student towards further study.

Theorem 8.4. *Let $g(x)$ be a continuous function and let $\{B(t); t \geq 0\}$ be a standard Brownian motion. For each fixed value of $t > 0$, there exists a random variable*

$$\mathcal{I}(g) = \int_0^t g(x)dB(x) \tag{8.72}$$

that is the limit of the approximating sums

$$\mathcal{I}_n(g) = \sum_{k=1}^{2^n} g\left(\frac{k}{2^n}t\right)\left[B\left(\frac{k}{2^n}t\right) - B\left(\frac{k-1}{2^n}t\right)\right] \tag{8.73}$$

as $n \to \infty$. The random variable $\mathcal{I}(g)$ is normally distributed with mean zero and variance

$$\mathrm{Var}[\mathcal{I}(g)] = \int_0^t g^2(u)du. \tag{8.74}$$

If $f(x)$ is another continuous function of x, then $\mathcal{I}(f)$ and $\mathcal{I}(g)$ have a joint normal distribution with covariance

$$E[\mathcal{I}(f)\mathcal{I}(g)] = \int_0^t f(x)g(x)dx. \tag{8.75}$$

The proof of the theorem is more tedious than difficult, but it does require knowledge of facts that are not included among our prerequisites. The theorem asserts that a

---

* This subsection is both more advanced and more abstract than those that have preceded it.

sequence of random variables, the Riemann approximations, converge to another random variable. The usual proof begins with showing that the expected mean square of the difference between distinct Riemann approximations satisfies the Cauchy criterion for convergence, and then goes on from there.

When $g(x)$ is differentiable, then an integration by parts may be validated, which shows that

$$\int_0^t g(x)dB(x) = g(t)B(t) - \int_0^t B(x)g'(x)dx, \tag{8.76}$$

and this approach may yield a concrete representation of the integral in certain circumstances. For example, if $g(x) = 1$, then $g'(x) = 0$, and

$$\int_0^t 1dB(x) = g(t)B(t) - 0 = B(t),$$

as one would hope. When $g(x) = t - x$, then $g'(x) = -1$, and

$$\int_0^t (t-x)dB(x) = \int_0^t B(x)dx. \tag{8.77}$$

The process on the right side of (8.77) is called *integrated Brownian motion*. Theorem 8.4, then, asserts that integrated Brownian motion is normally distributed with mean zero and variance

$$\mathrm{Var}\left[\int_0^t B(x)dx\right] = \int_0^t (t-x)^2 dx = \frac{t^3}{3}.$$

The calculus of the *Brownian integral* of Theorem 8.4 offers a fresh and convenient way to determine functionals of some Gaussian processes. For example, in the case of the Ornstein–Uhlenbeck process, we have the integral representation

$$V(t) = ve^{-\beta t} + \sigma \int_0^t e^{-\beta(t-u)}dB(u). \tag{8.78}$$

The second term on the right of (8.78) expresses the random component of the Ornstein–Uhlenbeck process as an exponentially weighted moving average of infinitesimal Brownian increments. According to Theorem 8.4, this random component has mean zero and is normally distributed. We use Theorem 8.4 to determine the covariance. For $0 < s < t$,

$$
\begin{aligned}
\mathrm{Cov}[V(s), V(t)] &= E\left[\{V(s) - v e^{-\beta s}\}\{V(t) - v e^{-\beta t}\}\right] \\
&= \sigma^2 E\left[\int_0^s e^{-\beta(s-u)}\,dB(u) \int_0^t e^{-\beta(t-w)}\,dB(w)\right] \\
&= \sigma^2 \int_0^t 1(u < s) e^{-\beta(s-u)} e^{-\beta(t-u)}\,du \\
&= \sigma^2 e^{-\beta(s+t)} \int_0^s e^{2\beta u}\,du \\
&= \frac{\sigma^2}{2\beta} e^{-\beta(s+t)} \left(e^{2\beta s} - 1\right) \\
&= \frac{\sigma^2}{2\beta} \left(e^{-\beta(t-s)} - e^{-\beta(t+s)}\right),
\end{aligned}
$$

in agreement with (8.61).

Example *The Position Process Revisited* Let us assume that $V(0) = v = 0$. The integral of the Ornstein–Uhlenbeck velocity process gives the particle's position $S(t)$ at time t. If we replace the integrand by its representation (8.78) $(v = 0)$, we obtain

$$
\begin{aligned}
S(t) = \int_0^t V(s)\,ds &= \sigma \int_0^t \int_0^s e^{-\beta(s-u)}\,dB(u)\,ds \\
&= \sigma \int_0^t \int_u^t e^{-\beta(s-u)}\,ds\,dB(u) \\
&= \sigma \int_0^t e^{\beta u} \int_u^t e^{-\beta s}\,ds\,dB(u) \qquad (8.79) \\
&= \frac{\sigma}{\beta} \int_0^t e^{\beta u} \left(e^{-\beta u} - e^{-\beta t}\right)\,dB(u) \\
&= \frac{\sigma}{\beta} \int_0^t \left(1 - e^{-\beta(t-u)}\right)\,dB(u).
\end{aligned}
$$

Theorem 8.4 applied to (8.79) tells us that the position $S(t)$ at time t is normally distributed with mean zero and variance

$$
\begin{aligned}
\mathrm{Var}[S(t)] &= \frac{\sigma^2}{\beta^2} \int_0^t \left[1 - e^{-\beta(t-u)}\right]^2 du \\
&= \frac{\sigma^2}{\beta^2} \int_0^t \left(1 - e^{-\beta w}\right)^2 dw \\
&= \frac{\sigma^2}{\beta^2} \left[t - \frac{2}{\beta}\left(1 - e^{-\beta t}\right) + \frac{1}{2\beta}\left(1 - e^{-2\beta t}\right)\right],
\end{aligned}
$$

in agreement with (8.65). Problem 8.5.4 calls for using Theorem 8.4 to determine the covariance between the velocity $V(t)$ and position $S(t)$.

The position process under an Ornstein–Uhlenbeck velocity behaves like a Brownian motion over large time spans, and we can see this more clearly from the Brownian integral representation in (8.79). If we carry out the first term in the integral (8.79) and recognize that the second part is $V(t)$ itself, we see that

$$
\begin{aligned}
S(t) &= \frac{\sigma}{\beta} \int_0^t \left(1 - e^{-\beta(t-u)}\right) dB(u) \\
&= \frac{\sigma}{\beta} \left[B(t) - \int_0^t e^{-\beta(t-u)} dB(u)\right] \\
&= \frac{1}{\beta}[\sigma B(t) - V(t)].
\end{aligned}
\tag{8.80}
$$

Let us introduce a rescaled position process that will allow us to better see changes in position over large time spans. Accordingly, for $N > 0$, let

$$
\begin{aligned}
S_N(t) &= \frac{1}{\sqrt{N}} S(Nt) \\
&= \frac{1}{\beta} \left[\frac{\sigma}{\sqrt{N}} B(Nt) + \frac{1}{\sqrt{N}} V(t)\right] \\
&= \frac{1}{\beta} \left[\sigma \tilde{B}(t) + \frac{1}{\sqrt{N}} V(t)\right],
\end{aligned}
\tag{8.81}
$$

where $\tilde{B}(t) = B(Nt)/\sqrt{N}$ remains a standard Brownian motion. (See Exercise 8.1.2.) Because the variance of $V(t)$ is always less than or equal to $\sigma^2/(2\beta)$, the variance of $V(t)/\sqrt{N}$ becomes negligible for large N. Equation (8.81), then, shows more clearly in what manner the position process becomes like a Brownian motion: For large N,

$$S_N(t) \approx \frac{\sigma}{\beta} \tilde{B}(t).$$

Exercises

8.5.1 An Ornstein–Uhlenbeck process $V(t)$ has $\sigma^2 = 1$ and $\beta = 0.2$. What is the probability that $V(t) \leq 1$ for $t = 1, 10$, and 100? Assume that $V(0) = 0$.

8.5.2 The velocity of a certain particle follows an Ornstein–Uhlenbeck process with $\sigma^2 = 1$ and $\beta = 0.2$. The particle starts at rest ($v = 0$) from position $S(0) = 0$. What is the probability that it is more than one unit away from its origin at time $t = 1$. What is the probability at times $t = 10$ and $t = 100$?

8.5.3 Let $\xi_1, \xi_2, \ldots$ be independent standard normal random variables and β a constant, $0 < \beta < 1$. A discrete analog to the Ornstein–Uhlenbeck process may be constructed by setting

$$V_0 = v \quad \text{and} \quad V_n = (1 - \beta)V_{n-1} + \xi_n \quad \text{for} \quad n \geq 1.$$

(a) Determine the mean value function and covariance function for $\{V_n\}$.

(b) Let $\Delta V = V_{n+1} - V_n$. Determine the conditional mean and variance of ΔV, given that $V_n = v$.

Problems

8.5.1 Let $\xi_1, \xi_2, \ldots$ be independent standard normal random variables and β a constant, $0 < \beta < 1$. A discrete analog to the Ornstein–Uhlenbeck process may be constructed by setting

$$V_0 = v \quad \text{and} \quad V_n = (1 - \beta)V_{n-1} + \xi_n \quad \text{for} \quad n \geq 1.$$

(a) Show that

$$V_n = (1 - \beta)^n v + \sum_{k=1}^{n} (1 - \beta)^{n-k} \xi_k.$$

Comment on the comparison with (8.78).

(b) Let $\Delta V_n = V_n - V_{n-1}$, $S_1 = v + V_1 + \cdots + V_n$, and $B_n = \xi_1 + \cdots + \xi_n$. Show that

$$V_n = v - \beta S_{n-1} + B_n.$$

Compare and contrast with (8.80).

8.5.2 Let $S(t)$ be the position process corresponding to an Ornstein–Uhlenbeck velocity $V(t)$. Assume that $S(0) = V(0) = 0$. Obtain the covariance between $S(t)$ and $V(t)$.

8.5.3 Verify the option valuation formulation (8.66).

   Hint: Use the result of Exercise 8.4.6.

8.5.4 In the Ehrenfest urn model (see Chapter 3, Section 3.3.2) for molecular diffusion through a membrane, if there are i particles in urn A, the probability that there will be $i + 1$ after one time unit is $1 - i/(2N)$, and the probability of $i - 1$ is $i/(2N)$, where $2N$ is the aggregate number of particles in both urns. Following Chapter 3, Section 3.3.2, let Y_n be the number of particles in urn A after the nth transition, and let $X_n = Y_n - N$. Let $\Delta X = X_{n+1} - X_n$ be the change in urn composition. The probability law is

$$\Pr\{\Delta X = \pm 1 | X_n = x\} = \frac{1}{2} \mp \frac{x}{2N}.$$

We anticipate a limiting process in which the time between transitions becomes small and the number of particles becomes large. Accordingly, let $\Delta t = 1/N$ and measure fluctuations of a rescaled process in units of order $1/\sqrt{N}$. The definition of the rescaled process is

$$V_N(t) = \frac{X_{[Nt]}}{\sqrt{N}}.$$

Note that in the duration $t = 0$ to $t = 1$ in the rescaled process, there are N transitions in the urns, and a unit change in the rescaled process corresponds to a fluctuation of order $\sqrt{N}$ in the urn composition. Let $\Delta V = V_N(t + 1/N) - V_N(t)$ be the displacement in the rescaled process over the time interval $\Delta t = 1/N$. Show that

$$E[\Delta V | V_N(t) = v] = \frac{1}{\sqrt{N}} \left(\frac{1}{2} - \frac{v}{2\sqrt{N}} \right) - \frac{1}{\sqrt{N}} \left(\frac{1}{2} + \frac{v}{2\sqrt{N}} \right)$$

$$= -v \left(\frac{1}{N} \right) = -v\Delta t,$$

and that $(\Delta V)^2 = 1/N$, whence

$$\mathrm{Var}[\Delta V | V_N(t) = v] = E\left[(\Delta V)^2 \right] - \{E[\Delta V]\}^2$$

$$= \frac{1}{N} + o\left(\frac{1}{N} \right) = \Delta t + o(\Delta t).$$

9 Queueing Systems

9.1 Queueing Processes

A queueing system consists of "customers" arriving at random times to some facility where they receive service of some kind and then depart. We use "customer" as a generic term. It may refer, e.g., to *bona fide* customers demanding service at a counter, to ships entering a port, to batches of data flowing into a computer subsystem, to broken machines awaiting repair, and so on. Queueing systems are classified according to

1. *The input process*, the probability distribution of the pattern of arrivals of customers in time;
2. *The service distribution*, the probability distribution of the random time to serve a customer (or group of customers in the case of batch service); and
3. *The queue discipline*, the number of servers and the order of customer service.

While a variety of input processes may arise in practice, two simple and frequently occurring types are mathematically tractable and give insights into more complex cases. First is the scheduled input, where customers arrive at fixed times $T, 2T, 3T, \ldots$. The second most common model is the "completely random" arrival process, where the times of customer arrivals form a Poisson process. Understanding the axiomatic development of the Poisson process in V may help one to evaluate the validity of the Poisson assumption in any given application. Many theoretical results are available when the times of customer arrivals form a renewal process. Exponentially distributed interarrival times, then, correspond to a Poisson process of arrivals as a special case.

We will always assume that the durations of service for individual customers are independent and identically distributed nonnegative random variables and are independent of the arrival process. The situation in which all service times are the same fixed duration D, is, then, a special case.

The most common queue discipline is *first come, first served*, where customers are served in the same order in which they arrive. All of the models that we consider in this chapter are of this type.

Queueing models aid the design process by predicting system performance. For example, a queueing model might be used to evaluate the costs and benefits of adding a server to an existing system. The models enable us to calculate system performance measures in terms of more basic quantities. Some important measures of system behavior are

1. *The probability distribution of the number of customers in the system.* Not only do customers in the system often incur costs, but in many systems, physical space for waiting customers

must be planned for and provided. Large numbers of waiting customers can also adversely affect the input process by turning potential new customers away. (See Section 9.4.1 on queueing with balking.)

2. *The utilization of the server(s).* Idle servers may incur costs without contributing to system performance.

3. *System throughput.* The long run number of customers passing through the system is a direct measure of system performance.

4. *Customer waiting time.* Long waits for service are annoying in the simplest queueing situations and directly associated with major costs in many large systems such as those describing ships waiting to unload at a port facility or patients awaiting emergency care at a hospital.

9.1.1   The Queueing Formula $L = \lambda W$

Consider a queueing system that has been operating sufficiently long to have reached an appropriate steady state, or a position of statistical equilibrium. Let

L = the average number of customers in the system;

λ = the rate of arrival of customers to the system; and

W = the average time spent by a customer in the system.

The equation $L = \lambda W$ is valid under great generality for such systems and is of basic importance in the theory of queues, since it directly relates two of our most important measures of system performance, the mean queue size and the mean customer waiting time in the steady state, i.e., mean queue size and mean customer waiting time evaluated with respect to a limiting or stationary distribution for the process.

The validity of $L = \lambda W$ does not rest on the details of any particular model, but depends only upon long run mass flow balance relations. To sketch this reasoning, consider a time T sufficiently long so that statistical fluctuations have averaged out. Then, the total number of customers to have entered the system is λT, the total number to have departed is $\lambda(T - W)$, and the net number remaining in the system L must be the difference

$$L = \lambda T - [\lambda(T - W)] = \lambda W.$$

Figure 9.1 depicts the relation $L = \lambda W$.

Of course, what we have done is by no means a proof, and indeed, we shall give no proof. We shall, however, provide several sample verifications of $L = \lambda W$ where L is the mean of the stationary distribution of customers in the system, W is the mean customer time in the system determined from the stationary distribution, and λ is the arrival rate in a Poisson arrival process.

Let L_0 be the average number of customers waiting in the system who are not yet being served, and let W_0 be the average waiting time in the system excluding service time. In parallel to $L = \lambda W$, we have the formula

$$L_0 = \lambda W_0 \tag{9.1}$$

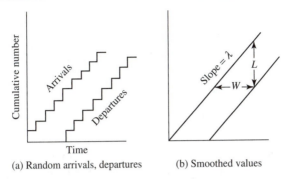

(a) Random arrivals, departures   (b) Smoothed values

Figure 9.1 The cumulative number of arrivals and departures in a queueing system. The smoothed values in (b) are meant to symbolize long run averages. The rate of arrivals per unit time is λ, the mean number in the system is L, and the mean time a customer spends in the system is W.

The total waiting time in the system is the sum of the waiting time before service and the service time. In terms of means, we have

$$W = W_0 + \text{mean service time.} \qquad (9.2)$$

9.1.2 A Sampling of Queueing Models

In the remainder of this chapter, we will study a variety of queueing systems. A standard shorthand is used in much of the queueing literature for identifying simple queueing models. The shorthand assumes that the arrival times form a renewal process, and the format $A/B/c$ uses A to describe the interarrival distribution, B to specify the individual customer service time distribution, and c to indicate the number of servers. The common cases for the first two positions are $G = GI$ for a general or arbitrary distribution, M (memoryless) for the exponential distribution, E_k (Erlang) for the gamma distribution of order k, and D for a deterministic distribution, a schedule of arrivals or fixed service times.

Some examples discussed in the sequel are the following:

The M/M/1 queue Arrivals follow a Poisson process; service times are exponentially distributed; and there is a single server. The number $X(t)$ of customers in the system at time t forms a birth and death process. (See Section 9.2.)

The M/M/∞ queue There are Poisson arrivals and exponentially distributed service times. Any number of customers are processed simultaneously and independently. Often self-service situations may be described by this model. In the older literature, this was called the "telephone trunking problem."

The M/G/1 queue In this model, there are Poisson arrivals but arbitrarily distributed service times. The analysis proceeds with the help of an embedded Markov chain.

More elaborate variations will also be set forth. *Balking* is the refusal of new customers to enter the system if the waiting line is too long. More generally, in a

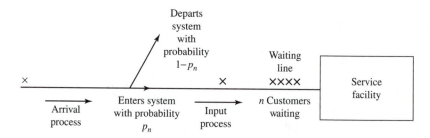

Figure 9.2 If n customers are waiting in a queueing system with balking, an arriving customer enters the system with probability p_n and does not enter with probability $1 - p_n$.

queueing system with balking, an arriving customer enters the system with a probability that depends on the size of the queue. Here it is important to distinguish between the *arrival process* and the *input process*, as shown in Figure 9.2. A special case is a queue with *overflow*, in which an arriving customer enters the queue if and only if there is at least one server free to begin service immediately.

In a *priority queue*, customers are allowed to be of different types. Both the service discipline and the service time distribution may vary with the customer type.

A *queueing network* is a collection of service facilities where the departures from some stations form the arrivals of others. The network is *closed* if the total number of customers is fixed, with these customers continuously circulating through the system. The machine repair model (see the example entitled "Repairman Models" in Chapter 6, Section 6.4) is an example of a closed queueing network. In an open queueing network, customers may arrive from, and depart to, places outside the network, as well as move from station to station. Queueing network models have found much recent application in the design of complex information processing systems.

Exercises

9.1.1 What design questions might be answered by modeling the following queueing systems?

The Customer	The Server
(a) Arriving airplanes	The runway
(b) Cars	A parking lot
(c) Broken TVs	Repairman
(d) Patients	Doctor
(e) Fires	Fire engine company

What might be reasonable assumptions concerning the arrival process, service distribution, and priority in these instances?

9.1.2 Consider a system, such as a barber shop, where the service required is essentially identical for each customer. Then, actual service times would tend to cluster near the mean service time. Argue that the exponential distribution would not be appropriate in this case. For what types of service situations might the exponential distribution be quite plausible?

9.1.3 Oil tankers arrive at an offloading facility according to a Poisson process whose rate is $\lambda = 2$ ships per day. Daily records show that there is an average of 3 ships unloading or waiting to unload at any instant in time. On average, what is the duration of time that a ship spends in port? Assume that a ship departs immediately after unloading.

Problem

9.1.1 Two dump trucks cycle between a gravel loader and a gravel unloader. Suppose that the travel times are insignificant relative to the load and unload times, which are exponentially distributed with parameters μ and λ, respectively. Model the system as a closed queueing network. Determine the long run gravel loads moved per unit time.

Hint: Refer to the example entitled "Repairman Models" in Section 6.4.

9.2   Poisson Arrivals, Exponential Service Times

The simplest and most extensively studied queueing models are those having a Poisson arrival process and exponentially distributed service times. In this case, the queue size forms a birth and death process (see Sections 6.3 and Sections 6.4 of Chapter 6), and the corresponding stationary distribution is readily found.

We let λ denote the intensity, or rate, of the Poisson arrival process and assume that the service time distribution is exponential with parameter μ. The corresponding density function is

$$g(x) = \mu e^{-\mu x} \quad \text{for } x > 0. \tag{9.3}$$

For the Poisson arrival process we have

$$\Pr\{\text{An arrival in } [t, t+h)\} = \lambda h + o(h) \tag{9.4}$$

and

$$\Pr\{\text{No arrivals in } [t, t+h)\} = 1 - \lambda h + o(h). \tag{9.5}$$

Similarly, the memoryless property of the exponential distribution as expressed by its constant hazard rate (see Chapter 1, Section 1.4.2) implies that

$$\Pr\{\text{A service is completed in } [t, t+h)|\text{Service in progress at time } t\}$$
$$= \mu h + o(h), \tag{9.6}$$

and

$$\Pr\{\text{Service not completed in } [t, t+h)|\text{Service in progress at time } t\}$$
$$= 1 - \mu h + o(h), \tag{9.7}$$

The service rate μ applies to a particular server. If k servers are simultaneously operating, the probability that one of them completes service in a time interval of duration h is $(k\mu)h + o(h)$, so that the system service rate is $k\mu$. The principle used here is the same as that used in deriving the infinitesimal parameters of the Yule process (Chapter 6, Section 6.1).

We let $X(t)$ denote the number of customers in the system at time t, counting the customers undergoing service as well as those awaiting service. The independence of arrivals in disjoint time intervals together with the memoryless property of the exponential service time distribution implies that $X(t)$ is a time homogeneous Markov chain, in particular, a birth and death process. (See Sections 6.3 and 6.4 of Chapter 6.)

9.2.1 The M/M/1 System

We consider first the case of a single server and let $X(t)$ denote the number of customers in the system at time t. An increase in $X(t)$ by one unit corresponds to a customer arrival, and in view of (9.4) and (9.7) and the postulated independence of service times and the arrival process, we have

$$\Pr\{X(t+h) = k+1|X(t) = k\} = [\lambda h + o(h)] \times [1 - \mu h + o(h)]$$
$$= \lambda h + o(h) \quad \text{for } k = 0, 1, \dots .$$

Similarly, a decrease in $X(t)$ by one unit corresponds to a completion of service, whence

$$\Pr\{X(t+h) = k-1|X(t) = k\} = \mu h + o(h) \quad \text{for } k = 1, 2, \dots .$$

Then, $X(t)$ is a birth and death process with birth parameters

$$\lambda_k = \lambda \quad \text{for } k = 0, 1, 2, \dots$$

and death parameters

$$\mu_k = \mu \quad \text{for } k = 1, 2, \dots .$$

Of course, no completion of service is possible when the queue is empty. We, thus, specify $\mu_0 = 0$.

Let

$$\pi_k = \lim_{t\to\infty} \Pr\{X(t) = k\} \quad \text{for } k = 0, 1, \ldots$$

be the limiting, or equilibrium, distribution of queue length. Section 6.4 of Chapter 6 describes a straightforward procedure for determining the limiting distribution π_k from the birth and death parameters λ_k and μ_k. The technique is to first obtain intermediate quantities θ_j defined by

$$\theta_0 = 1 \quad \text{and} \quad \theta_j = \frac{\lambda_0 \lambda_1 \cdots \lambda_{j-1}}{\mu_1 \mu_2 \cdots \mu_j} \quad \text{for } j \geq 1, \tag{9.8}$$

and then

$$\pi_0 = \frac{1}{\sum_{j=0}^{\infty} \theta_j} \quad \text{and} \quad \pi_k = \theta_k \pi_0 = \frac{\theta_k}{\sum_{j=0}^{\infty} \theta_j} \quad \text{for } k \geq 1. \tag{9.9}$$

When $\sum_{j=0}^{\infty} \theta_j = \infty$, then $\lim_{t\to\infty} \Pr\{X(t) = k\} = 0$ for all k, and the queue length grows unboundedly in time.

For the $M/M/1$ queue at hand we readily compute $\theta_0 = 1$ and $\theta_j = (\lambda/\mu)^j$ for $j = 1, 2, \ldots$. Then,

$$\sum_{j=0}^{\infty} \pi_j = \sum_{j=0}^{\infty} \left(\frac{\lambda}{\mu}\right)^j = \frac{1}{(1 - \lambda/\mu)} \quad \text{if } \lambda < \mu,$$

$$= \infty \quad \text{if } \lambda \geq \mu.$$

Thus, no equilibrium distribution exists when the arrival rate λ is equal to or greater than the service rate μ. In this case, the queue length grows without bound.

When $\lambda < \mu$, a *bona fide* limiting distribution exists, given by

$$\pi_0 = \frac{1}{\sum \theta_j} = 1 - \frac{\lambda}{\mu} \tag{9.10}$$

and

$$\pi_k = \pi_0 \theta_k = \left(1 - \frac{\lambda}{\mu}\right) \left(\frac{\lambda}{\mu}\right)^k \quad \text{for } k = 0, 1, \ldots. \tag{9.11}$$

The equilibrium distribution (9.11) gives us the answer to many questions involving the limiting behavior of the system. We recognize the form of (9.11) as that of a geometric distribution, and then reference to Chapter 1, Section 1.3.3 gives us the mean queue length in equilibrium to be

$$L = \frac{\lambda}{\mu - \lambda}. \tag{9.12}$$

The ratio $\rho = \lambda/\mu$ is called the *traffic intensity*,

$$\rho = \frac{\text{arrival rate}}{\text{system service rate}} = \frac{\lambda}{\mu}. \tag{9.13}$$

As the traffic intensity approaches one, the mean queue length $L = \rho/(1-\rho)$ becomes infinite. Again using (9.10), the probability of being served immediately upon arrival is

$$\pi_0 = 1 - \frac{\lambda}{\mu},$$

the probability, in the long run, of finding the server idle. The server utilization, or long run fraction of time that the server is busy, is $1 - \pi_0 = \lambda/\mu$.

We can also calculate the distribution of waiting time in the stationary case when $\lambda < \mu$. If an arriving customer finds n people in front of him, his total waiting time T, including his own service time, is the sum of the service times of himself and those ahead, all distributed exponentially with parameter μ, and since the service times are independent of the queue size, T has a gamma distribution of order $n+1$ with scale parameter μ,

$$\Pr\{T \le t | n \text{ ahead}\} = \int_0^t \frac{\mu^{n+1} \tau^n e^{-\mu\tau}}{\Gamma(n+1)} d\tau. \tag{9.14}$$

By the law of total probability, we have

$$\Pr\{T \le t\} = \sum_{n=0}^{\infty} \Pr\{T \le t | n \text{ ahead}\} \times \left(\frac{\lambda}{\mu}\right)^n \left(1 - \frac{\lambda}{\mu}\right),$$

since $(\lambda/\mu)^n (1 - \lambda/\mu)$ is the probability that in the stationary case a customer on arrival will find n ahead in line. Now, substituting from (9.14), we obtain

$$\Pr\{T \le t\} = \sum_{n=0}^{\infty} \int_0^t \frac{\mu^{n+1} \tau^n e^{-\mu\tau}}{\Gamma(n+1)} \left(\frac{\lambda}{\mu}\right)^n \left(1 - \frac{\lambda}{\mu}\right) d\tau$$

$$= \int_0^t \mu e^{-\mu\tau} \left(1 - \frac{\lambda}{\mu}\right) \sum_{n=0}^{\infty} \frac{\tau^n \lambda^n}{\Gamma(n+1)} d\tau$$

$$= \int_0^t \left(1 - \frac{\lambda}{\mu}\right) \mu \exp\left\{-\tau\mu\left(1 - \frac{\lambda}{\mu}\right)\right\} d\tau = 1 - \exp[-t(\mu - \lambda)],$$

which is also an exponential distribution.

The mean of this exponential waiting time distribution is the reciprocal of the exponential parameter, or

$$W = \frac{1}{\mu - \lambda}. \tag{9.15}$$

Reference to (9.12) and (9.15) verifies the fundamental queueing formula $L = \lambda W$.

A queueing system alternates between durations when the servers are busy and durations when the system is empty and the servers are idle. An *idle period* begins the instant the last customer leaves, and endures until the arrival of the next customer. When the arrival process is Poisson of rate λ, then an idle period is exponentially distributed with mean

$$E[I_1] = \frac{1}{\lambda}.$$

A busy period is an uninterrupted duration in which the system is not empty. When arrivals to a queue follow a Poisson process, then the successive durations X_k from the commencement of the kth busy period to the start of the next busy period form a renewal process (see Figure 9.3). Each X_k is composed of a busy portion B_k and an idle portion I_k. Then, the renewal theorem (see "A Queueing Model" in Chapter 7, Section 7.5.3) applies to tell us that $p_0(t)$, the probability that the system is empty at time t, converges to

$$\lim_{t \to \infty} p_0(t) = \pi_0 = \frac{E[I_1]}{E[I_1] + E[B_1]}.$$

We substitute the known quantities $\pi_0 = 1 - \lambda/\mu$ and $E[I_1] = 1/\lambda$ to obtain

$$1 - \frac{\lambda}{\mu} = \frac{1/\lambda}{1/\lambda + E[B_1]},$$

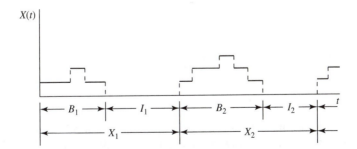

Figure 9.3 The busy periods B_k and idle periods I_k of a queueing system. When arrivals form a Poisson process, then $X_k = B_k + I_k$, $k = 1, 2, \ldots$, are independent, identically distributed non-negative random variables, and thus form a renewal process.

which gives

$$E[B_1] = \frac{1}{\mu - \lambda}$$

for the mean length of a busy period.

In Section 9.3, in studying the $M/G/1$ system, we will reverse this reasoning, calculate the mean busy period directly, and then use renewal theory to determine the server idle fraction π_0.

9.2.2   The M/M/∞ System

When an unlimited number of servers are always available, then all customers in the system at any instant are simultaneously being served. With the departure rate of a single customer being μ, the departure rate of k customers is $k\mu$, and we obtain the birth and death parameters

$$\lambda_k = \lambda \quad \text{and} \quad \mu_k = k\mu \quad \text{for } k = 0, 1, \ldots.$$

The auxiliary quantities of (9.8) are

$$\theta_k = \frac{\lambda_0 \lambda_1 \cdots \lambda_{k-1}}{\mu_1 \mu_2 \cdots \mu_k} = \frac{1}{k!} \left(\frac{\lambda}{\mu} \right)^k \quad \text{for } k = 0, 1, \ldots,$$

which sum to

$$\sum_{k=0}^{\infty} \theta_k = \sum_{k=0}^{\infty} \frac{1}{k!} \left(\frac{\lambda}{\mu} \right)^k = e^{\lambda/\mu},$$

whence

$$\pi_0 = \frac{1}{\sum_{k=0}^{\infty} \theta_k} = e^{-\lambda/\mu}$$

and

$$\pi_k = \theta_k \pi_0 = \frac{(\lambda/\mu)^k e^{-\lambda/\mu}}{k!} \quad \text{for } k = 0, 1, \ldots, \tag{9.16}$$

a Poisson distribution with mean queue length

$$L = \frac{\lambda}{\mu}.$$

Since a customer in this system begins service immediately upon arrival, customer waiting time consists only of the exponentially distributed service time, and the mean waiting time is $W = 1/\mu$. Again, the basic queueing formula $L = \lambda W$ is verified.

The $M/G/\infty$ queue will be developed extensively in the next section.

9.2.3 The M/M/s System

When a fixed number s of servers are available and the assumption is made that a server is never idle if customers are waiting, then the appropriate birth and death parameters are

$$\lambda_k = \lambda \quad \text{for } k = 1, 2, \ldots,$$

$$\mu_k = \begin{cases} k\mu & \text{for } k = 0, 1, \ldots, s, \\ s\mu & \text{for } k > s. \end{cases}$$

If $X(t)$ is the number of customers in the system at time t, then the number undergoing service is $\min\{X(t), s\}$, and the number waiting for service is $\max\{X(t) - s, 0\}$. The system is depicted in Figure 9.4.

The auxiliary quantities are given by

$$\theta_k = \frac{\lambda_0 \lambda_1 \cdots \lambda_{k-1}}{\mu_1 \mu_2 \cdots \mu_k} = \begin{cases} \dfrac{1}{k!} \left(\dfrac{\lambda}{\mu} \right)^k & \text{for } k = 0, 1, \ldots, s, \\[3mm] \dfrac{1}{s!} \left(\dfrac{\lambda}{\mu} \right)^s \left(\dfrac{\lambda}{s\mu} \right)^{k-s} & \text{for } k \geq s, \end{cases}$$

and when $\lambda < s\mu$, then

$$\sum_{j=0}^{\infty} \theta_j = \sum_{j=0}^{s-1} \frac{1}{j!} \left(\frac{\lambda}{\mu} \right)^j + \sum_{j=s}^{\infty} \frac{1}{s!} \left(\frac{\lambda}{\mu} \right)^s \left(\frac{\lambda}{s\mu} \right)^{j-s}$$

$$= \sum_{j=0}^{s-1} \frac{1}{j!} \left(\frac{\lambda}{\mu} \right)^j + \frac{(\lambda/\mu)^s}{s!(1 - \lambda/s\mu)} \quad \text{for } \lambda < s\mu.$$

(9.17)

The traffic intensity in an $M/M/s$ system is $\rho = \lambda/(s\mu)$. Again, as the traffic intensity approaches one, the mean queue length becomes unbounded. When $\lambda < s\mu$, then

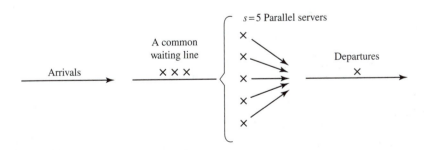

Figure 9.4 A queueing system with s servers.

from (9.10) and (9.17),

$$\pi_0 = \left\{ \sum_{j=0}^{s-1} \frac{1}{j!} \left(\frac{\lambda}{\mu} \right)^j + \frac{(\lambda/\mu)^s}{s!(1-\lambda/s\mu)} \right\}^{-1},$$

and

$$\pi_k = \begin{cases} \dfrac{1}{k!} \left(\dfrac{\lambda}{\mu} \right)^k \pi_0 & \text{for } k = 0, 1, \dots, s, \\[2ex] \dfrac{1}{s!} \left(\dfrac{\lambda}{\mu} \right)^s \left(\dfrac{\lambda}{s\mu} \right)^{k-s} \pi_0 & \text{for } k \geq s. \end{cases} \tag{9.18}$$

We evaluate L_0, the mean number of customers in the system waiting for, and not undergoing, service. Then,

$$\begin{aligned}
L_0 &= \sum_{j=s}^{\infty} (j-s)\pi_j = \sum_{k=0}^{\infty} k\pi_{s+k} \\
&= \pi_0 \sum_{k=0}^{\infty} k \frac{1}{s!} \left(\frac{\lambda}{\mu} \right)^s \left(\frac{\lambda}{s\mu} \right)^k \\
&= \frac{\pi_0}{s!} \left(\frac{\lambda}{\mu} \right)^s \sum_{k=0}^{\infty} k \left(\frac{\lambda}{s\mu} \right)^k \\
&= \frac{\pi_0}{s!} \left(\frac{\lambda}{\mu} \right)^s \frac{(\lambda/s\mu)}{(1-\lambda/s\mu)^2}.
\end{aligned} \tag{9.19}$$

Then,

$$W_0 = \frac{L_0}{\lambda},$$

$$W = W_0 + \frac{1}{\mu},$$

and

$$L = \lambda W = \lambda \left(W_0 + \frac{1}{\mu} \right) = L_0 + \frac{\lambda}{\mu}.$$

Exercises

9.2.1 Customers arrive at a tool crib according to a Poisson process of rate $\lambda = 5$ per hour. There is a single tool crib employee, and the individual service times are exponentially distributed with a mean service time of 10 min. In the long run, what is the probability that two or more workers are at the tool crib being served or waiting to be served?

9.2.2 On a single graph, plot the server utilization $1 - \pi_0 = \rho$ and the mean queue length $L = \rho/(1 - \rho)$ for the $M/M/1$ queue as a function of the traffic intensity $\rho = \lambda/\mu$ for $0 < \rho < 1$.

9.2.3 Customers arrive at a checkout station in a market according to a Poisson process of rate $\lambda = 1$ customer per minute. The checkout station can be operated with or without a bagger. The checkout times for customers are exponentially distributed, and with a bagger the mean checkout time is 30 s, while without a bagger this mean time increases to 50 s. Compare the mean queue lengths with and without a bagger.

Problems

9.2.1 Determine explicit expressions for π_0 and L for the $M/M/s$ queue when $s = 2$. Plot $1 - \pi_0$ and L as a function of the traffic intensity $\rho = \lambda/2\mu$.

9.2.2 Determine the mean waiting time W for an $M/M/2$ system when $\lambda = 2$ and $\mu = 1.2$. Compare this with the mean waiting time in an $M/M/1$ system whose arrival rate is $\lambda = 1$ and service rate is $\mu = 1.2$. Why is there a difference when the arrival rate per server is the same in both cases?

9.2.3 Determine the stationary distribution for an $M/M/2$ system as a function of the traffic intensity $\rho = \lambda/2\mu$, and verify that $L = \lambda W$.

9.2.4 The problem is to model a queueing system having finite capacity. We assume arrivals according to a Poisson process of rate λ, with independent exponentially distributed service times having mean $1/\mu$, a single server, and a finite system capacity N. By this we mean that if an arriving customer finds that there are already N customers in the system, then that customer does not enter the system and is lost.

Let $X(t)$ be the number of customers in the system at time t. Suppose that $N = 3$ (2 waiting, 1 being served).

(a) Specify the birth and death parameters for $X(t)$.

(b) In the long run, what fraction of time is the system idle?

(c) In the long run, what fraction of customers are lost?

9.2.5 Customers arrive at a service facility according to a Poisson process having rate λ. There is a single server, whose service times are exponentially distributed with parameter μ. Let $N(t)$ be the number of people in the system at time t. Then, $N(t)$ is a birth and death process with parameters $\lambda_n = \lambda$ for $n \geq 0$ and $\mu_n = \mu$ for $n \geq 1$. Assume $\lambda < \mu$. Then, $\pi_k = (1 - \lambda/\mu)(\lambda/\mu)^k, k \geq 0$, is a stationary distribution for $N(t)$; cf. equation (9.11).

Suppose the process begins according to the stationary distribution. That is, suppose $\Pr\{N(0) = k\} = \pi_k$ for $k = 0, 1, \ldots$. Let $D(t)$ be the number of people completing service up to time t. Show that $D(t)$ has a Poisson distribution with mean λt.

Hint: Let $P_{kj}(t) = \Pr\{D(t) = j | N(0) = k\}$ and $P_j(t) = \Sigma \, \pi_k P_{kj}(t) = \Pr\{D(t) = j\}$. Use a first step analysis to show that $P_{0j}(t + \Delta t) = \lambda(\Delta t)P_{1j}(t) + [1 - \lambda(\Delta t)] P_{0j}(t) + o(\Delta t)$, and for $k = 1, 2, \ldots,$

$$P_{kj}(t + \Delta t) = \mu(\Delta t)P_{k-1,j-1}(t) + \lambda(\Delta t)P_{k+1,j}(t)$$
$$+ [1 - (\lambda + \mu)(\Delta t)]P_{kj}(t) + o(t).$$

Then, use $P_j(t) = \sum_k \pi_k P_{kj}(t)$ to establish a differential equation. Use the explicit form of π_k given in the problem.

9.2.6 Customers arrive at a service facility according to a Poisson process of rate λ. There is a single server, whose service times are exponentially distributed with parameter μ. Suppose that "gridlock" occurs whenever the total number of customers in the system exceeds a capacity C. What is the smallest capacity C that will keep the probability of gridlock, under the limiting distributing of queue length, below 0.001? Express your answer in terms of the traffic intensity $\rho = \lambda/\mu$.

9.2.7 Let $X(t)$ be the number of customers in an $M/M/\infty$ queueing system at time t. Suppose that $X(0) = 0$.

(a) Derive the forward equations that are appropriate for this process by substituting the birth and death parameters into Chapter 6, equation (6.24).

(b) Show that $M(t) = E[X(t)]$ satisfies the differential equation $M'(t) = \lambda - \mu M(t)$ by multiplying the jth forward equation by j and summing.

(c) Solve for $M(t)$.

9.3 General Service Time Distributions

We continue to assume that the arrivals follow a Poisson process of rate λ. The successive customer service times $Y_1, Y_2, \ldots$, however, are now allowed to follow an arbitrary distribution $G(y) = \Pr\{Y_k \leq y\}$ having a finite mean service time $v = E[Y_k]$. The long run service rate is $\mu = 1/v$. Deterministic service times of an equal fixed duration are an important special case.

9.3.1 The M/G/1 System

If arrivals to a queue follow a Poisson process, then the successive durations X_k from the commencement of the kth busy period to the start of the next busy period form a renewal process. (A busy period is an uninterrupted duration when the queue is not empty. See Figure 9.3.) Each X_k is composed of a busy portion B_k and an idle portion I_k. Then $p_0(t)$, the probability that the system is empty at time t, converges to

$$\lim_{t \to \infty} p_0(t) = \pi_0 = \frac{E[I_1]}{E[X_1]}$$
$$= \frac{E[I_1]}{E[I_1] + E[B_1]} \tag{9.20}$$

by the renewal theorem (see "A Queueing Model" in Chapter 7, Section 7.5.3).

The idle time is the duration from the completion of a service that empties the queue to the instant of the next arrival. Because of the memoryless property that characterizes the interarrival times in a Poisson process, each idle time is exponentially distributed with mean $E[I_1] = 1/\lambda$.

The busy period is composed of the first service time Y_1, plus busy periods generated by all customers who arrive during this first service time. Let A denote this random number of new arrivals. We will evaluate the conditional mean busy period given that $A = n$ and $Y_1 = y$. First,

$$E[B_1|A = 0, Y_1 = y] = y,$$

because when no customers arrive, the busy period is composed of the first customer's service time alone. Next, consider the case in which $A = 1$, and let B' be the duration from the beginning of this customer's service to the next instant that the queue is empty. Then,

$$\begin{aligned} E[B_1|A = 1, Y_1 = y] &= y + E[B'] \\ &= y + E[B_1], \end{aligned}$$

because upon the completion of service for the initial customer, the single arrival begins a busy period B' that is statistically identical to the first, so that $E[B'] = E[B_1]$. Continuing in this manner we deduce that

$$E[B_1|A = n, Y_1 = y] = y + nE[B_1]$$

and then, using the law of total probability, that

$$\begin{aligned} E[B_1|Y_1 = y] &= \sum_{n=0}^{\infty} E[B_1|A = n, Y_1 = y] \Pr\{A = n|Y_1 = y\} \\ &= \sum_{n=0}^{\infty} \{y + nE[B_1]\} \frac{(\lambda y)^n e^{-\lambda y}}{n!} \\ &= y + \lambda y E[B_1]. \end{aligned}$$

Finally,

$$\begin{aligned} E[B_1] &= \int_0^{\infty} E[B_1|Y_1 = y] dG(y) \\ &= \int_0^{\infty} \{y + \lambda y E[B_1]\} dG(y) \qquad\qquad (9.21) \\ &= \nu\{1 + \lambda E[B_1]\}. \end{aligned}$$

Since $E[B_1]$ appears on both sides of (9.21), we may solve to obtain

$$E[B_1] = \frac{v}{1 - \lambda v}, \quad \text{provided that } \lambda v < 1. \tag{9.22}$$

To compute the long run fraction of idle time, we use (9.22) and

$$
\begin{aligned}
\pi_0 &= \frac{E[I_1]}{E[I_1] + E[B_1]} \\
&= \frac{1/\lambda}{1/\lambda + v/(1 - \lambda v)} \\
&= 1 - \lambda v \quad \text{if } \lambda v < 1.
\end{aligned}
\tag{9.23}
$$

Note that (9.23) agrees, as it must, with the corresponding expression (9.10) obtained for the $M/M/1$ queue where $v = 1/\mu$. For example, if arrivals occur at the rate of $\lambda = 2$ per hour and the mean service time is 20 min, or $v = \frac{1}{3}$ h, then in the long run, the server is idle $1 - 2\left(\frac{1}{3}\right) = \frac{1}{3}$ of the time.

The Embedded Markov Chain

The number $X(t)$ of customers in the system at time t is not a Markov process for a general $M/G/1$ system, because if one is to predict the future behavior of the system, one must know, in addition, the time expended in service for the customer currently in service. (It is the memoryless property of the exponential service time distribution that makes this additional information unnecessary in the $M/M/1$ case.)

Let X_n, however, denote the number of customers in the system immediately after the departure of the nth customer. Then, $\{X_n\}$ is a Markov chain. Indeed, we can write

$$
X_n = \begin{cases} X_{n-1} - 1 + A_n & \text{if } X_{n-1} > 0, \\ A_n & \text{if } X_{n-1} = 0, \end{cases}
\tag{9.24}
$$

$$= (X_{n-1} - 1)^+ + A_n,$$

where A_n is the number of customers that arrive during the service of the nth customer and where $x^+ = \max\{x, 0\}$. Since the arrival process is Poisson, the number of customers A_n that arrive during the service of the nth customer is independent of earlier arrivals, and the Markov property follows instantly. We calculate

$$
\begin{aligned}
\alpha_k = \Pr\{A_n = k\} &= \int_0^\infty \Pr\{A_n = k | Y_n = y\} \, dG(y) \\
&= \int_0^\infty \frac{(\lambda y)^k e^{-\lambda y}}{k!} \, dG(y),
\end{aligned}
\tag{9.25}
$$

and then, for $j = 0, 1, \ldots,$

$$P_{ij} = \Pr\{X_n = j | X_{n-1} = i\} = \Pr\{A_n = j - (i-1)^+\}$$

$$= \begin{cases} \alpha_{j-i+1} & \text{for } i \geq 1, j \geq i+1, \\ \alpha_j & \text{for } i = 0. \end{cases} \tag{9.26}$$

The Mean Queue Length in Equilibrium L

The embedded Markov chain is of special interest in the $M/G/1$ queue because in this particular instance, the stationary distribution $\{\pi_j\}$ for the Markov chain $\{X_n\}$ equals the limiting distribution for the queue length process $\{X(t)\}$. That is, $\lim_{t\to\infty} \Pr\{X(t) = j\} = \lim_{n\to\infty} \Pr\{X_n = j\}$. We will use this helpful fact to evaluate the mean queue length L.

The equivalence between the stationary distribution for the Markov chain $\{X_n\}$ and that for the non-Markov process $\{X(t)\}$ is rather subtle. It is not the consequence of a general principle and should not be assumed to hold in other circumstances without careful justification. The equivalence in the case at hand is sketched in an appendix to this section.

We will calculate the expected queue length in equilibrium $L = \lim_{t\to\infty} E[X(t)]$ by calculating the corresponding quantity in the embedded Markov chain, $L = \lim_{n\to\infty} E[X_n]$. If $X = X_\infty$ is the number of customers in the system after a customer departs and X' is the number after the next departure, then in accordance with (9.24),

$$X' = X - \delta + N, \tag{9.27}$$

where N is the number of arrivals during the service period and

$$\delta = \begin{cases} 1 & \text{if } X > 0, \\ 0 & \text{if } X = 0. \end{cases}$$

In equilibrium, X has the same distribution as does X', and in particular,

$$L = E[X] = E[X'], \tag{9.28}$$

and taking expectation in (9.27) gives

$$E[X'] = E[X] - E[\delta] + E[N],$$

and, by (9.28) and (9.23), then

$$E[N] = E[\delta] = 1 - \pi_0 = \lambda \nu. \tag{9.29}$$

Squaring (9.27) gives

$$(X')^2 = X^2 + \delta^2 + N^2 - 2\delta X + 2N(X - \delta),$$

and since $\delta^2 = \delta$ and $X\delta = X$, then

$$(X')^2 = X^2 + \delta + N^2 - 2X + 2N(X - \delta). \tag{9.30}$$

Now N, the number of customers that arrive during a service period, is independent of X, and hence of δ, so that

$$E[N(X - \delta)] = E[N]E[X - \delta], \tag{9.31}$$

and because X and X' have the same distribution, then

$$E\big[(X')^2\big] = E\big[X^2\big]. \tag{9.32}$$

Taking expectations in (9.30) we deduce that

$$E\big[(X')^2\big] = E\big[X^2\big] + E[\delta] + E\big[N^2\big] - 2E[X] + 2E[N]E[X - \delta],$$

and then substituting from (9.29) and (9.32), we obtain

$$0 = \lambda v + E\big[N^2\big] - 2L + 2\lambda v\{L - \lambda v\},$$

or

$$L = \frac{\lambda v + E\big[N^2\big] - 2(\lambda v)^2}{2(1 - \lambda v)}. \tag{9.33}$$

It remains to evaluate $E\big[N^2\big]$, where N is the number of arrivals during a service time Y. Conditioned on $Y = y$, the random variable N has a Poisson distribution with a mean (and variance) equal to λy [see equation (9.26)], whence $E\big[N^2|Y = y\big] = \lambda y + (\lambda y)^2$. Using the law of total probability, then, gives

$$\begin{aligned}
E[N^2] &= \int_0^\infty E[N^2|Y = y]dG(y) \\
&= \lambda \int_0^\infty y\,dG(y) + \lambda^2 \int_0^\infty y^2\,dG(y) \\
&= \lambda v + \lambda^2\big(\tau^2 + v^2\big),
\end{aligned} \tag{9.34}$$

where τ^2 is the variance of the service time distribution $G(y)$. Substituting (9.34) into (9.33) gives

$$\begin{aligned}
L &= \frac{2\lambda v + \lambda^2\tau^2 - (\lambda v)^2}{2(1 - \lambda v)} \\
&= \rho + \frac{\lambda^2\tau^2 + \rho^2}{2(1 - \rho)},
\end{aligned} \tag{9.35}$$

where $\rho = \lambda v$ is the traffic intensity.

Finally, $W = L/\lambda$, which simplifies to

$$W = v + \frac{\lambda(\tau^2 + v^2)}{2(1 - \rho)}. \tag{9.36}$$

The results (9.35) and (9.36) express somewhat surprising facts. They say that for a given average arrival rate λ and mean service time v, we can decrease the expected queue size L and waiting time W by decreasing the variance of service time. Clearly, the best possible case in this respect corresponds to constant service times, for which $\tau^2 = 0$.

Appendix

We sketch a proof of the equivalence between the limiting queue size distribution and the limiting distribution for the embedded Markov chain in an $M/G/1$ model. First, beginning at $t = 0$ let η_n denote those instants when the queue size $X(t)$ increases by one (an arrival), and let ξ_n denote those instants when $X(t)$ decreases by one (a departure). Let $Y_n = X(\eta_n-)$ denote the queue length immediately prior to an arrival and let $X_n = X(\xi_n+)$ denote the queue length immediately after a departure. For any queue length i and any time t, the number of visits of Y_n to i up to time t differs from the number of visits of X_n to i by at most one unit. Therefore, in the long run the average visits per unit time of Y_n to i must equal the average visits of X_n to i, which is π_i, the stationary distribution of the Markov chain $\{X_n\}$. Thus, we need only show that the limiting distribution of $\{X(t)\}$ is the same as that of $\{Y_n\}$, which is $X(t)$ just prior to an arrival. But because the arrivals are Poisson, and arrivals in disjoint time intervals are independent, it must be that $X(t)$ is independent of an arrival that occurs at time t. It follows that $\{X(t)\}$ and $\{Y_n\}$ have the same limiting distribution, and therefore $\{X(t)\}$ and the embedded Markov chain $\{X_n\}$ have the same limiting distribution.

9.3.2 The $M/G/\infty$ System

Complete results are available when each customer begins service immediately upon arrival independently of other customers in the system. Such situations may arise when modeling customer self-service systems. Let $W_1, W_2, \ldots$ be the successive arrival times of customers, and let $V_1, V_2, \ldots$ be the corresponding service times. In this notation, the kth customer is in the system at time t if and only if $W_k \leq t$ (the customer arrived prior to t) and $W_k + V_k > t$ (the service extends beyond t).

The sequence of pairs $(W_1, V_1), (W_2, V_2), \ldots$ forms a *marked Poisson process* (see Chapter 5, Section 5.6.2), and we may use the corresponding theory to quickly obtain results in this model. Figure 9.5 illustrates the marked Poisson process. Then $X(t)$, the number of customers in the system at time t, is also the number of points (W_k, V_k) for which $W_k \leq t$ and $W_k + V_k > t$. That is, it is the number of points (W_k, V_k) in the unbounded trapezoid described by

$$A_t = \{(w, v) : 0 \leq w \leq t \text{ and } v > t - w\}. \tag{9.37}$$

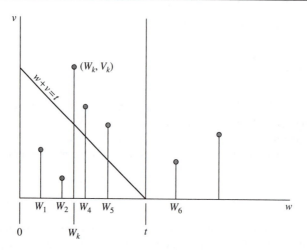

Figure 9.5 For the $M/G/\infty$ queue, the number of customers in the system at time t corresponds to the number of pairs (W_k, V_k) for which $W_k \le t$ and $W_k + V_k > t$. In the sample illustrated here, the number of customers in the system at time t is 3.

According to Chapter 5, Theorem 5.8, the number of points in A_t follows a Poisson distribution with mean

$$
\begin{aligned}
\mu(A_t) &= \iint_{A_t} \lambda(\mathrm{d}w)\,\mathrm{d}G(v) \\
&= \lambda \int_0^t \left\{ \int_{t-w}^\infty \mathrm{d}G(v) \right\} \mathrm{d}w \\
&= \lambda \int_0^t [1 - G(t - w)]\mathrm{d}w \\
&= \lambda \int_0^t [1 - G(x)]\mathrm{d}x.
\end{aligned}
\tag{9.38}
$$

In summary,

$$
\begin{aligned}
p_k(t) &= \Pr\{X(t) = k\} \\
&= \frac{\mu(A_t)^k e^{-\mu(A_t)}}{k!} \quad \text{for } k = 0, 1, \ldots,
\end{aligned}
$$

where $\mu(A_t)$ is given by (9.38). As $t \to \infty$, then

$$\lim_{t \to \infty} \mu(A_t) = \lambda \int_0^\infty [1 - G(x)]dx = \lambda\nu,$$

where ν is the mean service time. Thus, we obtain the limiting distribution

$$\pi_k = \frac{(\lambda\nu)^k e^{-\lambda\nu}}{k!} \quad \text{for } k = 0, 1, \ldots.$$

Exercises

9.3.1 Suppose that the service distribution in a single server queue is exponential with rate μ; i.e., $G(\nu) = 1 - e^{-\mu\nu}$ for $\nu \geq 0$. Substitute the mean and variance of this distribution into (9.35) and verify that the result agrees with that derived for the $M/M/1$ system in (9.12).

9.3.2 Consider a single-server queueing system having Poisson arrivals at rate λ. Suppose that the service times have the gamma density

$$g(y) = \frac{\mu^\alpha y^{\alpha-1} e^{-\mu y}}{\Gamma(\alpha)} \quad \text{for } y \geq 0,$$

where $\alpha > 0$ and $\mu > 0$ are fixed parameters. The mean service time is α/μ and the variance is α/μ^2. Determine the equilibrium mean queue length L.

9.3.3 Customers arrive at a tool crib according to a Poisson process of rate $\lambda = 5$ per hour. There is a single tool crib employee, and the individual service times are random with a mean service time of 10 min and a standard deviation of 4 min. In the long run, what is the mean number of workers at the tool crib either being served or waiting to be served?

9.3.4 Customers arrive at a checkout station in a market according to a Poisson process of rate $\lambda = 1$ customer per minute. The checkout station can be operated with or without a bagger. The checkout times for customers are random. With a bagger the mean checkout time is 30 s, while without a bagger this mean time increases to 50 s. In both cases, the standard deviation of service time is 10 s. Compare the mean queue lengths with and without a bagger.

9.3.5 Let $X(t)$ be the number of customers in an $M/G/\infty$ queueing system at time t. Suppose that $X(0) = 0$. Evaluate $M(t) = E[X(t)]$, and show that it increases monotonically to its limiting value as $t \to \infty$.

Problems

9.3.1 Let $X(t)$ be the number of customers in an $M/G/\infty$ queueing system at time t, and let $Y(t)$ be the number of customers who have entered the system and completed service by time t. Determine the joint distribution of $X(t)$ and $Y(t)$.

9.3.2 In operating a queueing system with Poisson arrivals at a rate of $\lambda = 1$ per unit time and a single server, you have a choice of server mechanisms. Method A has a mean service time of $\nu = 0.5$ and a variance in service time of $\tau^2 = 0.2$, while Method B has a mean service time of $\nu = 0.4$ and a variance of $\tau^2 = 0.9$. In terms of minimizing the waiting time of a typical customer, which method do you prefer? Would your answer change if the arrival rate were to increase significantly?

9.4 Variations and Extensions

In this section, we consider a few variations on the simple queueing models studied so far. These examples do not exhaust the possibilities but serve only to suggest the richness of the area.

Throughout we restrict ourselves to Poisson arrivals and exponentially distributed service times.

9.4.1 Systems with Balking

Suppose that a customer who arrives when there are n customers in the systems enters with probability p_n and departs with probability $q_n = 1 - p_n$. If long queues discourage customers, then p_n would be a decreasing function of n. As a special case, if there is a finite waiting room of capacity C, we might suppose that

$$
p_n = \begin{cases} 1 & \text{for } n < C, \\ 0 & \text{for } n \geq C, \end{cases}
$$

indicating that once the waiting room is filled, no more customers can enter the system.

Let $X(t)$ be the number of customers in the system at time t. If the arrival process is Poisson at rate λ and a customer who arrives when there are n customers in the system enters with probability p_n, then the appropriate birth parameters are

$$
\lambda_n = \lambda p_n \quad \text{for } n = 0, 1, \ldots.
$$

In the case of a single server, then $\mu_n = \mu$ for $n = 1, 2, \ldots$, and we may evaluate the stationary distribution π_k of queue length by the usual means.

In systems with balking, not all arriving customers enter the system, and some are lost. The *input rate* is the rate at which customers actually enter the system in the stationary state and is given by

$$
\lambda_I = \lambda \sum_{n=0}^{\infty} \pi_n p_n.
$$

The rate at which customers are lost is $\lambda \sum_{n=0}^{\infty} \pi_n q_n$, and the fraction of customers lost in the long run is

$$\text{fraction lost} = \sum_{n=0}^{\infty} \pi_n q_n.$$

Let us examine in detail the case of an $M/M/s$ system in which an arriving customer enters the system if and only if a server is free. Then,

$$\lambda_k = \begin{cases} \lambda & \text{for } k = 0, 1, \ldots, s-1, \\ 0 & \text{for } k = s, \end{cases}$$

and

$$\mu_k = k\mu \quad \text{for } k = 0, 1, \ldots, s.$$

To determine the limiting distribution, we have

$$\theta_k = \frac{1}{k!} \left(\frac{\lambda}{\mu} \right)^k \quad \text{for } k = 0, 1, \ldots, s,$$

and then

$$\pi_k = \frac{\dfrac{1}{k!} \left(\dfrac{\lambda}{\mu} \right)^k}{\displaystyle\sum_{j=0}^{s} \frac{1}{j!} \left(\frac{\lambda}{\mu} \right)^j} \quad \text{for } k = 0, 1, \ldots, s. \tag{9.39}$$

The long run fraction of customers lost is $\pi_s q_s = \pi_s$, since $q_s = 1$ in this case.

9.4.2 Variable Service Rates

In a similar vein, one can consider a system whose service rate depends on the number of customers in the system. For example, a second server might be added to a single-server system whenever the queue length exceeds a critical point ξ. If arrivals are Poisson and service rates are memoryless, then the appropriate birth and death parameters are

$$\lambda_k = \lambda \quad \text{for } k = 0, 1, \ldots, \quad \text{and} \quad \mu_k = \begin{cases} \mu & \text{for } k \leq \xi, \\ 2\mu & \text{for } k > \xi. \end{cases}$$

More generally, let us consider Poisson arrivals $\lambda_k = \lambda$ for $k = 0, 1, \ldots$, and arbitrary service rates μ_k for $k = 1, 2, \ldots$. The stationary distribution in this case is given by

$$\pi_k = \frac{\pi_0 \lambda^k}{\mu_1 \mu_2 \cdots \mu_k} \quad \text{for } k \geq 1, \tag{9.40}$$

where

$$\pi_0 = \left\{ 1 + \sum_{k=1}^{\infty} \frac{\lambda^k}{\mu_1 \mu_2 \cdots \mu_k} \right\}^{-1}. \tag{9.41}$$

9.4.3 A System with Feedback

Consider a single-server system with Poisson arrivals and exponentially distributed service times, but suppose that some customers, upon leaving the server, return to the end of the queue for additional service. In particular, suppose that a customer leaving the server departs from the system with probability q and returns to the queue for additional service with probability $p = 1 - q$. Suppose that all such decisions are statistically independent, and that a returning customer's demands for service are statistically the same as those of a customer arriving from outside the system. Let the arrival rate be λ and the service rate be μ. The queue system is depicted in Figure 9.6.

Let $X(t)$ denote the number of customers in the system at time t. Then, $X(t)$ is a birth and death process with parameters $\lambda_n = \lambda$ for $n = 0, 1, \dots$ and $\mu_n = q\mu$ for $n = 1, 2, \dots$. It is easily deduced that the stationary distribution in the case that $\lambda < q\mu$ is

$$\pi_k = \left(1 - \frac{\lambda}{q\mu} \right) \left(\frac{\lambda}{q\mu} \right)^k \quad \text{for } k = 0, 1, \dots. \tag{9.42}$$

9.4.4 A Two-Server Overflow Queue

Consider a two-server system where server i has rate μ_i for $i = 1, 2$. Arrivals to the system follow a Poisson process of rate λ. A customer arriving when the system is empty goes to the first server. A customer arriving when the first server is occupied goes to the second server. If both servers are occupied, the customer is lost. The flow is depicted in Figure 9.7.

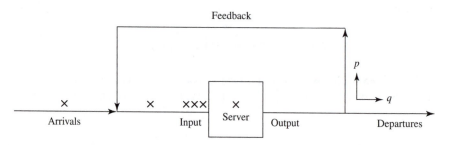

Figure 9.6 A queue with feedback.

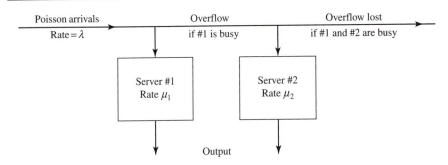

Figure 9.7 A two-server overflow model.

The system state is described by the pair $(X(t), Y(t))$, where

$$X(t) = \begin{cases} 1 & \text{if Server \#1 is busy,} \\ 0 & \text{if Server \#1 is idle.} \end{cases}$$

and

$$Y(t) = \begin{cases} 1 & \text{if Server \#2 is busy,} \\ 0 & \text{if Server \#2 is idle.} \end{cases}$$

The four states of the system are $\{(0,0), (1,0), (0,1), (1,1)\}$, and transitions among these states occur at the rate given in the following table:

From State	To State	Transition Rate	Description
$(0,0)$	$(1,0)$	λ	Arrival when system is empty
$(1,0)$	$(0,0)$	μ_1	Service completion by #1 when #2 is free
$(1,0)$	$(1,1)$	λ	Arrival when #1 is busy
$(1,1)$	$(1,0)$	μ_2	Service completion by #2 when #1 is busy
$(1,1)$	$(0,1)$	μ_1	Service completion by #1 when #2 is busy
$(0,1)$	$(1,1)$	λ	Arrival when #2 is busy and #1 is free
$(0,1)$	$(0,0)$	μ_2	Service completion by #2 when #1 is free

The process $(X(t), Y(t))$ is a finite-state, continuous-time Markov chain (see Chapter 6, Section 6.6), and the transition rates in the table furnish the infinitesimal matrix of the Markov chain:

$$\mathbf{A} = \begin{array}{c} \\ (0,0) \\ (0,1) \\ (1,0) \\ (1,1) \end{array} \begin{array}{cccc} (0,0) & (0,1) & (1,0) & (1,1) \\ \left\| \begin{array}{cccc} -\lambda & 0 & \lambda & 0 \\ \mu_2 & -(\lambda+\mu_2) & 0 & \lambda \\ \mu_1 & 0 & -(\lambda+\mu_1) & \lambda \\ 0 & \mu_1 & \mu_2 & -(\mu_1+\mu_2) \end{array} \right\| \end{array}.$$

From Chapter 6, equations (6.68) and (6.69), we find the stationary distribution $\pi = (\pi_{(0,0)}, \pi_{(0,1)}, \pi_{(1,0)} \pi_{(1,1)})$ by solving $\pi\mathbf{A} = 0$, or

$$-\lambda\pi_{(0,0)} \quad +\mu_2\pi_{(0,1)} \quad +\mu_1\pi_{(1,0)} \qquad\qquad = 0,$$
$$-(\lambda+\mu_2)\pi_{(0,1)} \qquad\qquad +\mu_1\pi_{(1,1)} = 0,$$
$$\lambda\pi_{(0,0)} \qquad -(\lambda+\mu_1)\pi_{(1,0)} \quad +\mu_2\pi_{(1,1)} = 0,$$
$$\lambda\pi_{(0,1)} \quad +\lambda\pi_{(1,0)} - (\mu_1+\mu_2)\pi_{(1,1)} = 0,$$

together with

$$\pi_{(0,0)} + \pi_{(0,1)} + \pi_{(1,0)} + \pi_{(1,1)} = 1.$$

Tedious but elementary algebra yields the solution:

$$\pi_{(0,0)} = \frac{\mu_1\mu_2(2\lambda+\mu_1+\mu_2)}{D},$$
$$\pi_{(0,1)} = \frac{\lambda^2\mu_1}{D}, \tag{9.43}$$
$$\pi_{(1,0)} = \frac{\lambda\mu_2(\lambda+\mu_1+\mu_2)}{D},$$
$$\pi_{(1,1)} = \frac{\lambda^2(\lambda+\mu_2)}{D},$$

where

$$D = \mu_1\mu_2(2\lambda+\mu_1+\mu_2) + \lambda^2\mu_1 + \lambda\mu_2(\lambda+\mu_1+\mu_2)$$
$$+ \lambda^2(\lambda+\mu_2).$$

The fraction of customers that are lost, in the long run, is the same as the fraction of time that both servers are busy, $\pi_{(1,1)} = \lambda^2(\lambda+\mu_2)/D$.

9.4.5 Preemptive Priority Queues

Consider a single-server queueing process that has two classes of customers, *priority* and *nonpriority*, forming independent Poisson arrival processes of rates α and β, respectively. The customer service times are independent and exponentially distributed with parameters γ and δ, respectively. Within classes there is a first come, first served discipline, and the service of priority customers is never interrupted. If a priority customer arrives during the service of a nonpriority customer, then the latter's service is immediately stopped in favor of the priority customer. The interrupted customer's service is resumed when there are no priority customers present.

Let us introduce some convenient notation. The system arrival rate is $\lambda = \alpha + \beta$, of which the fraction $p = \alpha/\lambda$ are priority customers and $q = \beta/\lambda$ are nonpriority customers. The system mean service time is given by the appropriately weighted means

$1/\gamma$ and $1/\delta$ of the priority and nonpriority customers, respectively, or

$$\frac{1}{\mu} = p\left(\frac{1}{\gamma}\right) + q\left(\frac{1}{\delta}\right) = \frac{1}{\lambda}\left(\frac{\alpha}{\gamma} + \frac{\beta}{\delta}\right), \tag{9.44}$$

where μ is the system service rate. Finally, we introduce the traffic intensities $\rho = \lambda/\mu$ for the system, and $\sigma = \alpha/\gamma$ and $\tau = \beta/\delta$ for the priority and nonpriority customers, respectively. From (9.44) we see that $\rho = \sigma + \tau$.

The state of the system is described by the pair $(X(t), Y(t))$, where $X(t)$ is the number of priority customers in the system and $Y(t)$ is the number of nonpriority customers. Observe that the priority customers view the system as simply an $M/M/1$ queue. Accordingly, we have the limiting distribution from (9.11) to be

$$\lim_{t\to\infty} \Pr\{X(t) = m\} = (1 - \sigma)\sigma^m \quad \text{for } m = 0, 1, \ldots \tag{9.45}$$

provided $\sigma = \alpha/\gamma < 1$.

Reference to (9.12) and (9.15) gives us, respectively, the mean queue length for priority customers

$$L_{\mathrm{p}} = \frac{\alpha}{\gamma - \alpha} = \frac{\sigma}{1 - \sigma} \tag{9.46}$$

and the mean wait for priority customers

$$W_{\mathrm{p}} = \frac{1}{\gamma - \alpha}. \tag{9.47}$$

To obtain information about the nonpriority customers is not as easy, since these arrivals are strongly affected by the priority customers. Nevertheless, $(X(t), Y(t))$ is a discrete-state, continuous-time Markov chain, and the techniques of Chapter 6, Section 6.6 enable us to describe the limiting distribution, when it exists. The transition rates of the $(X(t), Y(t))$ Markov chain are described in the following table:

From State	To State	Transition Rate	Description
(m, n)	$(m+1, n)$	α	Arrival of priority customer
(m, n)	$(m, n+1)$	β	Arrival of nonpriority customer
$(0, n)$ $n \geq 1$	$(0, n-1)$	δ	Completion of nonpriority service
(m, n) $m \geq 1$	$(m-1, n)$	γ	Completion of priority service

Let

$$\pi_{m,n} = \lim_{t\to\infty} \Pr\{X(t) = m, Y(t) = n\}$$

be the limiting distribution of the process. Reasoning analogous to that of Chapter 6, equations (6.68) and (6.69) (where the theory was derived for a finite-state Markov chain) leads to the following equations for the stationary distribution:

$$(\alpha + \beta)\pi_{0,0} = \gamma\pi_{1,0} \quad + \delta\pi_{0,1}, \tag{9.48}$$

$$(\alpha + \beta + \gamma)\pi_{m,0} = \gamma\pi_{m+1,0} \qquad \qquad + \alpha\pi_{m-1,0},$$
$$m \geq 1, \tag{9.49}$$

$$(\alpha + \beta + \delta)\pi_{0,n} = \gamma\pi_{1,n} \quad + \delta\pi_{0,n+1} + \beta\pi_{0,n-1},$$
$$n \geq 1, \tag{9.50}$$

$$(\alpha + \beta + \gamma)\pi_{m,n} = \gamma\pi_{m+1,n} \qquad + \beta\pi_{m,n-1} + \alpha\pi_{m-1,n},$$
$$m, n \geq 1. \tag{9.51}$$

The transition rates leading to equation (9.8) are shown in Figure 9.8.

In principle, these equations, augmented with the condition $\sum_m \sum_n \pi_{m,n} = 1$, may be solved for the stationary distribution, when it exists. We will content ourselves with determining the mean number L_n of nonpriority customers in the system in steady state, given by

$$L_n = \sum_{m=0}^{\infty} \sum_{n=0}^{\infty} n\pi_{m,n}. \tag{9.52}$$

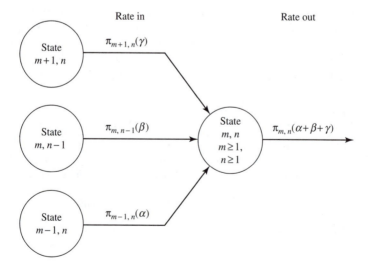

Figure 9.8 In equilibrium, the rate of flow into any state must equal the rate of flow out. Illustrated here is the state (m, n) when $m \geq 1$ and $n \geq 1$, leading to equation (9.51).

We introduce the notation

$$M_m = \sum_{n=0}^{\infty} n\pi_{m,n} = \sum_{n=1}^{\infty} n\pi_{m,n}, \tag{9.53}$$

so that

$$L_n = M_0 + M_1 + \cdots. \tag{9.54}$$

Using (9.45), let

$$p_m = \Pr\{X(t) = m\} = \sum_{n=0}^{\infty} \pi_{m,n} = (1 - \sigma)\sigma^m \tag{9.55}$$

and

$$\pi_n = \Pr\{Y(t) = n\} = \sum_{m=0}^{\infty} \pi_{m,n}. \tag{9.56}$$

We begin by summing both sides of (9.48) and (9.49) for $m = 0, 1, \ldots$ to obtain

$$(\alpha + \beta)\pi_0 + \gamma \sum_{m=1}^{\infty} \pi_{m,0} = \gamma \sum_{m=1}^{\infty} \pi_{m,0} + \delta\pi_{0,1} + \alpha\pi_0,$$

which simplifies to give

$$\beta\pi_0 = \delta\pi_{0,1}. \tag{9.57}$$

Next, we sum (9.50) and (9.51) over $m = 0, 1, \ldots$ to obtain

$$(\alpha + \beta)\pi_n + \delta\pi_{0,n} + \gamma \sum_{m=1}^{\infty} \pi_{m,n} = \gamma \sum_{m=1}^{\infty} \pi_{m,n} + \delta\pi_{0,n+1} + \beta\pi_{n-1} + \alpha\pi_n,$$

which simplifies to

$$\beta\pi_n + \delta\pi_{0,n} = \beta\pi_{n-1} + \delta\pi_{0,n+1},$$

and inductively with (9.57), we obtain

$$\beta\pi_n = \delta\pi_{0,n+1} \quad \text{for } n = 0, 1, \ldots. \tag{9.58}$$

Summing (9.58) over $n = 0, 1, \ldots$ and using $\Sigma \pi_n = 1$, we get

$$\beta = \delta \sum_{n=0}^{\infty} \pi_{0,n+1} = \delta \Pr\{X(t) = 0, Y(t) > 0\},$$

or

$$\Pr\{X(t) = 0, Y(t) > 0\} = \sum_{n=1}^{\infty} \pi_{0,n} = \frac{\beta}{\delta} = \tau. \tag{9.59}$$

Since (9.55) asserts that $\Pr\{X(t) = 0\} = 1 - (\alpha/\gamma) = 1 - \sigma$, we have

$$\pi_{0,0} = \Pr\{X(t) = 0, Y(t) = 0\} = \Pr\{X(t) = 0\} - \Pr\{X(t) = 0, Y(t) > 0\}$$

$$= 1 - \frac{\alpha}{\gamma} - \frac{\beta}{\delta} = 1 - \sigma - \tau \quad \text{when } \sigma + \tau < 1. \tag{9.60}$$

With these preliminary results in hand, we turn to determining $M_m = \sum_{n=1}^{\infty} n\pi_{m,n}$. Multiplying (9.50) by n and summing, we derive

$$(\alpha + \beta + \delta)M_0 = \gamma M_1 + \delta \sum_{n=1}^{\infty} n\pi_{0,n+1} + \beta \sum_{n=1}^{\infty} n\pi_{0,n-1}$$

$$= \gamma M_1 + \delta M_0 - \delta \sum_{n=0}^{\infty} \pi_{0,n+1} + \beta M_0 + \beta \sum_{n=1}^{\infty} \pi_{0,n-1}$$

$$= \gamma M_1 + \delta M_0 - \delta \left(\frac{\beta}{\delta}\right) + \beta M_0 + \beta(1 - \sigma),$$

where the last line results from (9.55) and (9.59). After simplification and rearrangement, the result is

$$M_1 = \sigma M_0 + \frac{\beta}{\gamma}\sigma. \tag{9.61}$$

We next multiply (9.51) by n and sum to obtain

$$(\alpha + \beta + \gamma)M_m = \gamma M_{m+1} + \beta \sum_{n=1}^{\infty} n\pi_{m,n-1} + \alpha M_{m-1}$$

$$= \gamma M_{m+1} + \beta M_m + \beta \sum_{n=1}^{\infty} \pi_{m,n-1} + \alpha M_{m-1}.$$

Again, referring to (9.55) and simplifying, we see that

$$(\alpha + \gamma)M_m = \gamma M_{m+1} + \alpha M_{m-1} + \beta(1 - \sigma)\sigma^m$$

$$\text{for } m = 1, 2, \ldots. \tag{9.62}$$

Equation (9.61) and (9.62) can be solved inductively to give

$$M_m = M_0\sigma^m + \frac{\beta}{\gamma}m\sigma^m \quad \text{for } m = 0, 1, \ldots,$$

which we sum to obtain

$$L_n = \sum_{m=0}^{\infty} M_m = \frac{1}{1-\sigma}\left[M_0 + \frac{\beta}{\gamma}\frac{\sigma}{(1-\sigma)}\right]. \tag{9.63}$$

This determines L_n in terms of M_0. To obtain a second relation, we multiply (9.58) by n and sum to obtain

$$\beta L_n = \delta \sum_{n=0}^{\infty} n\pi_{0,n+1} = \delta M_0 - \delta \sum_{n=0}^{\infty} \pi_{0,n+1}$$

$$= \delta M_0 - \delta\left(\frac{\beta}{\delta}\right) \quad \text{[see (9.59)]},$$

or

$$M_0 = \frac{\beta}{\delta}(L_n + 1) = \tau(L_n + 1). \tag{9.64}$$

We substitute (9.64) into (9.63) and simplify, yielding

$$L_n = \frac{1}{1-\sigma}\left[\tau(L_n+1) + \frac{\beta}{\gamma}\frac{\sigma}{1-\sigma}\right],$$

$$\left(1 - \frac{\tau}{1-\sigma}\right)L_n = \frac{1}{1-\sigma}\left[\tau + \frac{\beta}{\gamma}\frac{\sigma}{1-\sigma}\right],$$

and finally,

$$L_n = \left(\frac{\tau}{1-\sigma-\tau}\right)\left[1 + \left(\frac{\delta}{\gamma}\right)\frac{\sigma}{1-\sigma}\right]. \tag{9.65}$$

The condition that L_n be finite (and that a stationary distribution exist) is that

$$\rho = \sigma + \tau < 1.$$

That is, the system traffic intensity ρ must be less than one.

Since the arrival rate for nonpriority customers is β, we know that the mean waiting time for nonpriority customers is given by $W_n = L_n/\beta$.

Some simple numerical studies of (9.46) and (9.65) yield surprising results concerning adding priority to an existing system. Let us consider first a simple $M/M/1$ system with traffic intensity ρ whose mean queue length is given by (9.12) to be $L = \rho/(1-\rho)$. Let us propose modifying the system in such a way that a fraction $p = \frac{1}{2}$ of the customers have priority. We assume that priority is independent of service time. These assumptions lead to the values $\alpha = \beta = \frac{1}{2}\lambda$ and $\gamma = \delta = \mu$, whence

$\sigma = \tau = \rho/2$. Then, the mean queue lengths for priority and nonpriority customers are given by

$$L_p = \frac{\sigma}{1-\sigma} = \frac{\rho/2}{1-(\rho/2)} = \frac{\rho}{2-\rho}$$

and

$$L_n = \left(\frac{\rho/2}{1-\rho}\right)\left[1 + \frac{\rho/2}{1-(\rho/2)}\right] = \frac{\rho}{(2-\rho)(1-\rho)}.$$

The mean queue lengths L, L_p, and L_n were determined for several values of the traffic intensity ρ. The results are listed in the following table:

ρ	L	L_p	L_n
0.6	1.50	0.43	1.07
0.8	4.00	0.67	3.34
0.9	9.00	0.82	8.19
0.95	19.00	0.90	18.10

It is seen that the burden of increased queue length, as the traffic intensity increases, is carried almost exclusively by the nonpriority customers!

Exercises

9.4.1 Consider a two-server system in which an arriving customer enters the system if and only if a server is free. Suppose that customers arrive according to a Poisson process of rate $\lambda = 10$ customers per hour, and that service times are exponentially distributed with a mean service time of 6 min. In the long run, what is the rate of customers served per hour?

9.4.2 Customers arrive at a checkout station in a small grocery store according to a Poisson process of rate $\lambda = 1$ customer per minute. The checkout station can be operated with or without a bagger. The checkout times for customers are exponentially distributed, and with a bagger the mean checkout time is 30 s, while without a bagger this mean time increases to 50 s. Suppose the store's policy is to have the bagger help whenever there are two or more customers in the checkout line. In the long run, what fraction of time is the bagger helping the cashier?

9.4.3 Consider a two-server system in which an arriving customer enters the system if and only if a server is free. Suppose that customers arrive according to a Poisson process of rate $\lambda = 10$ customers per hour, and that service times are exponentially distributed. The servers have different experience in the job, and the newer server has a mean service time of 6 min, while the older has a mean service time of 4 min. In the long run, what is the rate of customers served per hour? Be explicit about any additional assumptions that you make.

9.4.4 Suppose that incoming calls to an office follow a Poisson process of rate $\lambda = 6$ per hour. If the line is in use at the time of an incoming call, the secretary has a HOLD button that will enable a single additional caller to wait. Suppose that the lengths of conversations are exponentially distributed with a mean length of 5 min, that incoming calls while a caller is on hold are lost, and that outgoing calls can be ignored. Apply the results of Section 9.4.1 to determine the fraction of calls that are lost.

Problems

9.4.1 Consider the two-server overflow queue of Section 9.4.4 and suppose the arrival rate is $\lambda = 10$ per hour. The two servers have rates 6 and 4 per hour. Recommend which server should be placed first. That is, choose between

$$\begin{matrix} \mu_1 = 6, \\ \mu_2 = 4 \end{matrix} \quad \text{and} \quad \begin{matrix} \mu_1 = 4, \\ \mu_2 = 6, \end{matrix}$$

and justify your answer. Be explicit about your criterion.

9.4.2 Consider the preemptive priority queue of Section 9.4.5 and suppose that the arrival rate is $\lambda = 4$ per hour. Two classes of customers can be identified, having mean service times of 12 min and 8 min, and it is proposed to give one of these classes priority over the other. Recommend which class should have priority. Be explicit about your criterion and justify your answer. Assume that the two classes appear in equal proportions and that all service times are exponentially distributed.

9.4.3 *Balking* refers to the refusal of an arriving customer to enter the queue. *Reneging* refers to the departure of a customer in the queue before obtaining service. Consider an $M/M/1$ system with reneging such that the probability that a specified single customer in line will depart prior to service in a short time interval $(t, t + \Delta t]$ is $r_n(\Delta t) + o(\Delta t)$ when n is the number of customers in the system. (Note that $r_0 = r_1 = 0$.) Assume Poisson arrivals at rate λ and exponential service times with parameter μ, and determine the stationary distribution when it exists.

9.4.4 A small grocery store has a single checkout counter with a full-time cashier. Customers arrive at the checkout according to a Poisson process of rate λ per hour. When there is only a single customer at the counter, the cashier works alone at a mean service rate of α per hour. Whenever there is more than one customer at the checkout, however, a "bagger" is added, increasing the service rate to β per hour. Assume that service times are exponentially distributed and determine the stationary distribution of the queue length.

9.4.5 A ticket office has two agents answering incoming phone calls. In addition, a third caller can be put on HOLD until one of the agents becomes available. If all three phone lines (both agent lines plus the hold line) are busy, a potential caller gets a busy signal, and is assumed lost. Suppose that the calls and attempted calls occur according to a Poisson process of rate λ, and that the length of a telephone conversation is exponentially distributed with parameter μ. Determine the stationary distribution for the process.

9.5 Open Acyclic Queueing Networks

Queueing networks, composed of groups of service stations, with the departures of some stations forming the arrivals of others, arise in computer and information processing systems, manufacturing job shops, service industries such as hospitals and airport terminals, and in many other contexts. A remarkable result often enables the steady-state behavior of these complex systems to be analyzed component by component.

9.5.1 The Basic Theorem

The result alluded to in the preceding paragraph asserts that the departures from a queue with Poisson arrivals and exponentially distributed service times in statistical equilibrium also form a Poisson process. We give the precise statement as Theorem 9.1. The proof is contained in an appendix at the end of this section. See also Problem 9.2.5.

Theorem 9.1. *Let $\{X(t), t \geq 0\}$ be a birth and death process with constant birth parameters $\lambda_n = \lambda$ for $n = 0, 1, \ldots$, and arbitrary death parameters μ_n for $n = 1, 2, \ldots$. Suppose there exists a stationary distribution $\pi_k \geq 0$ where $\sum_k \pi_k = 1$ and that $\Pr\{X(0) = k\} = \pi_k$ for $k = 0, 1, \ldots$. Let $D(t)$ denote the number of deaths in $(0, t]$. Then*

$$\Pr\{X(t) = k, D(t) = j\} = \Pr\{X(t) = k\}\Pr\{D(t) = j\}$$

$$= \pi_k \frac{(\lambda t)^j e^{-\lambda t}}{j!} \quad \text{for } k, j \geq 0.$$

Remark The stipulated conditions are satisfied, e.g., when $X(t)$ is the number of customers in an $M/M/s$ queueing system that is in steady state wherein $\Pr\{X(0) = j\} = \pi_j$, the stationary distribution of the process. In this case, a stationary distribution exists provided that $\lambda < s\mu$, where μ is the individual service rate.

To see the major importance of this theorem, suppose that $X(t)$ represents the number of customers in some queueing system at time t. The theorem asserts that the departures form a Poisson process of rate λ. Furthermore, the number $D(t)$ of departures up to time t is independent of the number $X(t)$ of customers remaining in the system at time t.

We caution the reader that the foregoing analysis applies only if the processes are in statistical equilibrium where the stationary distribution $\pi_k = \Pr\{X(t) = k\}$ applies. In contrast, under the condition that $X(0) = 0$, then neither will the departures form a Poisson process, nor will $D(t)$ be independent of $X(t)$.

9.5.2 Two Queues in Tandem

Let us use Theorem 9.1 to analyze a simple queueing network composed of two single-server queues connected in series as shown in Figure 9.9.

Let $X_k(t)$ be the number of customers in the kth queue at time t. We assume steady state. Beginning with the first server, the stationary distribution (9.11) for a single server queue applies, and

$$\Pr\{X_1(t) = n\} = \left(1 - \frac{\lambda}{\mu_1}\right)\left(\frac{\lambda}{\mu_1}\right)^n \quad \text{for } n = 0, 1, \dots.$$

Theorem 9.1 asserts that the departure process from the first server, denoted by $D_1(t)$, is a Poisson process of rate λ that is statistically independent of the first queue length $X_1(t)$. These departures form the arrivals to the second server, and therefore the second system has Poisson arrivals and is thus an $M/M/1$ queue as well. Thus, again using (9.11),

$$\Pr\{X_2(t) = m\} = \left(1 - \frac{\lambda}{\mu_2}\right)\left(\frac{\lambda}{\mu_2}\right)^m \quad \text{for } m = 0, 1, \dots.$$

Furthermore, because the departures $D_1(t)$ from the first server are independent of $X_1(t)$, it must be that $X_2(t)$ is independent of $X_1(t)$. We, thus, obtain the joint distribution

$$\Pr\{X_1(t) = n \quad \text{and} \quad X_2(t) = m\} = \Pr\{X_1(t) = n\}\Pr\{X_2(t) = m\}$$
$$= \left(1 - \frac{\lambda}{\mu_1}\right)\left(\frac{\lambda}{\mu_1}\right)^n\left(1 - \frac{\lambda}{\mu_2}\right)\left(\frac{\lambda}{\mu_2}\right)^m$$
$$\text{for } n, m = 0, 1, \dots.$$

We again caution the reader that the foregoing analysis applies only when the network is in its limiting distribution. In contrast, if both queues are empty at time $t = 0$,

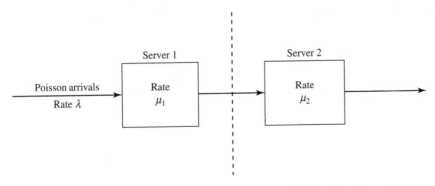

Figure 9.9 Two queues in series in which the departures from the first form the arrivals for the second.

then neither will the departures $D_1(t)$ form a Poisson process nor will $D_1(t)$ and $X_1(t)$ be independent.

9.5.3 Open Acyclic Networks

The preceding analysis of two queues in series applies to more general systems. An *open* queueing network (see Figure 9.10) has customers arriving from and departing to the outside world. (The repairman model in Chapter 6, Section 6.4 is a prototypical *closed* queueing network.) Consider an open network having K service stations, and let $X_k(t)$ be the number of customers in queue k at time t. Suppose

1. The arrivals from outside the system to distinct servers form independent Poisson processes.
2. The departures from distinct servers independently travel instantly to other servers, or leave the system, with fixed probabilities.
3. The service times for the various servers are *memoryless* in the sense that

$$\Pr\{\text{Server } \#k \text{ completes a service in } (t, t + \Delta t] | X_k(t) = n\}$$
$$= \mu_{kn}(\Delta t) + o(\Delta t) \quad \text{for } n = 1, 2, \ldots, \tag{9.66}$$

and does not otherwise depend on the past.
4. The system is in statistical equilibrium (steady state).
5. The network is *acyclic* in that a customer can visit any particular server at most once. (The case where a customer can visit a server more than once is more subtle, and is treated in the next section.)

Then,

(a) $X_1(t), X_2(t), \ldots, X_K(t)$ are independent processes, where

$$\Pr\{X_1(t) = n_1, X_2(t) = n_2, \ldots, X_K(t) = n_K\}$$
$$= \Pr\{X_1(t) = n_1\} \Pr\{X_2(t) = n_2\} \cdots \Pr\{X_K(t) = n_K\}. \tag{9.67}$$

(b) The departure process $D_k(t)$ associated with the kth server is a Poisson process, and $D_k(t)$ and $X_k(t)$ are independent.
(c) The arrivals to the kth station form a Poisson process of rate λ_k.
(d) The departure rate at the kth server equals the rate of arrivals to that server.

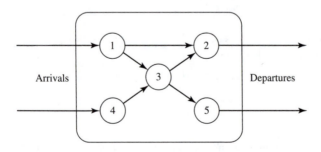

Figure 9.10 An open queueing network.

Let us add some notation so as to be able to express these results more explicitly. Let

λ_{0k} = rate of arrivals to station k from outside the system,

λ_k = rate of total arrivals to station k,

P_{kj} = probability that a customer leaving station k next visits station j.

Then, the arrivals to station k come from outside the system or from some other station j. The departure rate from j equals the arrival rate to j, the fraction P_{jk} of which go to station k, whence

$$\lambda_k = \lambda_{0k} + \sum_j \lambda_j P_{jk}. \tag{9.68}$$

Since the network is acyclic, (9.68) may be solved recursively, beginning with stations having only outside arrivals. The simple example that follows will make the procedure clear.

The arrivals to station k form a Poisson process of rate λ_k. Let

$$\psi_k(n) = \pi_{k0} \times \frac{\lambda_k^n}{\mu_{k1}\mu_{k2}\cdots\mu_{kn}} \quad \text{for } n = 1, 2, \ldots, \tag{9.69}$$

where

$$\psi_k(0) = \pi_{k0} = \left\{ 1 + \sum_{n=1}^{\infty} \left(\frac{\lambda_k^n}{\mu_{k1}\mu_{k2}\cdots\mu_{kn}} \right) \right\}^{-1}. \tag{9.70}$$

Referring to (9.40) and (9.41) we see that (9.69) and (9.70) give the stationary distribution for a queue having Poisson arrivals at rate λ_k and memoryless service times at rates μ_{kn} for $n = 1, 2, \ldots$. Accordingly, we may now express (9.67) explicitly as

$$\Pr\{X_1(t) = n_1, X_2(t) = n_2, \ldots, X_K = n_k\} \tag{9.71}$$
$$= \psi_1(n_1)\psi_2(n_2)\cdots\psi_K(n_K).$$

Example Consider the three-station network as shown in Figure 9.11.

The first step in analyzing the example is to determine the arrival rates at the various stations. In equilibrium, the arrival rate at a station must equal its departure rate, as asserted in (d). Accordingly, departures from state 1 occur at rate $\lambda_1 = 4$, and since these departures independently travel to stations 2 and 3 with respective probabilities $P_{12} = \frac{1}{3}$ and $P_{13} = \frac{2}{3}$, we determine the arrival rate $\lambda = \left(\frac{1}{3}\right)4$. At station 3 the arrivals include both those from station 1 and those from station 2. Thus, $\lambda_3 = \left(\frac{2}{3}\right)4 + \left(\frac{1}{3}\right)4 = 4$.

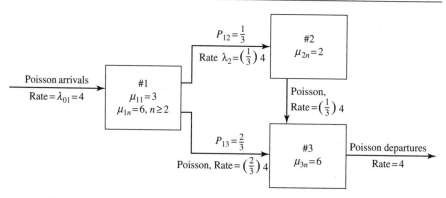

Figure 9.11 A three-station open acyclic network. Two servers, each of rate 3, at the first station give rise to the station rates $\mu_{11} = 3$ and $\mu_{1n} = 6$ for $n \geq 2$. Stations 2 and 3 each have a single server of rate 2 and 6, respectively.

Having determined the arrival rates at each station, we turn to determining the equilibrium probabilities. Station 1 is an $M/M/2$ system with $\lambda = 4$ and $\mu = 3$. From (9.18), or (9.69) and (9.70), we obtain

$$\Pr\{X_1(t) = 0\} = \pi_0 = \left\{ 1 + \left(\frac{4}{3} \right) + \frac{(4/3)^2}{2(1/3)} \right\}^{-1} = 0.2$$

and

$$\Pr\{X_1(t) = n\} = \begin{cases} \left(\dfrac{4}{3} \right)(0.2) & \text{for } n = 1, \\ (0.4) \left(\dfrac{2}{3} \right)^n & \text{for } n \geq 2. \end{cases}$$

Station 2 is an $M/M/1$ system with $\lambda = \frac{4}{3}$ and $\mu = 2$. From (9.11) we obtain

$$\Pr\{X_2(t) = n\} = \left(\frac{1}{3} \right) \left(\frac{2}{3} \right)^n \quad \text{for } n = 0, 1, \dots.$$

Similarly station 3 is an $M/M/1$ system with $\lambda = 4$ and $\mu = 6$, so that (9.11) yields

$$\Pr\{X_3(t) = n\} = \left(\frac{1}{3} \right) \left(\frac{2}{3} \right)^n \quad \text{for } n = 0, 1, \dots.$$

Finally, according to Property (a), the queue lengths $X_1(t)$, $X_2(t)$, and $X_3(t)$ are independent, so that

$$\Pr\{X_1(t) = n_1, X_2(t) = n_2, X_3(t) = n_3\}$$
$$= \Pr\{X_1(t) = n_1\} \Pr\{X_2(t) = n_2\} \Pr\{X_3(t) = n_3\}.$$

9.5.4 Appendix: Time Reversibility

Let $\{X(t), -\infty < t < +\infty\}$ be an arbitrary countable-state Markov chain having a stationary distribution $\pi_j = \Pr\{X(t) = j\}$ for all states j and all times t. Note that the time index set is the whole real line. We view the process as having begun indefinitely far in the past, so that it now is evolving in a stationary manner. Let $Y(t) = X(-t)$ be the same process, but with time reversed. The stationary process $\{X(t)\}$ is said to be *time reversible* if $\{X(t)\}$ and $\{Y(t)\}$ have the same probability laws. Clearly, $\Pr\{X(0) = j\} = \Pr\{Y(0) = j\} = \pi_j$, and it is not difficult to show that both processes are Markov. Hence, in order to show that they share the same probability laws it suffices to show that they have the same transition probabilities. Let

$$P_{ij}(t) = \Pr\{X(t) = j | X(0) = i\},$$
$$Q_{ij}(t) = \Pr\{Y(t) = j | Y(0) = i\}.$$

The process $\{X(t)\}$ is reversible if

$$P_{ij}(t) = Q_{ij}(t) \tag{9.72}$$

for all states i, j and all times t. We evaluate $Q_{ij}(t)$ as follows:

$$\begin{aligned}
Q_{ij}(t) &= \Pr\{Y(t) = j | Y(0) = i\} \\
&= \Pr\{X(-t) = j | X(0) = i\} \\
&= \Pr\{X(0) = j | X(t) = i\} \quad \text{(by stationarity)} \\
&= \frac{\Pr\{X(0) = j, X(t) = i\}}{\Pr\{X(t) = i\}} \\
&= \frac{\pi_j P_{ji}(t)}{\pi_i}.
\end{aligned}$$

In conjunction with (9.72) we see that the process $\{X(t)\}$ is reversible if

$$P_{ij}(t) = Q_{ij}(t) = \frac{\pi_j P_{ji}(t)}{\pi_i},$$

or

$$\pi_i P_{ij}(t) = \pi_j P_{ji}(t), \tag{9.73}$$

for all states i, j and all times t.

As a last step, we determine a criterion for reversibility in terms of the infinitesimal parameters

$$a_{ij} = \lim_{t \downarrow 0} \frac{1}{t} \Pr\{X(t) = j | X(0) = i\}, \quad i \neq j.$$

It is immediate that (9.73) holds when $i = j$. When $i \neq j$,

$$P_{ij}(t) = a_{ij}t + o(t), \tag{9.74}$$

which substituted into (9.73) gives

$$\pi_i[a_{ij}t + o(t)] = \pi_j[a_{ji}t + o(t)],$$

and after dividing by t and letting t vanish, we obtain the criterion

$$\pi_i a_{ij} = \pi_j a_{ji} \quad \text{for all } i \neq j. \tag{9.75}$$

When the transition probabilities are determined by the infinitesimal parameters, we deduce that the process $\{X(t)\}$ is time reversible whenever (9.75) holds.

All birth and death processes satisfying Chapter 6, (6.21) and having stationary distributions are time reversible! Because birth and death processes have

$$a_{i,i+1} = \lambda_i,$$
$$a_{i,i-1} = \mu_i,$$

and

$$a_{i,j} = 0 \quad \text{if } |i - j| > 1,$$

in verifying (9.75) it suffices to check that

$$\pi_i a_{i,i+1} = \pi_{i+1} a_{i+1,i},$$

or

$$\pi_i \lambda_i = \pi_{i+1} \mu_{i+1} \quad \text{for } i = 0, 1, \dots. \tag{9.76}$$

But [see Chapter 6, equations (6.36) and (6.73)],

$$\pi_i = \pi_0 \left(\frac{\lambda_0 \lambda_1 \cdots \lambda_{i-1}}{\mu_1 \mu_2 \cdots \mu_i} \right) \quad \text{for } i = 1, 2, \dots,$$

whence (9.76) becomes

$$\pi_0 \left(\frac{\lambda_0 \lambda_1 \cdots \lambda_{i-1}}{\mu_1 \mu_2 \cdots \mu_i} \right) \lambda_i = \pi_0 \left(\frac{\lambda_0 \lambda_1 \cdots \lambda_i}{\mu_1 \mu_2 \cdots \mu_{i+1}} \right) \mu_{i+1},$$

which is immediately seen to be true.

9.5.5 Proof of Theorem 9.1

Let us consider a birth and death process $\{X(t)\}$ having the constant birth rate $\lambda_k = \lambda$ for $k = 0, 1, \ldots$ and arbitrary death parameters $\mu_k > 0$ for $k = 1, 2, \ldots$. This process corresponds to a memoryless server queue having Poisson arrivals. A typical evolution is illustrated in Figure 9.12. The arrival process for $\{X(t)\}$ is a Poisson process of rate λ. The reversed time process $Y(t) = X(-t)$ has the same probabilistic laws as does $\{X(t)\}$, so the arrival process for $\{Y(t)\}$ also must be a Poisson process of rate λ. But the arrival process for $\{Y(t)\}$ is the departure process for $\{X(t)\}$ (see Figure 9.12). Thus, it must be that these departure instants also form a Poisson process of rate λ. In particular, if $D(t)$ counts the departures in the $X(\cdot)$ process over the duration $(0, t]$, then

$$\Pr\{D(t) = j\} = \frac{(\lambda t)^j e^{-\lambda t}}{j!} \quad \text{for } j = 0, 1, \ldots. \tag{9.77}$$

Moreover, looking at the reversed process $Y(-t) = X(t)$, the "future" arrivals for $Y(-t)$ in the Y duration $[-t, 0)$ are independent of $Y(-t) = X(t)$. (See Figure 9.12.) These future arrivals for $Y(-t)$ are the departures for $X(\cdot)$ in the interval $(0, t]$. Therefore, these departures and $X(t) = Y(-t)$ must be independent. Since $\Pr\{X(t) = k\} = \pi_k$, by the assumption of stationarity, the independence of $D(t)$ and $X(t)$ and (9.77) give

$$\Pr\{X(t) = k, D(t) = j\} = \Pr\{X(t) = k\}\Pr\{D(t) = j\}$$
$$= \frac{\pi_k e^{-\lambda t}(\lambda t)^j}{j!},$$

and the proof of Theorem 9.1 is complete.

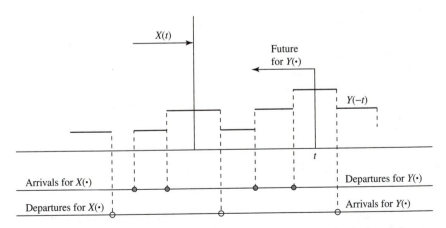

Figure 9.12 A typical evolution of a queueing process. The instants of arrivals and departures have been isolated on two time axes below the graph.

Exercises

9.5.1 Consider the three-server network pictured here:

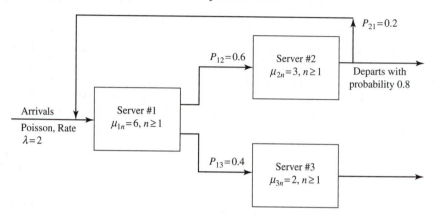

In the long run, what fraction of time is server #2 idle while, simultaneously, server #3 is busy? Assume that all service times are exponentially distributed.

9.5.2 Refer to the network of Exercise 9.5.1. Suppose that server #2 and server #3 share a common customer waiting area. If it is desired that the total number of customers being served and waiting to be served not exceed the waiting area capacity more than 5% of the time in the long run, how large should this area be?

Problem

9.5.1 Suppose three service stations are arranged in tandem so that the departures from one form the arrivals for the next. The arrivals to the first station are a Poisson process of rate $\lambda = 10$ per hour. Each station has a single server, and the three service rates are $\mu_1 = 12$ per hour, $\mu_2 = 20$ per hour, and $\mu_3 = 15$ per hour. In-process storage is being planned for station 3. What capacity C_3 must be provided if in the long run, the probability of exceeding C_3 is to be less than or equal to 1%? That is, what is the smallest number $C_3 = c$ for which $\lim_{t \to \infty} \Pr\{X_3(t) > c\} \le 0.01$?

9.6   General Open Networks

The preceding section covered certain memoryless queueing networks in which a customer could visit any particular server at most once. With this assumption, the departures from any service station formed a Poisson process that was independent of the number of customers at that station in steady state. As a consequence, the numbers

$X_1(t), X_2(t), \ldots, X_K(t)$ of customers at the K stations were independent random variables, and the product form solution expressed in (9.67) prevailed.

The situation where a customer can visit a server more than once is more subtle. On the one hand, many flows in the network are no longer Poisson. On the other hand, rather surprisingly, the product form solution of (9.67) remains valid.

Example To begin our explanation, let us first reexamine the simple feedback model of Section 9.4.3. The flow is depicted in Figure 9.13. The arrival process is Poisson, but the input to the server is not. (The distinction between the arrival and input processes is made in Figures 9.2 and 9.6.) The output process, as shown in Figure 9.13, is not Poisson, nor is it independent of the number of customers in the system. Recall that each customer in the output is fed back with probability p and departs with probability $q = 1 - p$. In view of this non-Poisson behavior, it is remarkable that the distribution of the number of customers in the system is the same as that in a Poisson $M/M/1$ system whose input rate is λ/q and whose service rate is μ, as verified in (9.42).

Example Let us verify the product form solution in a slightly more complex two-server network, depicted in Figure 9.14.

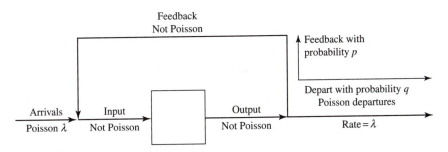

Figure 9.13 A single server with feedback.

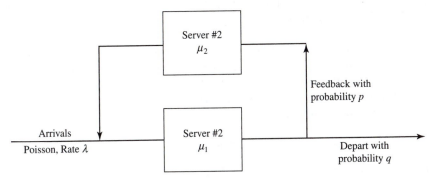

Figure 9.14 A two-server feedback system. For example, server #2 in this system might be an inspector returning a fraction p of the output for rework.

If we let $X_j(t)$ denote the number of customers at station i at time t, for $i = 1, 2$, then $\mathbf{X}(t) = [X_1(t), X_2(t)]$ is a Markov chain whose transition rates are given in the following table:

From State	To State	Transition Rate	Description
(m, n)	$(m+1, n)$	λ	Arrival of new customer
(m, n) $n \geq 1$	$(m+1, n-1)$	μ_2	Input of feedback customer
(m, n) $m \geq 1$	$(m-1, n)$	$q\mu_1$	Departure of customer
(m, n) $m \geq 1$	$(m-1, n+1)$	$p\mu_1$	Feedback to server #2

Let $\pi_{m,n} = \lim_{t \to \infty} \Pr\{X_1(t) = m, X_2(t) = n\}$ be the stationary distribution of the process. Reasoning analogous to that of (6.68) and (6.69) of Chapter 6 (where the theory was developed for finite-state Markov chains) leads to the following equations for the stationary distribution:

$$\lambda \pi_{0,0} = q\mu_1 \pi_{1,0}, \tag{9.78}$$

$$(\lambda + \mu_2)\pi_{0,n} = p\mu_1 \pi_{1,n-1} + q\mu_1 \pi_{1,n}, \quad n \geq 1, \tag{9.79}$$

$$(\lambda + \mu_1)\pi_{m,0} = \lambda \pi_{m-1,0} + q\mu_1 \pi_{m+1,0} + \mu_2 \pi_{m-1,1}, \quad m \geq 1, \tag{9.80}$$

$$(\lambda + \mu_1 + \mu_2)\pi_{m,n} = \lambda \pi_{m-1,n} + p\mu_1 \pi_{m+1,n-1} + q\mu_1 \pi_{m+1,n} \tag{9.81}$$
$$+ \mu_2 \pi_{m-1,n+1}, m, \quad n \geq 1.$$

The mass balance interpretation as explained following (6.69) in Chapter 6 may help motivate (9.78) through (9.81). For example, the left side in (9.78) measures the total rate of flow out of state $(0, 0)$ and is jointly proportional to $\pi_{0,0}$, the long run fraction of time the process is in state $(0, 0)$, and λ, the (conditional) transition rate out of $(0, 0)$. Similarly, the right side of (9.78) measures the total rate of flow into state $(0, 0)$.

Using the product form solution in the acyclic case, we will "guess" a solution and then verify that our guess indeed satisfies (9.78) through (9.81). First we need to determine the input rate, call it λ_1, to server #1. In equilibrium, the output rate must equal the input rate, and of this output, the fraction p is returned to join the new arrivals after visiting server #2. We have

Input Rate = New Arrivals + Feedback.

which translates into

$$\lambda_1 = \lambda + p\lambda_1,$$

or

$$\lambda_1 = \frac{\lambda}{1-p} = \frac{\lambda}{q}. \tag{9.82}$$

The input rate to server #2 is

$$\lambda_2 = p\lambda_1 = \frac{p\lambda}{q}. \tag{9.83}$$

The solution that we guess is to treat server #1 and server #2 as independent $M/M/1$ systems having input rates λ_1 and λ_2, respectively (even though we know from our earlier discussion that the input to server #2, while of rate λ_2, is not Poisson). That is, we attempt a solution of the form

$$\pi_{m,n} = \left(1 - \frac{\lambda_1}{\mu_1}\right)\left(\frac{\lambda_1}{\mu_1}\right)^m \left(1 - \frac{\lambda_2}{\mu_2}\right)\left(\frac{\lambda_2}{\mu_2}\right)^n$$

$$= \left(1 - \frac{\lambda}{q\mu_1}\right)\left(\frac{\lambda}{q\mu_1}\right)^m \left(1 - \frac{p\lambda}{q\mu_2}\right)\left(\frac{p\lambda}{q\mu_2}\right)^n \quad \text{for } m, n \geq 1.$$

It is immediate that

$$\sum_{m=0}^{\infty}\sum_{n=0}^{\infty}\pi_{m,n} = 1,$$

provided that $\lambda_1 = (\lambda/q) < \mu_1$ and $\lambda_2 = p\lambda/q < \mu_2$.

We turn to verifying (9.78) through (9.81). Let $\theta_{m,n} = (\lambda/q\mu_1)^m \times (p\lambda/q\mu_2)^n$. It suffices to verify that $\theta_{m,n}$ satisfies (9.78) through (9.81), since $\pi_{m,n}$ and $\theta_{m,n}$ differ only by the constant multiple $\pi_{0,0} = (1 - \lambda_1/\mu_1) \times (1 - \lambda_2/\mu_2)$. Thus, we proceed to substitute $\theta_{m,n}$ into (9.78) through (9.81) and verify that equality is obtained. We verify (9.78):

$$\lambda = q\mu_1\left(\frac{\lambda}{q\mu_1}\right) = \lambda.$$

We verify (9.79):

$$(\lambda + \mu_2)\left(\frac{p\lambda}{q\mu_2}\right)^n = p\mu_1\left(\frac{\lambda}{q\mu_1}\right)\left(\frac{p\lambda}{q\mu_2}\right)^{n-1} + q\mu_1\left(\frac{\lambda}{q\mu_1}\right)\left(\frac{p\lambda}{q\mu_2}\right)^n,$$

or after dividing by $(p\lambda/q\mu_2)^n$ and simplifying,

$$\lambda + \mu_2 = \left(\frac{p\lambda}{q}\right)\left(\frac{q\mu_2}{p\lambda}\right) + \lambda = \lambda + \mu_2.$$

We verify (9.80):

$$(\lambda + \mu_1)\left(\frac{\lambda}{q\mu_1}\right)^m = \lambda\left(\frac{\lambda}{q\mu_1}\right)^{m-1} + q\mu_1\left(\frac{\lambda}{q\mu_1}\right)^{m+1} + \mu_2\left(\frac{\lambda}{q\mu_1}\right)^{m-1}\left(\frac{p\lambda}{q\mu_2}\right),$$

which, after dividing by $(\lambda/q\mu_1)^m$, becomes

$$(\lambda + \mu_1) = \lambda\left(\frac{q\mu_1}{\lambda}\right) + q\mu_1\left(\frac{\lambda}{q\mu_1}\right) + \mu_2\left(\frac{q\mu_1}{\lambda}\right)\left(\frac{p\lambda}{q\mu_2}\right),$$

or

$$\lambda + \mu_1 = q\mu_1 + \lambda + p\mu_1 = \lambda + \mu_1.$$

The final verification, that $\theta_{m,n}$ satisfies (9.81), is left to the reader as Exercise 9.6.1.

9.6.1 The General Open Network

Consider an open queueing network having K service stations, and let $X_k(t)$ denote the number of customers at station k at time t. We assume that

1. The arrivals from outside the network to distinct servers form independent Poisson processes, where the outside arrivals to station k occur at rate λ_{0k}.
2. The departures from distinct servers independently travel instantly to other servers, or leave the system, with fixed probabilities, where the probability that a departure from station j travels to station k is P_{jk}.
3. The service times are *memoryless,* or *Markov,* in the sense that

$$\Pr\{\text{Server \#}k \text{ completes a service in } (t, t + \Delta t] | X_k(t) = n\}$$
$$= \mu_{kn}(\Delta t) + o(\Delta t) \quad \text{for } n = 1, 2, \ldots, \tag{9.84}$$

and does not otherwise depend on the past.
4. The system is in statistical equilibrium (stationary).
5. The system is completely open in that all customers in the system eventually leave.

Let λ_k be the rate of input at station k. The input at station k is composed of customers entering from outside the system, at rate λ_{0k}, plus customers traveling from (possibly) other stations. The input to station k from station j occurs at rate $\lambda_j P_{jk}$, whence, as in (9.68),

$$\lambda_k = \lambda_{0k} + \sum_{j=1}^{K} \lambda_j P_{jk} \quad \text{for } k = 1, \ldots, K. \tag{9.85}$$

Condition 5 above, that all entering customers eventually leave, ensures that (9.85) has a unique solution.

With $\lambda_1, \ldots, \lambda_K$ given by (9.86), the main result is the product form solution

$$\Pr\{X_1(t) = n_1, X_2(t) = n_2, \ldots, X_k(t) = n_K\}$$
$$= \psi_1(n_1)\psi_2(n_2) \cdots \psi_K(n_K), \tag{9.86}$$

where

$$\psi_k(n) = \frac{\pi_{k0}\lambda_k^n}{\mu_{k1}\mu_{k2}\cdots\mu_{kn}} \quad \text{for } n = 1, 2, \ldots, \tag{9.87}$$

and

$$\psi_k(0) = \pi_{k0} = \left\{ 1 + \sum_{n=1}^{\infty} \frac{\lambda_k^n}{\mu_{k1}\mu_{k2}\cdots\mu_{kn}} \right\} - 1. \tag{9.88}$$

Example The example of Figure 9.13 (see also Section 9.4.3) corresponds to $K = 1$ (a single service station) for which $P_{11} = p < 1$. The external arrivals are at rate $\lambda_{01} = \lambda$, and (9.86) becomes

$$\lambda_1 = \lambda_{01} + \lambda_1 P_{11}, \quad \text{or} \quad \lambda_1 = \lambda + \lambda_1 p,$$

which gives $\lambda_1 = \lambda/(1-p) = \lambda/q$. Since the example concerns a single server, then $\mu_{1n} = \mu$ for all n, and (9.88) becomes

$$\psi_1(n) = \pi_{10}\left(\frac{\lambda_1}{\mu}\right)^n = \pi_{10}\left(\frac{\lambda}{q\mu}\right)^n,$$

where

$$\pi_{10} = \left(1 - \frac{\lambda}{q\mu}\right),$$

in agreement with (9.42).

Example Consider next the two-server example depicted in Figure 9.14. The data given there furnish the following information:

$$\lambda_{01} = \lambda, \quad \lambda_{02} = 0,$$
$$P_{11} = 0, \quad P_{12} = p,$$
$$P_{21} = 1, \quad P_{22} = 0,$$

which substituted into (9.86) gives

$$\lambda_1 = \lambda + \lambda_2(1),$$
$$\lambda_2 = 0 + \lambda_1(p),$$

which readily yields

$$\lambda_1 = \frac{\lambda}{q} \quad \text{and} \quad \lambda_2 = \frac{p\lambda}{q},$$

in agreement with (9.82) and (9.83). It is readily seen that the product solution of (9.86) through (9.88) is identical with (9.84), which was directly verified as the solution in this example.

Exercise

9.6.1 In the case $m \geq 1, n \geq 1$, verify that $\theta_{m,n}$ as given following (9.84) satisfies the equation for the stationary distribution (9.81).

Problem

9.6.1 Consider the three-server network pictured here:

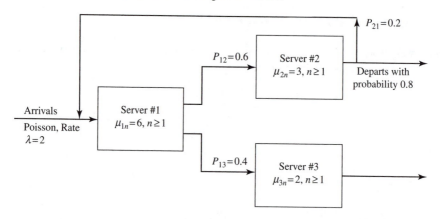

In the long run, what fraction of the time is server #2 idle while, simultaneously, server #3 is busy? Assume that the system satisfies assumptions (1) through (5) of a general open network.

10 Random Evolutions

In the previous chapters, we have examined stochastic models whose path functions are either of the jump variety or related to Brownian motion—which is continuous but has infinite velocity. The aim of this chapter is to introduce a class of continuous parameter processes, which move in a piecewise linear fashion and whose slopes jump at the times of a Poisson process. The transition probabilities satisfy a system of linear partial differential equations. In the simplest case, the components of the system satisfy the one-dimensional telegraph equation, which was studied by Mark Kac[1] and Sidney Goldstein[2] in the 1950s.

10.1 Two-State Velocity Model

We begin with the simplest case of random evolution, based on a set of two real numbers $v_0 = 1, v_1 = -1$, which are interpreted as velocities. Meanwhile, we introduce a probability space $(\Omega, \mathcal{F}, \Pr)$ on which is defined a sequence of independent random variables with the common exponential distribution

$$\Pr[e_n > t] = e^{-\lambda t}, \quad 0 < t < \infty, \quad n = 1, 2, \dots \tag{10.1}$$

and $\lambda > 0$ is a parameter, interpreted as the *rate*.

An increasing sequence of times is defined by forming the sums

$$\tau_n := e_1 + \cdots + e_n, \quad n = 1, 2, \dots \tau_0 := 0. \tag{10.2}$$

As we showed in Theorem 5.4, p. 242, τ_n has a gamma distribution with parameters (n, λ) so that

$$\Pr[\tau_n \in dt] = \frac{(\lambda t)^{n-1}}{(n-1)!} \lambda e^{-\lambda t} dt. \tag{10.3}$$

Closely associated to τ_n is the *counting process*, defined as

$$N(t) := \#\{k : \tau_k \leq t\}. \tag{10.4}$$

---

[1] Kac, M. (1974). *Rocky Mountain Journal of Mathematics*, 4, 497–520.
[2] Goldstein, S. (1951). *Quarterly Journal of Mechanics and Applied Mathematics*, 4, 129–156.

An Introduction to Stochastic Modeling

Proposition 10.1. $N(t)$ has a Poisson distribution with parameter λ.

Proof.

$$
\begin{aligned}
\Pr[N(t) = k] &= \Pr[\tau_k \le t < \tau_{k+1}] \\
&= \Pr[\tau_{k+1} > t] - \Pr[\tau_k > t] \\
&= \int_t^\infty \frac{(\lambda s)^k}{k!} \lambda e^{-\lambda s}\, ds - \int_t^\infty \frac{(\lambda s)^{k-1}}{(k-1)!} \lambda e^{-\lambda s}\, ds \\
&= -\int_t^\infty \frac{d}{ds}\left[\frac{(\lambda s)^k}{k!} e^{-\lambda s} \right] ds \\
&= \frac{(\lambda t)^k}{k!} e^{-\lambda t},
\end{aligned}
$$

which was to be proved.   ∎

The two-state velocity process is defined by

$$
V(t) = \begin{cases} 1 & \text{for } \tau_{2k} < t \le \tau_{2k+1} \\ -1 & \text{otherwise.} \end{cases}
$$

Equivalently, we can write $V(t) = (-1)^{N(t)}$. This process is a special case of the two-state continuous Markov chain. The transition probabilities $P_{ij}(t)$ of a two-state continuous Markov chain have been computed in Chapter 7. Alternatively, we can exploit the symmetry of the problem and the initial conditions $V(0) = +1$. Then, $P_{00}(t) = \Pr[N(t) \text{even}]$; the values of the process are ± 1 so that the transition matrix can be computed in terms of the oddness/evenness of the state variable. We have

$$
\begin{aligned}
P_{1,1}(t) = \Pr[V(t) = 1 | V(0) = 1] &= \Pr[N(t) \text{ even}] \\
&= \sum_{k=0}^\infty \frac{(\lambda t)^{2k}}{(2k)!} e^{-\lambda t} \\
&= e^{-\lambda t} \cosh \lambda t \\
&= \frac{1}{2}\left(1 + e^{-2\lambda t}\right).
\end{aligned}
$$

To compute $\Pr[V(t) = -1 | V(0) = 1]$, use the fact that the sum of each row of the matrix $P_{ij}(t)$ is equal to one. Thus, $P_{1,-1}(t) = \frac{1}{2}\left(1 - e^{-2\lambda t}\right)$. Symmetry considerations suggest that $P_{1,-1}(t) = P_{-1,1}(t), P_{-1,-1}(t) = P_{1,1}(t)$.

From these considerations, we conclude that the transition matrix is

$$
P(t) = \begin{pmatrix} \frac{1}{2}\left(1 + e^{-2\lambda t}\right) & \frac{1}{2}\left(1 - e^{-2\lambda t}\right) \\ \frac{1}{2}\left(1 - e^{-2\lambda t}\right) & \frac{1}{2}\left(1 + e^{-2\lambda t}\right) \end{pmatrix}.
$$

The properties of the matrices $P_{ij}(t)$ are listed as follows:

Proposition 10.2

1. $P(t+s) = P(t)P(s)$ $t, s > 0$

2. $\lim_{t \to 0} P(t) = I$, $I = \begin{pmatrix} 1 & 0 \\ 0 & 1 \end{pmatrix}$

3. $P'(t) = QP(t) = P(t)Q$ $Q := \begin{pmatrix} -\lambda & \lambda \\ \lambda & -\lambda \end{pmatrix}$

4. $\lim_{t \to \infty} P(t) = \begin{pmatrix} \dfrac{1}{2} & \dfrac{1}{2} \\ \dfrac{1}{2} & \dfrac{1}{2} \end{pmatrix}$

Exercise 10.1
Suppose that $P(t)$ is the family of transition matrices that correspond to a two-state Markov chain with the values v_1, v_2 and

$$P[V(t) = v_1 | V(0) = v_1] = 1 - tb + o(t),$$
$$P[V(t) = v_2 | V(0) = v_2] = 1 - ta + o(t), t \downarrow 0,$$

where $a > 0, b > 0$.

(a) Show that the transition matrix is given by the explicit formula

$$P(t) = \frac{1}{a+b} \begin{pmatrix} (a + be^{-\mu t}) & (b - be^{-\mu t}) \\ (a - ae^{-\mu t}) & (b + ae^{-\mu t}) \end{pmatrix},$$

where $\mu = a + b$.

(b) Show that

$$E[V(t)|V(0) = v_1] = \frac{av_1 + bv_2}{a+b} + e^{-\mu t} \frac{bv_1 - bv_2}{a+b}.$$

(c) Show that

$$E[V(t)|V(0) = v_2] = \frac{av_1 + bv_2}{a+b} + e^{-\mu t} \frac{av_2 - av_1}{a+b}.$$

(d) Show that $E[V(t)^2|V(0 = v_1] = \frac{av_1^2 + bv_2^2}{a+b} + be^{-\mu t} \frac{v_1^2 - v_2^2}{a+b}$.

(e) Show that $E[V(t)^2|V(0) = v_2] = \frac{av_1^2 + bv_2^2}{a+b} + ae^{-\mu t} \frac{v_2^2 - v_1^2}{a+b}$.

(f) Re-work Problem 6.3.3 of Chapter 6, p. 304.

Many of the above properties of two-state Markov chains extend to the case of several states. The *transition matrix* consists of a set of nonnegative functions $P_{ij}(t)$

so that

$$\sum_{k=1}^{N} P_{ik}(t) = 1, \quad \lim_{t \to 0} P_{ij}(t) = 0 \quad (i \neq j), \quad \lim_{t \to 0} P_{ij}(t) = 1 \quad (i = j).$$

$$P_{ij}(s+t) = \sum_{k=1}^{N} P_{ik}(t) P_{kj}(s) \quad \text{or} \quad P(s+t) = P(t)P(s).$$

We use these to conclude continuity of $t \to P(t)$. $P(t+h) = P(t)P(h)$ proves that the right-hand limit $P^{+}(t)$ exists and is equal to $P(t)$. Furthermore, $P(h)$ has an inverse for small h so that we can write $P(t) = P(t-h)P(h), P(t-h) = P(t)P(h)^{-1}$, which implies that the left-hand limit exists with $P(t^{-}) = P(t)$.

We can also use these ideas to prove the differentiability of $t \to P(t)$. Writing $P(t+s) = P(t)P(s)$, we have for a $\delta > 0$

$$\int_{t}^{t+b} P(u)\, du = R P(t), \quad R := \int_{0}^{\delta} P(u)\, du.$$

If δ is sufficiently small, then the matrix R has an inverse and we can write for $0 < h < \delta$

$$P(t) = R^{-1} \int_{t}^{t+h} P(u)\, du.$$

Any function of this form has a derivative, given by

$$P'(t) = R^{-1}[P(t+h) - P(t)].$$

In particular, we can set $t = 0$ to express $P'(0) = R^{-1}[P(h) - I]$; this proves the existence of the rates $P'_{ij}(0)$ for all i,j—assuming only the continuity of $P_{ij}(t)$ at $t = 0$. The set of numbers $Q_{ij} = P'_{ij}(0)$ is the *infinitesimal matrix*.

10.1.1 Two-State Random Evolution

Beginning with $V(t)$, a Markov chain with two states, ± 1, we define the associated random evolution by

$$X(t) = x + \int_{0}^{t} V(s)\, ds, \tag{10.5}$$

where we assume unit rates: $q_1 = 1 = q_2$. A major step is to determine a set of partial differential equations, corresponding to the backward equations studied in Chapter 6, pp. 295–296. In order to determine these equations, we let $\mathbf{f} = (f_1, f_{-1})$ be a pair of bounded and differentiable functions on $\mathbf{R}$ and define

$$\mathbf{u}(x,t) = E\left(\mathbf{f}\left(x + \int_0^t V(s)ds\right)\right).$$

Fix t and consider the integrand separately on the sets $\tau_1 \leq t$ and $\tau_1 > t$. Thus,

$$\mathbf{u}(x,t) = E\left(\mathbf{f}(x + \int_0^t V(s)ds)I_{\tau_1 \leq t}\right) + E\left(\mathbf{f}(x + \int_0^t V(s)ds)I_{\tau_1 > t}\right)$$

$$= \int_0^t \mathbf{u}(x+s)\lambda e^{-\lambda s}ds + e^{-\lambda t}\mathbf{f}(x+t).$$

The event $N(t) = 0$ has probability $e^{-\lambda t} = 1 - \lambda t + o(t), t \to 0$. On this set, we have $V(t) = 1, X(t) = x + t$.

The event $N(t) = 1$ has probability $\lambda t e^{-\lambda t} = \lambda t + o(t), t \to 0$. On this event, we have $V(t) = -1, X(t) = x - t$.

The event $N(t) \geq 1$ has probability $= o(t), t \to 0$ and can be ignored. Combining these yields the result

$$T_t f(x, v) = (1 - \lambda t)f_i(x + t) + \lambda t f_{-i}(x + O(t)) + o(t).$$

When we subtract $f_i(x)$ from both sides and take the limit when $t \to 0$, we obtain

$$\lim_{t \downarrow 0} t^{-1}[T_t f(x, i) - f(x, i)] = v_i f' - v_i + \lambda[f_{-i} - f_i]. \tag{10.6}$$

This information can be efficiently summarized in terms of a matrix-valued partial differential equation. We are implicitly using the Markov property $T_{t+s}f(x) = T_t(T_s f(x))$ to extend (10.6) from $s = 0$ to all $s > 0$.

Corollary 10.1. Let $\mathbf{f} = (f_1, f_{-1})$ be bounded and differentiable. Then, we have the system of backward equations

$$\frac{\partial}{\partial t}\begin{pmatrix} u_1 \\ u_{-1} \end{pmatrix} = \begin{pmatrix} u_1' + \lambda(u_{-1} - u_1) \\ -u_{-1}' + \lambda(u_1 - u_{-1}) \end{pmatrix}, \tag{10.7}$$

where $' = d/dx$.

10.1.2   The Telegraph Equation

The telegraph equation with rate λ is, by definition, the following partial differential equation for a function u of two variables (t, x):

$$\frac{\partial^2 u}{\partial t^2} + 2\lambda \frac{\partial u}{\partial t} = \frac{\partial^2 u}{\partial x^2}, \tag{10.8}$$

where $\lambda > 0$ is a positive constant. The telegraph equation is related to random evolution in much the same way that the heat equation is related to Brownian motion.

The history of the telegraph equation in probability theory reveals a different source: in 1951, Goldstein demonstrated that a two-step random walk, when properly scaled, has a distributional limit that satisfies the telegraph equation. Our approach will derive the telegraph equation directly from a continuous-time model, without taking limits. To see this in detail, recall the backward equations that govern the evolution of the distributions:

$$
\begin{aligned}
\frac{\partial u_1}{\partial t} &= \frac{\partial u_1}{\partial x} + \lambda(u_{-1} - u_1) \\
\frac{\partial u_{-1}}{\partial t} &= -\frac{\partial u_{-1}}{\partial x} + \lambda(u_1 - u_{-1}).
\end{aligned}
\tag{10.9}
$$

On the one hand, the unique solution of the system (10.9) can be written as the operator $\mathbf{f} \to \mathbf{u} = E[\mathbf{f}(\cdot + \int_0^t V)]$. Choosing $\mathbf{f}$ to be the indicator of an interval, then $\mathbf{u}$ is the probability that $X(t)$ falls in that interval. This gives a justification for studying the solutions of the system (10.9).

We now proceed to demonstrate that the all solutions of (10.9) satisfy the telegraph equation.

Set $U = u_1 + u_{-1}, V = u_1 - u_{-1}$. With this notation, we have

$$\frac{\partial U}{\partial t} = \frac{\partial V}{\partial x}, \tag{10.10}$$

$$\frac{\partial V}{\partial t} = \frac{\partial U}{\partial x} - 2\lambda V. \tag{10.11}$$

We differentiate (10.10) with respect to t to obtain

$$
\begin{aligned}
\frac{\partial^2 U}{\partial t^2} &= \frac{\partial^2 V}{\partial x \partial t} \\
&= \frac{\partial}{\partial x}\left(\frac{\partial U}{\partial x} - 2\lambda V\right) \\
&= \frac{\partial^2 U}{\partial x^2} - 2\lambda \frac{\partial V}{\partial x} \\
&= \frac{\partial^2 U}{\partial x^2} - 2\lambda \frac{\partial U}{\partial t},
\end{aligned}
$$

which proves that U satisfies the telegraph equation. On the other hand, from (10.11), we have

$$\frac{\partial^2 V}{\partial t^2} = \frac{\partial}{\partial t}\left(\frac{\partial U}{\partial x} - 2\lambda V\right)$$

$$= \frac{\partial^2 U}{\partial t \partial x} - 2\lambda \frac{\partial V}{\partial t}$$

$$= \frac{\partial}{\partial x}\left(\frac{\partial V}{\partial x}\right) - 2\lambda \frac{\partial V}{\partial t}$$

$$= \frac{\partial^2 V}{\partial x^2} - 2\lambda \frac{\partial V}{\partial t},$$

which proves that V also satisfies the telegraph equation. But $u_1 = (U+V)/2, u_{-1} = (U-V)/2$ so that both components u_1, u_2 satisfy the telegraph equation $u_{tt} + 2\lambda u_t = u_{xx}$. We summarize these computations as follows:

Proposition 10.3. Let (u_1, u_{-1}) be a solution of the system (10.9). Then, each component u_j satisfies the telegraph equation (10.8) for $j = \pm 1$.

The converse is false: if we are given two solutions v_1, v_{-1} of the telegraph equation, it is not necessary that they be the components of a solution of a first-order system of the form (10.9). For example, if v_1 is a solution of the telegraph equation, then the first equation of (10.9) implies that $v_{-1} = v_1 + (v_1)_t - (v_1)_x$, proving that v_{-1} cannot be arbitrary.

10.1.3   Distribution Functions and Densities in the Two-State Model

In order to discuss the explicit form of the distribution function of the two-state process, we appeal to some elementary analysis. We have shown that the distribution of the two-state process is a solution of the telegraph equation. On the other hand, if we have a solution of the telegraph equation with the same initial conditions (Cauchy data), then we can be assured that we have found the explicit distribution function. We also present a computation of the distribution functions, which avoids the Fourier transform. For notational simplicity, we take $\lambda = 1$.

Solution Using Fourier Analysis

In order to solve the telegraph equation, we use the method of characteristic functions, otherwise known as the Fourier transform and defined by

$$\hat{P}_{i,j}(t, \mu) := \int_{-\infty}^{\infty} e^{\sqrt{-1}\mu y} P_{ij}(t, 0, dy). \tag{10.12}$$

For simplicity, we take $\lambda = 1$. Here, μ is a real parameter that measures the spatial frequencies which are present in P. The measure P has a density inside the interval of length $2t$ and two weights at the ends of this interval.

The Fourier transform of the telegraph equation is

$$\hat{P}'' + 2\hat{P}' + \mu^2 \hat{P} = 0. \tag{10.13}$$

Exercise 10.2

Define the Fourier transform of an integrable function f by $\hat{f}(y) = \int_R f(x)e^{-i\mu x}\,dx$. Show that if $f(x, t)$ is a soluton of the telegraph equation, then $\hat{f} = g$ is a solution of the ordinary differential equation $g_{tt} + 2g_t + \mu^2 g = 0$. Assume as much smoothness and decay as you need.

The general solution of (10.13) is obtained by first finding the characteristic exponents, solutions of the algebraic equation

$$r^2 + 2r + \mu^2 = 0, \quad r = -1 \pm \sqrt{1 - \mu^2}$$

in case $|\mu| < 1$ and with a corresponding formula if $|\mu| > 1$.

The general solution of (10.13) for $|\mu| < 1$ is written in terms of hyperbolic functions:

$$\hat{P}_{ij}(\mu) = A_{ij}(\mu)e^{-t}\cosh t\sqrt{1 - \mu^2} + B_{ij}(\mu)e^{-t}\sinh t\sqrt{1 - \mu^2}.$$

For $|\mu| > 1$, the hyperbolic functions can be replaced by suitable trigonometric functions. In this way, we can obtain the representation of the distribution function in terms of its Fourier transform.

Exercise 10.3

Let $g(x)$ be a solution of the differential equation $g'' + 2g' + \mu^2 g = 0$, where μ is a real parameter. Find the general solution in case (1)$|\mu| < 1$, (2)$|\mu| = 1$, and (3)$|\mu| > 1$.

In detail, we have the following results:

$$\hat{P}_{1,1}(t, \mu) = e^{-t}\left[\cosh t\sqrt{1 - \mu^2} + \frac{i\mu}{\sqrt{1 - \mu^2}}\sinh t\sqrt{1 - \mu^2}\right]$$

$$\hat{P}_{1,-1}(t, \mu) = e^{-t}\frac{\sinh t\sqrt{1 - \mu^2}}{\sqrt{1 - \mu^2}}$$

$$\hat{P}_{-1,1}(t, \mu) = e^{-t}\frac{\sinh t\sqrt{1 - \mu^2}}{\sqrt{1 - \mu^2}}$$

$$\hat{P}_{-1,-1}(t, \mu) = e^{-t}\left[\cosh t\sqrt{1 - \mu^2} - \frac{i\mu}{\sqrt{1 - \mu^2}}\sinh t\sqrt{1 - \mu^2}\right].$$

These formulas may be inverted to obtain the densities dP/dy if we apply a classical formula for the modified Bessel function[3]

$$\frac{\sinh t\sqrt{1-\mu^2}}{\sqrt{1-\mu^2}} = \frac{1}{2}\int_{-t}^{t} I_0(\sqrt{t^2-x^2})e^{i\mu x}\,dx, \tag{10.14}$$

where I_0 is the modified Bessel function. Formula (10.14) makes it clear that $\hat{P}(t,\cdot)$ is the Fourier transform of a measure of compact support. Hence, the distribution $P_{ij}(t,x,y) = 0$ for $|y-x| > t$, whereas $|y-x| \le t$ gives the values

$$\frac{dP_{1,1}}{dy} = e^{-t}\left[\frac{t+y-x}{2}\frac{I_1(\sqrt{t^2-(x-y)^2}}{\sqrt{t^2-(y-x)^2}} + \delta(y-(x+t))\right]$$

$$\frac{dP_{1,-1}}{dy} = \frac{e^{-t}}{2}I_0\left(\sqrt{t^2-(x-y)^2}\right)$$

$$\frac{dP_{-1,1}}{dy} = \frac{e^{-t}}{2}I_0\left(\sqrt{t^2-(x-y)^2}\right)$$

$$\frac{dP_{-1,-1}}{dy} = e^{-t}\left[\frac{t-(y-x)}{2}\frac{1(\sqrt{t^2-(x-y)^2}}{\sqrt{t^2-(y-x)^2}} + \delta(y-(x-t))\right].$$

Probabilistic Approach

For a more probabilistic approach, begin with the entire function

$$I_0(2\sqrt{z}) = \sum_{n=0}^{\infty}\frac{z^n}{(n!)^2}.$$

We will group the exponential random variables e_n into odd and even indices, thus

$$U_n = e_1 + e_3 + \cdots + e_{2n-1}, \quad V_n = e_2 + e_4 + \cdots + e_{2n}.$$

For each $n \ge 1$, the random variables $U_n, V_n, e_{2n+1}, e_{2n+2}$ are independent, and the densities are given by

$$\Pr[U_n \in du] = \frac{u^{n-1}}{(n-1)!}e^u\,du = \Pr[V_n \in du], \quad n \ge 1$$

The random velocity process $V(t)$ satisfies

$$V(t) = U_n - V_n + (t - U_n - V_n) = t - 2V_n$$
$$\text{on the set } U_n + V_n \le t < U_n + V_n + e_{2n+1}$$

[3] Bateman, Tables of Integral Transforms, 1, 57.

and

$$V(t) = U_{n+1} - V_n + (t - U_{n+1} - V_n) = 2U_{n+1} - t$$
$$\text{on the set } U_{n+1} + V_n \leq t < U_{n+1} + V_n + e_{2n+2}.$$

These events are disjoint and exhaust the sample space. Hence, the distribution functions can be obtained by summation of the respective probabilities. Thus,

$$\Pr[X(t) \leq y, V(t) = 1] = \sum_{n=0}^{\infty} \Pr[e_{2n+1} > t - U_n - V_n, t - 2V_n \leq y, U_n + V_n \leq t]$$

$$\Pr[X(t) \leq y, V(t) = -1] = \sum_{n=0}^{\infty} \Pr[e_{2n+2} > t - U_{n+1} - V_n, t - 2V_n \leq y, U_{n+1} + V_n \leq t].$$

The term with $n = 0$ in the first sum is $\Pr[e_1 > t] = e^{-t}$. For the terms with $n \geq 1$, note that the conditional probability of $e_{2n+1} > t - U_n - V_n$ given U_n and V_n is equal to $e^{-(t-U_n-V_n)}$. Therefore, the required probability is given by the series of double integrals

$$\sum_{n=1}^{\infty} \int \int_{0 \leq u+v<t, v \geq \frac{1}{2}(t-y)} e^{-(t-u-v)} \frac{u^{n-1}e^{-u}}{(n-1)!} \frac{v^{n-1}e^{-v}}{(n-1)!}$$

$$= e^{-t} \int \int_{0 \leq u+v<t, v \geq \frac{1}{2}(t-y)} \sum_{n=1}^{\infty} \left(\frac{u^{n-1}v^{n-1}}{(n-1)!^2} \right) du \, dv$$

$$= e^{-t} \int \int_{0 \leq u+v<t, v \geq \frac{1}{2}(t-y)} \sum_{n=1}^{\infty} \int \int_{0 \leq u+v<t, v \geq \frac{1}{2}(t-y)} I_0(2\sqrt{uv}) du \, dv.$$

The second sum is handled in a similar manner. For $n = 0$, we have

$$\Pr[e_2 > t - U_1, 2U_1 \leq t + y, U_1 \leq t] = e^{-t} \frac{t+y}{2}.$$

For $n \geq 1$, we have

$$\sum_{n=1}^{\infty} \int \int_{0 \leq u+v<t, v \geq \frac{1}{2}(t-y)} e^{-(t-u-v)} \frac{u^n e^{-u}}{n!} \frac{v^{n-1}e^{-v}}{(n-1)!} du \, dv$$

$$= e^{-t} \int \int_{0 \leq u+v<t, v \geq \frac{1}{2}(t-y)} \sum_{n=1}^{\infty} \left(\sum_{n=1}^{\infty} \sum_{n=1}^{\infty} \frac{u^n v^{n-1}}{(n-1)!^2} \right) du \, dv$$

$$= e^{-t} \int\limits_{0 \le u+v < t, v \ge \frac{1}{2}(t-y)} \int \sum_{n=1}^{\infty} \int\limits_{0}^{\frac{1}{2}(t+y)} \frac{(t-u)^n}{n!} \frac{u^n}{n!}.$$

The term $n = 0$ is also given correctly by this formula so that we can perform the sum for $n \ge 0$ in the form

$$e^{-t} \int\limits_{0}^{\frac{1}{2}(t+y)} \sum_{n=0}^{\infty} \frac{u^n(t-u)^n}{n!^2} \, du = e^{-t} \int\limits_{0}^{\frac{1}{2}(t+y)} I_0(2\sqrt{u(t-u)}) \, du.$$

The above calculations show that the distribution functions of $X(t)$ are given by

$$\Pr[X(t) \le y, V(t) = 1] = e^{-t} \left[1 + \int\limits_{0 \le u+v < t, v \ge \frac{1}{2}(t-y)} \int I_0(2\sqrt{uv}) \, du \, dv \right]$$

$$\Pr[X(t) \le y, V(t) = -1] = \int\limits_{0 \le \frac{1}{2}(t+y)} I_0(2\sqrt{u(t-u)}) \, du.$$

10.1.4 Passage Time Distributions

In Chapter 8, Section 8.2.2, p. 407, we established the distributional properties of the first hitting time of Brownian motion $B(t), t > 0$ defined as

$$\tau_y := \inf\{t > 0 : B(t) = y\}. \tag{10.15}$$

An explicit formula was derived for the density of this random variable. This formula shows, in particular, that the moments of order less than 1/2 are finite but higher moments are infinite. We will now discuss the corresponding properties for two-state random evolution.

In order to discuss the hitting time of the two-state random evolution process, we use a discrete variable $j \in \{1, -1\}$ as well as the continuous parameter $y \in \mathbf{R}$, which defines the position on the real line. We define

$$T_{y,j} = \inf\{t > 0 : (X(t), V(t)) = (y,j)\}. \tag{10.16}$$

In order to obtain some perspective, think of a two-state process where the velocities do not sum to zero; in that case, the process tends to infinity and cannot be expected to hit all points.

To generate a computational algorithm, consider the *Laplace transform*, defined as

$$\phi_i(\alpha, x, y) = E\left(e^{-\alpha T_{y,1}} | X(0) = x, V(0) = i\right), \quad i = \pm 1. \tag{10.17}$$

For $x < y$, they satisfy the following system of integral equations

$$\phi_1(\alpha, x, y) = e^{-(1+\alpha)(y-x)} + \int_0^{y-x} e^{-(1+\alpha)s}\phi_{-1}(\alpha, x+s, y)\,ds \qquad (10.18)$$

$$\phi_{-1}(\alpha, x, y) = \int_0^\infty e^{-(1+\alpha)s}\phi_1(\alpha, x-s, y)\,ds. \qquad (10.19)$$

When we differentiate (10.18) with respect to x and simplify, we obtain the following first-order system:

$$\frac{\partial \phi_1}{\partial x} + (\phi_{-1} - \phi_1) = \alpha\,\phi_1 \qquad (10.20)$$

$$-\frac{\partial \phi_{-1}}{\partial x} + (\phi_1 - \phi_{-1}) = \alpha\,\phi_{-1}. \qquad (10.21)$$

Both components ϕ_1, ϕ_{-1} are solutions of the single second-order equation $\partial^2\phi/\partial x^2 = \alpha(\alpha+2)\phi$ and are bounded functions when $x < y$. Hence, one cannot use the exponent that is positive. Substituting into (10.20), this can be solved to obtain

$$\phi_1(\alpha, x, y) = e^{-(y-x)\sqrt{\alpha(\alpha+2)}} \quad x < y \qquad (10.22)$$

$$\phi_{-1}(\alpha, x, y) = e^{-(y-x)\sqrt{\alpha(\alpha+2)}}[1 + \alpha + \sqrt{\alpha(\alpha+2)}]^{-1}. \qquad (10.23)$$

These Laplace transforms can be inverted in terms of the modified Bessel functions if we apply the formula of (Bateman, *Higher Transcendental Functions*, vol. 2, p. 200, no. 18):

$$\Pr[T_{y,1} \in ds | V(0) = 1, X(0) = x] = (y-x)e^{-s}\frac{I_1(\sqrt{s^2 - (y-x)^2})}{\sqrt{s^2 - (y-x)^2}}. \quad s < y-x \qquad (10.24)$$

For $s > y - x$, the density function is zero, whereas there is a point mass of weight e^{y-x} at the point $s = y - x$, corresponding to those polygonal paths that have suffered no changes of direction.

The above formulas can be applied to study the *recurrence* of the Kac–Goldstein process. In general, we say that a point is recurrent if and only if $P[\tau_y < \infty | X(0) = y, V(0) = i] = 1$. In this case, we have

$$P[\tau_y < \infty | X(0) = x, V(0) = i] = \lim_{\alpha \to 0} \phi(\alpha, x, y) = 1$$

for all $y < x$. Taking $y \to x$ yields the conclusion.

Proposition 10.4. The passage time distributions of the two-state random evolution are given by (10.24). Every point is recurrent.

10.2   N-State Random Evolution

We now generalize the notion of random evolution to systems based on a Markov chain in continuous-time with an arbitrary finite number of states. We first discuss the random velocity model—a Markov chain with finitely many states. The spectral properties of the Markov chain are used to prove the law of large numbers and the central limit theorem.

10.2.1   Finite Markov Chains and Random Velocity Models

A continuous-time finite-state Markov chain is associated with a one-parameter family of matrices $P(t) = P_{ij}(t), 1 \leq i,j \leq N$, which has the properties

$$P_{ij}(t) \geq 0, \quad \sum_{j=1}^{N} P_{ij}(t) = 1, 1 \leq i \leq N$$

$$P_{ij}(t+s) = \sum_{k=1}^{N} P_{ik}(t)P_{kj}(s), \quad \lim_{t \to 0} P(t) = I.$$

From the results in Chapter 6, Section 6.6, we recall that $t \to P(t)$ is continuous at every $t > 0$ and the derivative $P'(t)$ exists, especially at $t = 0$. This defines the *infinitesimal rates* of the Markov chain:

$$q_{ij} := \lim_{t \to 0} \frac{P_{ij}(t) - \delta_{ij}}{t},$$

which satisfies $\sum_{j=1}^{N} q_{ij} = 0, q_{ii} \leq 0$. The row sums of Q are zero, and the diagonal elements of Q are negative or zero, so we can define $q_i := -q_{ii}, 1 \leq i \leq N$ where $q_i \geq 0$. The case $q_i = 0$ is trivial and is excluded, so we can divide by q_j to obtain a stochastic matrix

$$p_{ij} = \frac{q_{ij}}{q_j}. \quad 1 \leq i \neq j \leq N, \quad p_{jj} = 0$$

The matrix p_{ij} is a stochastic matrix, since

$$p_{ij} \geq 0, \quad \sum_{j=1}^{N} p_{ij} = 1.$$

10.2.2   Constructive Approach of Random Velocity Models

Given a Q matrix, as above, and a finite set of real numbers

$$\Lambda = \{v_1 < v_2 < \cdots < v_N\},$$

let W be the set of all piecewise constant, right continuous functions $t \to V(t, \omega)$.
We construct a family of random process $\Pr_x, x \in \Lambda$ on W as follows:
A pair of stochastic sequences $(e_n, Z_n)_{n \geq 1} \in R \times \Lambda$ is defined by:

$$\Pr_x(e_1 > t) = e^{-q_x}, \quad \Pr_x[Z_1 = z|e_1] = q_{xz}/q_x.$$

Assuming that $e_1, Z_1, \ldots e_N, Z_N$ have been defined, the conditional distributions of e_{N+1}, Z_{N+1} are postulated as follows:

$$\Pr[e_{N+1} > t|e_1, \ldots, e_N, Z_N] = e^{-tq_{Z_N}}, \Pr[Z_{N+1} = z|e_1, \ldots, e_N, Z_N] = q_{Z_N, z}/q_{Z_n}.$$

It is directly verified that

- The random variables $Z_n, n \geq 1$ form a discrete-time Markov chain.
- If q_x is constant for $x \in \Lambda$, then the random variables $e_n, n \geq 1$ are independent and identically distributed: $\Pr_x[e_n > t] = e^{-q_x t}$.

10.2.3 Random Evolution Processes

Having constructed the random velocity model, we can define the random evolution process on $\mathbf{R} \times \Lambda$ by setting

$$Z_x(t, \omega) = \left(x + \int_0^t V(s, \omega) \, ds, V(t, \omega) \right). \tag{10.25}$$

Although the second component $(V(t))$ has the Markov property, this is not true for the first component alone. However, the joint process $Z(t)$ enjoys the Markov property, written in the form

$$\Pr[Z. \in A|Z_r, r \leq s)$$

that depends only upon Z_s.

The next theorem is the *backward equation* for the general random evolution process. It gives the time evolution of the probabilities of general sets, through the equation $P(Z(t) \in A|Z(0) = x) = E(1_A(Z(t))|Z(0) = x)$, where 1_A is the indicator function of the set A: $1_A = 1$ on A, $1_A = 0$ on A^c.

Theorem 10.1. *Let* $\mathbf{f} = (f_1, f_2, \ldots, f_N)$ *be an n-tuple of differentiable functions.* $u(t, x, v) := E[\mathbf{f}(Z(t))|Z(0) = (y, v)]$. *Then, u satisfies the system of partial differential equations*

$$\frac{\partial u}{\partial t}(t, x, v_i) = v \frac{\partial u}{\partial x}(t, x, v_i) + \sum_{j=1}^N q_{ij} u(t, x, v_j), \quad v = v_i, x \in \mathbf{R}, t > 0, i \leq i \leq N.$$

$$\tag{10.26}$$

This is proved using a corresponding system of integral equations.

Proposition 10.5. If $\mathbf{f} = (f_1, \ldots, f_N)$ is an n-tuple of differentiable functions and $v = v_i, x \in \mathbf{R}, t > 0, 1 \le i \le N$, then

$$u(t, x, v) = e^{-tq_x} f_v(x + vt) + \sum_{j \ne i} q_{ij} \int_0^t e^{-sq_i} u(t - s, x + v_i s) \, ds. \qquad (10.27)$$

This is proved in the same manner as in the case $N = 2$ studied in Section 10.1. At a fixed moment $t > 0$, either the first jump has not occurred and this event has probability q_x or the first jump occurs at some time τ, which is distributed on the interval $[0, t]$ according to the exponential distribution with density $q_x e^{-sq_x}$.

10.2.4 Existence-Uniqueness of the First-Order System (10.26)

Quite apart from the probabilistic model, it is important to know the properties of the system (10.26).

Proposition 10.6. If $\mathbf{f} = (f_1, \ldots, f_N)$ is an n-tuple of differentiable functions, the equation (10.26) has a solution $u(t, \cdot)$ that is unique within the class of bounded functions and satisfies $u(0, \cdot) = f$.

Proof of Existence. Let

$$u_0(t, x, v_i) = e^{-tq_i} f_i(x + v_i t)$$

$$\qquad\qquad\qquad\qquad (10.28)$$

$$u_{n+1}(t, x, v_i) = e^{-tq_i} f_i(x + v_i t) + \sum_{j \ne i} q_{ij} \int_0^t e^{-sq_i} u_n(t - s, x + v_i s, v_j) \quad n \ge 0.$$

Since u_0 is in $C(\Lambda)$, mathematical induction shows that $u_n(t, \cdot) \in C(\Lambda)$ for all $n \ge 0$. Now,

$$u_{n+1}(t, x, v_i) - u_n(t, x, v_i)$$

$$= \sum_{j \ne i} q_{ij} \int_0^t e^{-sq_i} \left(u_n(t - s, x + v_i s, v_j) - u_{n-1}(t - s, x + v_i s, v_j) \right) ds.$$

Let

$$\phi_n(t) = \sup_{a, v, s \le t} |u_{n+1}(s, a, v) - u_n(s, a, v)|.$$

Upon iteration, this becomes the factorial estimate

$$\phi_n(t) \le \frac{(Qt)^n}{n!} \phi_0(t),$$

which shows that the series $u_0 + \sum_0^\infty (u_{n+1} - u_n)$ converges uniformly in a to a continuous limit u_∞. Using this uniform convergence, it follows that u_∞ is a solution of (10.26). ∎

Proof of Uniqueness. Let u_1, u_2 be two solutions and set $U := u_1 - u_2$. It satisfies

$$U(t, x, v_i) = \sum_{j \neq i} q_{ij} \int_0^t e^{-q_i s} U(t - s, x + v_i s, v_j) \, ds.$$

Letting $\Phi(t) = \sup_{a,t} U(a, t)$, we find that $\Phi(t) \leq Q \int_0^t \Phi(s) \, ds$ whose only solution is $\Phi(t) = 0$. ∎

10.2.5 Single Hyperbolic Equation

The system of first-order partial differential equations

$$\frac{\partial u_i}{\partial t} = v_i \frac{\partial u_i}{\partial x} + \sum_{j=1}^N q_{ij} u_j \quad 1 \leq i \leq N \tag{10.29}$$

can be related to a single partial differential equation of the Nth order, satisfied by each of the component functions u_i, $1 \leq i \leq N$. In case $N = 2$ and $v_1 = 1 = -v_2, q_1 = q = q_2$, this PDE includes the telegraph equation with rate q, which has been shown in the previous section.

Lemma 10.1. *Let Q, V be N-dimensional real matrices. A polynomial P is defined by*

$$P(\lambda, \mu) = \det(Q + \lambda V - \mu) = \sum_{k+l \leq N} a_{kl} \lambda^k \mu^l$$

for suitable constants a_{kl}. A differential operator on functions is defined by

$$\mathcal{H} := \det(Q + V \partial_x - \partial_t) = \sum_{k+l \leq N} a_{kl} \partial_t^k \partial_x^l, \quad \partial_t := \frac{\partial}{\partial t}, \partial_x := \frac{\partial}{\partial x} \tag{10.30}$$

meaning that we compute $P(\lambda, \mu)$ and make the substitution $\lambda \to \partial_x, \mu \to \partial_t$. In full detail, $\mathcal{H} = \sum_{k+l \leq N} a_{kl} \partial_x^k \partial_t^l$.
Let $u = (u_i)$ be a solution of the first-order linear system

$$\partial_t u = V \partial_x u + Q u.$$

Then, for each i, $1 \leq i \leq N$, we have $\mathcal{H} u_i = 0$.

Proof. For any matrix A, we have the identity $\det A = (\text{adj } A) \times A$; apply this to the case $A = Q + \lambda V - \mu$. A is a linear function of (λ, μ) and the classical adjoint adj A is also a polynomial in λ, μ, of degree $N - 1$. Applying this to $A = Q + \lambda V - \mu$, we have

$$P(\lambda, \mu) = (\text{adj } A)(Q + \lambda V - \mu).$$

Making the substitutions for λ, μ, we have for each i, $1 \leq i \leq N$

$$\mathcal{H}u_i = (\text{adj } A)(Q + V\partial_x - \partial_t)\mathbf{u} = 0.$$

Hence for each i, we have $\mathcal{H}u_i = 0$, as required. ∎

Returning to equation (10.29), it follows that each component $u = u_i$ satisfies the (scalar) PDE

$$\det \begin{pmatrix} q_{11} + v_1\partial_x - \partial_t & q_{12} & \cdots & q_{1n} \\ q_{21} & q_{22} + v_2\partial_x - \partial_t & \cdots & \cdots \\ \vdots & \vdots & \ddots & \\ q_{n1} & q_{n2} & \cdots & q_{nn} + v_n\partial_x - \partial_t \end{pmatrix} u_i = 0. \quad (10.31)$$

In case $N = 2$, $q_1 = q = q_2$, $v_1 = 1 = -v_2$, this is the statement that both components satisfy the telegraph equation with rate q:

$$\det \begin{pmatrix} -1 - \partial/\partial t + \partial/\partial x & 1 \\ 1 & -1 - \partial/\partial t - \partial/\partial x \end{pmatrix} u = 0.$$

Example The most general two-state random evolution has $v_1 \neq v_2$, $q_1 \neq q_2$. In this case, the second-order PDE (10.31) is

$$\left(\frac{\partial^2}{\partial t^2} - (v_1 + v_2)\frac{\partial^2}{\partial t \partial x} + v_1 v_2 \frac{\partial^2}{\partial x^2} + (q_1 + q_2)\frac{\partial}{\partial t} - (q_1 v_2 + q_2 v_1)\frac{\partial}{\partial x} \right) u = 0. \quad (10.32)$$

In the special case $q_1 = q_2 = 1$, $v_1 = 1 = -v_2$, we obtain the telegraph equation with rate 1: $u_{tt} + 2u_t - u_{xx} = 0$.

Exercise 10.4
Show that (10.32) follows from (10.31), when we take $N = 2$.
If we write the PDE (10.31) in the form

$$\mathcal{H}u = \sum_{k+l \leq 2} a_{kl} D_t^k D_x^l u = 0, \quad (10.33)$$

then $a_{20} = 1$, $a_{11} = (v_1 + v_2)$, $a_{02} = v_1 v_2$, $a_{10} = q_1 + q_2$, $a_{01} = v_1 q_2 + v_2 q_1$, $a_0 = 0$. Then, a short calculation shows that

(i) The polynomial $\lambda^2 + a_{11}\lambda + a_{02} = 0$ has two distinct real roots $v_1 < v_2$.

(ii) $a_{10} > 0$.

(iii) $v_1 < a_{01}/a_{10} < v_2$.

Conversely, one may characterize the coefficients as follows: suppose that we are given a second-order constant coefficient linear PDE written in the form (10.33) and satisfying the conditions (i)-(ii)-(iii). Then, there exist constants $v_1 < v_2, q_1 > 0, q_2 > 0$ such that (10.32) holds.

Exercise 10.5

Prove that any PDE of the form (10.33) satisfies conditions (i)-(ii)-(iii), when one writes it in the form (10.33).

Exercise 10.6

Suppose that we are given constants $v_1 < v_2, q_1 > 0, q_2 > 0$ satisfying the conditions (i)-(ii)-(iii). Then, we can define a_{10}, a_{01} so that (10.33) holds.

Example Let $N = 3$ and (v_i, q_{ij}) be otherwise arbitrary. Then, the system (10.29) becomes the single third-order equation

$$
\begin{aligned}
L := {}& D_t^3 - (v_1 + v_2 + v_3)D_t^2 D_x + (v_1 v_2 + v_1 v_3 + v_2 v_3)D_t D_x^2 - v_1 v_2 v_3 D_x^3 \\
& -(q_{11} + q_{22} + q_{33})D_t^2 + q_{11}(v_2 + v_3) + q_{22}(v_1 + v_3) + q_{33}(v_1 + v_2))D_t D_x \\
& -(q_{11}v_2 v_3 + q_{22}v_1 v_3 + q_{33}v_1 v_2)D_x^2 \\
& +(q_{11}q_{33} - q_{13}q_{31} + q_{11}q_{22} + q_{12}q_{21} + q_{22}q_{33} - q_{23}q_{32})D_t \\
& +(v_1(q_{23}q_{32} - q_{22}q_{33}) + v_2(q_{13}q_{31} - q_{11}q_{33}) + v_3(q_{12}q_{21} - q_{11}q_{22})u = 0.
\end{aligned}
$$

Writing $\mathcal{H}$ in the form (10.33), it is easy to see that the coefficients obey the following necessary conditions:

(i') The polynomial $\lambda^3 + a_{21}\lambda^2 + a_{12}\lambda + a_{03}$ has distinct real roots.

(ii') $a_{20} > 0, v_1 + v_2 < a_{11}/a_{20} < v_2 + v_3, ; a_{02}/a_{20} \in [\min_{i \neq j} v_i v_j, \max_{i \neq j} v_i v_j]$.

(iii') There exists $\delta > 0$ such that $\delta < a_{10} < q_1 q_2 + q_1 q_3 + q_2 q_3$. In case q_i is constant q, then $\delta = (9q^2/4)$.

Exercise 10.7

Prove the necessary conditions (i'), (ii'), and (iii').

10.2.6 Spectral Properties of the Transition Matrix

In this section, we return briefly to study N-state Markov chains. The results will be used to obtain the central limit theorem and the law of large numbers for the random evolution process.

To study the asymptotic properties of the transition matrix $P(t)$, we need to obtain information about its eigenvalues. Clearly, the complex number $\gamma = 0$ is an eigenvalue with eigenvector $(1, 1, \ldots, 1)^T$. The next lemma gives further information.

Lemma 10.2. *If γ is any eigenvalue of the matrix Q, then $\mathrm{Re}\,\gamma \leq 0$.*
If γ is a purely imaginary eigenvalue of the matrix Q, then $\gamma = 0$.

Proof. If γ is any eigenvalue, then there exists a set of complex numbers (c_k), not all zero, so that

$$\sum_{k=1}^{N} q_{ik}c_k = \gamma c_i. \quad 1 \leq i \leq N$$

Moving the term $q_{ii}c_i$ to the other side and taking the modulus, we have

$$|\gamma + q_i||c_i| = \left|\sum_{k \neq i} q_{ik}c_k\right| \quad 1 \leq i\, leN$$

$$\leq \max_k \left|\sum_{k \neq i} q_{ik}c_k\right|$$

$$= \max_k |c_k|\, q_i.$$

Now, choose i to maximize $|c_i|$. This allows one to cancel a common factor, resulting in the inequality $|\gamma + q_i| \leq q_i$ which states that γ lies in a circle of radius q_i, centered at $-q_i$. In particular $\mathrm{Re}\,\gamma \leq 0$, with equality if and only if $\gamma = 0$.

In order to treat the law of large numbers and the central limit theorem, we need to develop the properties of the matrix $Q + i\xi V$, where ξ is real, V is a real diagonal matrix, and Q is the infinitesimal matrix of an irreducible continuous-time Markov chain. In particular, the matrix Q has a simple eigenvalue $\gamma = 0$, and all other eigenvalues lie in the strict left half-plane $\mathrm{Re}(\gamma) < 0$. The detailed behavior is in the next proposition. ∎

Proposition 10.7. With the above assumptions, we have the following behavior: there exist eigenvalues $\gamma_1(\xi), \ldots, \gamma_N(\xi)$ of $Q + i\xi - V$ and $\delta > 0$ so that

$$\mathrm{Re}\,\gamma_k(\xi) \leq -\delta < 0 \quad \text{if} -\infty < \xi < \infty,\, 2 \leq k \leq N, \tag{10.34}$$

$$\mathrm{Re}\,\gamma_1(\xi) \leq -\delta < 0 \quad \text{if}|\xi| > \delta, \tag{10.35}$$

$$\gamma_1(\xi) = \bar{\phi}\xi + \frac{\sigma^2\xi^2}{2} + O(\xi^3), \quad \xi \to 0 \quad \bar{\phi} := \sum_j v_j \pi_j. \tag{10.36}$$

Proof. To prove the first two statements, we recall that $P(\gamma, \lambda) = \sum_{k+l \leq N} a_{kl}\lambda^k \gamma^l$ and that $\gamma(\lambda)$ is obtained by solving $P(\gamma, \lambda) = 0$. Let $v(\lambda) := \lambda\gamma(1/\lambda)$. Then, $B(v, \lambda) := \sum a_{k+l \leq N} v^k \lambda^{n-k-l} = 0$. Setting $\lambda = 0$, it follows that $\sum a_{k+l=N} v^k = 0$. This shows that $\lim_{v \to 0} v(\lambda) = iv_j$ for some j. Since (v_j) are distinct, it follows that

$v(\lambda) = iv_j + \lambda\phi(1/\lambda)$ for some smooth function $\phi(\lambda)$. Translating back into the λ language, this is written as

$$\gamma(\lambda) = i\lambda v_j + \phi(1/\lambda),$$

where $\lambda \to \infty$. If we substitute this expansion into the original equation $\det(Q + \lambda V - \gamma) = 0$, it follows that $\phi(0) = q_{jj}$, which was to be proved.   ∎

The eigenvalue equation for $\gamma(\xi)$ is

$$(Q + i\xi V)c = \gamma c$$

$$\sum_{k=1}^{N} q_{jk}c_k + i\xi v_j c_j = \gamma c_j \quad 1 \le j \le N$$

$$\sum_{k \ne j} q_{jk}c_k = (\gamma - i\xi v_j - q_j)c_j$$

$$|\gamma - i\xi v_j + q_j|c_j \le \max_j |c_j| \sum_{k \ne j} q_{jk} = |c_j|q_j.$$

Choose j so that $|c_j| = \max_k |c_k|$, which leads to the inequality

$$|\gamma + q_j - i\xi v_j| \le q_j.$$

This is the equation of a disk centered at $(-q_j, v_j\xi)$ of radius q_j. A glance at the γ plane shows that every point in the disk satisfies the inequality $\operatorname{Re}(\gamma) \ge q_j - |v_j| > 0$. But $\operatorname{Re}\gamma(0) < 0$ for $2 \le k \le N$ from which (10.34) and (10.35) follow.

To prove (10.36), recall that the eigenvectors and eigenvalues have asymptotic expansions about a simple eigenvalue, e.g., $\gamma = 0$. Thus,

$$\mathbf{e}(\xi) = \mathbf{e}_0 + \mathbf{e}_1\xi + O(|\xi|^2) \tag{10.37}$$

$$\gamma(\xi) = \gamma_0 + \gamma_1\xi + \gamma_2\xi^2 + O(|\xi|^3). \tag{10.38}$$

The coefficients need to be chosen so that $(Q + i\xi V)\mathbf{e} = \gamma(\xi)\mathbf{e}$. This requires that

$$Q\mathbf{e}_0 = \gamma_0\mathbf{e}_0 \tag{10.39}$$

$$Q\mathbf{e}_1 + iV\mathbf{e}_0 = \gamma_1\mathbf{e}_0 + \gamma_0\mathbf{e}_1 \tag{10.40}$$

$$Q\mathbf{e}_2 + iV\mathbf{e}_1 = \gamma_2\mathbf{e}_0 + \gamma_1\mathbf{e}_1 + \gamma_0\mathbf{e}_2 \tag{10.41}$$

and so forth. Equation (10.39) is solved by taking $\gamma_0 = 0, \mathbf{e}_0 = 1$. To solve the second equation, take the inner product of each side with the stationary distribution π_j, solution of $\sum_j \pi_j q_{jk} = 0$. The final term is already zero from the choice of γ_0. We are left with two terms involving Ve_0 and γ_1, which yield the first nonzero term in the expansion (10.38) of the eigenvalue. The next term is obtained by computing the inner product of e_0 and Qe_2.

10.2.7 Recurrence Properties of Random Evolution

It is well known (see Section 4.3.3) that a random walk in one dimension is recurrent if and only if the common distribution has mean value zero. The same holds true for one-dimensional Brownian motion: the process is recurrent if and only if the drift (mean value) is zero.

When we come to random evolution a similar criterion is valid: the process is recurrent if and only if the overall process has mean zero. As in the case of irreducible Markov chains, returning once is equivalent to returning infinitely often, which we make precise below.

Let $\Lambda = \mathbf{R} \times \{1, 2, \dots N\}$ be the state space of a random evolution process $Z(t) = (X(t), Y(t))$. If $z = (a, i) \in \Lambda$, $w = (b, j) \in \Lambda$, we write the *hitting time* as the extended random variable

$$T_w = \inf\{t > 0 : Z(t) = w\}, \quad T_w = +\infty \quad \text{otherwise}$$

$$\pi(z, w) := P_z[T_w < \infty]$$

is the hitting probability of w starting at z. In these terms, recurrence means that $\pi(z, z) = 1$, for all pairs $z \in \Lambda$.

The proof hinges on a system of integral equations satisfied by the Laplace transform of the hitting time distribution, defined by

$$u_\alpha(z, w) = E_z\left[e^{-\alpha T_w}\right] \quad z, w \in \Lambda, \alpha > 0. \tag{10.42}$$

Lemma 10.3. *If* $I(z) = I(w)$ *and* $(b - a)/v_i > 0$, *then*

$$u_\alpha(z, w) = e^{-(\alpha+q_i)(b-a)/v_i} + \sum_{k \neq i} q_{ik} \int_0^{(b-a)/v_i} e^{-(\alpha+q_i)s} u_\alpha(z_s, w)ds, \tag{10.43}$$

where $z_s := (a + v_i s)$, $I(z) = i$.
If $I(z) \neq I(w)$ *or* $(b - a)/v_i \leq 0$, *then*

$$u_\alpha(z, w) = \sum_{k \neq i} q_{ik} \int_0^\infty e^{-(\alpha+q_i)s} u_\alpha(z_s, w)ds. \tag{10.44}$$

In the first case, the random process hits w (with a positive probability) before changing directions. In the second case, the process changes direction before hitting w, with probability one.

Letting $\alpha \to 0$, we have the system of integral equations for the hitting probabilities.

Lemma 10.4. *If* $I(z) = I(w)$ *and* $(b-a)v_i > 0$, *then* $\pi(z, w) = e^{-(q_i(b-a)/v_i}$ $+ \sum_{k \neq i} \int_0^{(b-a)/v_i} e^{-q_i s} \pi(z_s, w) ds$.
If $I(z) \neq I(w)$ *or* $(b-a)/v_i \leq 0$, *then* $\pi(z, w) = \sum_{k \neq i} \int_0^\infty e^{-q_i s} \pi(z_s, w) ds$.

These integral equations allow us to deduce the smoothness properties of the hitting probabilities.

Lemma 10.5. *For fixed* $w = (b, j)$, *the mapping* $z \to \pi(z, w)$ *is continuous everywhere, with the possible exception of the place* $z = w$. *The mapping is infinitely differentiable for* $z \neq w$ *provided that* $v_i \neq 0$.
This is proved by changing variables in the integrals, which represent π. *The proof is left to the reader.*

A simple example shows that the statement of the lemma cannot be improved in general. To see this, take $N = 2$, $Q = \begin{pmatrix} -1 & 1 \\ 1 & -1 \end{pmatrix}$, and $v_1 > 0 > v_2, v_1 + v_2 > 0$. Direct computation using lemma (10.4) yields the formulas

$$\pi(x, 1; 0, 2) = \frac{v_2}{v_1 + v_2}(1 - e^{\mu x})1_{(-\infty, 0)}(x)$$

$$\pi(x, 2; 0, 2) = \frac{1}{v_1 + v_2}(v_2 + v_1 e^{\mu x})1_{(-\infty, 0)}(x),$$

where $\mu = (v_1 + v_2)/v_1 v_2 > 0$. This reflects the fact that if we start at the left of zero and are moving to the right, then we almost certainly hit zero, moving to the right; if we start at the left of zero and are moving to the left, the probability of hitting zero from the left is nearly zero.

The corresponding local statements follow immediately.

Lemma 10.6. *If* (b, j) *is fixed, then the hitting probabilities are harmonic functions of* (x, i):

$$v_i \pi_i'(x) + \sum_{k=1}^N q_{ik} \pi_k(x) = 0, \quad 1 \leq i \leq N \tag{10.45}$$

10.3 Weak Law and Central Limit Theorem

The normal distribution plays a fundamental role in the theory of probability and stochastic processes. Brownian motion furnishes a family of normally distributed random variables with mean zero and variance proportional to the time t. Many other nonnormal distributions are well approximated by the normal distribution if one takes sufficiently many independent components. This leads to the central limit theorem—the subject of deep analysis on the one hand and numerical confidence levels on the other hand.

In the case of random evolutions, there is a counterpart of these limit theorems. Consider a finite-state Markov chain $V(t)$ with real numbers $v_1 < v_2 < \cdots < v_N$ and form

$$M_t := \int_0^t (V(s) - m)\, ds. \qquad (10.46)$$

With the proper choice of m, we will have $E_x[M_t] = 0$ $\text{Var}[M_t^2] \sim \sigma^2 t$. The weak law of large numbers asserts the weak convergence of M_t/t to the constant m. The central limit theorem asserts the weak convergence of the distributions of $(M_t - mt)/\sqrt{t}$ to a normal distribution whose variance will be computed.

Based on the analogy with Brownian motion, we can formulate and prove the analogs of the classical weak law of large numbers and the central limit theorem. The setup is based on a continuous-parameter finite-state Markov chain $V(t), t \geq 0$ with one ergodic class and no transient states. A real-valued function ϕ is written $\phi(v_i), 1 \leq i \leq N$. For maximum flexibility, we consider a *continuous additive functional* defined by

$$X(t) = \int_0^t \phi(V(s))\, ds,$$

where ϕ is a real-valued function. The weak law of large numbers states that

$$\lim_{T \to \infty} \frac{1}{T} \int_0^T \phi(V(s))\, ds = \sum_1^N \pi(k)\phi(v_k), \qquad (10.47)$$

where the convergence is in probability, i.e., for every $\delta > 0$

$$P\left[\omega : \left| \frac{1}{T} \int_0^T \phi(V(s)\, ds - \sum_1^N \pi(k)\phi(v_k) \right| > \delta \right] \to 0 \quad N \to \infty.$$

(10.47) is equivalent to the condition that for every bounded and continuous f

$$\lim_{T \to \infty} Ef\left(\frac{1}{T} \int_0^T \phi(s)\, ds \right) = f\left(\sum_1^N \pi(k)\phi(v_k) \right) = 0.$$

For example, suppose that $\phi(v) = 1$ for $v = v_1$ and $\phi(v) = 0$ otherwise. Then, the left side of (10.47) is the limiting fraction of time in the interval $[0, T]$ that the Markov chain spends at v_1. The right side of (10.47) is the stationary measure of the point v_1. In other words,

THE TIME AVERAGE EQUALS THE SPACE AVERAGE.

We can obtain an intuitive idea of the validity of (10.47) by computing expectations. Then,

$$E[X(t)|V(0) = v_i] = \int_0^t E[\phi(V(s))|V(0) = v_i]\,ds$$

$$= \int_0^t \sum_{j=1}^N \phi(v_j)P_{ij}(s)\,ds.$$

Recalling that $\lim_{t\to\infty} P_{ij}(t) = \pi_j$, we obtain

$$\lim_{t\to\infty} t^{-1}E[X(t)|V(0) = v_i] = \sum_{j=1}^N \phi(v_j) \lim_{t\to\infty} t^{-1} \int_0^t P_{ij}(s)\,ds$$

$$= \sum_{j=1}^N \phi(v_j)\pi_j := \bar{\phi}.$$

This calculation shows that the mean value of the time average tends to the space average, as expressed through the stationary distribution. The weak law of large numbers shows that the *mean value* can be omitted in the previous sentence.

The central limit theorem is a refinement of the weak law of large numbers. Using the same notation as above, it states that

$$P\left[\frac{X(t) - \bar{\phi}t}{\sqrt{t}} < x\right] \to \Phi\left(\frac{x}{\sigma}\right), \quad t \to \infty$$

where $\sigma > 0$ and Φ is the standard normal distribution function, defined by the integral

$$\Phi(x) = \int_{-\infty}^x \frac{e^{-y^2/2}}{\sqrt{2\pi}}\,dy, \quad -\infty < x < \infty.$$

The proof will be organized in a non-probabilistic fashion, using the Fourier transform of the various random variables. In what follows, we will use the indices (j, k) in place of (i, j), since we want to save the letter i for $\sqrt{-1}$.

$$WLLN: \ E\left[e^{i\xi X(t)/t}1_k(V(t))|V(0) = j\right] \to e^{i\xi\bar{\phi}}\pi_k, \quad \xi \in R, \quad 1 \le j, k \le N$$

$$CLT: \ E\left[e^{i\xi\frac{X(t)-\bar{\phi}t}{\sqrt{t}}}1_k(V(t))|V(0) = j\right] \to e^{-\gamma_2\xi^2}\pi_k, \quad \xi \in R, \quad 1 \le j, k \le N$$

The constant $\gamma_2 > 0$ indicates the amount of randomness in the original Markov chain.

The proof will be broken into several stages, each of which involves elementary calculations. The common hypothesis is that the matrix Q has one single ergodic class and no transient states. In particular, zero is a simple eigenvalue and all other eigenvalues of Q are strictly in the left half-plane.

Step 1: (Solution of Poisson's equation) The matrix equation $QH = HQ = -I + \Pi$ has the solution $H_{ij} = \int_0^\infty (P_{ij}(t) - \pi_j)\,dt$ where the convergent of the integral is exponentially fast, and π_j is the stable probability distribution, solution of $\pi Q = 0$.

Step 2: (Quadratic forms in the Q matrix) The matrix Q satisfies the identity

$$< Qv, v >_\pi := \sum_{i,j=1}^N \pi_i q_{ij} v_i v_j = -\frac{1}{2} \sum_{i,j=1}^N \pi_i q_{ij}(v_i - v_j)^2 \leq 0$$

with equality iff $v = c(1, 1, \ldots, 1)$ for some constant c.

Let $\hat{P}_{jk}(t, \xi)$ be the Fourier transform, defined by

$$\hat{P}_{jk}(t, \xi) = E\left[e^{i\xi X(t)}\, 1_{V(t)=k} | V(0) = j \right].$$

Step 3: The Fourier transform can be written as the matrix exponential

$$\hat{P}_{jk}(t, \xi) = E\left[e^{t(Q+i\xi\phi)} \right]_{jk}. \tag{10.48}$$

From the elementary theory of matrices, it is known that the solutions of the equation $(Q + i\xi\phi)e(\xi) = \gamma(\xi)e(\xi)$ have expansions as analytic functions of ξ. The solution $\gamma 1$ tends to zero while the other branches satisfy $\mathrm{Re}(\gamma_j) \leq -\delta < 0$ for some $\delta > 0$ and all ξ, $-\infty < \xi < \infty$.

Step 4: There exists a solution of the eigenvalue problem $(Q + i\xi\phi)e = \gamma(\xi)e(\xi)$ with the expansions about $\xi = 0$

$$\gamma(\xi) = \bar{\phi}\xi + \sigma_2\xi^2 + O(\xi^3), \quad \xi \to 0 \tag{10.49}$$

$$e(\xi) = 1 + e_1(\xi) + O(\xi^2), \quad \xi \to 0. \tag{10.50}$$

Proof of the WLLN. We have the matrix exponential representation

$$E\left(\exp(iX(t)/t)\right)_{jk} = \exp t(Q + i\xi/t) = e^{t\gamma(\xi/t)} + O(e^{-t\delta}), \quad t \to \infty.$$

When $t \to \infty$, the right-hand side converges to $e^{it\bar{\phi}\xi}$. The continuity theorem proves that the distribution functions converge to a distribution concentrated at the point $x = \bar{\phi} := \sum_j \pi_j \phi_j$. ∎

Proof of the CLT. We have the matrix exponential representation

$$E\left(\exp(i(X(t) - mt)/\sqrt{t})]_{jk}\right) = \exp t(Q + i\xi/t)_{j,k} = \sum_{l=1}^N e^{t\gamma_l(\xi/t)}.$$

Applying Step 4, we have

$$\lim_{t \to \infty} E\left[\exp\left[i\xi \frac{X(t) - t\bar{\phi}}{\sqrt{t}}\right]1_{v(t)=j}\right] = \pi_j e^{\gamma_2 \xi^2}.$$

We outline the main steps to prove Step 1. The other proofs are left to the reader. ∎

Proposition 10.8. If zero is a simple eigenvalue of Q, then $\pi_k = \lim_{t \to \infty} p_{jk}(t)$ exists, is independent of j, and satisfies $\pi Q = 0$. The convergence is exponentially fast and the integral

$$H_{ij} = \int_0^\infty (p_{ij}(t) - \pi_j)\, dt \tag{10.51}$$

is convergent. It satisfies the equations $H1 = 0$, $< Hv, 1 >= 0$, for any vector v and satisfies the Poisson equation

$$QH = HQ = -I + \Pi,$$

where Π is the projection onto the constant vectors, defined by $\Pi f = \sum_1^N f_k \pi_k$.

Proof. The exponential convergence is guaranteed by the location of the eigenvalues in the left half-plane, coupled with the Jordan canonical form. From the differential equation $P'(t) = QP(t) = P(t)Q$, we take $t \to \infty$ to obtain that the limiting matrix R satisfies $QR = RQ = 0$. In particular, the columns of R are eigenvectors with eigenvalue zero, hence each column of R is a constant: $R_{ij} = \pi_j$. Furthermore, the equation $RQ = 0$ shows that the row vector π_i is a left eigenvector, solution of $\pi Q = 0$. Finally, we come to the H matrix:

$$\sum_{k=1}^N p_{ik}(s)H_{kj} = \sum_{k=1}^N p_{ik}(s)\int_0^\infty (p_{kj}(t) - \pi_j)\, dt$$

$$= \int_0^\infty (p_{ij}(s+t) - \pi_j)\, dt$$

$$= \int_s^\infty (p_{ij}(u) - \pi_j)\, du.$$

Therefore,

$$\sum_{k=1}^n q_{ik}H_{kj} = \frac{d}{ds}\int_s^\infty (p_{ij}(u) - \pi_j)du|_{s=0} = \pi_j - \delta_{ij}.$$

∎

10.4   Isotropic Transport in Higher Dimensions

In this section, we consider models of random motion in several dimensions. The finite Markov chain models of the previous sections are not well suited for this purpose, since one needs a continuum of directions as soon as the dimension is two or greater.

10.4.1   The Rayleigh Problem of Random Flights

The classical theory of probability provides a starting point for the higher-dimensional case, by means of the *Rayleigh problem of random flights*. This was first posed by Karl Pearson in 1905 in the following terms: "A man starts from a point O and walks l yards in a straight line; he then turns through any angle whatever and walks another l yards in a second straight line. He repeats this process n times. It is required to find the probability that after n stretches he is at a distance between r and $r + dr$ from his starting point O."

To solve Pearson's problem, we assume that each step is a random variable and that the collection of steps forms a sequence of R^2-valued independent and isotropic random variables with the same distribution. This is written

$$S_n = X_1 + \cdots + X_n, \quad \Pr[x_1 \in dr \times d\theta] = f(r)dr\,d\theta, \tag{10.52}$$

where $f(r)$ is a nonnegative f on $[0, \infty)$ normalized so that $\int_0^\infty f(r)\,dr = 1/2\pi$.

The behavior of the sum is most effectively studied by means of the Fourier transform

$$
\begin{aligned}
\Phi_n(\lambda) : &= E\big[e^{i\lambda \dot{S}_n}\big] \\
&= E\big[e^{i\lambda \cdot X_1}\big]^n \\
&= \left[\int_0^\infty \int_0^{2\pi} e^{i|\lambda|r\cos\theta} f(r)dr\,d\theta\right]^n \\
&= \left[2\pi \int_0^\infty J_0(|\lambda|rf(r)\,dr\right]^n 2.
\end{aligned}
$$

Thus,

$$
\begin{aligned}
\frac{\mathrm{Prob}[S_n \in (r, r+dr)]}{dx} &= \frac{1}{(2\pi)^2} \int_{R^2} \Phi_n(\lambda)e^{-i\lambda \cdot x}d\lambda_1\,d\lambda_2 \\
&= \frac{1}{(2\pi)^2} \int_0^{2\pi}\int_0^\infty \Phi(\lambda)e^{-i\lambda r\cos\phi}\lambda\,d\lambda d\phi \\
&= \int_0^\infty J_0(r\lambda)\Phi_n(r\lambda)\lambda d\lambda.
\end{aligned}
$$

The solution of Pearson's problem is given formally by

$$\frac{P[S_n \in r, r + dr)]}{dr} = \int_0^\infty J_0(r\lambda) \Phi_n(\lambda) r \lambda d\lambda. \tag{10.53}$$

The exact computation of the integral defining $\Phi_n(\lambda)$ may be difficult, depending on the form of the density function $f(r)$. Moreover, the inversion of the resultant nth power may be formidable. Nevertheless, it is straightforward to obtain the analog of the law of large numbers and the central limit theorem for this model. We recall the asymptotic behavior of the Bessel function J_0:

$$
\begin{aligned}
J_0(x) &= \frac{1}{2\pi} \int_0^{2\pi} e^{iz\cos\theta} \\
&= \frac{1}{2\pi} \int_0^{2\pi} \left[1 + ix\cos\theta - \frac{x^2}{2}\cos^2\theta + O(|x|^3) \right] d\theta \\
&= 1 - \frac{x^2}{4} + O(|x|^3), \quad x \to 0
\end{aligned}
$$

When we replace λ by λ/n, the integral $= 1 + O(1/n^2)$, which, when taken to the nth power, tends to 1. This gives the law of large numbers in the form

$$\lim_{n\to\infty} E\left[e^{i\lambda \cdot S_n/n} \right] = 1 \quad \lambda \in R^2. \tag{10.54}$$

To obtain the appropriate form of the central limit theorem, we replace λ by $\lambda/\sqrt{n}$, which yields an integral of the form $1 - (\pi|\lambda|^2/2n) \int_0^\infty r^2 f(r) dr + O(1/n^{3/2})$. When we take this to the nth power, we obtain the central limit theorem in the form

$$\lim_{n\to\infty} E\left[e^{i\lambda S_n/\sqrt{n}} \right] = e^{-\lambda^2\sigma^2/2}, \tag{10.55}$$

where $\sigma^2 = \int_0^\infty \pi r^2 f(r) dr$.

These results are the counterparts of the asymptotic results obtained for the two-state velocity model, studied in Section 10.1, which is the one-dimensional continuous-time analog of the Rayleigh problem of random flights. The variance parameter σ^2 depends on the form of the radial density function $f(r)$.

Exercise 10.8

Suppose that we have an exponential distribution with radial density function $f(r) = (2\pi a)^{-1} e^{-r/a}$. Then, $\sigma^2 e^{-r/a} dr = a^2$.

10.4.2 Three-Dimensional Rayleigh Model

The three-dimensional counterpart of Pearson's problem can be solved by reduction to a well-studied one-dimensional problem—as is often the case in three-dimensional spherically symmetric models.

Let $\{X_n\}$ be a sequence of independent and uniformly distributed random variables on the unit sphere in three-dimensional space; $S_n := X_1 + \cdots + X_n$. The radial density function $f(r)$ is defined as the quotient

$$f_n(r) = \frac{P[|S_n| \in (r, r+dr)]}{dr}. \tag{10.56}$$

This can be expressed in terms of the density function of a sum of independent and uniformly distributed random variables on the interval $[-1, 1]$, as follows.

Proposition 10.9. The density function is expressed in the form

$$f_n(r) = -r g_n'(r),$$

where

$$g_n(r) = \frac{P[[Y_1 + \cdots + Y_n] \in (r, r+dr)]}{dr}$$

and where $\{Y_n, n \geq 1\}$ is a sequence of real-valued and independent random variables with the uniform distribution on $[-1/2, 1/2]$.

Proof. We have

$$E\left[e^{i\lambda \cdot S_n}\right] = E\left[e^{i\lambda \cdot X_1}\right]^n$$

$$= \left[\frac{1}{4\pi} \int_0^{2\pi} \int_0^{\pi} e^{i\lambda \cos\theta} \sin\theta \, d\theta \, d\phi\right]^n$$

$$= \left(\frac{\sin\lambda}{\lambda}\right)^n$$

$$\frac{P[S_n \in dx]}{dx} = \left(\frac{1}{2\pi}\right)^3 \int_{R^3} \left(\frac{\sin\lambda}{\lambda}\right)^n e^{-i\lambda \cdot x} d^3 x$$

$$= \left(\frac{1}{2\pi}\right)^3 \int_0^{\infty} \left(\frac{\sin\lambda}{\lambda}\right)^n \lambda^2 d\lambda \int_0^{2\pi} \int_0^{\pi} e^{-i\lambda r \cos\theta} \sin\theta \, d\theta \, d\phi$$

$$= \frac{1}{2\pi^2} \int_0^{\infty} \left(\frac{\sin\lambda}{\lambda}\right)^n \left(\frac{\sin\lambda r}{\lambda r}\right) \lambda^2 d\lambda$$

$$f_n(r) = \frac{P[S_n \in (r, r+dr)]}{dr}$$

$$= 4\pi^2 \frac{1}{2\pi^2} \int_0^\infty \left(\frac{\sin\lambda}{\lambda}\right)^n \left(\frac{\sin\lambda r}{\lambda r}\right) \lambda^2 \, d\lambda$$

$$= \frac{2}{\pi} \int_0^\infty \left(\frac{\sin\lambda}{\lambda}\right)^n r\lambda \sin(r\lambda) \, d\lambda.$$

On the other hand, the density of the sum $\bar{S}_n := Y_1 + \cdots + Y_n$ is computed as

$$E\left[e^{it\bar{S}_n}\right] = E\left[e^{itX_1}\right]^n$$

$$= \left(\frac{1}{2}\int_{-1}^1 e^{ity} \, dy\right)^n$$

$$= \left(\frac{\sin t}{t}\right)$$

so that

$$g_n(x) = \frac{P[S_n \in dx]}{dx}$$

$$= \frac{2}{\pi} \int_0^\infty \left(\frac{\sin t}{t}\right)^n \cos tx \, dt$$

$$g_n'(x) = -\frac{2}{\pi} \int_0^\infty \left(\frac{\sin t}{t}\right)^n t \sin tx \, dt,$$

which completes the proof that $f_n(r) = -rg_n'(r)$. The functions $g_n(r)$ are polynomials on the interval $0 \le r \le n$ and can be directly computed. ∎

The Rayleigh model can be generalized to an arbitrary number of dimensions, where the Bessel function J_0 is replaced by $J_{(p-2)/2}$ in dimension p. It can also be carried out in the case of a continuous parameter t: a skater moves along a straight line at constant velocity for an exponentially distributed amount of time, at the end of which he or she chooses a new direction at random according to a uniform distribution on the unit sphere. In fact, the displacement after n changes of direction will be given by a discrete-time model with an exponential distribution of radial displacement. These considerations can be carried out on a surface or higher dimensional manifold.

11 Characteristic Functions and Their Applications

The moment generating function (m.g.f.) of a random variable X is defined as the average of the exponential function:

$$M_X(t) := E\left(e^{tX}\right) = \int_R e^{tx} F(\mathrm{d}x) = \sum_{n=0}^{\infty} \frac{t^n}{n!} E(X^n).$$

For example, if X is normally distributed with mean zero and variance 1, then

$$M_X(t) = (2\pi)^{-1/2} \int_R e^{tx} e^{-x^2/2} \, \mathrm{d}x = (2\pi)^{-1/2} e^{t^2/2} \int_R e^{-(t-x)^2/2} \, \mathrm{d}x = e^{t^2/2}.$$

From this, the moments are computed by expanding both sides in powers of t, which yields $E(X^{2n}) = (2n)!/2^n n!$ for even moments and zero for odd moments.

The m.g.f. is a useful computational device, which can be used to tabulate the moments of a large class of probability distributions, both discrete and continuous. However, the m.g.f. is not defined for all random variables, e.g., a Cauchy distribution, where $M_X(t) = (1/\pi) \int_R e^{tx}/\left(1+x^2\right) \, \mathrm{d}x = +\infty$, for $t \neq 0$. Accordingly, we now introduce a universal label for an arbitrary distribution function, known as the *characteristic function* and defined as follows. We recall the complex exponential function, which satisfies $e^{it} = \cos t + i \sin t$, where t is a real number and $i = \sqrt{-1}$.

Exercise 11.1
If X is normally distributed with mean zero and variance 1, then $E(X^{2n}) = (2n)!/2^n n!$ for even moments and zero for odd moments.

11.1 Definition of the Characteristic Function

The characteristic function of a random variable X is defined as

$$\phi_X(t) := E\left(e^{itX}\right) = E\left(\cos tX\right) + iE\left(\sin tX\right), \qquad -\infty < t < \infty$$

The expectation is finite for all real values of t. In case X is a discrete random variable with discrete density f_X, we can write

$$\phi_X(t) = \sum_x e^{itx} f_X(x), \tag{11.1}$$

An Introduction to Stochastic Modeling
© 2011 Elsevier Inc. All rights reserved.

whereas if X is continuous with density f_X, then

$$\phi_X(t) = \int_{-\infty}^{\infty} e^{itx} f_X(x)\,dx. \tag{11.2}$$

If all the moments of X are finite, then the characteristic function can be computed by expanding the complex exponential function in its power series. We illustrate with the normal distribution $n(\mu, \sigma^2)$, for which we know the central moments. In this case,

$$
\begin{aligned}
E\left[e^{itX}\right] &= e^{it\mu} E\left[e^{it(X-\mu)}\right] \\
&= e^{it\mu} \sum_{m=0}^{\infty} \frac{(it)^m}{m!} E(X-\mu)^m \\
&= e^{it\mu} \sum_{k=0}^{\infty} \frac{\left(-t^2\right)^k}{(2k)!} E(X-\mu)^{2k} \\
&= e^{it\mu} \sum_{k=0}^{\infty} \frac{\left(-t^2\right)^k}{(2k)!} \frac{\sigma^{2k}(2k)!}{2^k k!} \\
&= e^{it\mu} \sum_{k=0}^{\infty} \frac{\left(-\sigma^2 t^2/2\right)^k}{k!} \\
&= e^{it\mu} e^{-\sigma^2 t^2/2}.
\end{aligned}
$$

which shows that the characteristic function has a normal-type dependence in the variable t.

If X is an arbitrary random variable, the characteristic function is a bounded continuous function of t with $|\phi_X(t)| \le 1, \phi_X(0) = 1$. This clearly holds for a normally distributed random variable and is easily proved in general. Additional smoothness properties of the characteristic function depend on the existence of higher moments, which is satisfied by the normal distribution, but not for an arbitrary random variable.

Exercise 11.2
Let X be a real-valued random variable. Show that the characteristic function is continuous in t.

Exercise 11.3
Let X be a real-valued random variable. Show that $|\phi_X(t)| \le 1$ and $\phi_X(0) = 1$.

11.1.1 Two Basic Properties of the Characteristic Function

In general, the characteristic function defines a *homomorphism*, converting sums of independent random variables into products. The precise statement is the following:

Theorem 11.1. *If X_1, X_2 are independent random variables, then*

$$\phi_{X_1+X_2}(t) = \phi_{X_1}(t)\phi_{X_2}(t)$$

The proof is a one-liner: if X_1, X_2 are independent, then

$$\phi_{X_1+X_2}(t) = E\left[e^{it(X_1+X_2)}\right] = E\left[e^{itX_1}e^{itX_2}\right] = E\left[e^{itX_1}\right]E\left[e^{itX_2}\right] = \phi_{X_1}(t)\phi_{X_2}(t)$$

where we have first used the properties of the exponential function followed by independence in the last step.

The other important property of the characteristic function is that it serves as a label for the distribution function of the random variable. This is formalized as follows.

Theorem 11.2. *If X_1, X_2 are random variables with $\phi_{X_1}(t) = \phi_{X_2}(t)$ for all t, then $F_{X_1}(x) = F_{X_2}(x)$ for all x.*

A theorem of this type can be proved by first proving an inversion formula, where we explicitly display the density/distribution in terms of the characteristic function.

11.2 Inversion Formulas for Characteristic Functions

We first illustrate the proof of Theorem 11.2 in case of discrete random variables, where the characteristic function is written as an infinite series:

$$\phi_X(t) = \sum_{x\in R} e^{itx} f_X(x).$$

We fix $y \in R$, multiply by the complex exponential e^{-ity}, and average on the interval $-L \le t \le L$ with the result

$$e^{-ity}\phi_X(t) = \sum_{x\in R} e^{it(x-y)} f_X(x)$$

$$\frac{1}{2L}\int_{-L}^{L} e^{-ity}\phi_X(t)\,dt = \sum_{x\in R} \left(\frac{1}{2L}\int_{-L}^{L} e^{it(x-y)}\,dt\right) f_X(x)$$

$$= \sum_{x\in R} \frac{\sin L(x-y)}{L(x-y)} f_X(x). \tag{11.3}$$

(If $x = y$, then the integral on the right side has the value $2L$, which agrees with the limiting value of the indicated quotient.) In particular, if X takes only integer values $0, \pm 1, \pm 2, \ldots$ and y is an integer, then we can take $L = \pi$ and note that all of the terms on the right side are zero, except in the case that $x = y$, an integer. From this, we obtain the **inversion formula for integer-valued random variables**:

$$X \in \mathbf{Z} \Longrightarrow f_X(y) = \frac{1}{2\pi} \int_{-\pi}^{\pi} e^{-ity} \phi_X(t)\, dt \qquad y = 0, \pm 1, \pm 2, \ldots \tag{11.4}$$

Example If X has a binomial distribution $B(n, p)$, then

$$\phi_X(t) = \sum_{k=0}^{n} e^{itk} \binom{n}{k} p^k q^{n-k} = \left(q + p e^{it} \right)^n \tag{11.5}$$

The inversion formula (11.4) takes the form

$$\binom{n}{y} p^y q^{n-y} = \frac{1}{2\pi} \int_{-\pi}^{\pi} e^{-ity} \left(q + p e^{it} \right)^n dt, \qquad y = 0, 1, \ldots, n$$

Formula (11.4) will be used to prove the local limit theorem of de Moivre and Laplace.

Example If X has a Poisson distribution $\mathcal{P}(\lambda)$, then

$$\phi_X(t) = \sum_{k=0}^{\infty} e^{itk} \frac{\lambda^k}{k!} e^{-\lambda} = e^{\lambda(e^{it}-1)}$$

and the inversion formula (11.4) takes the form

$$\frac{\lambda^y}{y!} e^{-\lambda} = \frac{1}{2\pi} \int_{-\pi}^{\pi} e^{-ity} e^{\lambda(e^{it}-1)}\, dt, \qquad y = 0, 1, 2, \ldots \tag{11.6}$$

This will be used to do the proof of Stirling's formula.

In the case of a more general discrete random variable, we can take the limit $L \to \infty$ in (11.3). The terms on the right side are bounded by an absolutely convergent series and tend to zero, save for $x = y$, so that we obtain the **inversion formula for discrete random variables**:

$$X \in \mathbf{D} \Longrightarrow f_X(y) = \lim_{L \to \infty} \frac{1}{2L} \int_{-L}^{L} e^{-ity} \phi_X(t)\, dt, \qquad y \in \mathbf{R} \tag{11.7}$$

where $\mathbf{D}$ is the set of possible values of X, with $\sum_{x \in \mathbf{D}} f_X(x) = 1$. Formula (11.7) shows explicitly that ϕ_X determines f_X and thus F_X, since $F_X(x) = \sum_{z \leq x} f_X(x)$. Hence, we have proved Theorem 11.2 in the case of general discrete random variables.

We can make a similar argument in the continuous case, when X has a density:

$$\phi_X(t) = \int_{-\infty}^{\infty} e^{itx} f_X(x)\, dx.$$

Again, we multiply by e^{-ity} and integrate over the real line tempered with the factor $e^{-\sigma^2 t^2/2}$— to ensure convergence of the improper integral; explicitly

$$\int_{-\infty}^{\infty} \phi_X(t) e^{-ity} e^{-\sigma^2 t^2/2}\, dt = \int_{-\infty}^{\infty} \left(\int_{-\infty}^{\infty} e^{it(x-y)} e^{-\sigma^2 t^2/2}\, dt \right) f_X(x)\, dx. \tag{11.8}$$

The inner integral is $\sqrt{2\pi}/\sigma \times$ the characteristic function of a normal density with mean zero and variance $1/\sigma^2$. In the special case, where ϕ_X is integrable over the real line, we can take the limit $\sigma \to 0$ to obtain the **Fourier inversion formula for integrable characteristic functions**:

$$\int_{-\infty}^{\infty} |\phi(t)|\, dt < \infty \implies \frac{1}{2\pi} \int_{-\infty}^{\infty} \phi_X(t) e^{-ity}\, dt = f_X(y) \tag{11.9}$$

valid at all continuity points of f_X. More generally, if the limit $\bar{f}_X$ of f_X exists in some averaged sense at $x = y$, then we can take the limit in equation (11.8) to obtain the inversion formula

$$f_X(y+0) = f_X(y-0) \implies \frac{1}{2\pi} \lim_{\sigma \to 0} \int_{-\infty}^{\infty} \phi_X(t) e^{-ity} e^{-\sigma^2 t^2/2}\, dt = \bar{f}_X(y) \tag{11.10}$$

If, e.g., f_X has a simple jump at y, then the right side of (11.10) needs to be interpreted as the average of the left and right limits at y.

Example In the case of the bilateral exponential density $f(x) = \frac{1}{2} e^{-|x|}$, the characteristic function is computed directly as $\phi(t) = 1/(1+t^2)$. This is an integrable function on the real line, so that the inversion formula (11.9) applies, to yield

$$\frac{1}{2} e^{-|y|} = \frac{1}{2\pi} \int_{-\infty}^{\infty} \frac{1}{1+t^2} e^{-ity}\, dt, \qquad t \in \mathbf{R}$$

As a by-product, we can change the roles of y and t to obtain the characteristic function of the Cauchy density $f_X(x) = 1/\pi \left(1+x^2\right)$, namely $\phi_X(t) = e^{-|t|}$. Since this is integrable, we also have the inversion formula (11.9) without any limiting procedure.

Example In the case of the uniform density $f_X(x) = (1/(b-a))1_{[a,b]}(x)$, the characteristic function is $\phi_X(t) = (e^{itb} - e^{ita})/(it(b-a))$, which is not integrable over the real line. Hence, we must use the general form (11.10) of the inversion formula.

Example In the case of a triangular density, e.g., $f(x) = 1 - |x|$ for $|x| \le 1$ and zero elsewhere, we may justify the inversion formula (11.9) by noting that f is the convolution of two uniform densities on $\left[-\frac{1}{2}, \frac{1}{2}\right]$, for which $\phi(t) = O(1/t), t \to \infty$. Hence, by Theorem 11.1, the characteristic function of f is $O(1/t^2), t \to \infty$. Hence, we can apply (11.9) to obtain the Fourier inversion formula.

11.2.1 Fourier Reciprocity/Local Non-Uniqueness[*]

The previous example can be re-written as a pair of Fourier integrals: Let $\phi(t) = 1 - |t|$ for $|t| < 1$ and zero elsewhere.

$$f(x) := \int_{-\infty}^{\infty} \phi(t)e^{itx}\,dt = \int_{-1}^{1} (1-|t|)e^{itx}\,dt = 2\frac{1-\cos x}{x^2} \qquad x \ne 0, f(0) = 1.$$

Applying the inversion formula for integrable characteristic functions, we have

$$\phi(t) = \frac{1}{2\pi} \int_{-\infty}^{\infty} e^{-itx}2\frac{1-\cos x}{x^2}\,dx, \quad t \in \mathbf{R}$$

which shows that $\phi(t)$ is a characteristic function. Now, we *periodize* by defining

$$\Phi(t) := \sum_{k \in Z} \phi(t - 2k\pi)$$

which is a 2π periodic function on the line and which agrees with $\phi(t)$ for $|t| < 1$. Its Fourier coefficients are computed by

$$\frac{1}{2\pi} \int_{-\pi}^{\pi} \Phi(t)e^{-int}\,dt = \frac{1}{2\pi} \int_{-\pi}^{\pi} \sum_{k \in Z} \phi(t - 2k\pi)e^{-int}\,dt$$

$$= \frac{1}{2\pi} \sum_{k \in Z} \int_{-(2k-1)\pi}^{(2k+1)\pi} \phi(y)e^{-in(y+2k\pi)}\,dy$$

$$= \frac{1}{2\pi} \int_{-\infty}^{\infty} \phi(y)e^{-iny}\,dy$$

$$= \frac{1}{\pi} \frac{1-\cos k}{k^2}$$

[*] This section can be omitted without loss of continuity.

leading to the absolutely convergent Fourier series:

$$\Phi(t) = \frac{1}{\pi} \sum_{k \in Z} \frac{1 - \cos k}{k^2} e^{ikt}, \qquad t \in \mathbf{R}.$$

This allows one to define an integer-valued random variable by the distribution

$$p_k = \frac{1 - \cos k}{\pi k^2}, \qquad 0 \neq k \in Z, \quad p_0 = \frac{1}{2\pi}$$

which is clearly nonnegative and sums to $\Phi(0) = 1$. Since this distribution is concentrated on the integers, its characteristic function must be a periodic function, namely $\Phi(t)$. Clearly $\Phi(t) = \phi(t)$ for $|t| < 1$, but the equality fails outside of the interval $[-1, 1]$. In summary,

THERE EXIST TWO DISTINCT CHARACTERISTIC FUNCTIONS WHICH AGREE ON THE INTERVAL $[-1, 1]$.

11.2.2 Fourier Inversion and Parseval's Identity

The ideas used to prove the inversion formula (11.9) can be extended to treat the Fourier transform of an absolutely integrable function ψ, where we define

$$\hat{\psi}(t) = \int_{-\infty}^{\infty} \psi(x) e^{itx} \, dx \tag{11.11}$$

If ψ is the probability density of a random variable X, then $\hat{\psi} = \phi_X$, the characteristic function of X. In the more general case, we can apply the same transformations to $\hat{\psi}$ as above, namely multiply (11.11) by $e^{-ity} e^{-\sigma^2 t^2 / 2}$ and integrate, to obtain

$$\int_{-\infty}^{\infty} \hat{\psi}(t) e^{-ity} e^{-\sigma^2 t^2 / 2} \, dt = \int_{-\infty}^{\infty} \left(\int_{-\infty}^{\infty} e^{it(x-y)} e^{-\sigma^2 t^2 / 2} \, dt \right) \psi(x) \, dx$$

$$= 2\pi \int_{-\infty}^{\infty} \frac{e^{-(x-y)^2 / 2\sigma^2}}{\sqrt{2\pi\sigma^2}} \psi(x) \, dx$$

If $\hat{\psi}$ is also absolutely integrable, then we can take the limit $\sigma \to 0$ and obtain the Fourier inversion formula

$$\psi(y) = \frac{1}{2\pi} \int_{-\infty}^{\infty} \hat{\psi}(t) e^{-ity} \, dt. \tag{11.12}$$

Applied to a random variable $y = X(\omega)$ and taking the expectation, we obtain a useful corollary.

Proposition 11.1 (Parseval's identity). Suppose that $\psi, \hat{\psi}$ are absolutely integrable. Then, we have the inversion formula (11.12) and for any random variable X, we have Parseval's identity:

$$E\psi(X) = \frac{1}{2\pi} \int_{-\infty}^{\infty} \hat{\psi}(t)\,\phi_X(-t)\,dt \tag{11.13}$$

This will be used in the proof of the continuity theorem, below.

11.3   Inversion Formula for General Random Variables

In the general case, the random variable X is neither discrete nor continuous. It is still possible to obtain an inversion formula in this general case by inserting an additional integration. We begin with the symbolic formula

$$\phi_X(t) = \int_{-\infty}^{\infty} e^{itx} F_X(dx)$$

The distribution function F_X may be purely discrete, purely continuous, or a combination of both types. We use the above steps to write

$$\int_{-\infty}^{\infty} \phi_X(t) e^{-ity} e^{-\sigma^2 t^2/2}\,dt = 2\pi \int_{-\infty}^{\infty} \frac{e^{-(x-y)^2/2\sigma^2}}{\sqrt{2\pi\sigma^2}} F_X(dx)$$

Now, we integrate on the left side over the interval $a \leq y \leq b$ to obtain

$$\int_{-\infty}^{\infty} \phi_X(t) \left(\frac{e^{-itb} - e^{-ita}}{-it} \right) e^{-\sigma^2 t^2/2}\,dt = 2\pi \int_{-\infty}^{\infty} \left(\int_a^b \frac{e^{-(x-y)^2/2\sigma^2}}{\sqrt{2\pi\sigma^2}}\,dy \right) F_X(dx)$$

$$\tag{11.14}$$

The integrand on the left side of (11.14) is defined by continuity at $t = 0$. The integral inside the parentheses on the right side can be written in terms of the normal distribution function Φ as $\Phi((x-a)/\sigma) - \Phi((x-b)/\sigma)$. Using the properties that $\Phi(+\infty) = 1, \Phi(0) = \frac{1}{2}, \Phi(-\infty) = 0$, we see that

$$\lim_{\sigma \to 0} [\Phi((x-a)/\sigma) - \Phi((x-b)/\sigma)] = 1 \qquad a < x < b$$

$$\lim_{\sigma \to 0} [\Phi((x-a)/\sigma) - \Phi((x-b)/\sigma)] = \frac{1}{2} \qquad x = a \text{ or } x = b$$

$$\lim_{\sigma \to 0} [\Phi((x-a)/\sigma) - \Phi((x-b)/\sigma)] = 0 \qquad x < a \text{ or } x > b$$

Therefore, we obtain the **general form of the inversion formula**

$$\frac{1}{2\pi} \lim_{\sigma \to 0} \int_{-\infty}^{\infty} \phi_X(t) \left(\frac{e^{-itb} - e^{-ita}}{-it} \right) e^{-\sigma^2 t^2/2} \, dt$$

$$= P[a < X < b] + \frac{1}{2}P[X = a] + \frac{1}{2}P[X = b], \tag{11.15}$$

which completes the proof of Theorem 11.2 in the most general case.

Corollary 11.1. If the distribution function of the random variable X is continuous at the points $x = a, x = b$, then

$$\frac{1}{2\pi} \lim_{\sigma \to 0} \int_{-\infty}^{\infty} \phi_X(t) \left(\frac{e^{-itb} - e^{-ita}}{-it} \right) e^{-\sigma^2 t^2/2} \, dt = P[a < X < b] \tag{11.16}$$

Proof. Indeed, in this case $P[X = a] = 0 = P[X = b]$. ∎

11.4   The Continuity Theorem

In order to use the characteristic function to prove limit theorems, we need to know that convergence of a sequence of characteristic functions implies convergence of the corresponding distribution functions, in an appropriate sense. The general result of this type is known as the *continuity theorem*, which is stated as follows.

Theorem 11.3. *Let $(X_n, n \geq 1)$ be a sequence of random variables with characteristic functions $\phi_n(t)$. If for each real number t, we have*

$$\lim_{n \to \infty} \phi_n(t) = \phi_X(t)$$

for some random variable X, then

$$\lim_{n \to \infty} P[a \leq X_n \leq b] = P[a \leq X \leq b]$$

provided that $P[X = a] = 0 = P[X = b]$; in particular, this occurs if X has a continuous distribution function.

Example Apply the continuity theorem to the (suitably normalized) binomial distribution with $p = \frac{1}{2}$.

Solution. If we have the binomial distribution with $p = \frac{1}{2}$, then the characteristic function of $X_n = (S_n - n/2)/\sqrt{n/4}$ is $\phi_n(t) = \cos(t/\sqrt{n})^n$. When $n \to \infty$, we have

$$\lim_n \phi_n(t) = e^{-t^2/2}$$

which is the characteristic function of the standard normal distribution. Applying the continuity theorem shows that we have a limiting normal distribution.

If the limiting random variable X has a continuous distribution function, then $P[X = a] = 0 = P[X = b]$, so that we can assert that the probabilities of all intervals converge to the corresponding probabilities for the limiting random variable. In this case, we can assert that the probability of any interval, e.g., $[a, b), (a, b]$ or (a, b) converges to the same limit.

The next example illustrates what can happen if the limiting distribution is not continuous.

Example Let the random variable $X_n := (-1)^n/n$, so that $X_n \to 0$ when $n \to \infty$. The distributions satisfy

$$P[0 \le X_n \le 1] = 1 \quad \text{if } n \text{ is even,} \qquad P[0 \le X_n \le 1] = 0 \quad \text{if } n \text{ is odd}$$

so that $\lim_n P[0 \le X_n \le 1]$ does not exist. This illustrates the possible limiting behavior when the limiting random variable has a discontinuous distribution function.

11.4.1 Proof of the Continuity Theorem[*]

The proof uses the notion of *upper limit* and *lower limit* of a sequence of real numbers. We begin with the *test functions* $\psi_{\pm}^{\epsilon}(x)$, depending on an additional parameter $\epsilon > 0$ and defined as follows: $\psi_{+}^{\epsilon}(x) = 1$ on the interval $[a, b]$ and $\psi(x) = 0$ if $x \le a - \epsilon$ or $x \ge b + \epsilon$. Otherwise, ψ_{+}^{ϵ} is a linear function, which interpolates between these values: $\psi_{+}(x) = (x - a + \epsilon)/\epsilon$ for $a - \epsilon \le x \le a$ and $\psi_{+}^{\epsilon}(x) = (b + \epsilon - x)/\epsilon$ for $b \le x \le b + \epsilon$. In the same manner, we define ψ_{-}^{ϵ}, which is piecewise linear, equal to 1 if $a + \epsilon \le x \le b - \epsilon$ and is zero for $x \le a$ and $x \ge b$. In particular, we have the double system of inequalities

$$\psi_{-}^{\epsilon}(x) \le 1_{[a,b]}(x) \le \psi_{+}^{\epsilon}(x) \tag{11.17}$$

On the other hand, both $\psi_{\pm}^{\epsilon}$ have trapezoidal profiles and can, thus, be expressed as the difference of two triangular profiles, both of which have integrable characteristic functions, from Equation (11.2). Hence, the Fourier inversion formula (11.12) and Parseval's identity (11.13) apply to both ψ_{-}^{ϵ} and ψ_{+}^{ϵ}. Applying both sides of (11.17) to X_n and taking the expectation, we have

$$E\left(\psi_{-}^{\epsilon}(X_n)\right) \le P[a \le X_n \le b] \le E\left(\psi_{+}^{\epsilon}(X_n)\right). \tag{11.18}$$

But for each n, we can use the Parseval's identity (11.13) to write

$$E\left(\psi_{\pm}^{\epsilon}(X_n)\right) = \frac{1}{2\pi} \int\limits_{-\infty}^{\infty} \phi_n(t) \hat{\psi}_{\pm}^{\epsilon}(-t) \, dt.$$

---

[*] This section may be skipped on the first reading.

Taking the limit $n \to \infty$, we see that the right side converges, hence we have

$$\lim_n E\left(\psi_\pm^\epsilon(X_n)\right) = \frac{1}{2\pi} \int\limits_{-\infty}^{\infty} \phi(t)\hat{\psi}_\pm^\epsilon(-t)\,dt = E\left(\psi_\pm^\epsilon(X)\right)$$

where we have used Parseval's identity again. Referring to (11.18), we have the double system of inequalities

$$\limsup_n P[a \le X_n \le b] \le E\left(\psi_+^\epsilon(X)\right) \le P[a - \epsilon \le X \le b + \epsilon],$$

$$\liminf_n P[a \le X_n \le b] \ge E\left(\psi_-^\epsilon(X)\right) \ge P[a + \epsilon \le X \le b - \epsilon].$$

But the upper and lower limits do not depend on ϵ. Taking $\epsilon \to 0$, we obtain

$$P[a < X < b] \le \liminf_n P[a \le X_n \le b] \le \limsup_n P[a \le X_n \le b] \le P[a \le X \le b]$$

If $P[X = a] = 0 = P[X = b]$, then the two extreme members are equal and we have proved the required result.

Exercise 11.4
Show that the above proof applies equally well to compute $\lim_n P[a < X_n < b]$ or $\lim_n P[a \le X_n < b]$ or $\lim_n P[a < X_n \le b]$.

11.5 Proof of the Central Limit Theorem

The main application of the continuity theorem is to prove the classical CLT:

Theorem 11.4. *Let $Y_n, n \ge 1$ be a sequence of independent and identically distributed random variables with mean μ and variance σ^2 with $0 < \sigma^2 < \infty$. Denoting $S_n := Y_1 + \cdots + Y_n$, then for every pair of reals $a < b$*

$$\lim_n P\left[a \le \frac{S_n - n\mu}{\sigma\sqrt{n}} \le b\right] = \int\limits_a^b \frac{e^{-u^2/2}}{\sqrt{2\pi}}\,du. \tag{11.19}$$

Proof. This is proved by reducing to the case $\mu = 0, \sigma = 1$ as follows. Letting $Y_i' := (Y_i - \mu)/\sigma$, $S_n' = Y_1' + \cdots + Y_n'$, it is immediate that Y_i' has mean zero and variance 1. Furthermore $(S_n - n\mu)/\sigma\sqrt{n} = S_n'/\sqrt{n}$.

Assuming that $\mu = 0, \sigma = 1$, we have the characteristic function $\phi(t) = \phi_{Y_1}(t)$, a twice differentiable function with

$$|\phi(t)| \le 1, \quad \phi(0) = 1, \quad \phi'(0) = 0, \quad \phi''(0) = -1.$$

From Taylor's formula with remainder

$$\phi(s) = 1 - \frac{s^2}{2} + \epsilon_1(s), \qquad \lim_{s \to 0} \frac{\epsilon_1(s)}{s^2} = 0 \qquad (11.20)$$

$$e^{-\frac{s^2}{2}} = 1 - \frac{s^2}{2} + \epsilon_2(s), \qquad \lim_{s \to 0} \frac{\epsilon_2(s)}{s^2} = 0. \qquad (11.21)$$

The characteristic function of the normalized sum is

$$\phi_n(t) := E\left[e^{it\frac{S_n}{\sqrt{n}}}\right] = E\left[e^{it\frac{Y_1}{\sqrt{n}}}\right]^n = \phi\left(\frac{t}{\sqrt{n}}\right)^n$$

The characteristic function of the standard normal distribution is $e^{-t^2/2}$, so that the difference is written using the identity $A^n - B^n = (A - B)\left(A^{n-1} + \cdots + B^{n-1}\right)$:

$$\phi_n(t) - e^{-t^2/2} = \phi\left(\frac{t}{\sqrt{n}}\right)^n - \left(e^{-t^2/2n}\right)^n$$

$$= \left[\phi\left(\frac{t}{\sqrt{n}}\right) - e^{-t^2/2n}\right]\left[A^{n-1} + \cdots + B^{n-1}\right]$$

where $A = \phi(t/\sqrt{n}), B = e^{-t^2/2n}$. Each of the n terms on the right is less than 1 in modulus, so that we can write

$$\left|\phi_n(t) - e^{-t^2/2}\right| \le n\left|\phi\left(\frac{t}{\sqrt{n}}\right) - e^{-t^2/2n}\right|.$$

Setting $s = t/\sqrt{n}$ in (11.20) and (11.21) with t fixed, we have

$$\phi\left(\frac{t}{\sqrt{n}}\right) = 1 - \frac{t^2}{2n} + \epsilon_1\left(\frac{t}{\sqrt{n}}\right), \qquad e^{-t^2/2n} = 1 - \frac{t^2}{2n} + \epsilon_2\left(\frac{t}{\sqrt{n}}\right).$$

Subtracting these two expressions, the first two terms cancel and we are left with terms of the form $n\epsilon(t/\sqrt{n})$, which tend to zero when $n \to \infty$ and t is fixed. We have proved that $\phi_n(t)$ converges to the standard normal characteristic function, which has a continuous distribution function. Hence, by the continuity theorem, the probabilities of all intervals converge, as required. ∎

11.6 Stirling's Formula and Applications

Often, we encounter the factorial function of a large integer argument. The numerical evaluation of these expressions can be cumbersome, which leads one to search for an *asymptotic formula*, meaning a simpler formula, which provides a good approximation for large arguments.

Stirling's formula is the following limiting statement involving $n!$:

$$\lim_{n} \frac{n!}{n^{n+\frac{1}{2}}e^{-n}} = \sqrt{2\pi} \tag{11.22}$$

where $e = 2.71828\cdots$ is the base of the natural logarithms. This is also written in the form

$$n! \sim n^{n+\frac{1}{2}}e^{-n}\sqrt{2\pi}, \qquad n \to \infty \tag{11.23}$$

where the tilde sign means that the ratio of the two terms tends to 1 when $n \to \infty$. The Stirling's approximation (11.23) is already extremely accurate for small values of n; for example, if $n = 5$, then $n! = 120$, whereas the Stirling's approximation gives 118.019, an error of less than 2%. For $n = 10$, we have the exact value of 3,628,800, whereas Stirling's approximation is 3,598,690, an error of less than 1%.

We will prove Stirling's formula by representing the reciprocal of $n!$ in terms of a Poisson distribution, which we can estimate. No previous knowledge of the Poisson distribution is assumed.

11.6.1 *Poisson Representation of n!*

The Poisson distribution with parameter $\lambda > 0$ is defined by the sequence

$$p(k; \lambda) = \frac{\lambda^k}{k!}e^{-\lambda}, \qquad k = 0, 1, 2, \ldots \tag{11.24}$$

It is immediate that $p(k; \lambda) > 0$ and $\sum_{k=0}^{\infty} p(k; \lambda) = 1$, so that we have a probability distribution on the nonnegative integers. The characteristic function is the following trigonometric series, which can be summed in closed form and which defines a 2π-periodic function:

$$\hat{p}(\theta; \lambda) = \sum_{k=0}^{\infty} p(k; \lambda)e^{ik\theta} = e^{\lambda(e^{i\theta}-1)}, \qquad \theta \in \mathbf{R}. \tag{11.25}$$

For each $\lambda > 0$, the series (11.25) converges uniformly on $\mathbf{R}$, as well as the series obtained by multiplying by $e^{-ik\theta}$. Hence, we can integrate term-by-term on any period interval to obtain

$$p(k; \lambda) = \frac{1}{2\pi} \int_{-\pi}^{\pi} \hat{p}(\theta, \lambda)e^{-ik\theta}d\theta, \qquad \lambda > 0, k = 0, 1, 2, \ldots \tag{11.26}$$

Now, we are free to take $\lambda = k$, to obtain the useful representation of the reciprocal factorial function:

$$p(k; k) = \frac{k^k}{k!} e^{-k} = \frac{1}{2\pi} \int\limits_{-\pi}^{\pi} e^{k(e^{i\theta} - 1 - i\theta)} \, d\theta, \qquad k = 0, 1, 2, \ldots. \tag{11.27}$$

11.6.2 Proof of Stirling's Formula

We, now, make the substitution $\psi = \theta \sqrt{k}$ to obtain the integral formula

$$\frac{k^{k+\frac{1}{2}}}{k!} e^{-k} = \frac{1}{2\pi} \int\limits_{-\pi\sqrt{k}}^{\pi\sqrt{k}} e^{k\left(e^{i\psi/\sqrt{k}} - 1 - i\psi/\sqrt{k}\right)} \, d\psi, \qquad k = 1, 2, \ldots. \tag{11.28}$$

When $k \to \infty$, the integrand on the right side tends to $e^{-\psi^2/2}$ and is pointwise dominated by $e^{-\delta\psi^2}$, where $\delta := \inf_{0 < |\theta| \leq \pi} (1 - \cos\theta)/\theta^2 = 2/\pi^2$, since the modulus of the exponential is the exponential of the real part, namely $k(\cos\theta - 1)$, which is bounded above by $-k\delta\theta^2$ for $|\theta| \leq \pi$. Hence, by the dominated convergence theorem, we have

$$\lim_{k\to\infty} \frac{k^{k+\frac{1}{2}}}{k!} e^{-k} = \frac{1}{2\pi} \int\limits_{-\infty}^{\infty} e^{-\psi^2/2} \, d\psi = \frac{1}{\sqrt{2\pi}} \tag{11.29}$$

which is the statement of Stirling's formula, where we have used the normalization of the standard normal density: $\int_{-\infty}^{\infty} e^{-\psi^2/2} \, d\psi = \sqrt{2\pi}$.

Exercise 11.5
Prove the limiting relation $\lim_{z\to 0} \left(e^{iz} - 1 - iz\right)/z^2 = -1/2$.

Exercise 11.6
If $z = \alpha + i\beta$ is an arbitrary complex number, show that $\left|e^{\alpha + i\beta}\right| = e^{\alpha}$.

Exercise 11.7
Prove the upper and lower bounds $2\theta/\pi \leq \sin\theta \leq \theta$ for $0 \leq \theta \leq \pi/2$.
Hint: Look at the graphs of these three functions.

Exercise 11.8
Prove that $\int_R e^{-x^2/2} dx = \sqrt{2\pi}$.
Hint: Square both sides and use polar coordinates.

11.7   Local deMoivre–Laplace Theorem

We define the characteristic function and its normalized version by

$$F(\theta) := q + p e^{i\theta}, \qquad F_0(\theta) := e^{-ip\theta} F(\theta) = q e^{-ip\theta} + p e^{iq\theta} \tag{11.30}$$

The binomial probability is the Fourier coefficient of the characteristic function, which can be represented as a suitable integral on $(-\pi, \pi)$.

Lemma 11.1. *For $n = 0, 1, 2, \ldots$ and $k = 0, 1, \ldots, n$, let $x = x(k, n) = (k - np)/\sqrt{npq}$. Then,*

$$P_{kn} =: \binom{n}{k} p^k q^{n-k} = \frac{1}{2\pi} \int_{-\pi}^{\pi} F(\theta)^n e^{-ik\theta} \, d\theta \tag{11.31}$$

$$= \frac{1}{2\pi} \int_{-\pi}^{\pi} F_0(\theta)^n e^{-i\theta(k-np)} \, d\theta \tag{11.32}$$

$$= \frac{1}{2\pi \sqrt{npq}} \int_{-\pi\sqrt{npq}}^{\pi\sqrt{npq}} F_0\left(\frac{\psi}{\sqrt{npq}}\right)^n e^{-ix\psi} \, d\psi \tag{11.33}$$

Proof. From the binomial theorem

$$F(\theta)^n = \left(q + p e^{i\theta}\right)^n = \sum_{j=0}^{n} \binom{n}{k} p^j q^{n-j} e^{ij\theta},$$

Multiply both sides by $e^{-ik\theta}$ and integrate on $(-\pi, \pi)$, from which we conclude

$$P_{k,n} = \binom{n}{k} p^k q^{n-k} = \frac{1}{2\pi} \int_{-\pi}^{\pi} F(\theta)^n e^{-ik\theta} \, d\theta. \tag{11.34}$$

Formula (11.32) comes from the definition of $F_0(\theta)$. The final formula (11.33) comes from the definition of $x = x(k, n)$ and the substitution of $\psi = \theta \sqrt{npq}$, which completes the proof.

Now, we note

$$F_0(0) = 1, F_0'(0) = 0, F_0''(0) = -pq$$

$$|F_0(\theta)|^2 = q^2 + p^2 + 2pq\cos\theta = 1 - 2pq(1 - \cos\theta)$$

$$\leq \left(1 - 4pq\theta^2/\pi^2\right) \leq e^{-4pq\theta^2/\pi^2}, \quad |\theta| < \pi$$

$$|F_0(\theta)| \leq e^{-2pq\theta^2/\pi^2} \qquad |\theta| < \pi$$

$$\lim_{n\to\infty} F_0\left(\frac{\psi}{\sqrt{npq}}\right)^n = e^{-\psi^2/2} \qquad \psi \in \mathbf{R}$$

For any real number x, let $k \to \infty, n \to \infty$ so that $(k - np)/\sqrt{npq} \to x$. Then, by the dominated convergence theorem,

$$\lim_{n \to \infty} \sqrt{npq}\, P_{kn} = \frac{1}{2\pi} \int_{-\infty}^{\infty} e^{-\psi^2/2} e^{-ix\psi}\, d\psi = \frac{1}{\sqrt{2\pi}} e^{-x^2/2}$$

which is the statement of the *local central limit theorem of de Moivre and Laplace*.
   This can be rewritten in a more intuitive form:

$$P(k, n) = p^k q^{n-k} \binom{n}{k} \sim \frac{e^{-x^2/2}}{\sqrt{2\pi npq}}, \quad (n \to \infty)$$

■

Further Reading

Elementary Textbooks

Breiman, L. (1969). *Probability and stochastic processes with a view toward applications.* Boston: Houghton Mifflin.

Cinlar, E. (1975). *Introduction to stochastic processes.* Englewood Cliffs, NJ: Prentice-Hall.

Cox, D. R., & Miller, H. D. (1965). *The theory of stochastic processes.* New York: John Wiley & Sons.

Hoel, R. G., Port, S. C., & Stone, C. J. (1972). *Introduction to stochastic processes.* Boston: Houghton Mifflin.

Kemeny, J. G., & Snell, J. L. (1960). *Finite markov chains.* New York: Van Nostrand Reinhold.

Intermediate Textbooks

Bhat, U. N. (1972). *Elements of applied stochastic processes.* New York: John Wiley & Sons.

Breiman, L. (1968). *Probability.* Reading, MA: Addison-Wesley.

Dynkin, E. B., & Yushkevich, A. A. (1969). *Markov processes: Theorems and problems.* New York: Plenum.

Feller, W. (1966–1986). *An introduction to probability theory and its applications* (3rd ed., Vols. 1–2). New York: John Wiley & Sons.

Karlin, S., & Taylor, H. M. (1975). *A first course in stochastic processes.* New York: Academic Press.

Karlin, S., & Taylor, H. M. (1981). *A second course in stochastic processes.* New York: Academic Press.

Kemeny, J. G., Snell, J. L., & Knapp, A. W. (1966). *Denumerable markov chains.* New York: Van Nostrand Reinhold.

Ross, S. M. (1983). *Stochastic processes.* New York: John Wiley & Sons.

Ross, S. M. (1993). *Introduction to probability models* (5th ed.,). New York: Academic Press.

Renewal Theory

Kingman, J. F. C. (1972). *Regenerative phenomena.* New York: John Wiley & Sons.

Queueing Processes

Kleinrock, L. (1976). *Queueing systems. Theory* (Vol 1), *Computer applications* (Vol 2). New York: John Wiley & Sons, Interscience.

Branching Processes

Athreya, K. B., & Ney, P. (1970). *Branching processes*. New York: Springer-Verlag.

Harris, T. (1963). *The theory of branching processes*. New York: Springer-Verlag.

Stochastic Models

Bartholomew, D. J. (1967). *Stochastic models for social processes*. New York: John Wiley & Sons.

Bartlett, M. S. (1960). *Stochastic population models in ecology and epidemiology*. New York: John Wiley & Sons.

Goel, N. S., & Richter-Dyn, N. (1974). *Stochastic models in biology*. New York: Academic Press.

Point Processes

Lewis, P. A. (1972). *Stochastic point processes: Statistical analysis, theory, and applications*. New York: John Wiley & Sons, Interscience.

Answers to Exercises

Chapter 1

1.2.1 Because B and B^c are disjoint events whose union is the whole sample space, the law of total probability (Section 1.2.1) applies to give the desired formula.

1.2.3 (b) $f(x) = \begin{cases} 0 & \text{for } x \leq 0; \\ 3x^2 & \text{for } 0 < x < 1; \\ 0 & \text{for } x \geq 1. \end{cases}$

    (c) $E[X] = \frac{3}{4}$.

    (d) $\Pr\left\{\frac{1}{4} \leq X \leq \frac{3}{4}\right\} = \frac{26}{64}$.

1.2.4 (b) $E[Z] = \frac{9}{8}$.

    (c) $\text{Var}[Z] = \frac{55}{64}$.

1.2.7 (a) $F_X(x) = \begin{cases} 0 & \text{for } x < 0; \\ x^R & \text{for } 0 \leq x \leq 1; \\ 1 & \text{for } 1 < x. \end{cases}$

    (b) $E[X] = R/(1+R)$.

    (c) $\text{Var}[X] = R/\left[(R+2)(R+1)^2\right]$.

1.2.8 $f(v) = A(1-v)^{A-1}$ for $0 \leq v \leq 1$;

    $E[V] = 1/(A+1)$;

    $\text{Var}[V] = A/\left[(A+2)(A+1)^2\right]$.

1.2.9 $F_X(x) = \begin{cases} 0 & \text{for } x < 0; \\ \frac{1}{2}x^2 & \text{for } 0 \leq x \leq 1; \\ 1 - \frac{1}{2}(2-x)^2 & \text{for } 1 < x \leq 2; \\ 1 & \text{for } x > 2. \end{cases}$

    $E[X] = 1$; $\text{Var}[X] = \frac{1}{6}$.

1.3.1 $\Pr\{X = 3\} = \frac{10}{32}$.

1.3.2 $\Pr\{0 \text{ defective}\} = 0.3151$.

    $\Pr\{0 \text{ or } 1 \text{ defective}\} = 0.9139$.

1.3.3 $\Pr\{N = 10\} = 0.0315$.

1.3.4 $\Pr\{X = 2\} = 2e^{-2} = 0.2707$.

    $\Pr\{X \leq 2\} = 5e^{-2} = 0.6767$.

1.3.5 $\Pr\{X \geq 8\} = 0.1334$.

1.3.6 (a) Mean $= \frac{n+1}{2}$; Variance $= \frac{n^2-1}{12}$.

(b) $\Pr\{Z = m\} = \begin{cases} \frac{m+1}{n^2} & \text{for } m = 0, \ldots, n; \\ \frac{2n+1-m}{n^2} & \text{for } m = n+1, \ldots, 2n. \end{cases}$

(c) $\Pr\{U = k\} = \frac{1+2(n-k)}{(n+1)^2}$ for $k = 0, \ldots, n.$

1.4.1 $\Pr\{X > 1.5\} = e^{-3} = 0.0498.$
$\Pr\{X = 1.5\} = 0.$

1.4.2 Median $= \frac{1}{\lambda} \log 2$; Mean $= \frac{1}{\lambda}.$

1.4.3 Exponential distribution with parameter $\lambda/2.54.$

1.4.4 Mean $= 0$; Variance $= 1.$

1.4.5 $\alpha^* = \frac{\sigma_Y^2 - \rho \sigma_X \sigma_Y}{\sigma_X^2 + \sigma_Y^2 - 2\rho \sigma_X \sigma_Y}$ for $\rho \neq \pm 1.$

1.4.6 **(a)** $f_Y(y) = e^{-y}$ for $y \geq 0.$

   (b) $f_W(w) = \frac{1}{n} \left(\frac{1}{w} \right)^{(n-1)/n}$ for $0 < w < 1.$

1.4.7 R has the gamma density $f_R(r) = \lambda^2 r e^{-\lambda r}$ for $r > 0.$

1.5.1 $\Pr\{X \geq 1\} = 0.6835938$
$\Pr\{X \geq 2\} = 0.2617188$
$\Pr\{X \geq 3\} = 0.0507812$
$\Pr\{X \geq 4\} = 0.0039062.$

1.5.2 Mean $= \frac{5}{7}.$

1.5.3 $E[X] = \frac{1}{\lambda}.$

1.5.4 **(a)** $E[X_A] = \frac{1}{2}; E[X_B] = \frac{1}{3};$

   (b) $E[\min\{X_A, X_B\}] = \frac{1}{5};$

   (c) $\Pr\{X_A < X_B\} = \frac{2}{5};$

   (d) $E[X_B - X_A | X_A < X_B] = \frac{1}{3}.$

1.5.5 **(a)** $\Pr\{\text{Naomi is last}\} = \frac{1}{2};$

   (b) $\Pr\{\text{Naomi is last}\} = \frac{282}{2500} = 0.1128;$

   (c) $c = 2 + \sqrt{3}.$

Chapter 2

2.1.1 $\Pr\{N = 3, X = 2\} = \frac{1}{16};$
$\Pr\{X = 5\} = \frac{1}{48};$
$E[X] = \frac{7}{4}.$

2.1.2 $\Pr\{\text{two nickel heads} | N = 4\} = \frac{3}{7}.$

2.1.3 $\Pr\{X \geq 1 | X \geq 1\} = 0.122184.$
$\Pr\{X > 1 | \text{Ace of spades}\} = 0.433513.$

2.1.4 $\Pr\{X = 2\} = 0.2204.$

2.1.5 $E[X | X \text{ is odd}] = \lambda \left(\frac{e^\lambda + e^{-\lambda}}{e^\lambda - e^{-\lambda}} \right).$

2.1.6 $\Pr\{U = u, Z = z\} = \rho^2 (1 - \rho)^z,$ $0 \leq u \leq z;$
$\Pr\{U = u | Z = n\} = \frac{1}{n+1},$ $0 \leq u \leq n.$

2.2.1 Pr{Game ends in a 4} $= \frac{1}{4}$.

2.2.3 Pr{Win} $= 0.468984$.

2.3.1

k	Pr{$Z = k$}
0	0.16406
1	0.31250
2	0.25781
3	0.16667
4	0.07552
5	0.02083
6	0.00260

$E[Z] = \frac{7}{4}$; $\text{Var}[Z] = 1.604167$.

2.3.2 $E[Z] = \frac{3}{2}$; $\text{Var}[Z] = \frac{9}{8}$;
Pr{$Z = 2$} $= 0.29663$.

2.3.3 $E[Z] = \mu^2$; $\text{Var}[Z] = \mu(1 + \mu)\sigma^2$.

2.3.4 Pr{$X = 2$} $= 0.2204$;
$E[X] = 2.92024$.

2.3.5 $E[Z] = 6$; $\text{Var}[Z] = 26$.

2.4.1 Pr{$X = 2$} $= \frac{1}{4}$.

2.4.2 Pr{System operates} $= \frac{1}{2}$.

2.4.3 $\Pr\left\{U > \frac{1}{2}\right\} = 1 - \frac{1}{2}(1 + \log 2) = 0.1534$.

2.4.4 $f_Z(z) = \frac{1}{(1+z)^2}$ for $0 < z < \infty$.

2.4.5 $f_{U,V}(u, v) = e^{-(u+v)}$ for $u > 0, v > 0$.

2.5.1

x	$\frac{1}{2}$	1	2
Pr{$X > x$}	0.61	0.37	0.14
$\frac{1}{x}E[X]$	2	1	$\frac{1}{2}$.

2.5.2 Pr{$X \geq 1$} $= E[X] = p$.

Chapter 3

3.1.1 0.

3.1.2 0.12, 0.12.

3.1.3 0.03.

3.1.4 0.02, 0.02.

3.1.5 0.025, 0.0075.

3.2.1 (a) $\mathbf{P}^2 = \begin{Vmatrix} 0.47 & 0.13 & 0.40 \\ 0.42 & 0.14 & 0.44 \\ 0.26 & 0.17 & 0.57 \end{Vmatrix}$.

   (b) 0.13.
   (c) 0.16.

3.2.2

n	0	1	2	3	4
$\Pr\{X_n = 0 \mid X_0 = 0\}$	1	0	$\frac{1}{2}$	$\frac{1}{4}$	$\frac{3}{8}$

3.2.3 0.264, 0.254.

3.2.4 0.35.

3.2.5 0.27, 0.27.

3.2.6 0.42, 0.416.

3.3.1

$$
\mathbf{P} = \begin{array}{c} \\ -1 \\ 0 \\ 1 \\ 2 \\ 3 \end{array}
\begin{array}{ccccc}
-1 & 0 & 1 & 2 & 3 \\
\left\|\begin{array}{ccccc}
0 & 0 & 0.3 & 0.3 & 0.4 \\
0 & 0 & 0.3 & 0.3 & 0.4 \\
0.3 & 0.3 & 0.4 & 0 & 0 \\
0 & 0.3 & 0.3 & 0.4 & 0 \\
0 & 0 & 0.3 & 0.3 & 0.4
\end{array}\right\|
\end{array}
$$

3.3.2 $P_{ii} = \left(\frac{i}{N}\right)p + \left(\frac{N-i}{N}\right)q;$

$P_{i,i+1} = \left(\frac{N-i}{N}\right)p;$

$P_{i,i-1} = \left(\frac{i}{N}\right)q.$

3.3.3

$$
\mathbf{P} = \begin{array}{c} \\ -1 \\ 0 \\ 1 \\ 2 \\ 3 \end{array}
\begin{array}{ccccc}
-1 & 0 & 1 & 2 & 3 \\
\left\|\begin{array}{ccccc}
0 & 0 & 0.1 & 0.4 & 0.5 \\
0 & 0 & 0.1 & 0.4 & 0.5 \\
0.1 & 0.4 & 0.5 & 0 & 0 \\
0 & 0.1 & 0.4 & 0.5 & 0 \\
0 & 0 & 0.1 & 0.4 & 0.5
\end{array}\right\|
\end{array}
$$

3.3.4

$$
\begin{array}{c} \\ -2 \\ -1 \\ 0 \\ 1 \\ 2 \\ 3 \end{array}
\begin{array}{cccccc}
-2 & -1 & 0 & 1 & 2 & 3 \\
\left\|\begin{array}{cccccc}
0 & 0 & 0.2 & 0.3 & 0.4 & 0.1 \\
0 & 0 & 0.2 & 0.3 & 0.4 & 0.1 \\
0 & 0 & 0.2 & 0.3 & 0.4 & 0.1 \\
0.2 & 0.3 & 0.4 & 0.1 & 0 & 0 \\
0 & 0.2 & 0.3 & 0.4 & 0.1 & 0 \\
0 & 0 & 0.2 & 0.3 & 0.4 & 0.1
\end{array}\right\|
\end{array}
$$

3.3.5

$$
\mathbf{P} = \begin{array}{c} \\ 0 \\ 1 \\ 2 \end{array}
\begin{array}{ccc}
0 & 1 & 2 \\
\left\|\begin{array}{ccc}
0 & 1 & 0 \\
\frac{1}{2} & 0 & \frac{1}{2} \\
0 & 1 & 0
\end{array}\right\|
\end{array}
$$

3.4.1 $v_{03} = 10.$

3.4.2 (a) $u_{10} = \frac{1}{4};$

(b) $v_1 = \frac{5}{2}.$

3.4.3 (a) $u_{10} = \frac{40}{105};$

(b) $v_1 = \frac{10}{3}.$

3.4.4 $v_0 = 6$.

3.4.5 $u_{\text{H,TT}} = \frac{1}{3}$.

3.4.6 (a) $u_{10} = \frac{6}{23}$;

    (b) $v_1 = \frac{50}{23}$.

3.4.7 $w_{11} = \frac{20}{11}$; $w_{12} = \frac{25}{11}$

    $v_1 = \frac{45}{11}$.

3.4.8 $w_{11} = 1.290$; $w_{12} = 0.323$

    $v_1 = 1.613$.

3.4.9 $u_{10} = \frac{9}{22} = 0.40909\ldots$;

$$P_{10}^{(2)} = 0.17;$$

$$P_{10}^{(4)} = 0.2658;$$

$$P_{10}^{(8)} = 0.35762\ldots;$$

$$P_{10}^{(16)} = 0.40245\ldots.$$

3.5.1 0.71273.

3.5.2 (a) 0.8044; 0.99999928....

    (b) 0.3578; 0.00288....

3.5.3 $\mathbf{P}^2 = \begin{Vmatrix} 0.58 & 0.42 \\ 0.49 & 0.51 \end{Vmatrix}$.

$$\mathbf{P}^3 = \begin{Vmatrix} 0.526 & 0.474 \\ 0.553 & 0.447 \end{Vmatrix}.$$

$$\mathbf{P}^4 = \begin{Vmatrix} 0.5422 & 0.4578 \\ 0.5341 & 0.4659 \end{Vmatrix}.$$

$$\mathbf{P}^5 = \begin{Vmatrix} 0.53734 & 0.46266 \\ 0.53977 & 0.46023 \end{Vmatrix}.$$

3.5.4

$$\mathbf{P} = \begin{array}{c} \\ 0 \\ 1 \\ 2 \\ 3 \end{array} \begin{array}{cccc} 0 & 1 & 2 & 3 \\ \begin{Vmatrix} \frac{1}{2} & \frac{1}{2} & 0 & 0 \\ \frac{1}{2} & 0 & \frac{1}{2} & 0 \\ \frac{1}{2} & 0 & 0 & \frac{1}{2} \\ 0 & 0 & 0 & 1 \end{Vmatrix} \end{array}.$$

3.5.5 $P_{\text{GG}}^{(8)} = 0.820022583$.

3.5.6 2.73.

3.5.7 $u_{10} = 0.3797468$.

3.5.8 $p_0 = \alpha$, $r_0 = 1 - \alpha$;

    $p_i = \alpha(1 - \beta)$, $q_i = \beta(1 - \alpha)$,

    $r_j = \alpha\beta + (1 - \alpha)(1 - \beta)$, for $i \geq 1$.

3.5.9 $p_0 = 1$, $q_0 = 0$,

    $p_i = p$, $q_i = q$, $r_i = 0$ for $i \geq 1$.

3.6.1 (a) $u_{35} = \frac{3}{5}$;

(b) $u_{35} = \left[1 - \left(\frac{q}{p}\right)^3\right] \Big/ \left[1 - \left(\frac{q}{p}\right)^5\right]$.

3.6.2 $u_{10} = 0.65$.

3.6.3 $v = 2152.777\ldots$.

3.6.4 $v_1 = 2.1518987$.

3.7.1 $\mathbf{W} = \left\|\begin{matrix} \frac{20}{11} & \frac{25}{11} \\ \frac{10}{11} & \frac{40}{11} \end{matrix}\right\|$.

(a) $u_{10} = \frac{9}{22}$;

(b) $w_{11} = \frac{20}{11}$; $w_{12} = \frac{25}{11}$.

3.7.2 $\mathbf{W} = \left\|\begin{matrix} \frac{100}{79} & \frac{70}{79} \\ \frac{30}{79} & \frac{100}{79} \end{matrix}\right\|$.

(a) $u_{10} = \frac{30}{79}$;

(b) $w_{11} = \frac{100}{79}$; $w_{12} = \frac{70}{79}$.

3.8.1 $M(n) = 1$, $V(n) = n$.

3.8.2 $\mu = b + 2c$; $\sigma^2 = b + 4c - (b + 2c)^2$.

3.8.3

n	1	2	3	4	5
u_n	0.5	0.625	0.695	0.742	0.775

3.8.4 $M(n) = \lambda^n$, $V(n) = \lambda^n\left(\frac{1-\lambda^n}{1-\lambda}\right)$, $\lambda \neq 1$.

3.9.1

n	1	2	3	4	5
u_n	0.333	0.480	0.564	0.619	0.658
u_∞	$= 0.82387$.				

3.9.2 $\varphi(s) = p_0 + p_2 s^2$.

3.9.3 $\varphi(s) = p + q s^N$.

3.9.4 $\frac{\varphi(s) - \varphi(0)}{1 - \varphi(0)}$.

Chapter 4

4.1.1 $\pi_0 = \frac{10}{21}$, $\pi_1 = \frac{5}{21}$, $\pi_2 = \frac{6}{21}$.

4.1.2 $\pi_0 = \frac{31}{66}$, $\pi_1 = \frac{16}{66}$, $\pi_2 = \frac{19}{66}$.

4.1.3 $\pi_1 = \frac{3}{13}$.

4.1.4 2.94697.

4.1.5 $\pi_0 = \frac{10}{29}$, $\pi_1 = \frac{5}{29}$, $\pi_2 = \frac{5}{29}$, $\pi_3 = \frac{9}{29}$.

4.1.6 $\pi_0 = \frac{5}{14}$, $\pi_1 = \frac{6}{14}$, $\pi_2 = \frac{3}{14}$.

4.1.7 $\pi_0 = \frac{140}{441}$, $\pi_1 = \frac{40}{441}$, $\pi_2 = \frac{135}{441}$, $\pi_3 = \frac{126}{441}$.

4.1.8 $\pi_u = \frac{4}{17}$.

4.1.9 $\pi_0 = \frac{2}{7}$, $\pi_1 = \frac{3}{7}$, $\pi_2 = \frac{2}{7}$.

4.1.10 $\pi_{\text{late}} = \frac{17}{40}$.

4.2.1 $\pi_s = \frac{8}{9}$.

4.2.2 One facility: $\Pr\{\text{Idle}\} = \frac{q^2}{1+p^2}$;

Two facilities: $\Pr\{\text{Idle}\} = \frac{1}{1+p+p^2}$.

4.2.3 **(a)**

p	0	0.02	0.04	0.06	0.08	0.10
AFI	0.10	0.11	0.12	0.13	0.14	0.16
AOQ	0	0.018	0.036	0.054	0.072	0.090

(b)

p	0	0.02	0.04	0.06	0.08	0.10
AFI	0.20	0.23	0.27	0.32	0.37	0.42
AOQ	0	0.016	0.032	0.048	0.064	0.080

4.2.4

p	0.05	0.10	0.15	0.20	0.25
R_1	0.998	0.990	0.978	0.962	0.941
R_2	0.998	0.991	0.981	0.968	0.952

4.2.5 $\pi_A = \frac{1}{5}$.

4.2.6 $\pi_0 = \frac{1}{3}$.

4.2.7 **(a)** 0.6831;

(b) $\pi_1 = \pi_2 = \frac{10}{21}$, $\pi_3 = \frac{1}{21}$;

(c) $\pi_3 = \frac{1}{21}$.

4.2.8 $\pi_3 = \frac{8}{51}$.

4.3.1 $\left\{ n \geq 1; P_{00}^{(n)} > 0 \right\} = \{5, 8, 10, 13, 15, 16, 18, 20, 21, 23, 24, 25, 26, 28, \ldots\}$

$d(0) = 1$, $P_{5,7}^{(37)} = 0$, $P_{i,j}^{(38)} > 0$ for all i, j.

4.3.2 Transient states: $\{0, 1, 3\}$.
Recurrent states: $\{2, 4, 5\}$.

4.3.3 **(a)** $\{0, 2\}, \{1, 3\}, \{4, 5\}$;

(b) $\{0\}, \{5\}, \{1, 2\}, \{3, 4\}$.

4.3.4 $\{0\}, d = 1$;

$\{1\}, d = 0$;

$\{2, 3, 4, 5\}, d = 1$.

4.4.1 $\pi_k = p^k / \left(1 + p + p^2 + p^3 + p^4\right)$ for $k = 0, \ldots, 4$.

4.4.2 **(a)** $\pi_0 = \frac{1449}{9999}$.

(b) $m_{10} = \frac{8550}{1449}$.

4.4.3 $\pi_0 = \pi_1 = 0.2$, $\pi_2 = \pi_3 = 0.3$.

4.5.1 $\lim P_{00}^{(n)} = \lim P_{10}^{(n)} = 0.4$;

$\lim P_{20}^{(n)} = \lim P_{30}^{(n)} = 0$;

$\lim P_{40}^{(n)} = 0.4$.

4.5.2 (a) $\frac{3}{11}$,   (e) $\frac{3}{11}$,

      (b) 0,   (f) X,

      (c) $\frac{2}{33}$,   (g) $\frac{1}{3}$,

      (d) $\frac{2}{9}$,   (h) $\frac{4}{27}$.

Chapter 5

5.1.1 (a) e^{-2};

      (b) e^{-2}.

5.1.2 $(p_k/p_{k-1}) = \lambda/k, \quad k = 0, 1, \ldots$.

5.1.3 $\Pr\{X = k | N = n\} = \binom{n}{k} p^k (1-p)^{n-k}, \quad p = \frac{\alpha}{\alpha+\beta}$.

5.1.4 (a) $\frac{(\lambda t)^k e^{-\lambda t}}{k!}, \quad k = 0, 1, \ldots$;

      (b) $\Pr\{X(t) = n + k | X(s) = n\} = \frac{[\lambda(t-s)]^k e^{-\lambda(t-s)}}{k!}$,

      $E[X(t)X(s)] = \lambda^2 ts + \lambda s$.

5.1.5 $\Pr\{X = k\} = (1-p)p^k \quad$ for $\quad k = 0, 1, \ldots$ where $p = 1/(1+\theta)$.

5.1.6 (a) e^{-12};

      (b) Exponential, parameter $\lambda = 3$.

5.1.7 (a) $2e^{-2}$;

      (b) $\frac{64}{3}e^{-6}$;

      (c) $\binom{6}{2}\left(\frac{1}{3}\right)^2\left(\frac{2}{3}\right)^4$;

      (d) $\frac{32}{3}e^{-4}$.

5.1.8 (a) $5e^{-2}$;

      (b) $4e^{-4}$;

      (c) $\frac{1-3e^{-2}}{1-e^{-2}}$.

5.1.9 (a) 4;

      (b) 6;

      (c) 10.

5.2.1

k	0	1	2
(a)	0.290	0.370	0.225
(b)	0.296	0.366	0.221
(c)	0.301	0.361	0.217

5.2.2 Law of rare events, e.g., (a) Many potential customers who could enter store, small probability for each to actually enter.

5.2.3 The number of distinct pairs is large; the probability of any particular pair being in sample is small.

5.2.4 $\Pr\{\text{Three pages error free}\} \approx e^{-12}$.

5.3.1 e^{-6}.

5.3.2 (a) $e^{-6} - e^{-10}$;

      (b) $4e^{-4}$.

5.3.3 $\frac{1}{4}$.

5.3.4 $\binom{5}{2}\left(\frac{1}{3}\right)^2\left(\frac{2}{3}\right)^3 = \frac{80}{243}$.

5.3.5 $\binom{n}{m}\left(\frac{t}{T}\right)^m\left(1-\frac{t}{T}\right)^{n-m}$, $\quad m = 0, 1, \ldots, n$.

5.3.6 $F(t) = \left(1 - e^{-\lambda t}\right)^n$.

5.3.7 $t + \frac{2}{\lambda}$.

5.3.8 $\binom{12}{5}\left(\frac{1}{2}\right)^5\left(\frac{1}{2}\right)^7$.

5.3.9 $\Pr\{W_r \le t\} = 1 - \sum_{k=0}^{r-1} \frac{(\lambda t)^k e^{-\lambda t}}{k!}$.

5.4.1 $\frac{1}{n+1}$.

5.4.2 $\frac{1}{4}$.

5.4.3 $\frac{5}{2}$.

5.4.4 See equation (5.23).

5.4.5 $\left[1 - \frac{1-e^{-\alpha}}{\alpha}\right]^5$.

5.5.1 0.9380.

5.5.2 0.05216.

5.5.3 0.1548.

5.6.1 0.0205.

5.6.2 Mean $= \frac{\lambda t}{\theta}$, Variance $= \frac{2\lambda t}{\theta^2}$.

5.6.3 $\frac{e^{-\lambda G(z)t} - e^{-\lambda t}}{1 - e^{-\lambda t}}$.

5.6.4 (a) $\frac{1}{9}$;

    (b) $\frac{11}{27}$.

5.6.5 $\Pr\{M(t) = k\} = \frac{\Lambda(t)^k e^{-\Lambda(t)}}{k!}$, $\quad$ where $\quad \Lambda(t) = \lambda \int_0^t [1 - G(u)]\mathrm{d}u$.

Chapter 6

6.1.1 $P_0(t) = e^{-t}$;

    $P_1(t) = \frac{1}{2}e^{-t} - \frac{1}{2}e^{-3t}$;

    $P_2(t) = 3\left[\frac{1}{2}e^{-t} + \frac{1}{2}e^{-3t} - \frac{1}{2}e^{-2t}\right]$

    $P_3(t) = 6\left[\frac{1}{8}e^{-t} + \frac{1}{4}e^{-3t} - \frac{1}{3}e^{-2t} - \frac{1}{24}e^{-5t}\right]$.

6.1.2 (a) $\frac{11}{6}$;

    (b) $\frac{25}{6}$;

    (c) $\frac{49}{36}$.

6.1.3 $X(t)$ is a Markov process (memoryless property of exponential distribution) for which the sojourn time in state k is exponentially distributed with parameter λk.

6.1.4 (a) $\Pr\{X = 0\} = (1 - \alpha h)^n = 1 - n\alpha h + o(h)$;

 (b) $\Pr\{X = 1\} = n\alpha h(1 - \alpha h)^n = n\alpha h + o(h)$;

 (c) $\Pr\{X \geq 2\} = 1 - \Pr\{X = 0\} - \Pr\{X - 1\} = o(h)$.

6.1.5 $E[X(t)] = \frac{1}{p}$, $\mathrm{Var}[X(t)] = \frac{1-p}{p^2}$, where $p = e^{-\beta t}$.

6.1.6 (a) $P_1(t) = e^{-5t}$;

 (b) $P_2(t) = 5\left[\frac{1}{2}e^{-3t} - \frac{1}{2}e^{-5t}\right]$

 (c) $P_3(t) = 15\left[\frac{1}{20}e^{-3t} - \frac{1}{16}e^{-5t} + \frac{1}{80}e^{-13t}\right]$.

6.2.1 $P_3(t) = e^{-5t}$;

 $P_2(t) = 5\left[\frac{1}{3}e^{-2t} - \frac{1}{3}e^{-5t}\right]$;

 $P_1(t) = 10\left[\frac{1}{6}e^{-5t} + \frac{1}{3}e^{-2t} - \frac{1}{2}e^{-3t}\right]$;

 $P_0(t) = 1 - P_1(t) - P_2(t) - P_3(t)$.

6.2.2 (a) $\frac{31}{30}$;

 (b) $\frac{29}{15}$;

 (c) $\frac{361}{900}$.

6.2.3 $P_3(t) = e^{-t}$;

 $P_2(t) = \left[e^{-t} - e^{-2t}\right]$;

 $P_1(t) = 2\left[\frac{1}{2}e^{-t} - e^{-2t} + \frac{1}{2}e^{-3t}\right]$;

 $P_0(t) = 1 - P_1(t) - P_2(t) - P_3(t)$.

6.2.4 $P_2(t) = 10e^{-4t}\left(1 - e^{-2t}\right)^3$.

6.3.1 $\lambda_n = \lambda, \mu_n = n\mu$ for $n = 0, 1, \ldots$.

6.3.2 Assume that $\Pr\{\text{Particular patient exits in } [t, t + h]|k \text{ patients}\} = \frac{1}{m_k}h + o(h)$.

6.4.2 $\pi_k = \binom{N}{k}p^k(1 - p)^{N-k}$, where $p = \frac{\alpha}{\alpha+\beta}$.

6.4.3 $\pi_k = \frac{\lambda^k e^{-\lambda}}{k!}$, where $\lambda = \frac{\alpha}{\beta}$.

6.4.4 (a) $\pi_0 = 1 \bigg/ \left(1 + \frac{\lambda}{\mu} + \frac{1}{2}\left(\frac{\lambda}{\mu}\right)^2\right)$;

 (b) $\pi_0 = 1 \bigg/ \left(1 + \frac{\lambda}{\mu} + \left(\frac{\lambda}{\mu}\right)^2\right)$.

6.4.5 $\pi_k = (k+1)(1 - \theta)^2\theta^k$.

6.4.6 $\pi_k = \frac{\theta^k e^{-\theta}}{k!}$, where $\theta = \frac{\lambda}{\mu}$.

6.5.1 Use $\log\frac{1}{1-x} = 1 + x + \frac{1}{2}x^2 + \cdots |x| < 1$.

6.5.2 Use $\frac{K}{K-i} \cong 1$ for $K \gg i$.

6.6.1 $1 - \left(\frac{\beta_A}{\alpha_A+\beta_A}\right)\left(\frac{\beta_B}{\alpha_B+\beta_B}\right)$.

6.6.2 $P_{00}(t) = \left\{ \frac{\mu}{\lambda+\mu} + \frac{\lambda}{\lambda+\mu} e^{-(\lambda+\mu)t} \right\}^2$.

6.7.1 $f(t) = \frac{2-\sqrt{2}}{4} e^{-(2+\sqrt{2})t} + \frac{2+\sqrt{2}}{4} e^{-(2-\sqrt{2})t}$.

6.7.2 $f_0'(t) = -\frac{\sqrt{2}}{4} e^{-(2-\sqrt{2})t} + \frac{\sqrt{2}}{4} e^{-(2+\sqrt{2})t}$;

$f_1'(t) = -\frac{2-\sqrt{2}}{4} e^{-(2-\sqrt{2})t} - \frac{2+\sqrt{2}}{4} e^{-(2+\sqrt{2})t}$.

Chapter 7

7.1.1 The age δ_t of the item in service at time t cannot be greater than t.

7.1.2 $F_2(t) = 1 - e^{-\lambda t} - \lambda t e^{-\lambda t}$.

7.1.3 (a) True;
   (b) False;
   (c) False.

7.1.4 (a) W_k has a Poisson distribution with parameter λk.

   (b) $\Pr\{N(t) = k\} = \sum_{n=0}^{t} \frac{e^{-\lambda k}(\lambda k)^n}{n!} - \sum_{n=0}^{t} \frac{e^{-\lambda(k+1)}(\lambda k + \lambda)^n}{n!}$.

7.2.1 $\frac{1}{a} + \frac{1}{b}$.

7.2.2 The system starts from an identical condition at those instants when first both components are OFF.

7.2.3

n	$M(n)$	$u(n)$
1	0.4	0.4
2	0.66	0.26
3	1.104	0.444
4	1.6276	0.5236
5	2.0394	0.41184
6	2.4417	0.4023
7	2.8897	0.44798
8	3.3374	0.44769
9	3.7643	0.42693
10	4.1947	0.4304

7.3.1 $e^{-\lambda s}; t + \frac{1}{\lambda}$.

7.3.2 The inter-recording time is a random sum with a geometric number of exponential terms and is exponential with parameter λp. (See the example following Chapter 2, (2.34)).

7.3.3 $\Pr\{N(t) = n, W_{N(t)+1} > t + s\}$

$= \left\{ \frac{(\lambda t)^n e^{-\lambda t}}{n!} \right\} e^{-\lambda s}$.

7.4.1 $M(t) \approx \frac{3}{2} t - \frac{7}{16}$.

7.4.2 $\frac{3}{8}$.

7.4.3 $T^* = \sqrt{\frac{\sqrt{273}-15}{8}}$.

7.4.4 $c(T)$ is a decreasing function of T.

7.4.5 $h(x) = 2(1 - x)$ for $0 \leq x \leq 1$.

7.4.6 (a) $\frac{\alpha}{\alpha+\beta}$;

    (b) $13\frac{\alpha}{\alpha+\beta}$;

    (c) $\frac{7}{\alpha+\beta}$.

7.5.1 $\frac{\mu\lambda}{1+\mu\lambda}$.

7.5.2 $\frac{2}{9}$.

7.5.3 $T^* = \infty$.

7.6.1–7.6.2

n	v_n	u_n
0	0.6667	1.33333
1	1.1111	0.88888
2	0.96296	1.03704
3	1.01235	0.98765
4	0.99588	1.00412
5	1.00137	0.99863
6	0.99954	1.00046
7	1.00015	0.99985
8	0.99995	1.00005
9	1.00002	0.99998
10	0.99999	1.00001

7.6.3 $E[X] = 1$; $\Sigma b_k = 1$.

Chapter 8

8.1.1 (a) 0.8413.
    (b) 4.846.

8.1.2 $\mathrm{Cov}[W(s), W(t)] = \min\{s, t\}$.

8.1.3 $\frac{\partial p}{\partial t} = \frac{1}{2}\varphi(z)t^{-\frac{3}{2}}\left[z^2 - 1\right]$;

    $\frac{\partial p}{\partial x} = \frac{1}{t}z\varphi(z)$.

8.1.4 (a) 0.
    (b) $3u^2 + 3uv + uw$.

8.1.5 (a) $e^{-|s-t|}$;
    (b) $t(1 - s)$ for $0 < t < s < 1$;
    (c) $\min\{s, t\}$.

8.1.6 (a) Normal, $\mu = 0, \sigma^2 = 4u + v$;
    (b) Normal, $\mu = 0, \sigma^2 = 9u + 4v + w$.

8.1.7 $\frac{s}{S}$.

8.2.1 (a) 0.6826.
    (b) 4.935.

8.2.2 If $\tan\theta = \sqrt{s/t}$, then $\cos\theta = \sqrt{t/(s+t)}$.

8.2.3 0.90.

8.2.4 Reflection principle: $\Pr\{M_n \geq a, S_n < a\} =$
$\Pr\{M_n \geq a, S_n > a\}(= \Pr\{S_n > a\})$.
Also, $\Pr\{M_n \geq a, S_n = a\} = \Pr\{S_n = a\}$.

8.2.5 $\Pr\{\tau_0 < t\} = \Pr\{B(u) \neq 0 \quad \text{for all } t \leq u < a\} = 1 - \vartheta(t, a)$.

8.2.6 $\Pr\{\tau_1 < t\} = \Pr\{B(u) = 0 \quad \text{for some } u, b < u < t\}$.

8.3.1 $\Pr\{R(t) < y | R(0) = x\}$
$= \Pr\{-y < B(t) < y | B(0) = x\}$
$= \Pr\{B(t) < y | B(0) = x\} - \Pr\{B(t) < -y | B(0) = x\}$.

8.3.2 0.3174, 0.15735.

8.3.3 0.83995.

8.3.4 0.13534.

8.3.5 No, No.

8.4.1 0.05.

8.4.2 0.5125, 0.6225, 0.9933.

8.4.3 0.25, 24.5, 986.6.

8.4.4 $\tau = 43.3 \quad \text{versus} \quad E[T] = 21.2$

8.4.5 0.3325.

8.4.6 (a) $E\left[(\xi - a)^+\right] = \int_a^\infty x\varphi(x)dx - a \int_a^\infty \varphi(x)dx$
$= \varphi(a) - a[1 - \Phi(a)]$.

(b) $(X - b)^+ = \sigma\left(\xi - \frac{b-\mu}{\sigma}\right)^+$.

8.5.1 0.8643, 0.7389, 0.7357.

8.5.2 0.03144, 0.4602, 0.4920.

8.5.3 (a) $E[V_n] = (1 - \beta)^n v$
$\text{Cov}[V_n, V_{n+k}] = (1 - \beta)^k$.

(b) $E[\Delta V | V_n = v] = -\beta v$
$\text{Var}[\Delta V | V_n = v] = 1$.

Chapter 9

9.1.1 (a) Probability waiting planes exceed available air space.
(b) Mean number of cars in lot.

9.1.2 The standard deviation of the exponential distribution equals the mean.

9.1.3 $1\frac{1}{2}$ days.

9.2.1 $\frac{25}{36}$.

9.2.3 $L = 1 \quad \text{versus} \quad L = 5$.

9.3.1 $v = \frac{1}{\mu}, \tau^2 = \frac{1}{\mu^2}, L = \frac{\rho}{1-\rho}$.

9.3.2 $L = \rho + \frac{\rho^2(1+\alpha)}{2\alpha(1-\rho)}$.

9.3.3 $3\frac{1}{4}$.

9.3.4 $\frac{27}{36}$ versus 3.

9.3.5 $M(t) = \lambda \int\limits_0^t [1 - G(y)]dy \to \lambda \nu$.

9.4.1 8 per hour.

9.4.2 $\frac{5}{16}$.

9.4.3 8.55.

9.4.4 0.0647.

9.5.1 $\frac{1}{5}\left(\frac{4}{5}\right) = \frac{4}{25}$.

9.5.2 $\Pr\{Z \le 20\} = 0.9520$.

Index